Wood: A Manual for its Use as a Shipbuilding Material

First Collected Edition

Teaparty Books

1983

Teaparty Books, P.O. Box 232
Kingston, MA 02364

First Printing

Library of Congress Catalog Card Number 82-74415
ISBN 0-9610602-0-4 (Hardbound)
ISBN 0-9610602-1-2 (Softbound)

WOOD:
A MANUAL FOR
ITS USE AS
A SHIPBUILDING MATERIAL

Volume I

DEPARTMENT OF THE NAVY

BUREAU OF SHIPS

1957

Bow view AM minesweeper.

WOOD: A MANUAL FOR ITS USE AS A SHIPBUILDING MATERIAL

Prepared by

Bureau of Ships
Department of the Navy

With the cooperation of

Forest Products Laboratory
Forest Service
United States Department of Agriculture

Issued July 1945
Second Printing October 1948
Extensively Revised October 1957

DEPARTMENT OF THE NAVY,
BUREAU OF SHIPS,
Washington 25, D. C.

During World War II and the Korean action, more than 40,000 wooden ships, boats, and landing craft were added to the Fleet. A continuing program of improvement in wood laminating, extreme service adhesives, and development of wood preservatives for protection against marine borers has been largely responsible for the satisfactory performance of these craft. During the past five years, many additional wooden ships and boats have been put into service, incorporating the most recent knowledge of wood as an engineering material.

Because of these many new developments in marine uses of wood, a need has arisen for a comprehensive, up-to-date collection of information based on recent investigations and tests. This new publication, "Wood: A Manual for Its Use as a Shipbuilding Material," has, therefore, been prepared and should prove most useful to shipyards, naval architects, Supervisors of Shipbuilding, Inspectors of Naval Materials, and officers and men concerned with all phases of wooden boat design, construction, or maintenance.

A. G. MUMMA,
Rear Admiral, USN,
Chief, Bureau of Ships.

ACKNOWLEDGMENTS

In an industry so old as shipbuilding, many men, most of them unknown, have contributed to our present day knowledge and advancement of the art and science. A notable early attempt to organize the improvement of shipbuilding was made by King James I in 1605 when he chartered a Shipwrights Guild with supervision over all shipbuilding in England, and which all English shipwrights were required to join. The Guild made significant progress under the leadership of Master Shipwright Phineas Pett who introduced the use of scale models and mathematically developed plans, making possible the building of two ships alike.

Later, the French took up the banner of progress and carried it well into the 1800's, surpassing the British. This era of shipbuilding witnessed the highest development of the wooden sailing ship in construction, speed, and beauty. During this time, the American forest contributed enormous quantities of shiptimbers, planking, decking, masts, and spars. American shipwrights achieved a reputation for skilled workmanship held to this day.

Shipbuilding began to emerge from an art to a science during the 19th century. Several schools of naval architecture were founded in England, the United States, and Germany, followed by such technical societies as the Institution of Engineers and Shipbuilders in Scotland, the British Institution of Naval Architects, and Institute of Marine Engineers, and in the United States, the Society of Naval Architects and Marine Engineers.

Thus, the modern science of wood shipbuilding has developed over the centuries. Most significant of the achievements during the 20th century has been the adopting of adhesive bonded wood as an engineering material in the construction of wooden hull ships. The bonding together of small pieces of wood into large structural members or panels in either straight or curved form results in a far more efficient use of wood than was formerly possible. This development is the first basic change in wooden shipbuilding in 2,000 years.

Previous editions of this manual have mentioned many individual shipbuilders, researchers, and other contributors who furnished information regarding modern wood ship and boat building practices. The present edition represents the combined efforts of virtually all American technical societies, research agencies, trade associations, and universities which foster progress in wood technology, naval architecture, and timber engineering. It is the Bureau's desire to give full recognition to these groups and to the part played by the lumber, laminating, plywood, adhesives, and wood preservation industries in modern wood shipbuilding. Without their cumulative contributions to the advancement of wood as an engineering material, this manual would not be possible.

The basic manuscript for "Wood: A Manual for Its Use as a Shipbuilding Material" was prepared by the Forest Products Laboratory, Forest Service, United States Department of Agriculture, at the request of the Bureau of Ships, Department of the Navy. A very large share of the credit for the material presented in the manual is due the Laboratory, which has been foremost in the field of forest products for almost 50 years. The Laboratory's individual staff members which have contributed to the manual are recognized authorities in their individual fields. Thanks are also due the staff of the Division of Forest Disease Research of the United States Forest Service, and the Timber Engineering Company Laboratory for their advice and guidance.

The manuscript was reviewed and edited for the Bureau of Ships by members of its staff. The final arrangement of the material for publication has been made under the direction of Mr. G. I. Dewey, now retired, who until 1948, supervised wood products development for the Bureau of Ships.

TABLE OF CONTENTS

CHAPTER 1. INTRODUCTION

1.0. PURPOSE AND SCOPE

When this publication was first issued in 1945 by the Department of the Navy under the title, "Wood: A Manual for Its Use in Wooden Vessels," its principal purpose was to serve the Nation in the construction of great numbers of wood hulls of all types then being built for World War II service. Hull-building techniques were then in a state of rapid development as new materials and methods were introduced. The main objective of the Manual at that time was to guide builders in making intelligent use of lumber, plywood, and allied materials under these rapidly changing conditions. The content of the book was carefully planned with this objective in mind. No attempt was made to present a comprehensive discussion of the many construction techniques involved in building a wood hull. Rather, the properties of wood and allied products were related to the techniques then in use in an effort to demonstrate how these properties affect the intelligent use of wood hull-building materials.

Since the Manual was first released, both materials and methods of hull construction have developed greatly. These developments, notably in the fields of gluing, laminating, preservative treatment, and fabrication, constitute one of the reasons that the Bureau of Ships undertook the complete rewriting of the Manual. This issue, therefore, contains much new information gained from research continuously in progress at the Forest Products Laboratory, the Timber Engineering Company, naval laboratories and elsewhere, much of it done in cooperation with the Bureau of Ships.

New information is also presented on hull maintenance, storage, and repair, a field about which little knowledge was available at the time the Manual was originally written. During the postwar years, however, problems of storage and maintenance have become prominent, and considerable research and study have been conducted on them. A summary of the information developed on this subject through research and experience at various Navy activities is presented herein.

The several chapters have been expanded to the extent that a single volume would no longer be convenient to handle. In view of the diverse uses to which the Manual may be put, as well as for convenience in handling, it has been divided into four volumes as follows:

- Volume I—Basic Wood Technology Applicable to Shipbuilding.
- Volume II—Techniques and Practices Applicable to Preservation and Storage.
- Volume III (Confidential)—Technical Data Applicable to Boat and Ship Design.
- Volume IV (Confidential)—Techniques Applicable to Boat and Ship Construction.

The name has been changed to "Wood: A Manual for Its Use as a Shipbuilding Material" to more accurately express the broadened scope.

Throughout recorded history, wood has been a major material for the construction of boats and ships. Its many natural advantages for such use have in recent years been extended by the results of scientific research. This Manual is intended to provided an up-to-date source of basic information about wood materials, the application of which in the construction and maintenance of wood-hulled boats and ships can improve its service under today's conditions.

1.1. NOMENCLATURE

So far as has been practicable, the shipbuilding terminology used in this manual has been kept in agreement with that of the Navy Department publication, "Nomenclature of Naval Vessels." For ship parts and techniques not defined in that publication, attempt has been made to use the names and phrases believed to be most generally employed throughout the United States. Terminology of a distinctly local character is avoided.

The more important wood terms used in this manual are defined in the Glossary, chapter 4.

The names given for the native woods discussed in this manual are those officially adopted for the various species by the United States Forest Service and given in "Check List of Native and Naturalized Trees of the United States (Including Alaska)." The common names of lumber species which are used by the trade and by shipbuilders are not always identical with those adopted as official by the Forest Service. To avoid possible confusion, table 1–1 covering commercial shipbuilding woods has been prepared to compare the standard commercial names, as prescribed in American Lumber Standards for softwood and by current trade usage for hardwood, with the corresponding botanical names and the names used in this Manual.

1.2. TRADE NAMES

The mention of any trade name or proprietary product in this Manual does not constitute an endorsement of the product by any Government agency.

TABLE 1–1.

NOMENCLATURE OF COMMERCIAL SHIP AND BOAT BUILDING WOODS [1]

HARDWOODS

Commercial name	Botanical name	Name used in this manual
Angelique, basra locus	*Dicorynia paraensis or guianensis*	Angelique.
Apitong, bagac, eng, gurjun, keruing, yang	*Dipterocarpus spp*	Apitong.
Ash:		
Black ash	*Fraxinus nigra*	Black ash.
Oregon ash	*Fraxinus latifolia*	Oregon ash.
White ash	*Fraxinus americana*	White ash.
White ash	*Fraxinus pennsylvanica*	Green ash.
Basswood	*Tilia americana*	American basswood.
Basswood	*Tilia heterophyilla*	White basswood.
Beech	*Fagus grandifolia*	America beech.
Birch	*Betula alleghaniensis*	Yellow birch.
Cherry	*Prunus serotina*	Black cherry.
Rock elm	*Ulmus thomasii*	Rock elm.
Greenheart	*Ocotea rodioei*	Greenheart.
Gum	*Liquidambar styraciflua*	Sweetgum.
Hickory	*Carya ovata*	Shagbark hickory.
Hickory	*Carya glabra*	Pignut hickory.
Hickory	*Carya laciniosa*	Shellbark hickory.
Hickory	*Carya tomentosa*	Mockernut hickory.
Iroko	*Chlorophora excelsa*	Iroko.
Ironbark:		
Gray ironbark	*Eucalyptus paniculata*	Gray ironbark.
Red ironbark	*Eucalyptus crebra*	Narrow-leaved red ironbark.
Red ironbark	*Eucalyptus siderophlia*	Broad-leaved red ironbark.
Red ironbark	*Eucalyptus sideroxylon*	Red ironbark.
Lignumvitae	*Guaiacum officinale*	Lignumvitae.
Locust	*Robinia pseudoacacia*	Black locust.
Mahogany	*Swietenia macrophylla*	Mahogany.
Mahogany, African	*Khaya ivorensis* *Khaya senegalensis.*	Khaya.
Mahogany, Philippine	*Shorea polysperma*	Tangile.
Mahogany, Philippine	*Shorea negrosensis*	Red lauan.
	Shorea teysmanniana	Tiaong.
	Pentacme contorta	White lauan.
Maple:		
Hard maple	*Acer saccharum*	Sugar maple.
Hard maple	*Acer nigrum*	Black maple.
Maria, Santa Maria	*Calophyllum spp*	Maria.

[1] Includes native and foreign species generally recognized as ship and boat building woods or which have been tested by the Navy and found suitable for certain ship and boat uses.

HARDWOODS—continued

Commercial name	Botanical name	Name used in this manual
Oak:		
Red oak	*Quercus velutina*	Black oak.
Red oak	*Quercus palustris*	Pin oak.
Red oak	*Quercus rubra*	Northern red oak.
Red oak	*Quercus coccinea*	Scarlet oak.
Red oak	*Quercus falcata*	Southern red oak.
White oak	*Quercus macrocarpa*	Bur oak.
White oak	*Quercus prinus*	Chestnut oak.
White oak	*Quercus virginiana*	Live oak.
White oak	*Quercus stellata*	Post oak.
White oak	*Quercus michauxii*	Swamp chestnut oak.
White oak	*Quercus alba*	White oak.
White oak	*Quercus lyrata*	Overcup oak.
Okoume (Gaboon)	*Aucoumea klaineana*	Okoume.
Pecan	*Carya cordiformis*	Bitternut hickory.
Pecan	*Carya illinoensis*	Pecan.
Pecan	*Carya aquatica*	Water hickory.
Pecan	*Carya myristicaeformis*	Nutmeg hickory.
Peroba do campo (ipé peroba)	*Paratecoma peroba*	Peroba do campo.
Teak	*Tectona grandis*	Teak.
Tupelo	*Nyssa sylvatica*	Black tupelo.
Walnut	*Juglans nigra*	Black walnut.
Yellow-poplar	*Liriodendron tulipifera*	Yellow-poplar.

SOFTWOODS

Cedar:		
Alaska cedar	*Chamaecyparis nootkatensis*	Alaska cedar.
Eastern red cedar	*Juniperus virginiana*	Eastern red cedar.
Incense cedar	*Libocedrus decurrens*	Incense-cedar.
Northern white cedar	*Thuja occidentalis*	Northern white cedar.
Port-Orford cedar	*Chamaecyparis lawsoniana*	Port-Orford cedar.
Southern white cedar	*Chamaecyparis thyoides*	Atlantic white cedar.
Western red cedar	*Thuja plicata*	Western red cedar.
Cypress, red	*Taxodium distichum*	Baldcypress.
Douglas fir	*Pseudostsuga menziesii* (formerly *taxifolia).*	Douglas fir.
Western larch	*Larix occidentalis*	Western larch.
Eastern larch	*Larix laricina*	Hackmatack, tamarack.
Pine:		
Idaho white pine	*Pinus monticola*	Western white pine.
Northern white pine	*Pinus strobus*	Eastern white pine.
Ponderosa pine	*Pinus ponderosa*	Ponderosa pine.
Sugar pine	*Pinus lambertiana*	Sugar pine.
Southern yellow pine	*Pinus palustris*	Longleaf pine.
Southern yellow pine	*Pinus echinata*	Shortleaf pine.
Southern yellow pine	*Pinus taeda*	Loblolly pine.
Southern yellow pine	*Pinus elliottii*	Slash pine.
Southern yellow pine	*Pinus caribaea*	Slash or Longleaf pine.
Redwood	*Sequoia sempervirens*	Redwood.
Spruce:		
Eastern spruce	*Picea glauca*	White spruce.
Eastern spruce	*Picea mariana*	Black spruce.
Sitka spruce	*Picea sitchensis*	Sitka spruce.

CHAPTER 2. BASIC INFORMATION ON WOOD SPECIES

2.0. GENERAL

The use of wood in boats and ships is as ancient as the shipbuilding art itself. Two characteristics inherent in most woods, buoyancy in water and workability with primitive tools, probably first led men to venture offshore on rafts or hollowed logs and thus to lay the foundation of navigation. The existence of bounteous forests containing wood well adapted to the construction of boats big and small undoubtedly contributed to its long dominance as the premier shipbuilding material. While wood has bowed to steel in the building of seagoing navies and merchant fleets, it possesses inherent properties that make it still unexcelled for the construction of certain types of naval vessels both large and small.

Wood consists mainly of hollow fibers, the walls of which are composed of cellulose and lignin, the latter, also acting as a cementing agent, binds them together. This lignin bond is so strong that breaks usually occur within the fiber walls rather than in the bond between the fibers.

Unlike metals, which have nearly uniform strength in all directions, wood has not, for instance, the same strength across the grain as parallel to the grain; its tensile strength may vary as much as 40 to 1, its crushing strength 7 to 1, and its modulus of elasticity 150 to 1. Not only do different species of wood differ in their properties, but trees of the same species and even parts of the same tree may vary, depending on the growth conditions prevailing when the wood was formed.

Wood is a hygroscopic material; that is, it takes on or gives off moisture until the amount of moisture it contains balances with that of the surrounding medium. Like many other hygroscopic materials, it shrinks as it loses moisture and swells as it absorbs moisture.

Plywood is made up by gluing plies or sheets of veneer together. Usually the grain of any one ply is at right angles to the grain of the adjacent ply or plies. The redistribution of material that results makes plywood much more uniform than solid wood as to strength and shrinkage. Synthetic-resin adhesives of the phenolic type contribute the water resistance necessary to make the joints in plywood durable for shipbuilding purposes.

The great water resistance of phenolic or resorcinol adhesives is also utilized in laminating ship timbers for frames, keels, stems, and other ship and boat parts which are more or less in direct contact with water. Laminates as distinct from plywood, do not redistribute wood in alternate crossed sheets; the grain of each layer, or lamination, runs in approximately the same direction as that of all others. Its principal advantages are that it permits the creation of large timbers from relatively small, thin pieces and, by careful selection and positioning of material, assures the manufacture of timbers in which the strength-reducing defects occurring in the tree or due to seasoning are controlled. Laminating has the added advantage that, while the thin laminations are being glued together, they can also be bent to desired curvatures for stems, frames, and other boat parts.

The properties of wood can also be modified by resin treatments and by combination of resin treatment and compression. Materials with marked stability against swelling and shrinking are thus made up of veneer impregnated with resins and pressed into solid pieces.

2.1. STRUCTURE AND GROWTH

2.10. BARK, WOOD, AND PITH. A cross section of a tree (figure 2–1) shows three main, well-defined features. One is bark, which consists of an outer corky, dead portion that varies greatly in thickness with age and with different species, and a thin, inner living portion. The second is wood, which in merchantable trees of most species consists of light-colored sapwood and darker heartwood. The third is the pith, a small, central, soft core that is formed when woody stems lengthen.

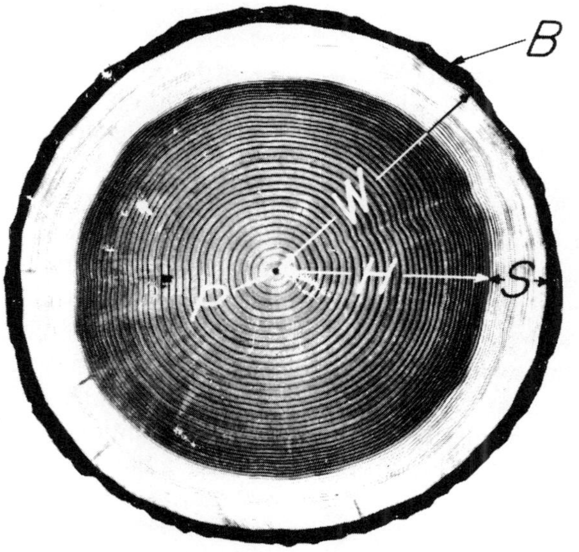

Figure 2–1. Cross section of a Douglas-fir log showing: *B*, bark; *W*, wood; *P*, pith; *H*, heartwood; *S*, sapwood.

2.11. CELLULAR STRUCTURE. Wood is composed of cells that are for the most part grown tightly together. Each year new cells are produced on the outer side of the sapwood by a thin layer of cells, called the cambium, located between the bark and the sapwood. Wood cells vary considerably in size and shape within a piece of wood and in different species. The principal kinds of cells and their usefulness in identifying wood species are discussed under section 2.14.

2.12. ANNUAL RINGS—SPRINGWOOD AND SUMMERWOOD. A cross section of a log grown in a temperate climate shows well-defined concentric layers of wood. These layers correspond closely to yearly increments of growth and for that reason are called annual rings. Many tropical timbers, however, show no well-defined annual rings because they grow more or less continuously throughout the year or growth may coincide with wet seasons.

Springwood is the wood formed on the inner side of the annual ring during the early part of each growing season. It is usually more porous, has thinner-walled cells, is softer, weaker, and, especially in the conifers, lighter in color than the summerwood, which is formed later in the season in the outer part of the annual ring. In most softwoods and some hardwoods—for example, oak—the summerwood is considerably denser than the springwood; in others, such as white pine, birch, maple, and yellow-poplar, the difference is much less pronounced, although the division between rings is plain enough. The width of the summerwood also varies considerably, in species such as Douglas fir, the southern yellow pines, and the oaks, according to environmental conditions and the vigor of the tree at the time the rings were formed.

2.13. SAPWOOD AND HEARTWOOD. In the living tree, the sapwood layer (figure 2–1) contains many living cells that serve mainly to transfer and store food. Most sapwood cells, however, are dead; they serve only as channels for the movement of sap and help to support the tree. The heartwood, in which all of the cells are dead, serves mainly to strengthen the trunk.

Year by year, as a tree increases in diameter by the addition of new layers of sapwood under the bark, the central heartwood area enlarges also. The change from sapwood to heartwood consists principally in the death of living cells in the adjacent sapwood and by the infiltration of coloring matter and various other materials into the cell walls and cell cavities; and in some woods tyloses (section 2.14), which plug the pores partially or completely, develop or increase in number. The circumference of the heartwood may be more or less irregular and need not closely follow the annual rings.

In many woods, the sapwood is markedly lighter in color than the heartwood. The pines, oaks, baldcypress, Douglas fir, and birch show this contrast. Some woods, as, for example, Port-Orford cedar and spruce, show little or no color difference between sapwood and heartwood.

Sapwood may, however, be discolored by sapstaining, wood-destroying, and other fungi, chemical stains within the wood, and color leached from the bark. The color of the heartwood may be uniform, or it may be streaked and variegated as in sweetgum. The outer part of the heartwood of old Douglas fir trees is often yellowish in color, in contrast to the more reddish inner part of the trunk. Occasional light-colored zones known as internal sapwood occur in heartwood of Douglas fir, Sitka spruce, Western redcedar, and some other species. Decay may discolor heartwood in various ways.

The sapwood of practically all trees is low in decay resistance. This is considered to be due mainly to lack of the toxic substances that are found in heartwood of durable species; some stain and mold fungi also favor the stored tree foods in sapwood. Unless given preservative treatment, sapwood should be excluded from structural members of boats and ships.

The thickness of the sapwood layer varies considerably among species and among logs of the same species. In Atlantic white-cedar, western red cedar, Douglas fir, and spruce, it is usually less than 1½ inches thick and consequently constitutes but a relatively small part of the lumber cut from these species. In southern yellow pine, white ash, birch, and maple, on the other hand, the sapwood band often comprises more than half the total volume of the log.

Within each species, the sapwood band is thickest in the most vigorous trees, especially those grown in the open. The proportion by volume of sapwood to heartwood generally increases toward the top of the tree as the trunk's diameter decreases.

Heartwood is fundamentally neither weaker nor stronger than sapwood; but some changes in physical characteristics other than color accompany heartwood formation. After the timber is cut, the heartwood in most species is more resistant to attack by certain insects and by decay, stain, and mold than is the sapwood. In the living tree, however, the sapwood is usually less subject to attack because it contains so much water, whereas specific fungi often infect the heartwood. The decay resistance of heartwood lumber varies markedly among species, some woods having heartwood that is not much superior to sapwood while in others the heartwood is highly resistant to decay.

As a rule, heartwood absorbs and loses moisture more slowly than sapwood and is more difficult to treat with preservatives. In resinous species, the heartwood usually contains more resin than the sapwood.

2.130. COLOR TESTS FOR DIFFERENTIATING HEARTWOOD AND SAPWOOD. The procedures used for differentiating between oak heartwood and sapwood, and between white oak and red oak are described in a Military Specification. Where it is difficult to tell sapwood and heartwood apart by visual inspection, one of two chemical solutions that have been found to give different coloration to heartwood than to sapwood may be used on some native species by applying them across the grain, preferably with a small camel's-hair brush.

One of these solutions works well on the natural wood of southern yellow pines and red pine and gives fairly satisfactory coloration with western white, sugar, and ponderosa pine; it is not effective with the larches, spruces, or Douglas fir. This solution consists of two separately prepared solutions, one of which is a 10 percent solution of sodium *nitrite* (sodium *nitrate* will not do). The other consists of 5 grams of benzidine dissolved in 23 cubic centimeters (25 grams) of 25 percent hydrochloric acid and 970 cubic centimeters of water. (Concentrated hydrochloric acid is diluted to the proper strength by pouring 25 cubic centimeters into a glass and adding enough water to make 38 cubic centimeters of liquid.) The sodium nitrite solution and the benzidine-hydrochloric acid solution should be mixed in equal amounts at the time the wood is tested. Within a few moments, sapwood will turn yellowish brown and heartwood red.

On white oak, chestnut oak, red oak, black oak, and scarlet oak, methyl orange solution is used as an indicator. The heartwood of white oak turns red, that of red oak turns pinkish-orange to orange, and the sapwood of both turns yellow. The indicator consists of 0.5 percent of methyl orange in a 25 percent solution of alcohol. The test may prove ineffective or erroneous, however, when applied to wood long exposed to weathering, to preservative-treated wood, or to wood infected with decay or stain fungi, molds, or the organisms present in "sour" logs.

2.14. HARDWOODS AND SOFTWOODS. All woods can be grouped in two general classes: Hardwoods, which come from trees with broad leaves, and softwoods, or conifers, which come from trees with needlelike or scalelike leaves. The terms "hardwoods" and "softwoods" are not descriptively exact, since some of the so-called softwoods, such as southern yellow pine and Douglas fir, are harder than some of the so-called hardwoods, such as basswood and cottonwood; but the terms have been in use so long that their meaning has been definitely established. Within each class there are considerable variations in structure.

Hardwoods as a class generally contain larger cells, constituting pores or vessels, scattered among the smaller cells, which are mostly fibers. Figure 2–2 shows the pores and other cells in a hardwood cube highly magnified. In many hardwoods the pores can be seen distinctly as small holes on smoothly cut cross sections and as fine grooves on planed longitudinal surfaces. In other hard-

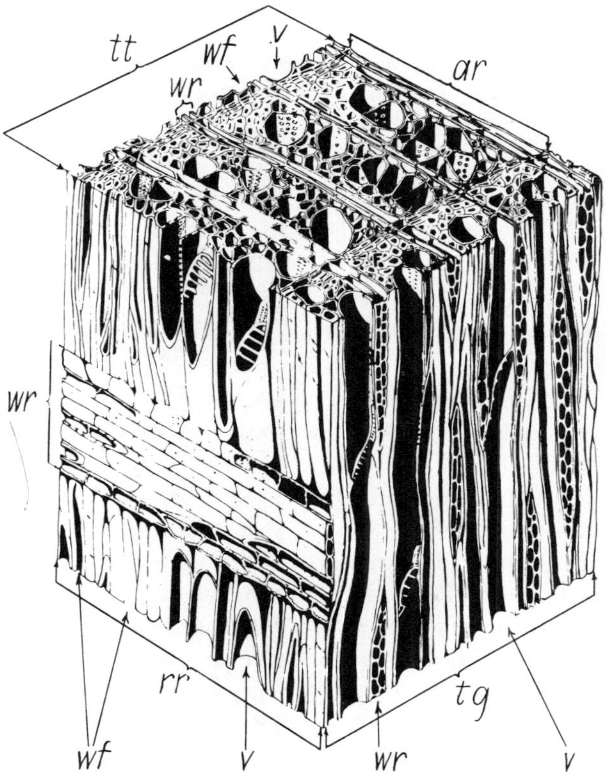

Figure 2–2. Drawing of a small cube of hardwood highly magnified: *tt*, transverse surface *rr*, radial surface; *tg*, tangential surface; *ar*, annual ring; *wr*, wood ray; *wf*, wood fibers; *v*, vessels or pores.

woods, the pores cannot be seen without magnification.

In certain hardwoods, the pores in the heartwood and inner sapwood become more or less plugged with hairlike ingrowths known as tyloses. These ingrowths are largely responsible for the watertightness of white oak.

The fibers in hardwoods are so small that they cannot readily be distinguished from one another with a hand lens. They average about $1/25$ inch in length. The thickness of the fiber walls greatly affects their strength.

In addition to the vertically arranged cells, there are horizontally elongated cells known as rays or wood rays, which extend radially from the bark inward. In oak, the rays are distinctly visible as broad lines on the cross section and as large "flakes" on radial surfaces. In all other native commercial species, they are much smaller.

In softwoods, the bulk of the wood is composed of fibrous cells called tracheids, averaging about $1/5$ inch in length. Figure 2–3 shows a drawing of a small softwood cube highly magnified. On account of the absence of pores, the softwoods are called nonporous woods.

On a smoothly cut cross section, the fibrous cells can be seen with a hand lens, resembling in their regularity the cells of a honeycomb, except that in the outer part of each annual ring they are flattened radially and thicker walled, producing the denser brand of summerwood. Rays are present in softwoods as well as in hardwoods, but they are always small and on the cross section are invisible without a lens.

Resin ducts are passages extending vertically between fibrous cells and radially within certain rays. They serve for the storage and conduction of resin and in native woods are present normally only in the pines, spruces, larches (tamaracks), and Douglas fir.

2.15. DISTINGUISHING BETWEEN WHITE AND RED OAK. In general the following characteristics should help in separating red from white oak. It is safest not to rely on only one characteristic.

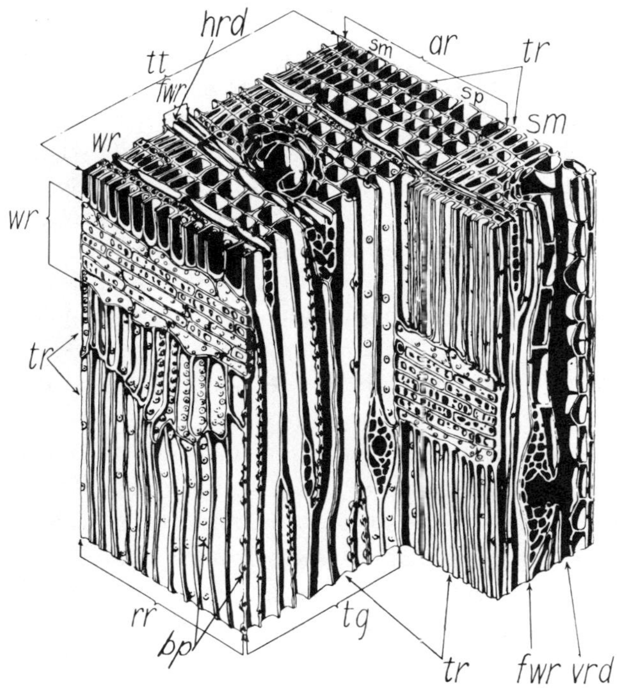

Figure 2–3. Drawing of a small cube of softwood highly magnified: *tt*, transverse surface; *rr*, radial surface; *tg*, tangential surface; *ar*, annual ring; *sp*, springwood; *sm*, summerwood; *tr*, tracheid, or fiber; *wr*, wood ray; *fwr*, fusiform wood ray containing horizontal resin duct; *hrd*, horizontal resin duct. The large hole near the center of the transverse section and the passage along the right edge are vertical resin ducts, *vrd*.

White Oaks	*Red Oaks*
1. Vessels in heartwood contain abundant tyloses	1. Vessels in heartwood contain few tyloses.
2. Color of heartwood is tan or light brown	2. Color of heartwood is reddish or pink.
3. Pores in the summerwood are very small and too numerous to be counted even with a hand lens (figure 2—4).	3. Pores in the summerwood are few, large, and open. They can be counted under a hand lens (figure 2–4).
4. Heartwood brushed with benzidine-sodium nitrite (section 2.130) turns dark brown or greenish brown.	4. Heartwood brushed with benzidine-sodium nitrite (section 2.130) turns light orange.

2.2. CHARACTERISTICS OF WOOD SPECIES

2.20. GENERAL. Of the great variety and number of wood species, relatively few have come into widespread use for ship and boat building. To a great extent, selection has been made on the basis of experience. Availability of certain woods in sufficient quantity and suitable dimensions, however, has also influenced their use, and the type of wood hull craft under construction has placed further limitations on the utility for shipbuilding of the many species native to a given region.

Broadly speaking, the principal considerations in selecting wood for ships and boats are (1) strength, (2) decay resistance, and (3) availability in the qualities, quantities, and sizes needed. The first two requirements are axiomatic. A boat must be strongly built to withstand the great stresses exerted against its structural members in heavy seas, grounding, dry docking, and under varying loads; and at least certain parts of it must be built of materials that are durable under use in conditions often highly favorable to decay fungi and certain physical hazards of wood. Other important properties of woods that affect their selection for boats are workability and water absorption.

Size and quantity requirements vary with the kinds and numbers of boats being built. Great amounts of large timbers are required to fabricate a single ship of the sawed-frame type. When large numbers of such hulls have been constructed, as under the stress of war emergency, availability of lumber often decides the species to be used. Even in Eastern yards close to southern yellow pine regions, Douglas fir planking from the west coast has been extensively used. Eastern white pine, once a favorite shipbuilding wood, has all but disappeared from New England shipyards because it is no longer available in the qualities, sizes, and quantities needed for decking, planking, and framing members.

The glue laminating process has greatly eased the shortages of our best shipbuilding woods in recent years by making possible large timbers made of thin lumber. Large solid timbers will undoubtedly remain scarce in the foreseeable future.

In smaller wood craft, such as motor launches and boats of V-bottom construction, the small sizes required permit a wider range of species selection on the basis of strength, durability, and such other factors as weight, easy working qualities, and freedom from defects.

The following summary of general characteristics of native and foreign woods available on the market presents brief descriptions of strength properties, decay resistance, and the sizes attained by mature trees as a general indication of what may be expected from these woods. Many of the species listed are currently in use for ship and boat parts. It must always be remembered, however, that variations exist even between trees of a single species which may have grown beside each other in the forest and that, in selecting wood for ship and boat building, sound judgment and knowledge are needed. The mere fact that white oak is an excellent species for keels and frames is no guarantee, for example, that every piece of white oak timber is suitable for such service.

2.21. HARDWOODS

2.2101. Apitong (*Dipterocarpus* spp.). Apitong, bagac, eng, gurjun, keruing, and yang are names commonly applied to the woods of some fifteen species of *Dipterocarpus* abundant in the Philippine Islands, Indo-Malaya, and New Guinea. The trees grow to 135 feet, often with 50 to 90 feet clear of branches, and diameters up to 90 inches. The wood of the different species looks very much alike. The sapwood is creamy yellow, gray, or reddish white; the heartwood is reddish purple or brown. The grain is usually straight. Apitong is slow to dry and somewhat refractory. The wood is moderately low in resistance to decay and insects, and moderately difficult to work.

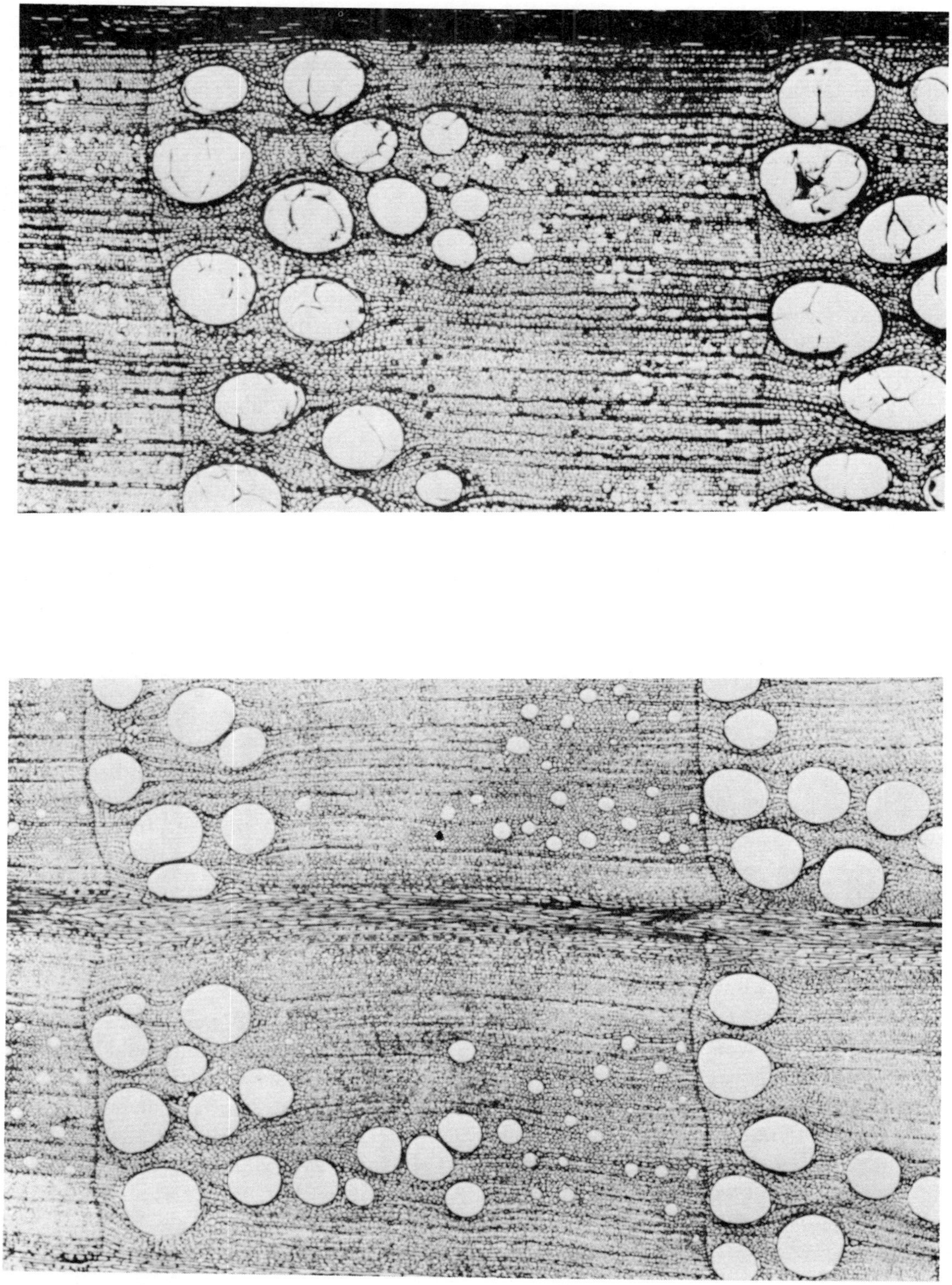

Figure 2–4. Appearance of pores under hand lens in red and white oak annual rings.

Uses include poles, framing lumber, flooring, bridge and wharf parts, boxes, and furniture. Indiscriminate use for boat and ship parts is not warranted.

2.2102. Ash (white ash, *Fraxinus americana*, green ash, *Fraxinus pennsylvanica*, black ash, *Fraxinus nigra*, and Oregon ash, *Fraxinus latifolia*). Ash has very limited use for boatbuilding. All species except Oregon ash grow in the eastern States. Commercial white ash consists mostly of white ash and green ash. Because of the demand, second-growth ash trees seldom exceed 80 feet in height and 18 inches in diameter. Heartwood of commercial white ash is brown; the sapwood is light colored or nearly white. Second-growth commercial white ash is heavy, strong, hard, and stiff, and has high resistance to shock. Such wide-ringed "tough ash" is used for handles, oars, vehicle parts, and sporting and athletic goods. Decay resistance is low. Ash trees growing in southern river bottoms produce buttresses that contain relatively light and weak wood sometimes solid as "cabinet ash" for cooperage, furniture, and containers. Oregon ash has somewhat lower strength properties than white ash, but it is used locally for the same purposes.

2.2103. Basswood (American basswood, *Tilia americana*, and white basswood, *Tilia heterophylla*). Basswood is not used in ship and boat parts. American and white basswood are the most important of several native basswood species. Forest-grown trees reach 90 feet with a diameter of 3 feet. No attempt is made to distinguish between the different species in lumber form. Basswood grows throughout the eastern half of the United States. Wide creamy-white or pale-brown sapwood merges gradually into yellowish-brown heartwood. The wood is soft and light in weight, straight-grained, easy to work, and stays in place well. Decay resistance is low. Basswood lumber is used largely for venetian blinds, sash and door frames, molding, woodenware, and boxes.

2.2104. Beech, American (*Fagus grandifolia*). American beech, which grows throughout most of the area east of the Great Plains, is not used for ship and boat parts. Mature trees attain a height of 120 feet with diameters of 2 feet or more. The wood varies in color from reddish brown to nearly white. Growth rings are not conspicuous. It is heavy, hard, stiff, of fine, uniform texture, has good shock-resisting ability, and resists abrasion. It is low in resistance to decay. The principal uses of beech are for boxes, baskets, crating, furniture, and handles.

2.2105. Birch, Yellow (*Betula lutea*). Yellow birch is not a shipbuilding wood. It cannot be distinguished from sweet birch (*B. lenta*) by the wood alone, and is very similar to sweet birch in strength properties, decay resistance, weight, stiffness, and hardness. It is the most abundant and commercially important of American birches, growing in the same regions as does sweet birch. It has white sapwood and light reddish-brown heartwood. It attains a little larger mature growth than does sweet birch. Yellow birch is used for the same purposes as sweet birch.

2.2106. Cherry, Black (*Prunus serotina*). Black cherry is not used for ship and boat parts. It occurs scatteringly throughout the Eastern States, and reaches 90 feet with a diameter of 3 feet. Greatest lumber production is in the Middle Atlantic States. The lustrous heartwood varies in color from light to dark reddish brown; sapwood is nearly white. The wood has a fairly uniform texture and machines very well. It is moderately heavy, strong, stiff, moderately hard, has high shock-resisting ability, and stays in place well after seasoning, being comparatively free from checking or warping. Principal use is furniture, though it is sometimes used for gunstocks.

2.2107. Elm, Rock (*Ulmus thomasii*). Rock elm, which grows in the North Central and Northern States, has some uses for ship and boat building. It is of smaller average size than American elm. The supply is very limited. Like American and slippery elm, it can be steam bent readily. Rock elm usually has interlocked grain and tends to warp. It is at least as decay resistant as American elm. In hardness and strength, rock elm is superior to the soft elms. It is also heavier, and has a high capacity to resist shock. For many uses, the different species of elm are employed indiscriminately. Where hardness or shock resistance is required, rock elm is preferred.

Cedar elm (*Elmus crassifolia*) and winged elm (*U. alata*), both southern species, have properties similar to those of rock elm, but both average somewhat smaller in tree size.

2.2108. Greenheart (*Ocotea rodiaei*). Greenheart, used for some ships parts, is native to tropical South America and the West Indies but is obtained largely from British Guiana. The tree grows to 125 feet with diameters up to 3 feet. Its heartwood is extremely decay-resistant. It is about as hard as black locust, and considerably stiffer and stronger. In shock resistance it compares with the beeches, birches, and oaks. It has long been a favorite wood of European shipbuilders. Its reputation for being highly resistant to marine borers, however, has not been supported by performance records in tropical waters. Its color ranges from pale greenish yellow to deep brownish purple, with little color difference between sapwood and heartwood. The most important use of greenheart is in marine construction, including ship and dock building, piling, and planking.

2.2109. Hickory, Pecan (bitternut hickory, *Carya cordiformis*, pecan, *Carya illinoensis*, water hickory, *Carya aquatica*, and nutmeg hickory, *Carya myristicaeformis*). The hickories are not used for shipbuilding. Bitternut hickory grows throughout the eastern half of the United States, the other three species of pecan hickory are largely in the Southern States. Mature trees are 2 feet or more in diameter and often more than 100 feet high. The wood resembles that of the true hickories, with white or nearly white sapwood and somewhat darker heartwood. The wood is heavy to very heavy and sometimes has very large shrinkage. Pecan hickory can be used for the same purposes as the true hickories, notably tool and implement handles. Some pecan hickory is giving satisfactory use as flooring. The lower grades are used for pallets.

2.2110. Hickory, True (shagbark, *Carya ovata*, pignut, *Carya glabra*, shellbark, *Carya laciniosa*, and mockernut, *Carya tomentosa*). True hickories, not used for shipbuilding, are found throughout most of the Eastern United States, reaching 110 feet and diameters of 2 feet. Commercial production is in the Middle Atlantic, Central, and Southern States. Sapwood is white, heartwood reddish. Hickory excels as a tough, heavy, hard, strong wood. Decay resistance is low and shrinkage considerable. Hickory is used for items like handles in which shock resistance is important. Lower-grade uses are pallets, blocking, and the like.

2.2111. Iroko (*Chlorophora excelsa*). Iroko is a large timber tree of tropical Africa, the wood of which resembles teak in a number of its properties, hence finds use for ship and boat parts. It comes mostly from the Ivory and Gold Coasts and Nigeria. Trees reach 150 feet with diameters up to 8 feet. It is low in shrinkage, comparing favorably in this respect with teak, but is not so strong as the latter species. The heartwood is resistant to decay and somewhat resistant to marine borer attack. The wood is hard, moderately heavy, and is worked with moderate ease. Heartwood is light to greenish yellow, darkening to brown upon exposure to light and air. Iroko is used in bridges, ships, and wharves.

2.2112. Ironbark (*Eucalyptus* sp.). Several species of *Eucalyptus*, termed red and gray ironbarks, are native to Australia. The trees grow to heights of 100 feet or more and furnish lumber for shipbuilding, poles, railroad ties, and structural purposes. Some is imported to the United States for specialty uses, such as ice sheathing for hulls. Heartwood is red to dark brown, sapwood light-colored. The heartwood has good decay resistance and its hardness makes it somewhat resistant to the crustacean forms of marine borers. The wood shrinks moderately, is very heavy and hard, and has good strength properties.

2.2113. Khaya (*Khaya* spp.). Very similar to mahogany and hence useful for shipbuilding, khaya is native to the West Coast of Africa from the Ivory Coast through the French Cameroons, and is known as "African mahogany." Trees grow to 150 feet with a trunk length of 80 to 100 feet and to 4 feet in diameter. Its color ranges from light pink through bright red to reddish brown, but is not quite so variable as that of mahogany. It is hard and decay resistant, with interlocked grain, but like mahogany, it has low shrinkage and seasons well. Khaya is used largely for furniture, fixtures, and interior finish.

2.2114. Lignumvitae (*Guaiacum* sp.). Lignumvitae is one of the hardest and heaviest woods known. The tree usually is small, about 30 feet high and 10 to 12 inches in diameter. Naturally impregnated with an oily substance, the wood has been extensively used for propellershaft bearings of ships. It is found principally in the coastal regions of Central America and the West Indies. Heartwood varies from olive brown or blue to dark brown or nearly black; sapwood is cream-colored. The heartwood is very resistant to decay and abrasion. Other uses include bearing parts that are permanently under water, such as water-power machinery.

2.2115. Locust, Black (*Robinia pseudoacacia*). Black locust is one of the heaviest and hardest woods native to the United States and has specialty uses in shipbuilding. Its growth range is the Appalachian Mountains from Pennslyvania to northern Georgia, southern Illinois and Indiana, and parts of Arkansas, Oklahoma, and Tennessee. Trees grow to 50 feet with diameters of 5 to 7 inches in 25 years. The wood ranks very high in shock-resisting ability and strength. For its density, it has very low shrinkage. Its heartwood is generally very decay resistant, some strains and varieties, such as shipmast and Flowerfield, being much more resistant than others. Black locust is subject to worm holes, cross grain, and bird pecks. Checking and splitting generally occur to only a moderate degree. Principal uses are treenails and insulator pins.

2.2116. Mahogany, "African." See Khaya.

2.2117. Mahogany, "Philippine." See Tangile.

2.2118. Mahogany (*Swietenia* sp.). Mahogany grows only in the West Indies, the extreme south of Florida, Mexico, Central America, and northern South America. Central American trees often attain 100 feet with diameters of 6 feet. Its merit for shipbuilding was early established. Color varies from deep red to reddish brown, with the sapwood pale yellow. It seasons readily and has very decay-resistant heartwood. That from the West Indies is uniformly heavy and hard, that from Central America more variable in weight, hardness, and strength. Its shrinkage is low and uniform. Mahogany is used principally for furniture, fixtures, interior woodwork, and shipbuilding.

2.2119. Maple, Hard (sugar maple, *Acer saccharum*, and black maple, *A. nigrum*). Hard maple is not a shipbuilding wood. It grows throughout the eastern United States and is commercially the most important and most abundant of the maples. Trees reach heights of 100 feet. Woods-run logs average 14 inches in diameter. Heartwood is light reddish brown and sapwood white with a slight reddish-brown tinge. The wood is heavy, hard, strong, and stiff, has good resistance to shock, and wears well under abrasion; decay resistance is low. It has moderate shrinkage and its grain is usually straight although sometimes curly, wavy, or bird's-eye grain occurs. Hard maple is used for flooring, furniture, and containers. Sugar maple is the source of maple sugar.

2.2120. Maria (*Calophyllum* spp.). Maria, or Santa Maria, grows in tropical America. Trees attain heights up to 150 feet with diameters up to 6 feet. Heartwood ranges from pink to rich reddish brown; sapwood is lighter. The grain is often interlocked. The wood is moderately durable, generally easy to work, but tends to "pick up" or show rough patches. It is used locally for general construction, bridges, shipbuilding, flooring, furniture, and rotary-cut veneer.

2.2121. Oak, Live (*Quercus virginiana*). Live oak grows principally along the south Atlantic and Gulf coasts. While the tree grows to heights of 50 or 60 feet, the main trunk, as a rule, divides into several large branches a few feet above ground. Mature trees sometimes exceed 6 feet in diameter. The wood has a distinctive dark brown color. Formerly important for shipbuilding, it is one of the heaviest woods native to the United States, is extremely dense and hard, and has good decay resistance. It is now cut very little and is not marketed through established hardwood lumber channels.

2.2122. Oak (*Quercus* sp.). The common commercial red oaks are black oak (*Quercus velutina*), pin oak (*Q. palustris*), Northern red oak (*Q. rubra*), scarlet oak (*Q. coccinea*), and Southern red oak (*Q. falcata*). Commercial white oaks are bur oak (*Q. macrocarpa*), chestnut oak (*Q. prinus*), post oak (*Q. stellata*), swamp chestnut oak (*Q. michauxii*), white oak (*Q. alba*), and overcup oak (*Q. lyrata*). The principal commercial species of oaks include about 11 of the more than 100 species, varieties, and hybrids native to the United States. Though they are generally similar in their properties, there are two essential differences. The white oaks differ from the red oaks in that their heartwood is more decay resistant; the heartwood pores, except in chestnut oak, are plugged with tyloses, which are much less numerous in most red oaks. This tylose growth accounts for the greater tightness of the white oaks against water and other liquids. Oak is generally heavy, hard, stiff, and strong, rapidly grown trees providing stronger, harder, tougher wood than those that have grown more slowly. The wood holds fastenings exceptionally well, has superior shock-resisting ability, and bends well. White oak is much preferred for most shipbuilding purposes; but if pressure-preservative treatment is practicable, red oak is suitable. The oaks are native to the eastern half of the United States, especially the Mississippi Valley and the South. Southern oak, particularly when swamp grown, is very difficult to dry without serious checking, honeycombing, and case-hardening, and machines with difficulty. Swamp chestnut oak and overcup oak are unsuitable for ship and boat parts because of their tendency to warp and their low strength. Oak is used principally for flooring, furniture, boxes and crates, and ship and boat parts.

2.2123. Okoume (*Aucoumea klaineana* Pierre). Okoume (called Gaboon in England) is available in large supply from West Africa and produces large, clear logs. Trees attain heights up to 130 feet with diameters up to 5 feet. Heartwood is salmon pink or pale pinkish brown, sapwood often grayish. The wood is usually straight-grained, uniform in structure, moderately light in weight, and fairly strong for its weight. It is low in resistance to decay and somewhat difficult to saw, requiring special saw teeth. Uses are paneling, furniture, boats and ships, and boxes.

2.2124. Peroba Do Campo (*Paratecoma peroba* (Record) Kuhlm). Peroba do campo, or ipe peroba, grows in Brazil. Trees attain 130 feet with diameters up to 60 inches. Sapwood is pale yellowish gray and heartwood pale olive with yellow, red, or brown shading. The wood is rated as strong and tough and highly decay resistant. Some difficulty is encountered in seasoning because of irregular grain. It shrinks more than mahogany but less than teak. The wood is not difficult to work, but tends to dull tools when much of a sulfur-yellow deposit is present in the pores. Uses include interior fittings, high-grade furniture, ship timbers, and boat decking.

2.2125. Sweetgum (*Liquidambar styraciflua*). Sweetgum has little use for shipbuilding. The heartwood is sold as red gum lumber. In color it ranges from a light to a deep reddish brown. The sapwood of sweetgum, commercially sold as sap gum lumber, is nearly white. Although sweetgum tends to warp badly in seasoning (because of interlocked grain), especially when plain sawed, drying and handling methods have been devised that produce satisfactory material. The wood is moderately heavy and strong, of fine, uniform texture, and finishes well. Heartwood is moderately decay resistant. Sweetgum lumber goes principally into boxes, crates, and furniture.

2.2126. Tangile (*Shorea polyspermal*). Tangile and several other hardwoods, including red lauan (*Shorea negrosensis*), white lauan (*Pentacme contorta*), tiaong (*Shorea teysmanniana*), obtained from the Philippine Islands, have been called "Philippine Mahogany," although they are not mahogany. They are used for ship and boat parts, though some, such as white lauan and tiaong, are limited to inside joinery. The trees yield large, clear boards. The heartwood varies from light to dark red or reddish brown, and is reported to be moderately decay resistant, moderately heavy, and somewhat difficult to season because of its interlocked grain. Tangile is also used for furniture and interior finish.

2.2127. Teak (*Tectona grandis*). Teak, valued especially for decking, grows chiefly in central and southern India, Burma, Thailand, and the East Indies. Mature trees attain diameters of more than 3 feet. Sapwood is white to pale yellowish brown, heartwood a dark golden yellow that turns almost black with age. The wood has a rough, oily feel to the touch, is straight-grained, coarse, and varies in weight from light to moderately heavy. It is strong, fairly hard, seasons slowly but without serious checking, has low shrinkage, and is easily worked but somewhat brittle. It is probably one of the most decay-resistant timbers in the world. It is not immune to attack by marine borers. Teak is extensively used for hull and deck planking, and fittings of ships and boats.

2.2128. Walnut, Black (*Juglans nigra*). Black walnut grows from New England to Texas. Heartwood is brown, sapwood nearly white. The wood is heavy, hard, strong, stiff, and has high resistance to decay. It stays in place well and is easily worked. Black walnut is used largely for furniture and interior finish.

2.2129. Yellow poplar (*Liriodendron tulipifera*). Yellow poplar, also known as tuliptree, grows throughout the eastern part of the United States, with the largest stands in the southern Appalachian Mountains. Trees reach 160 feet; woods-run logs average about 18 inches in diameter. The sapwood is white, the heartwood yellowish brown with a greenish or purplish tinge. The wood is moderately light in weight, straight-grained, moderately soft, and moderately weak. Its decay resistance is moderately low. It stays in place well and is easily worked. Uses are paneling, furniture, and interior cabin finish on ships and boats.

2.22. SOFTWOODS

2.2200. Baldcypress (*Taxodium distichum*). Baldcypress, a shipbuilding wood, grows along the southeastern coast of the United States and up the Mississippi Valley to Missouri. Lumber from trees grown in swamps near salt water is called red cypress or tidewater red cypress, while that from inland or upland growth is called yellow or white cypress. Mature trees are 70 to 150 feet high. Woods-run logs average about 18 inches. The heartwood of cypress near salt water varies from reddish to almost black, while that found farther inland is only slightly reddish or yellowish brown. Baldcypress is moderately light and moderately strong. Some is difficult to dry, but its heartwood is highly decay resistant. The wood is largely used where decay is an important factor, as in boats and buildings.

2.2201. Cedar, Alaska (*Chamaecyparis nootkatensis*). Alaska cedar, a much-used wood for boat and ship lumber, grows along the Pacific coast from Alaska to Oregon. The heartwood is a bright clear yellow, and the thin, scarcely distinguishable sapwood is a shade lighter. The wood is of a fine uniform texture and has low shrinkage. It is moderately light and moderately low in strength. It has excellent working and finishing properties, and heartwood ranks high in resistance to decay. A large part of the cut is used for small boats. Other principal uses are for interior finish, furniture, and cabinet work.

2.2202. Cedar, Atlantic White (*Chamaecyparis thyoides*). Atlantic white cedar, a tree of medium size, is native to the Eastern States. It is also widely known as southern white cedar and locally as juniper. It is very light in weight, soft, brittle, weak, and splits readily. Its low shrinkage with changes in moisture content and its high decay resistance, however, are qualities lending themselves to shipbuilding use, particularly boat planking. Atlantic white cedar lumber also goes largely into boxes and crates for fruits and vegetables.

2.2203. Cedar, Eastern Red (*Juniperus virginiana*). Eastern redcedar grows throughout the eastern United States. Trees may be up to 60 feet tall and 18 inches in diameter. Like all the cedars, eastern redcedar heartwood has exceptionally good decay resistance. Due to heavy use, trees are small in size. Sapwood is whitish, heartwood reddish brown. It is somewhat heavier, much harder, and tougher than western redcedar but not so stiff, and has slightly lower bending and compressive strength than the western species. Eastern redcedar lumber goes largely into chests, wardrobes, and closet lining.

2.2204. Cedar, Incense (*Libocedrus decurrens*). Incense cedar, little used for boatbuilding, grows in California, southwestern Oregon, and Nevada. At 100 years, the diameter is about 14 inches and the height, 65 feet. Sapwood is white or cream-colored, heartwood light brown, often tinged with red. The wood has a fine, uniform texture and a spicy odor. It is moderately weak, soft, low in shock-resisting ability, and lacks stiffness. It has small shrinkage and is easy to season with little checking or warping. Incense cedar is used principally for pencils, chests, construction lumber, posts, and ties. "Pecky" lumber is caused by a fungus in the tree, which is killed by seasoning. Because of its durability, incense cedar is well suited for fence posts.

2.2205. Cedar, Northern White (*Thuja occidentalis*). Northern white cedar, which grows in relatively small sizes, is used principally for small boat construction. Although its properties are similar to those of Atlantic white cedar, the decay resistance and low shrinkage of its heartwood make it well suited for boat planking.

2.2206. Cedar, Port Orford (*Chamaecyparis lawsoniana*). Port Orford cedar, a preferred boatbuilding species, is the heaviest and strongest of the cedars. It grows naturally along the coast of southern Oregon and northern California. Mature trees reach heights of 180 feet with diameters of 6 feet. It has a distinctive spicy odor. Heartwood is light yellow to pale brown, moderately light in weight, of moderate strength and hardness, and highly resistant to decay. The wood is of a fine and unusually uniform texture. It is used in boatbuilding principally for planking. Other important products are battery separators and venetian-blind slats.

2.2207. Cedar, Western Red (*Thuja plicata*). Western redcedar grows principally in the humid regions along the north Pacific coast and extends inland from Washington to Montana. In regions of best growth, trees reach 150 feet in height and 8 feet in diameter. It has narrow white sapwood and reddish-brown heartwood. The wood is easily worked, rather soft and weak, shrinks very little, and the heartwood has good resistance to decay. Uses are house siding, poles, shingles, and construction other than hulls.

2.2208. Douglas fir (*Pseudotsuga menziesii*, formerly *taxifolia*). Douglas fir is one of the principal shipbuilding woods grown in the United States. It is available in large sizes and quantities and is strong, moderately hard, heavy, and moderately decay resistant. Trees of the coastal region of Washington, Oregon, and California are larger than those growing inland east of the Cascades. Inland Douglas fir is often mixed with western larch and sold as "larch and fir," that from the Rocky Mountains as red fir. Most of the lumber sold as Douglas fir comes from the coast regions. Coast Douglas fir sometimes grows to 8-foot diameters. Old-growth Douglas fir has little sapwood. The species differs from the true firs in that the wood is heavier, stronger, more durable, darker colored, and pitchy. Uses other than shipbuilding are construction lumber, piling, plywood, millwork, poles, and containers.

2.2209. Larch, Western (*Larix occidentalis*). Western larch, though not much used for shipbuilding, is a promising wood for this use. It resembles Douglas fir in appearance except that it has narrower sapwood and its heartwood is russet brown in color rather than reddish. The wood is closely related to tamarack, or hackmatack. It is rated as moderately heavy, strong, stiff, and moderately good in shock resistance. Its resistance to decay is also moderate. Lumber from butt logs is often subject to shakes and its splitting resistance is low. Western larch and Douglas fir are logged together in the Inland Empire, where it is principally found, and are frequently graded, manufactured, and sold together as "larch and fir." It is a satisfactory construction wood, and in accordance with its properties should be suitable for wider use in ship construction.

2.2210. Eastern Larch, Tamarack (*Larix laricina*). Shipbuilders know tamarack as hackmatack. Tamarack grows chiefly in the New England and Lake States and is closely related to western larch. Mature trees are about 70 feet high and 20 inches in diameter. Heartwood is yellowish brown, the narrow sapwood nearly white. It has moderate strength properties. It has numerous small branches and relatively little clear material is obtained from the smaller trees. Moderately decay resistant, tamarack was formerly widely used for shipbuilding, especially for ship knees.

2.2211. Pine, Eastern White (*Pinus strobus*). The scarcity of virgin eastern white pine has sharply curtailed the use of this once popular shipbuilding wood. Second-growth stock is usually smaller and faster growing, hence weaker and more knotty than old growth. The species is found principally in the Lake States, the Northeastern States, and the Appalachian region. The wood is moderately light in weight, moderately low in strength, and usually straight-grained. The decay resistance of its heartwood is moderate, but second-growth trees usually contain much sapwood. As changes in moisture content affect its dimensions only slightly, and as it is easily worked, the species has long been popular among boat and ship builders for decking and planking as well as battens and templates.

2.2212. Pine, Ponderosa (*Pinus ponderosa*). Ponderosa pine, which has some uses for ship and boat building, grows from Washington to the Black Hills and southward through the Rocky Mountain and Pacific Coast regions. Trees grow to heights of 125 feet with diameters of 3 feet or more. The wood is somewhat similar to eastern white pine in strength and working properties but somewhat lower in decay resistance. It is used for containers, and for sash, doors, general millwork, patterns, and templates.

2.2213. Pine, Sugar (*Pinus lambertiana*). Sugar pine, a widely used pattern wood, grows in the inland empire region of Idaho, Montana, and eastern Washington and Oregon. It is the largest of the pines, sometimes reaching a height of 240 feet with a diameter of 10 feet. Although the lumber is similar in appearance and properties to eastern white pine, it can usually be distinguished from white pine by its more conspicuous resin ducts and lack of reddish color. It is largest species of the white pine group and yields the bulk of the exceptionally wide, thick stock of soft pine lumber. It is used interchangeably with white pine for many purposes. Its decay resistance is moderately low.

2.2214. Pine, Southern Yellow (*Pinus spp.*). Commercial southern yellow pine, when available in required sizes and grades, is a good shipbuilding wood. It includes a number of closely related species, chiefly longleaf (*Pinus palustris*), shortleaf (*P. echinata*), loblolly (*P. taeda*), and slash (*P. elliottii*). Growth range is from New Jersey south to the Gulf and west to Texas. Except for the fact that the pith often distinguishes longleaf from loblolly and shortleaf, the individual speeies cannot be positively identified by the wood alone. Grading rules of the Southern Pine Association classify all structural southern yellow pine as either "longleaf" or "shortleaf." Under these rules, "shortleaf" structural lumber includes loblolly, slash, or any other southern yellow pine species, but "longleaf" includes only longleaf pine (*Pinus palustris*). Structural lumber must conform with the density rule for the species. Structural southern yellow pine averages moderately hard and strong but is highly variable in these respects. Dense heartwood southern yellow pine is considered almost as decay resistant as white oak. The sapwood can be easily treated with preservatives to improve its naturally low decay resistance.

So-called North Carolina pine, which grows extensively in the Atlantic Coastal Plain, is largely second-growth loblolly pine. Arkansas soft pine is principally shortleaf pine that grows chiefly in or near the Ozark Mountains.

Dense southern yellow pine of structural grade is well suited for planking, keels, stems, stern posts, deck beams, and other structural members of ships. Southern yellow pine provides much of the Nation's general building and container lumber.

2.2215. Pine, Western White (*Pinus monticola*). Western white pine, sold commercially as Idaho white pine, is a good wood for ship patterns and templates. It grows principally in northern Idaho, eastern Washington, and western Montana. It reaches fairly large sizes in old-growth forests. It resembles eastern white pine so closely that the wood of the two species is indistinguishable. Like the eastern species, it is highly regarded for pattern making. It is light in color, moderately light in weight, and straight-grained. It swells and shrinks a little more than eastern white pine with changes in moisture content, but on the whole, compares favorably with the latter, having about the same decay resistance. It is easy to work and has low shrinkage. It is widely used for millwork and shop grades of lumber.

2.2216. Redwood (*Sequoia sempervirens*). Redwood, of limited use in ships and boats, grows in a narrow coastal belt in southern Oregon and northern California. Since the tree commonly reaches diameters of 5 to 10 feet, a high percentage of clear, large-sized lumber is obtainable. The wood is moderately light in weight and moderately strong. Its sapwood band is very narrow, almost white in color, and the heartwood varies from cherry to dark mahogany. The wood is exceptionally decay resistant, shrinks and swells little, stays in place well, and is easy to work. Redwood lumber is used principally for a wide range of building construction in the form of dimension, boards, timbers, siding, and millwork. Shipbuilding uses include hull planking of large and small ships and superstructure framing.

2.2217. Spruce, Eastern (*Picea* spp.). Eastern spruce gets some use in ship and boat building. Three species, red (*Picea rubens*), white (*P. glauca*), and black spruce (*P. Mariana*), are usually classed as eastern spruce. White and black spruce grow mainly in the Lake States and New England, while red spruce is found in New England and the Appalachian Mountains. Mature trees may be 60 to 80 feet in height and 18 inches in diameter. All three species have about the same properties and in commerce no distinction is made among them. The wood is light in color, with little difference in appearance of sapwood and heartwood. It dries easily, stays in place well, is moderately light in weight, easily worked, and has moderate shrinkage, strength, stiffness, toughness, and hardness. It is not resistant to decay. Eastern spruce is used principally for boxes and crates, pulpwood, and construction lumber.

2.2218. Spruce, Sitka (*Picea sitchensis*). Sitka spruce has some uses for shipbuilding, its low decay resistance being offset by high strength-to-weight ratio and moderate shrinkage. It grows along the Pacific Coast from Alaska to California. Old-growth trees may reach 200 feet in height with diameters of 6 feet. The heartwood is light pinkish brown and the sapwood creamy white. For its weight, which is moderately light, Sitka spruce is a strong wood. Compared with other woods without considering weight, it ranks as moderately hard, moderately stiff, moderately low in shock resistance, and moderately strong in bending. Sitka spruce lumber is used principally for boxes and crates. Furniture, planing-mill products, and general millwork are other major uses.

2.3. DEFECTS AND BLEMISHES

2.300. General. A defect is any irregularity occurring in wood that may lower its strength. A blemish is anything, not necessarily a defect, that mars the appearance of wood. Thus, decay is a defect because it impairs wood's strength; certain stains, on the other hand, do not affect strength appreciably, the chief objection to them being that they discolor the wood. Such defects as cross grain not only weaken wood but are responsible for much serious warpage and other effects of shrinking and swelling, and make wood surfaces less suited to gluing.

While defects such as decay must be guarded against when selecting wood for any type of ship or boat, others are important chiefly as their size relates to that of wood members in which they occur. A check or knot that may be acceptable in a 10 by 10 inch timber, for example, might be so large as to make a 1 by 6 inch frame member unfit for use. Likewise, such defects as compression failures, compression wood, and pitch pockets, which are often disregarded in large structural members, must not be permitted in relatively thin planking, beams, and frames, the strength of which is more critically affected by them.

The following is a discussion of the nature, occurrence, and objectionable features of defects and blemishes in lumber. Some of the defects described are common, but others occur in frequently or in only a few species.

2.301. Knots. A knot is a portion of a branch or limb embedded in a piece of lumber or veneer. Normally, a branch starts at the pith and increases in diameter from the pith of the tree outward as long as it is alive. Occasionally, however, branches start at some distance from the pith.

The fibers of a living branch interlace with those of the tree trunk, producing intergrown knots (figure 2–5). Many of the lower branches die, however, at some intermediate stage of the tree's growth, as a result of shading or other causes, and break off many years thereafter. After the death of a branch, the new wood formed in the tree trunk makes no further connection with it but grows around it, producing a dead knot, which may be either loose, so that it will drop out, or encased, so that it is held in position when the trunk is sawed into lumber.

Eventually, the dead branch breaks off, the stub heals over, and the distortion of grain in successive growth layers becomes less and less with increasing diameter of the trunk, until finally clear wood with normal grain is produced in the wood covering the knot.

A knot cut through transversely is known as a round knot, one cut through obliquely is known as an oval knot, and one cut through lengthwise is known as a spike knot (figure 2–5). A sound, tight knot is solid across its face, harder than the surrounding wood, shows no sign of decay, and is so fixed by growth or position that it will firmly retain its place in the piece. A watertight knot is one intergrown on at least one surface of the piece and sound on that surface.

Knots occur singly, in clusters, and as branch knots. A cluster of knots consists of two or more grouped together as a unit with the wood fibers deflected around the entire unit; a group of single knots does not constitute a knot cluster. Branch knots are two or more knots branching from a common center.

Knots are objectionable on account of the distortion and the discontinuity of the grain which they cause, which in turn weakens the wood, causes irregular shrinkage, and makes machining difficult. In resinous species, pitch often exudes more freely from knots than from the clear wood.

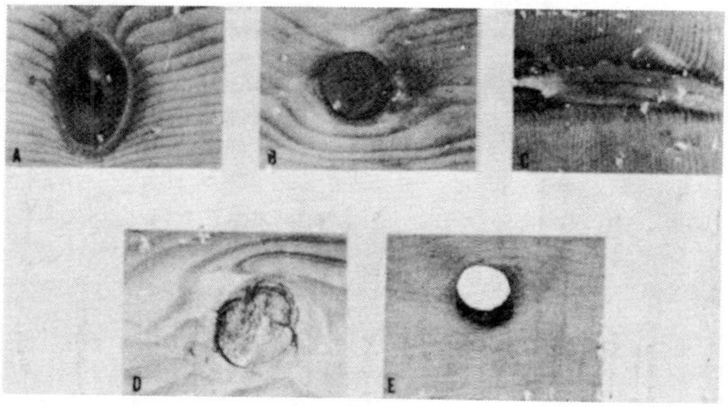

Figure 2–5. Types of knots. *A*, intergrown knots; *B*, encased knot; *C*, spike knot; *D*, decayed knot; *E*, knot hole.

When lumber dries, the knots tend to check or loosen because a knot shrinks differently from the wood around it. In ships and boats, knots are also objectionable because they do not assure permanent watertightness.

In determining the size of a knot, only the knot itself is measured, not the distorted grain around it, and specifications are based upon the size of the knot proper. A knot up to ½ inch in diameter is called a pin knot; a small knot is ½ to ¾ inch in diameter; a medium knot is ¾ to 1½ inches in diameter, and a large knot is more than 1½ inches in diameter.

Ship and boat lumber containing knots is, as a rule, graded according to (1) the size of the knot, (2) the condition of the knot whether it is sound or unsound, tight, or loose, and (3) the use to which the piece is to be put in the ship.

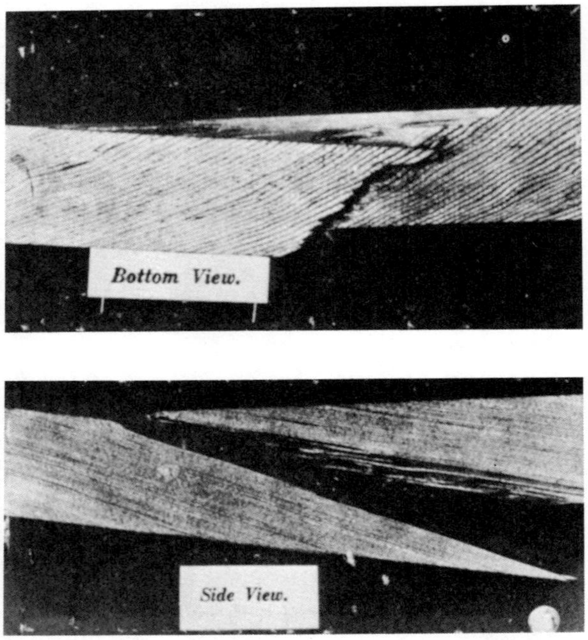

Figure 2–6. Typical break in diagonal-grained specimen (Douglas-fir).

2.302. Cross Grain. A piece of wood is called cross-grained when the fibers are not parallel with the major axis or the sides of the piece. Excessive slope of grain is objectionable because it reduces strengthen (chapter 4), causes warping in drying, makes smooth surfacing difficult when the wood is planed against the grain, and is responsible for much breakage in bending stock.

Slope of grain is conveniently measured in inches of deviation from straightness within a definite length (See *2.4(2)*).* A deviation, or "slope," of 1 inch in a length of 10 inches is steeper than a slope of 1 inch in a length of 15 inches. Slope-of-grain limitations are usually expressed as 1 in 15, 1 in 20, and so on.

Cross grain may be either of two major types, diagonal grain (figure 2–6) or spiral grain (figure 2–7).

Diagonal grain is deviation of the plane of the annual rings from parallelism with the longitudinal axis of a piece of wood. It is due to such natural causes as crook, bulges, butt swell, pitch pockets and bark pockets, blister grain, some types of curly grain, healing over of knots and injuries, and to the common practice of sawing tapered logs parallel to the pith instead of parallel to the bark.

For convenience, the slope of grain in the radial plane of a piece, no matter what its cause, is usually spoken of as diagonal grain. Since diagonal grain is usually easier to detect than spiral grain, examination should first be made for diagonal grain. If its slope is steeper than that allowed in the specification or other limiting requirement, an examination for spiral grain need not be made.

Spiral grain occurs when, due to an unknown cause, the fibers in a tree are not vertical but follow a spiral course similar to that of the stripes on a barber pole. Spiral grain in peeled posts, poles, or tree trunks is evidenced by inclined rather than vertical season checks, as in the dead tree to the left in figure 2–8. The slope of the spiral also fluctuates sometimes, especially near the butt of the tree, so that a block split radially will show a ruffled or fluted appearance, as in figure 2–9.

Truly straight-grained lumber cannot be produced from spiral-grained trees, but with wide flat-sawn boards or cants sawed some distance from the pith, the slope of the spiral can be reduced by proper attention in edging and ripping such pieces into smaller ones. On the other hand, false or artificial spiral grain is produced when straight-grained timber is not cut parallel to the fibers as seen on the tangential surface. It looks like natural spiral grain and should be so classed.

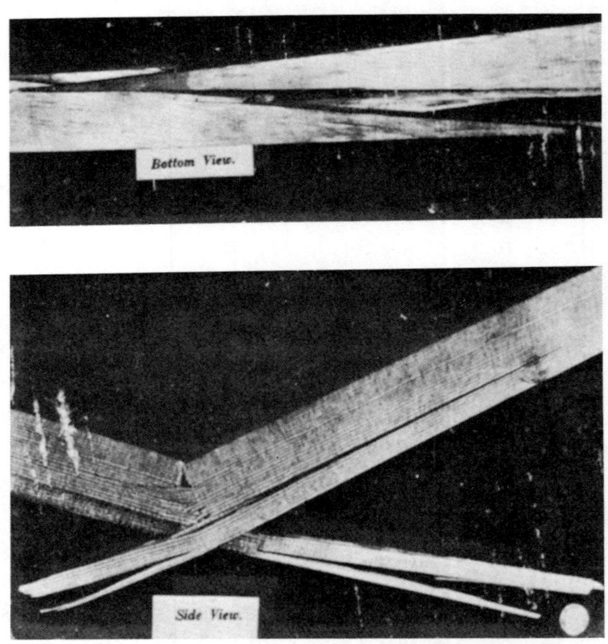

Figure 2–7. Typical break in spiral-grained specimen (Sitka spruce).

*Note: Italic numbers in parenthesis refer to references at the end of this chapter.

Figure 2–8. Straight grain and right-hand spiral indicated by direction of seasoning checks in dead tree trunks.

The principal indicators of the fiber direction on a truly tangential (plain-sawed or flat-grained) surface are checks, pores, resin ducts (fine brownish lines in the pines, spruces, Douglas fir, and tamarack), wood rays (most conspicuous in the oaks), the course taken by a scribe as explained in the following paragraphs, the direction in which a free-flowing ink or dye spreads, and the surface splits formed when a sharp pointed tool, such as a chisel, is pushed into the piece and pried against the wood.

The scribe test consists in drawing a sharply pointed steel scribe in the direction in which the grain seems to run. Enough pressure must be supplied so that the point will penetrate the wood slightly, and freedom of lateral movement must be enough to allow the point to follow the grain.

Figure 2–10 illustrates two scribes that have been found satisfactory. The point trails about 5 inches behind the vertical handle, thus affording it freedom of lateral movement. Scribe A is superior to B in that the handle is swiveled; the point, a phonograph needle or electric moisture-meter point, can be easily replaced; and the long, tapered head on the set screw that holds the needle in place can be used to line up the direction in which the scribe is pulled with the path it has recently traversed. Scribe B is made of a single piece of drill rod, tapered and hardened at the pointed end.

Figure 2–9. Spiral grain of varying or fluctuating slope indicated by split radial surface.

It has been found that the scribe points follow the grain best if inclined 10 to 20 degrees from the vertical toward the direction in which they are pulled.

In using the scribe, first make a trial scratch, pulling the scribe in the direction in which the grain appears to run; then make a second scratch

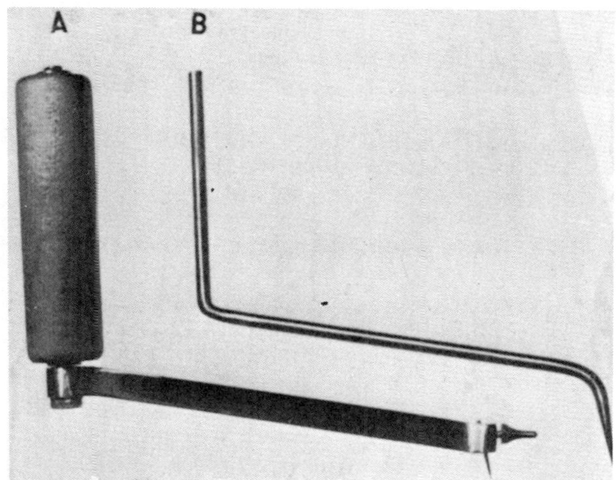

Figure 2–10. Scribes for determining the direction of fibers on the surface of wood. *A*, Swivel-handled with phonograph-needle point. *B*, one piece drill rod bent to the shape and point slightly hardened.

near the first one by pulling the scribe in the direction parallel to the first one unless it obviously did not follow the grain. Sometimes bands of summerwood deflect the scribe from the true course of the fibers; in such cases, a number of short scratches should be made on the springwood only.

It should be borne in mind that the scratches made by the scribe indicate the direction of the fibers on the surface only, and do not necessarily indicate the true slope of spiral grain unless the surface is tangential or practically so.

Sometimes a piece of wood has diagonal and spiral grain in combination. The true slope of grain of such pieces can be determined by measuring the slope of spiral grain and that of the diagonal grain separately, and then consulting table 2–1. In pieces of wood without radial or tangential surfaces, the true slope can be obtained by the grain slope on two adjacent surfaces, as determined with a scribe, and combining the two according to the data of table 2–1.

TABLE 2–1.

COMBINED SLOPE OF SPIRAL AND DIAGONAL GRAIN*

Slope of spiral or diagonal grain, whichever is lesser	1–10	1–11	1–12	1–13	1–14	1–15	1–16	1–17	1–18	1–19	1–20	1–21	1–22	1–23	1–24	1–25	1–26	1–27	1–28	1–29
1–10	7.1																			
1–11	7.4	7.8																		
1–12	7.7	8.1	8.5																	
1–13	7.9	8.4	8.8	9.2																
1–14	8.1	8.7	9.1	9.5	9.9															
1–15	8.3	8.9	9.4	9.8	10.2	10.6														
1–16	8.5	9.1	9.6	10.1	10.5	10.9	11.3													
1–17	8.6	9.2	9.8	10.3	10.8	11.2	11.7	12.0												
1–18	8.7	9.4	10.0	10.5	11.1	11.5	12.0	12.4	12.7											
1–19	8.8	9.5	10.1	10.7	11.3	11.8	12.2	12.7	13.1	13.4										
1–20	8.9	9.6	10.3	10.9	11.5	12.0	12.5	13.0	13.4	13.8	14.1									
1–21	9.0	9.7	10.4	11.1	11.6	12.2	12.7	13.2	13.7	14.1	14.5	14.9								
1–22	9.1	9.8	10.5	11.2	11.8	12.4	12.9	13.5	13.9	14.4	14.8	15.2	15.6							
1–23	9.2	9.9	10.6	11.3	12.0	12.6	13.1	13.7	14.2	14.6	15.1	15.5	15.9	16.3						
1–24	9.2	10.0	10.7	11.4	12.1	12.7	13.3	13.9	14.4	14.9	15.4	15.8	16.2	16.6	17.0					
1–25	9.3	10.1	10.8	11.5	12.2	12.9	13.5	14.1	14.6	15.1	15.6	16.1	16.5	16.9	17.3	17.7				
1–26	9.3	10.1	10.9	11.6	12.3	13.0	13.6	14.2	14.8	15.3	15.9	16.3	16.8	17.2	17.6	18.0	18.4			
1–27	9.4	10.2	11.0	11.7	12.4	13.1	13.8	14.4	15.0	15.5	16.1	16.6	17.1	17.5	17.9	18.3	18.7	19.1		
1–28	9.4	10.2	11.0	11.8	12.5	13.2	13.9	14.5	15.1	15.7	16.3	16.8	17.3	17.8	18.2	18.6	19.1	19.4	19.8	
1–29	9.5	10.3	11.1	11.9	12.6	13.3	14.0	14.7	15.3	15.9	16.5	17.0	17.5	18.0	18.5	18.9	19.4	19.8	20.1	20.5

Slope of spiral or diagonal grain, whichever is steeper (column headings)

How to Determine Combined Slope of Spiral and Diagonal Grain

First, determine slopes of spiral and of diagonal grain separately. To determine the combined or resultant slope, find the column headed by the steeper of the two slopes; and in this column, locate the figure in line with the lesser of the two slopes, as given in the left-hand column. This figure represents the length in inches (or other units) in which the grain deviates 1 inch (or other unit) with respect to the central axis of the piece.

Examples: If the slope of spiral grain is 1 in 20 and the slope of diagonal grain is 1 in 25, the combined slope is 1 in 15.6; or if the slope of diagonal grain is 1 in 18 and the slope of spiral grain is 1 in 22, the combined slope is 1 in 13.9.

*NOTE. Based on formula: Combined slope of grain $=\sqrt{(\text{slope of spiral grain})^2+(\text{slope of diagonal grain})^2}$.

2.303. Compression Wood. Leaning softwood trees often develop on their lower sides abnormal wood called compression wood (see 2.4 (5)). Such wood has relatively wide annual rings and unusually wide summerwood which, however, does not appear so dense and hornlike as normal summerwood. The resulting lack of contrast with springwood gives a lifeless appearance to compression wood, particularly when dry (figure 2–11). On lumber surfaces, compression wood usually has a yellowish or slightly brownish color when dry and a reddish color when wet. Streaks of compression wood are frequently interspersed in normal wood. Well-developed compression wood can usually be detected by ordinary visual inspection.

Compression wood shrinks excessively in the longitudinal direction. When compression wood is present in the same piece with normal wood, the unequal endwise shrinkage causes bowing and crooking. Occasionally, excessive and irregular shrinkage causes cross breaks.

Strength properties of compression wood are lower than those of normal dry wood (chapter 4). In bending, compression wood breaks with a brittle fracture, while normal wood splinters (figure 2–11).

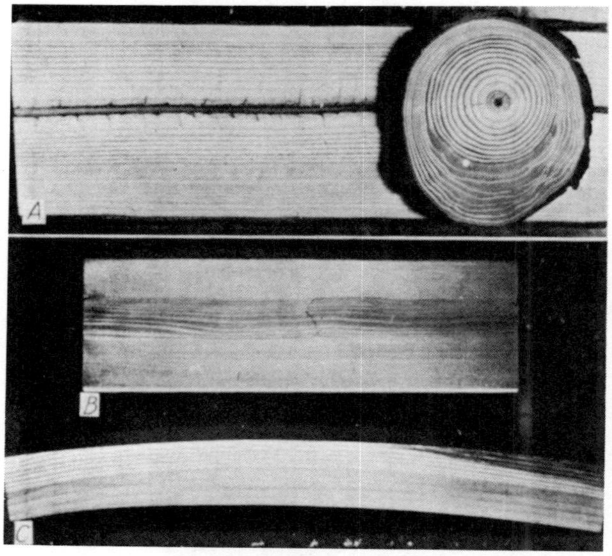

Figure 2–11. *A*, Longitudinal and cross sections through tree trunk with compression wood on lower side; *B*, cross break in compression wood and splits between compression wood and normal wood due to greater longitudinal shrinkage of compression wood; *C*, crook caused by longitudinal shrinkage of compression wood on one edge of a piece.

2.304. Tension Wood. In such hardwood species as oak, elm, and mahogany, tension wood includes numerous fibers with abnormal structure in their cell walls. In those fibers, the peculiar inner layers are only loosely attached to outer layers of the fibers and also show other unusual characteristics. Tension wood occurs predominantly on the upper side of leaning trees but sometimes is distributed entirely around the trees.

Tension wood frequently causes excessive longitudinal shrinkage and warping in hardwood lumber when moderate to large numbers of its abnormal fibers are present. In addition, tension wood fibers hold together tenaciously, so that sawed surfaces usually have projecting fiber bundles, sometimes nearly 1 inch in length, and planed, shaped, or turned surfaces have torn or raised grain (figure 2–13). Those abnormal fibers also cause veneer to buckle in drying, to the extent that it is difficult to make flat plywood. In addition, tension wood gives fibrous surfaces when the veneer is cut. These fibrous surfaces or raised grain on lumber or veneer make it possible to detect tension wood (see 2.4 (4)).

In white oak or elm bending lumber, however, tension wood is not a disadvantage, because it has somewhat lower compressive strength along the grain than normal wood of the same species. Such a property of the wood tends to permit shortening on the concave side without localized failures in the bent pieces.

2.305. Checks, Splits, and Shakes. A check is a longitudinal crack in wood, generally in the radial direction or across the annual rings (figure 2–14). Checks are usually due to uneven shrinkage in seasoning. Thick lumber checks more severely than thin lumber, and wide lumber more severely than narrow lumber.

Navy specifications define a small surface check as a perceptible opening not over 4 inches long; a medium surface check as not more than 1/32 inch wide and between 4 and 10 inches long; and a large surface check as more than 1/32 inch wide and more than 10 inches long.

A split is a longitudinal crack in wood caused usually by rough handling. Typically, it extends

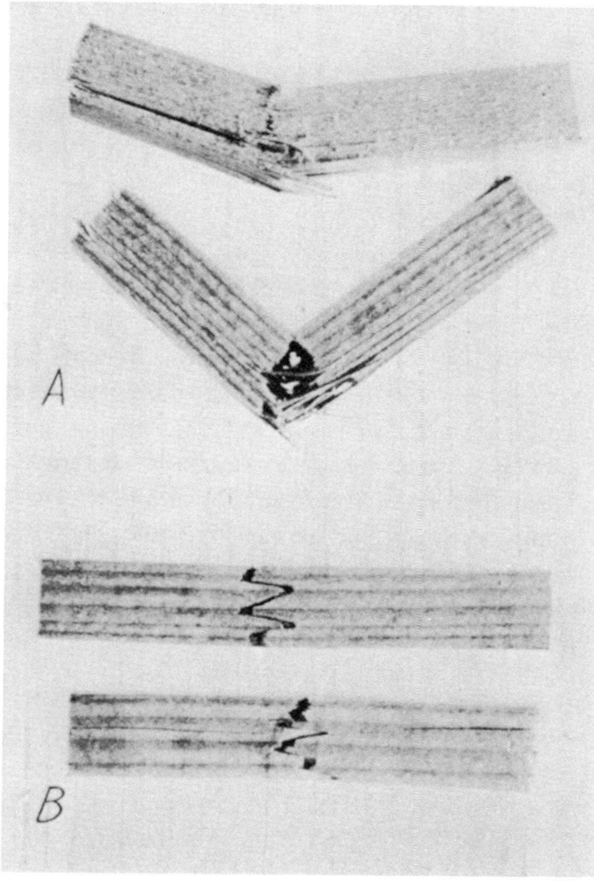

Figure 2–12. Typical bending failure represented by *A*, splintering normal wood, and *B* brittle compression wood.

Figure 2–13. Fibrous surface caused by tension wood in mahogany lumber.

through the thickness of a piece from side to side, regardless of whether the piece is edge-grained or flat-grained. Thick lumber does not split so readily as thin stock.

A shake is a longitudinal crack in wood along or between the annual rings (figure 2–15). Shakes, unlike checks and most splits, do not develop in seasoning or handling but originate in green timber. They may, however, become accentuated in seasoning. They are more prevalent near the pith or heart center of the tree and consequently occur chiefly in lower grades of lumber.

Depending on their size and the size of the piece of lumber, checks, splits, and shakes may seriously weaken wood members in resistance to longitudinal shear.

2.306. Compression Failures. Excessive bending of standing trees from wind or snow, felling trees across boulders, logs, or irregularities on the ground, or rough handling of logs or lumber may produce excessive compression stresses along the grain that causes minute buckling of the fibers. The deformation of the fibers so caused ranges from well-defined buckling visible to the unaided eye, as wrinkles across the face of a piece, to slight crinkling of the fiber walls visible only with a high-power microscope.

Compression failures can best be seen on wood so placed that light strikes along the fibers at an

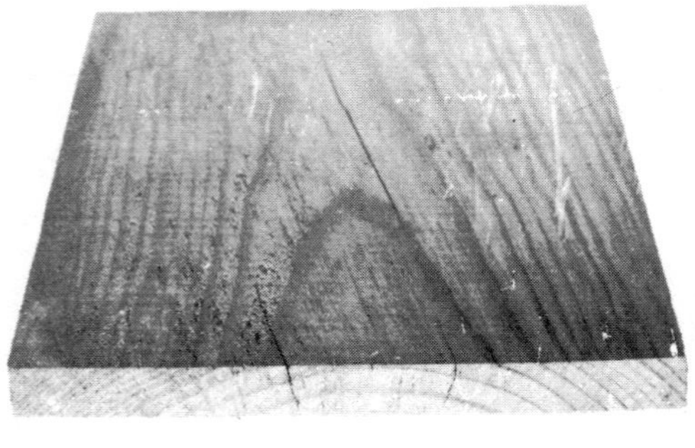

Figure 2–14. Checks in a flat-grain board.

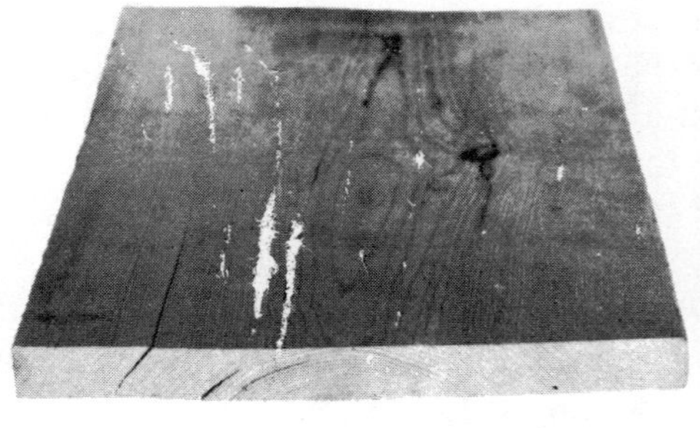

Figure 2–15. Shake in a flat-grain board.

angle of about 20 degrees to the surface. The wood should be viewed from the same direction but not at so small an angle as that of the light source. A concentrated light beam, such as a spotlight is best. Figure 2–16 shows how compression failures appear when viewed in this manner. A hand lens with a magnifying power up to 10 diameters aids in finding the less pronounced failures.

The readily visible compression failures usually are localized. Very small ones, however, commonly are widely distributed. Presence of the small, widely distributed ones frequently can be detected by fiber breakage on the end grain. In softwood lumber and such hardwoods as mahogany, fiber breakage shown in figure 2–17 can be seen on an ordinary saw cut across the grain when even microscopical compression failures are present (see 2.4 (3)).

2.307. Pitch Streaks. Pitch streaks are well-defined infiltrations of resin in the fibers that appear as streaks. They normally occur only in pine, Douglas fir, spruce, tamarack, and larch. As found in the better grades of lumber, practically all are small. They are chiefly objectionable because pitch may exude and damage paints, especially in warm weather.

2.308. Pitch Pockets. Pitch pockets are well-defined openings extending parallel to the annual rings, and containing more or less free resin and occasionally, bark. Like pitch streaks, they occur mainly in pine, Douglas fir, spruce, tamarack, and larch. They may weaken small members and resin may exude from them, especially when the wood becomes warm. Pitch pockets in large numbers and extent indicate a general fiber separation in the wood, and such occurrence may indicate presence of shake.

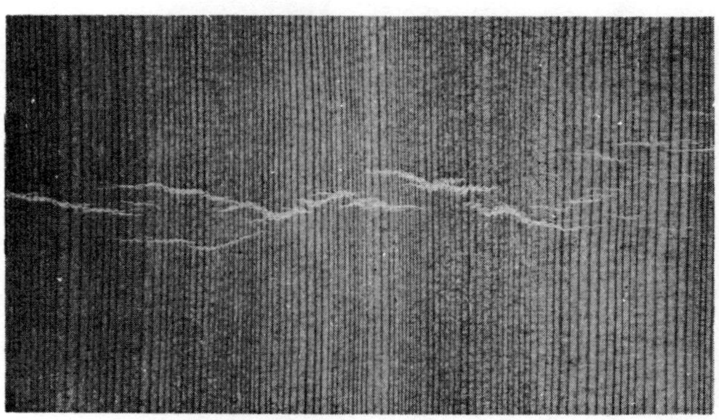

Figure 2–16. Pronounced compression failure.

Figure 2–17. Fiber breakage caused by compression failures.

2.309. Burls. A burl is a wartlike, small or large excrescence on a tree trunk with greatly distorted grain. Frequently, but not always, it contains the dark piths of one or many buds that rarely develop. Somewhat similar distorted grain may form over the end of a broken branch embedded in the trunk by subsequent growth of annual rings. In thick timbers, it is frequently a sign that a knot is embedded within the timber. Burls are usually objectionable because they cause local distortion of grain.

2.310. Pith Flecks. A pith fleck is a narrow streak resembling pith, usually brownish and up to several inches long on the surface of a piece. It is caused by larvae that burrow in the living cells of the tree. When present in considerable number, such streaks may have an appreciable weakening effect on thin lumber, but usually their effect is negligible. They occur chiefly in birch and soft maple.

2.311. Bark Pockets. A bark pocket is a patch of bark partially or wholly enclosed in the wood. There is usually some slight separation (or at least a lack of cohesion) of the wood substance that has definite weakening effect on thin material.

2.312. Indented Rings. Some softwood annual rings, as seen on the cross section, are indented along a radial line for many successive years. The grain distortion thus caused weakens thin material.

2.313. Strains and Decay.

2.3130. General. All forms of decay, and many kinds of stain as well, are caused by primitive plants called fungi that grow in the wood. Fungi are made up of fine threads (hyphae) invisible to the naked eye, unless they are massed together. A mass of hyphae is called mycelium.

Figure 2–18. Mycelium on oak, and characteristic cracking of typical brown rot.

Figure 2–18 shows mycelium on oak. Fungi produce great quantities of microscopic spores, which are carried by air currents or insects and act like seeds in starting new infections on wood. New infections also may start from hyphae that grow into sound wood from other infected wood or from soil. Fungi grow in wood only when it is moist, but they also need air and will not attack submerged or waterlogged wood.

Decay fungi soften wood and greatly weaken or destroy it. Staining and molding fungi are usually limited to sapwood, do not soften the wood, and have little effect on most strength properties.

2.3131. Stains and Molds. Sap stain and mold are not stages of decay. They are caused by a different class of fungus that generally infects the sapwood before the decay fungus does. The principal difference between the stain and mold fungi is that the molds discolor wood only superficially. It is helpful to recognize these discolorations, if only to avoid confusing them with decay.

The fungus-caused stains that will not surface off are objectionable for any members that are to have a natural finish. Furthermore, heavy stain may mean some reduction of shock resistance either by the discoloring fungi or by decay fungi that may also be present and are difficult to detect in stained wood. For this reason, heavily stained material should be discriminated against for members that are to be heavily stressed. Stained and moldy wood, like decayed wood, is more water-absorptive than uninfected wood and, therefore, more likely to decay unless protected from moisture.

Various other colored stains are of chemical or unknown origin. In most cases, these stains have little effect on strength or other properties of the wood.

2.31310. Blue Stain. Blue stain occurs mainly in the sapwood of logs and lumber (see 2.4 (6)), but is found in the heartwood of some coniferous trees, especially spruce. The color is predominantly bluish gray, although bluish-brown shades are found in hardwoods. If the stain is allowed to progress unhindered, the entire sapwood may become blue gray or even black.

2.31311. Brown Stain in Pines. Two types of light to dark brown stain are found most commonly in pines. One is caused by a fungus and the other is a chemical stain that occurs during seasoning. Neither is associated with any significant change in structure or reduction in strength of the wood.

2.31312. Yellow Stain in Oak. Oak heartwood occasionally becomes yellowish in color as a result of the action of a specific mold. The color is usually widespread and distinctive enough to distinguish it from yellow decay discolorations. Its effect on strength is unimportant.

2.31313. Mineral Stains. So-called mineral stains and streaks are common in hardwoods. Reddish brown to grayish brown colors, often with a tinge of green, occur in sweetgum, birch, and maple. In yellow poplar, a variety of colors are found. Red, purple, yellow, lavender, green, and black. Brownish mineral stains in yellow poplar are sometimes confused with decay. However, they differ from decay discolorations in that when they occur in streaks, their boundaries tend to contrast more sharply with the surrounding normal wood. Moreover, wood with even the darkest colored mineral stain is not softened.

2.31314. Iron Stain. A common bluish-black stain is due to the reaction of iron with the tannic acid present in many woods. This stain is often harmless, but may indicate a loss in decay resistance of certain species, due to alteration of natural preservatives in the wood. Section 3.090 gives a more detailed discussion of this stain.

2.3132. Decay. This discussion primarily concerns decay with which the shipbuilder should be familiar in order to avoid putting decayed material into the ship. Conditions affecting development of decay in ships after construction are discussed in section 3.09.

Decay fungi weaken wood by breaking down the cell walls, which are the main strength-giving components of the wood. Even before the decay has reached the stage of definite softening, the shock resistance of the wood is much reduced.

All tree species are subject to heartwood decay in the living tree. One of the fungi that causes heart rot in living oaks also is known to be common in decayed timber in ships. In cut products, heartwood of some species is much more resistant to decay than that of others, as described in section 3.08. Rotting of sapwood in living trees is rare; however, resistance of sapwood to decay in the cut products is uniformly low. Therefore, any decay that develops after felling and before use of lumber is likely to be found in sapwood.

It is usually a simple matter to recognize well-advanced or typical decay caused by prolonged action of a wood-destroying fungus. Depending on the causal fungus, the wood may be abnormally brown or white in color, greatly softened, pocketed, or brittle and cracked in collapsed areas.

Early, or incipient, decay is far from easy to recognize. In some cases, detection is practically impossible without examining the wood under a microscope. Incipient decay usually appears as a discoloration that rarely ends abruptly or evenly, but often fades out in one or more irregular streaks. Such streaks usually extend less than 3 or 4 feet beyond the typical decay, as measured along the grain of the wood. In the original piece of lumber or timber, incipient decay is often connected at one end with more advanced and recognizable decay. Furthermore, early decay is most easily distinguished when the wood is green or only partially seasoned. Therefore, a drastic rejection policy at the sawmill or shipyard is the most effective safeguard. As a general rule, an allowance should be made of 2 feet along the grain beyond the last visible evidence of incipient decay. Across the grain, an allowance of 2 or 3 inches is sufficient.

In inspection at later stages of manufacture, safe judgment depends first on thorough acquaintance with the normal appearance of the wood inspected. No known test can be substituted for expert knowledge of a wood species acquired through experience. In addition to visual inspection, probing with a pick or sharpened screw driver is the best way to detect early decay. By turning up a splinter, loss of toughness can sometimes be detected in wood that is not yet softened. A long splinter usually means sound wood, while an abrupt or square break means either that decay is present or that the wood was initially brash. If a large piece of wood is equally brash throughout, it indicates that the wood was week to start with.

2.31320. Types of Decay. The classification of decays into white rot and brown rot is sometimes made, but the color distinction does not hold true for incipient stages. A typical brown rot is shown in figure 2–18. Such classifications as pocket, stringy, cubical, or crumbly rot also are based on late stages of decay and have no interest to the shipbuilder. Brown, crumbly rot in the dry condition is sometimes called "dry rot," but the term is meaningless because all wood must be moist for active decay.

2.31321. Decay in Standing Softwood Trees. The most common decay in the heartwood of living coniferous trees honeycombs the heartwood with small white pockets in which the wood is reduced to a soft fibrous mass. The pockets are separated by firm and apparently sound wood.

Figure 2–19 shows decay of this type. Incipient stages of this decay cause, at the most, only slight discoloration in the cedars, and vary from pink to red brown in pines, light gray to brown in spruce, and reddish or olive purple to brown in Douglas fir.

Another common pocket rot in which the wood between pockets often remains firm is the pecky rot of cypress. Pecky cypress is successfully used in places of high decay hazard if the pockets are not too numerous and weakening.

Several fungi cause reddish brown crumbly rots in which the wood ultimately breaks into cubical fragments. Incipient stages may show an extremely faint yellow or brown discoloration that is very difficult to detect. Strength of the wood may be considerably reduced even in the early incipient stage.

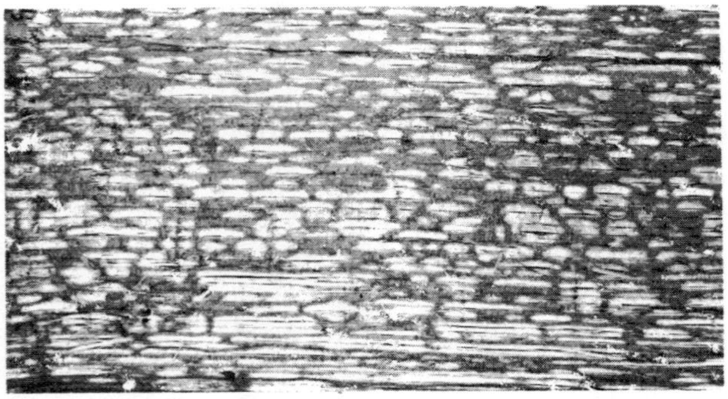

Figure 2–19. White pocket rot.

Some discolorations do not indicate decay infection, as already described in section 2.3131. These are often more conspicuous and general than the colors of incipient decays. A reddish color in Sitka spruce heartwood bordering on the sapwood is an example of this. Most reddish colors without brown or purple tinges are not connected with decay, especially if uniform over large surfaces. In Douglas fir, western redcedar, and western larch, light-colored zones that are restricted to certain bands of growth rings in the heartwood usually consist of "included" sapwood and are not indications of decay.

2.31322. Decay in Standing Hardwood Trees. In hardwoods, as in softwoods, rots are often of the white pocket or brown crumbly types. In such woods as sweetgum, grain figure is not an indication of decay. In freshly cut hardwood lumber, early stages of decay often have a water-soaked appearance. On drying, any abnormal shade of brown, such as yellow brown, greenish brown, and grayish brown, indicates decay.

2.31323. Detection of Incipient Decay. Although recognition of incipient decay is difficult, certain discolorations and other characteristics denote its presence and usually will serve to distinguish it from harmless discoloration. The following criteria are offered as a guide in this respect. It must be remembered, however, that no instructions can be followed successfully without thorough acquaintance with the normal appearance of wood of particular species.

Abnormal Color. (*a*) Any discoloration extending in zones from obviously decayed wood indicates infection.

(*b*) Any unusual shade of brown ordinarily indicates infection. Shades of brown caused by fungi are variable; in addition to "pure" browns, such shades as yellow brown, greenish brown, and grayish brown are common in hardwoods. In some cases, the decayed wood is bleached to a cream, straw, or yellow color. Distinctly reddish shades of brown are prevalent in coniferous woods, and in some species, notably Douglas fir, the earliest decay discoloration is often a shade of purple.

(*c*) Aside from color, an abnormal mottled appearance of the wood often indicates decay.

(d) Discoloration limited to certain growth rings is less likely to be decay than when it spreads across the rings.

(*e*) Infected wood commonly has a duller color than normal wood, and is sometimes described as having a "dead" appearance.

(*f*) Some decays, particularly in hardwoods, produce what are known as zone lines. These are very narrow black, orange, or yellow lines that tend to run in the direction of the grain. They are likely to be most prevalent at or near the border of the most conspicuously discolored areas. Sometimes, they are associated with little general discoloration, but their presence is certain evidence of decay.

Softness and Brashness. Abnormally soft or brash wood should be suspected of decay even if no abnormal color can be detected. A simple test for brashness in infected wood is described under section 2.3132.

Roughness of Surface. Wood containing even very early decay sometimes does not plane so smoothly as sound wood. The surface of veneer cut from wood with incipient decay is almost always rough.

2.3133. Prevention of Infections in Boat Lumber. A prime essential in preventing stains, molds, and decay between felling of the tree and use of the lumber is to saw the logs and season the lumber as rapidly as possible. None of the wood-attacking fungi can develop in wood that has a moisture content of 20 percent or less. Therefore, kiln drying of lumber green from the saw will prevent attack unless the wood again becomes wet. Fully air-seasoned lumber in all parts of the country also dries down to less than 20 percent moisture content.

Freshly sawed lumber should be either kiln dried or dried in accordance with good air-seasoning practice (chapter 11). Thereafter, the lumber should be protected from rain or other moisture until used. Storage directly on the ground should be particularly avoided. If the lumber cannot be dried rapidly, or is to be shipped in the green condition, a large measure of protection can be inexpensively provided by dipping or spraying the freshly cut lumber with water solutions of one of several standard fungicides (see 2.4 (9)). These products usually contain an organic mercury or chlorinated phenol, used either alone or in mixture with each other or with borax. The directions of the manufacturer or distributor of a particular product should be followed closely for maximum effectiveness. Although marketed primarily for stain and mold control, these treatments also are effective against decay for limited periods.

Logs cut in seasons when temperatures are higher than about 45 degrees F. should be sawed as soon as is practicable. If prompt utilization is impossible, temporary protection for logs can be obtained by treating all exposed surfaces with one of the same solutions mentioned for lumber. Somewhat stronger solutions usually are preferable for logs, and the incorporation of an insecticide is often necessary (see 2.4 (9)). Water storage (section 5.210) also is often used to protect logs in some regions.

2.314. Insect Damage. Insect damage may occur in standing trees, logs, and unseasoned or seasoned lumber (see 2.4 (1)). Damage in the standing tree is difficult to control, but otherwise insect damage can be largely eliminated by proper control methods.

Insect holes are generally classified as pin holes, grub holes, and powder post holes. No method is known for estimating, from the appearance of insect-damaged wood, the amount of reduction in strength, and when strength is of prime importance in members of small cross section, the safe procedure is to discard every piece that contains insect holes. Powder-post larvae, by their irregular burrows, may destroy most of the interior of a piece while the surface shows only small holes, and the strength of the piece may be reduced virtually to zero.

Pinholes are caused by ambrosia beetles (see 2.4 (7)), which attack living hardwood trees and green hardwood logs, as well as felled softwood trees, logs, and green lumber. These beetles leave telltale dark streaks in wood around the holes.

Grub holes may be caused by wood-boring grubs or round-headed borers in living trees or by ambrosia beetles in felled green timber or green lumber. The holes vary in diameter from less than ⅛ inch to more than ¼ inch. The hole itself may be either darkly stained or unstained; streaks are not present.

Powder-post beetles (see 2.4 (8)), attack hardwoods both in the freshly cut timber and in seasoned lumber and timbers. They chiefly damage the seasoned sapwood of hardwoods, favoring ash, oak, and, to a lesser degree, maple, elm, and yellow poplar.

Infested lumber may be treated to kill the beetles and prevent further damage by immersing it in a 5 percent solution of DDT in mineral spirits or a 2 percent solution of chlordane in mineral spirits. Another control is the steaming of infested wood in a kiln at 135° F. or higher at atmospheric pressure for 1½ hours, or the wood may be heated at 180° F. for at least one half hour.

2.4. REFERENCES

(*1*) FOREST PRODUCTS LABORATORY.

1955. "Wood Handbook," *U. S. Department of Agriculture (Agriculture Handbook No. 72)*, 523 pp., illus. (Revised.)

(*2*) KOEHLER, ARTHUR.

1943. "Guide to Determining Slope of Grain in Lumber and Veneer," *Forest Products Laboratory Report No. 1585*, 25 pp., illus.

(*3*) PILLOW, MAXON Y.

1949. "Studies of Compression Failures and Their Detection in Ladder Rails," *Forest Products Laboratory Report No. D1733.*

(*4*) ———

1953. "Effects of Tension Wood in Hardwood Lumber and Veneer," *Forest Products Laboratory Report No. R1943.*

(*5*) PILLOW, MAXON Y. and LUXFORD, R. F.

1937. "Structure, Occurence, and Properties of Compression Wood," *U. S. Department of Agriculture Technical Bulletin 546*, 32 pp. illus.

(*6*) SCHEFFER, T. C. and LINDGREN, R. M.

1940. "Stains of Sapwood and Sapwood Products and Their Controls," *U. S. Department of Agriculture Technical Bulletin 714*, 121 pp., illus.

(*7*) SNYDER, T. E.

1927. "Defects in Timber Caused by Insects," *U. S. Department of Agriculture Bulletin 1490*, 46 pp., illus.

(*8*) U. S. FOREST SERVICE.

1954. "Powder Post Beetles in Buildings," *U. S. Department of Agriculture Leaflet No. 358*, 8 pp., illus.

(*9*) VERRALL, A. F., and SCHEFFER, T. C.

1949. "Control of Stain, Mold and Decay in Green Lumber and Other Wood Products," *Proceedings Forest Products Research Society*, Vol. 3. Illus.

CHAPTER 3. FACTORS AFFECTING THE USE OF WOOD IN BOATS AND SHIPS

3.00. GENERAL. Most uses of wood in boat and ship construction involve the selection of a specific kind of lumber or plywood; the user may have erroneous concepts of the effect on strength of specific gravity or moisture content; he may hesitate to accept material from dead trees or turpentined trees; he may not appreciate the conditions under which wood can be used without decaying. These and many other factors that may arise affect the efficient use of wood in boats and ships.

This chapter contains information on many species, not all of which are used, or suitable for use, as shipbuilding woods. The information is presented as an aid in selecting suitable species for various purposes.

3.01. LIVE AND DEAD TREES. Timber from trees killed by insects, blight, wind, or fire is as good for any structural purpose as that from live trees, provided additional injury from further insect attack, staining, decay, or seasoning defects has not occurred.

If a tree stands on the stump too long after it is dead, the sapwood is likely to become decayed or to be severely attacked by wood-boring insects, and in time, the heartwood will be similarly affected. Such deterioration also occurs in logs that have been cut from live trees and improperly cared for subsequently. Because of variations in climatic and local weather conditions and other factors that affect deterioration, the length of the period during which dead timber may stand or lie in the forest without serious deterioration varies. Tests on wood from trees that had stood as long as 15 years after being killed by fire demonstrated that this wood was sound and as strong as wood from live trees. Logs of some of the more decay resistant species contain thoroughly sound heartwood after lying on the ground in the forest for many years. On the other hand, decay may cause great loss of strength within a very brief time, both in trees standing dead on the stump and in logs that have been cut from live trees and allowed to lie on the ground. Consequently, the important consideration is not whether the trees are alive or dead, but whether the products cut from those trees are free from decay and other defects that would make them unsuitable for use.

Lumber that is cut from dead trees should be more carefully inspected for decay, because it is present in them more frequently than in live trees. It is usually impossible to determine from the appearance of wood, however, whether the tree was alive or dead when it was cut.

3.02. WINTER CUTTING VERSUS SUMMER CUTTING. An old belief still given wide currency is that winter-cut lumber is more durable than summer-cut lumber. The belief is based on the erroneous assumption that in winter, " the sap is down," while in the summer, "the sap is up," in the living tree. Actually, tests have demonstrated conclusively that standing trees contain about as much sap in winter as in summer.

The only sound objection to summer-cut lumber is that logs are more likely to deteriorate if left exposed to high summer temperatures that may accelerate checking and attack by insects and decay fungi. Reasonable precautions, praticularly prompt sawing after felling and good piling and seasoning methods, remove the danger of such damage to summer-cut material.

3.03. TURPENTINING. Among the species of southern yellow pine, longleaf, and slash are frequently tapped for turpentine. If the turpentining is properly handled, the trees are not unduly damaged for lumber purposes. Comparative tests on mature turpentined and unturpentined trees have shown that: (1) Turpentined lumber is as strong as unturpentined lumber of the same weight, (2) the weight and shrinkage of wood are not affected by turpentining, and (3) practically speaking, properly handled turpentined trees contain neither more nor less resin than unturpentined trees.

3.04. WEIGHT OF WOOD

3.040. General. All wood substance, regardless of the species from which it comes, has about the same weight for a given volume. The differences in weight that occur between species arise from the cellular sturcture of wood. Depending upon the size of the cells, the thickness of the cell walls, and relatively minor factors, different species have greatly varying weights for a given volume. Thus, black ironwood, a very dense wood growing in Florida, is about 10 times as heavy per cubic foot as is balsa, a Latin American species. The difference in weight is due to the thickness of the cell walls in relation to the size of the cells of the two species.

Wood in the green state contains considerable moisture, varying from about 30 percent in the heartwood of some of the pines to more than 200 percent in the sapwood of some other species. The amount of moisture in a piece of wood accordingly affects the weight of the wood. Most species of wood, even when green, float in water. A few species, among them black ironwood, do not float even when dry. The buoyancy of wood is due to the air within its cells. When wood is kept submerged, the air within the cells is gradually replaced by water, and consequently the wood becomes heavier and finally sinks. Likewise, moisture absorbed within the cell walls, or wood tissue, makes wood heavier. A cubic foot of green wood is heavier than a cubic foot of dry wood. Table 3–1 gives average weights of a number of shipbuilding woods per cubic foot when green, at 15 percent moisture content, and when ovendry.

Table 3–2 provides a ready means for determining the weight per lineal feet of given cross sections of wood. Select from table 3–2 the factor corresponding to the given section and multiply it by the weight per cubic foot at 15 percent moisture content as shown in table 3–1.

TABLE 3–1

AVERAGE WEIGHT PER CUBIC FOOT FOR VARIOUS WOODS UNDER DIFFERENT MOISTURE CONTENT CONDITIONS, AND SPECIFIC GRAVITY BASED ON WEIGHT WHEN OVENDRY AND VOLUME WHEN GREEN

Species	Specific gravity	Weight per cubic foot [1]		
		Green	15 percent moisture content	Ovendry
HARDWOODS				
		Pounds	*Pounds*	*Pounds*
Angelique	0.60	67	46	44
Apitong	.58	------	44	41
Ash:				
Black	.45	[2] 55	35	33
Green	.53	------	41	38
Oregon	.50	------	38	36
White	.55	[2] 50	42	40
Basswood, American	.32	[2] 36	25	24
Beech, American	.56	[2] 54	44	42
Birch:				
Yellow	.55	[2] 60	43	41
Cherry, black	.47	[2] 46	36	33
Elm:				
Rock	.57	[2] 51	44	41
Greenheart	.81	------	61	56
Hickory:				
Mockernut	.64	[2] 68	50	49
Pignut	.66	[2] 70	52	50
Shagbark	.64	------	50	48
Shellbark	.62	------	49	48
Iroko	.55	------	40	36
Ironbark, gray	1.00	------	------	62
Khaya ("African mahogany")	.43	------	32	29
Lauan, red	.40	------	31	28
Lignumvitae (*Guaiacum* spp.)	1.20	------	[3] 76	------
Locust:				
Black	.66	------	50	46
Honey	.63	------	48	44
Mahogany (Swietenia spp.)	.45	------	34	30
Maple:				
Black	.52	------	40	38
Sugar	.56	[2] 58	43	41
Maria	.50	------	39	37
Oak:				
Red oaks:				
Black	.56	------	43	41
Northern red	.56	[2] 63	43	40
Pin	.58	------	44	42
Scarlet	.60	------	46	43
Southern red	.52	[2] 59	41	39

Footnotes at end of table.

TABLE 3-1.—Continued

AVERAGE WEIGHT PER CUBIC FOOT FOR VARIOUS WOODS UNDER DIFFERENT MOISTURE CONTENT CONDITIONS, AND SPECIFIC GRAVITY BASED ON WEIGHT WHEN OVENDRY AND VOLUME WHEN GREEN—Continued

Species	Specific gravity	Weight per cubic foot [1]		
		Green	15 per-cent mois-ture content	Oven-dry
HARDWOODS—continued				
Oak—Continued				
White oaks:		*Pounds*	*Pounds*	*Pounds*
Bur	.58	------	44	40
Chestnut	.57	------	45	43
Live	.81	------	63	59
White	.60	[2] 61	47	44
Okoume (Gaboon)	.35	------	27	25
Pecan	.60	------	46	43
Peroba do campo	.69	------	63	58
Sweetgum	.46	[2] 52	36	34
Tangile	.53	------	41	38
Teak	.58	------	43	39
Walnut, black	.51	[2] 60	39	37
Yellow-poplar	.40	[2] 46	31	28
SOFTWOODS				
Baldcypress	.42	[2] 58	32	29
Cedar:				
Alaska	.42	[2] 35	32	29
Atlantic White	.31	------	23	21
Eastern redcedar	.44	[2] 37	33	30
Incense	.35	[2] 32	26	24
Northern white	.20	------	21	19
Port-Orford	.40	[2] 37	30	28
Western redcedar	.31	[2] 31	23	21

TABLE 3-1.—Continued

AVERAGE WEIGHT PER CUBIC FOOT FOR VARIOUS WOODS UNDER DIFFERENT MOISTURE CONTENT CONDITIONS, AND SPECIFIC GRAVITY BASED ON WEIGHT WHEN OVENDRY AND VOLUME WHEN GREEN—Continued

Species	Specific gravity	Weight per cubic foot [1]		
		Green	15 per-cent mois-ture content	Oven-dry
SOFTWOODS—continued				
Douglas fir:		*Pounds*	*Pounds*	*Pounds*
Coast	.45	[2] 38	34	32
Intermediate	.41	[2] 34	31	29
Rocky mountain	.40	[2] 32	30	28
Larch, Eastern (Tamerack Hackmatack)	.49	[2] 36	30	28
Larch, western	.51	[2] 49	39	37
Pine:				
Eastern white	.34	------	25	23
Loblolly	.47	[2] 39	36	33
Longleaf	.54	[2] 44	41	38
Ponderosa	.38	[2] 33	28	26
Shortleaf	.46	[2] 38	35	33
Slash	.56	------	43	40
Sugar	.35	[2] 43	27	25
Western white	.36	[2] 36	27	25
Redwood	.38	[2] 44	28	25
Spruce:				
Red (Eastern)	.38	[2] 32	29	27
Sitka	.37	[2] 33	28	26
White (Eastern)	.37	------	28	27

[1] The weight per cubic foot includes both the weight of the dry wood substance and the water.

[2] Values given apply to heartwood.

[3] Weight at 12 percent moisture content.

TABLE 3.2

Weight Per Lineal Foot of Given Sections

Multiply factor from table by the weight per cubic foot of the kind of wood required drawn in table 3.1

Width	Thickness											
	⅛	¼	⅜	½	⅝	¾	⅞	1	1⅛	1¼	1⅜	1½
⅝	0.0005	0.0011	0.0016	0.0022	0.0027	0.0033	0.0038	0.0043	0.0049	0.0054	0.0060	0.0065
¾	.0006	.0013	.0020	.0026	.0033	.0039	.0046	.0052	.0059	.0065	.0072	.0078
⅞	.0008	.0015	.0023	.0030	.0038	.0045	.0053	.0061	.0068	.0076	.0083	.0091
1	.0009	.0017	.0026	.0034	.0043	.0052	.0061	.0069	.0080	.0087	.0095	.0104
1⅛	.0010	.0020	.0029	.0039	.0049	.0059	.0068	.0078	.0088	.0098	.0107	.0117
1¼	.0011	.0022	.0033	.0043	.0054	.0065	.0078	.0087	.0098	.0109	.0120	.0130
1⅜	.0012	.0024	.0036	.0048	.0060	.0072	.0083	.0095	.0107	.0119	.0131	.0143
1½	.0013	.0026	.0039	.0052	.0065	.0078	.0091	.0104	.0117	.0130	.0143	.0156
1⅝	.0014	.0028	.0042	.0056	.0070	.0084	.0099	.0113	.0127	.0141	.0155	.0169
1¾	.0015	.0030	.0046	.0061	.0076	.0091	.0106	.0121	.0136	.0152	.0167	.0182
1⅞	.0016	.0033	.0049	.0066	.0081	.0098	.0114	.0130	.0146	.0163	.0179	.0195
2	.0017	.0035	.0052	.0070	.0087	.0104	.0121	.0139	.0159	.0173	.0191	.0208
2⅛	0018	.0037	.0055	.0074	.0092	.0111	.0129	.0147	.0166	.0184	.0203	.0221
2¼	.0020	.0039	.0059	.0078	.0098	.0117	.0137	.0156	.0176	.0195	.0215	.0234
2⅜	.0021	.0041	.0062	.0082	.0103	.0123	.0144	.0165	.0185	.0206	.0226	.0247
2½	.0022	.0043	.0065	.0086	.0108	.0130	.0152	.0173	.0195	.0217	.0238	.0260
2⅝	.0023	.0046	.0068	.0092	.0114	.0137	.0159	.0182	.0205	.0228	.0250	.0273
2¾	.0024	.0048	.0072	.0096	.0119	.0143	.0167	.0191	.0215	.0238	.0263	.0286
2⅞	.0025	.0050	.0075	.0100	.0125	.0150	.0175	.0200	.0225	.0250	.0275	.0300
3	.0026	.0052	.0078	.0104	.0130	.0156	.0182	.0208	.0234	.0260	.0286	.0312
3⅛	.0027	.0054	.0082	.0108	.0135	.0162	.0190	.0217	.0244	.0271	.0293	.0325
3¼	.0028	.0056	.0085	.0112	.0141	.0169	.0197	.0225	.0253	.0282	.0310	.0338
3⅜	.0029	.0058	.0088	.0117	.0146	.0175	.0205	.0234	.0263	.0292	.0322	.0351
3½	.0030	.0061	.0091	.0121	.0152	.0182	.0212	.0243	.0273	.0303	.0334	.0364
3⅝	.0031	.0063	.0095	.0126	.0157	.0189	.0220	.0251	.0283	.0314	.0346	.0377
3¾	.0033	.0065	.0098	.0130	.0163	.0195	.0228	.0260	.0293	.0325	.0358	.0390
3⅞	.0034	.0067	.0101	.0134	.0168	.0201	.0235	.0269	.0302	.0336	.0369	.0403
4	.0035	.0069	.0104	.0139	.0173	.0208	.0243	.0277	.0312	.0347	.0381	.0416
4⅛	.0036	.0072	.0108	.0143	.0179	.0215	.0250	.0286	.0322	.0358	.0393	.0429
4¼	.0037	.0074	.0111	.0147	.0184	.0221	.0258	.0295	.0332	.0368	.0405	.0442
4⅜	.0038	.0076	.0114	.0152	.0192	.0228	.0265	.0303	.0341	.0379	.0417	.0445
4½	.0039	.0078	.0117	.0156	.0195	.0234	.0273	.0312	.0351	.0390	.0429	.0468
4⅝	.0040	.0080	.0120	.0160	.0200	.0240	.0281	.0321	.0361	.0401	.0441	.0481
4¾	.0041	.0082	.0124	.0165	.0206	.0247	.0288	.0329	.0370	.0412	.0453	.0494
4⅞	.0042	.0085	.0127	.0169	.0211	.0254	.0296	.0338	.0381	.0423	.0465	.0507
5	.0043	.0087	.0130	.0173	.0217	.0260	.0303	.0347	.0390	.0433	.0477	.0520
5⅛	.0044	.0089	.0133	.0178	.0222	.0267	.0311	.0355	.0400	.0444	.0489	.0533
5¼	.0046	.0091	.0137	.0182	.0228	.0273	.0319	.0364	.0410	.0455	.0501	.0546
5⅜	.0047	.0093	.0140	.0186	.0233	.0279	.0326	.0373	.0419	.0466	.0512	.0559
5½	.0048	.0095	.0143	.0191	.0238	.0286	.0334	.0381	.0429	.0477	.0524	.0572
5⅝	.0049	.0098	.0146	.0195	.0244	.0293	.0341	.0390	.0439	.0488	.0536	.0585
5¾	.0050	.0100	.0150	.0199	.0249	.0299	.0349	.0399	.0449	.0499	.0548	.0598
5⅞	.0051	.0102	.0153	.0204	.0255	.0306	.0356	.0407	.0458	.0509	.0560	.0611
6	.0052	.0104	.0156	.0208	.0260	.0312	.0364	.0416	.0468	.0520	.0572	.0624
6⅛	.0053	.0106	.0160	.0212	.0265	.0318	.0372	.0425	.0478	.0531	.0584	.0637
6¼	.0054	.0108	.0163	.0217	.0271	.0325	.0379	.0433	.0487	.0542	.0596	.0650
6⅜	.0055	.0111	.0166	.0221	.0276	.0332	.0387	.0442	.0497	.0553	.0608	.0663
6½	.0056	.0113	.0170	.0225	.0282	.0338	.0394	.0451	.0507	.0563	.0620	.0676
6⅝	.0057	.0115	.0173	.0230	.0287	.0345	.0402	.0459	.0517	.0574	.0632	.0689
6¾	.0059	.0117	.0176	.0234	.0293	.0351	.0410	.0468	.0527	.0585	.0644	.0702
6⅞	.0060	.0119	.0179	.0238	.0298	.0357	.0417	.0477	.0536	.0596	.0655	.0715
7	.0061	.0122	.0182	.0243	.0303	.0364	.0425	.0485	.0546	.0607	.0667	.0728

TABLE 3.2—Continued

Weight Per Lineal Foot of Given Sections—Continued

Multiply factor from table by the weight per cubic foot of the kind of wood required drawn in table 3.1

Width	Thickness											
	1⅝	1¾	1⅞	2	2⅛	2¼	2⅜	2½	2⅝	2¾	2⅞	3
¼	0. 0028	0. 0030	0. 0033	0. 0035	0. 0037	0. 0039	0. 0041	0. 0043	0. 0046	0. 0047	0. 0050	0. 0052
⅜	. 0042	. 0046	. 0050	. 0052	. 0055	. 0059	. 0062	. 0065	. 0068	. 0072	. 0075	. 0078
½	. 0056	. 0061	. 0065	. 0070	. 0074	. 0078	. 0083	. 0087	. 0091	. 0096	. 0100	. 0104
⅝	. 0071	. 0076	. 0081	. 0087	. 0092	. 0098	. 0103	. 0109	. 0114	. 0119	. 0125	. 0130
¾	. 0076	. 0091	. 0098	. 0104	. 0111	. 0117	. 0124	. 0130	. 0137	. 0143	. 0150	. 0156
⅞	. 0099	. 0106	. 0114	. 0122	. 0129	. 0137	. 0144	. 0152	. 0160	. 0167	. 0175	. 0182
1	. 0112	. 0121	. 0130	. 0139	. 0148	. 0156	. 0163	. 0174	. 0182	. 0191	. 0200	. 0208
1⅛	. 0127	. 0137	. 0147	. 0156	. 0166	. 0176	. 0186	. 0195	. 0205	. 0215	. 0225	. 0235
1¼	. 0141	. 0152	. 0163	. 0174	. 0185	. 0195	. 0206	. 0217	. 0228	. 0239	. 0250	. 0261
1⅜	. 0155	. 0167	. 0179	. 0191	. 0203	. 0215	. 0227	. 0239	. 0251	. 0263	. 0274	. 0287
1½	. 0169	. 0182	. 0195	. 0208	. 0221	. 0234	. 0247	. 0260	. 0274	. 0287	. 0300	. 0313
1⅝	. 0183	. 0197	. 0212	. 0226	. 0240	. 0254	. 0268	. 0282	. 0296	. 0310	. 0324	. 0339
1¾	. 0197	. 0213	. 0228	. 0243	. 0256	. 0273	. 0289	. 0304	. 0319	. 0334	. 0349	. 0365
1⅞	. 0212	. 0228	. 0244	. 0261	. 0277	. 0293	. 0309	. 0326	. 0342	. 0358	. 0374	. 0391
2	. 0226	. 0243	. 0261	. 0278	. 0295	. 0313	. 0330	. 0347	. 0365	. 0382	. 0399	. 0417
2⅛	. 0240	. 0258	. 0277	. 0295	. 0314	. 0332	. 0354	. 0369	. 0387	. 0406	. 0426	. 0443
2¼	. 0254	. 0278	. 0293	. 0313	. 0332	. 0352	. 0371	. 0391	. 0410	. 0430	. 0449	. 0469
2⅜	. 0267	. 0289	. 0309	. 0330	. 0350	. 0371	. 0392	. 0412	. 0433	. 0454	. 0474	. 0495
2½	. 0282	. 0304	. 0326	. 0347	. 0369	. 0391	. 0412	. 0434	. 0456	. 0477	. 0499	. 0521
2⅝	. 0296	. 0319	. 0342	. 0365	. 0387	. 0410	. 0433	. 0456	. 0479	. 0501	. 0524	. 0547
2¾	. 0310	. 0334	. 0358	. 0382	. 0406	. 0430	. 0454	. 0477	. 0501	. 0525	. 0549	. 0573
2⅞	. 0324	. 0349	. 0374	. 0399	. 0424	. 0449	. 0474	. 0499	. 0524	. 0549	. 0574	. 0599
3	. 0339	. 0365	. 0391	. 0417	. 0442	. 0469	. 0495	. 0521	. 0547	. 0573	. 0599	. 0625
3⅛	. 0353	. 0380	. 0407	. 0434	. 0461	. 0488	. 0515	. 0542	. 0570	. 0597	. 0624	. 0651
3¼	. 0367	. 0395	. 0423	. 0451	. 0480	. 0508	. 0536	. 0564	. 0593	. 0621	. 0649	. 0677
3⅜	. 0381	. 0410	. 0440	. 0469	. 0498	. 0527	. 0556	. 0586	. 0615	. 0645	. 0674	. 0703
3½	. 0395	. 0425	. 0456	. 0486	. 0517	. 0547	. 0577	. 0608	. 0638	. 0669	. 0699	. 0729
3⅝	. 0409	. 0440	. 0472	. 0504	. 0535	. 0566	. 0598	. 0629	. 0661	. 0692	. 0724	. 0752
3¾	. 0423	. 0456	. 0499	. 0521	. 0553	. 0586	. 0618	. 0651	. 0684	. 0716	. 0749	. 0781
3⅞	. 0437	. 0470	. 0505	. 0538	. 0572	. 0605	. 0639	. 0673	. 0707	. 0740	. 0774	. 0807
4	. 0451	. 0486	. 0521	. 0556	. 0590	. 0625	. 0660	. 0694	. 0729	. 0764	. 0799	. 0833
4⅛	. 0466	. 0501	. 0537	. 0573	. 0609	. 0645	. 0680	. 0716	. 0751	. 0788	. 0824	. 0860
4¼	. 0480	. 0516	. 0554	. 0590	. 0627	. 0664	. 0700	. 0738	. 0775	. 0812	. 0849	. 0896
4⅜	. 0494	. 0532	. 0570	. 0608	. 0646	. 0684	. 0722	. 0759	. 0796	. 0836	. 0874	. 0912
4½	. 0508	. 0547	. 0586	. 0625	. 0664	. 0703	. 0742	. 0781	. 0821	. 0860	. 0899	. 0938
4⅝	. 0522	. 0562	. 0602	. 0642	. 0683	. 0723	. 0763	. 0803	. 0843	. 0883	. 0923	. 0964
4¾	. 0536	. 0577	. 0619	. 0660	. 0701	. 0742	. 0783	. 0824	. 0866	. 0907	. 0948	. 0990
4⅞	. 0550	. 0592	. 0635	. 0677	. 0720	. 0762	. 0804	. 0846	. 0889	. 0931	. 0973	. 1016
5	. 0564	. 0608	. 0651	. 0695	. 0738	. 0781	. 0825	. 0868	. 0912	. 0955	. 0998	. 1042
5⅛	. 0578	. 0623	. 0668	. 0712	. 0756	. 0813	. 0845	. 0900	. 0934	. 0980	. 1023	. 1068
5¼	. 0592	. 0638	. 0684	. 0729	. 0775	. 0820	. 0866	. 0911	. 0957	. 1003	. 1048	. 1094
5⅜	. 0606	. 0653	. 0700	. 0747	. 0793	. 0840	. 0886	. 0933	. 0980	. 1027	. 1073	. 1120
5½	. 0620	. 0668	. 0716	. 0764	. 0812	. 0859	. 0907	. 0955	. 1003	. 1050	. 1098	. 1146
5⅝	. 0635	. 0684	. 0733	. 0781	. 0830	. 0879	. 0928	. 0977	. 1026	. 1074	. 1123	. 1172
5¾	. 0649	. 0699	. 0749	. 0799	. 0849	. 0899	. 0948	. 0998	. 1048	. 1098	. 1148	. 1198
5⅞	. 0663	. 0714	. 0765	. 0816	. 0867	. 0918	. 0969	. 1020	. 1071	. 1122	. 1173	. 1224
6	. 0677	. 0729	. 0782	. 0834	. 0886	. 0938	. 0990	. 1042	. 1094	. 1146	. 1198	. 1250
6⅛	. 0691	. 0744	. 0798	. 0851	. 0904	. 0957	. 1010	. 1062	. 1117	. 1170	. 1223	. 1276
6¼	. 0705	. 0759	. 0814	. 0868	. 0922	. 0977	. 1031	. 1085	. 1140	. 1194	. 1248	. 1302
6⅜	. 0719	. 0774	. 0830	. 0886	. 0941	. 0996	. 1051	. 1106	. 1162	. 1218	. 1273	. 1328
6½	. 0733	. 0790	. 0847	. 0903	. 0960	. 1016	. 1072	. 1128	. 1185	. 1242	. 1298	. 1354
6⅝	. 0748	. 0805	. 0863	. 0920	. 0978	. 1035	. 1093	. 1150	. 1203	. 1265	. 1323	. 1380
6¾	. 0762	. 0820	. 0879	. 0938	. 0996	. 1055	. 1113	. 1172	. 1231	. 1289	. 1348	. 1406
6⅞	. 0776	. 0835	. 0895	. 0955	. 1015	. 1074	. 1134	. 1193	. 1254	. 1313	. 1373	. 1432
7	. 0790	. 0850	. 0912	. 0972	. 1033	. 1094	. 1154	. 1215	. 1276	. 1337	. 1398	. 1458

TABLE 3.2—Continued

WEIGHT PER LINEAL FOOT OF GIVEN SECTIONS—Continued

Multiply factor from table by the weight per cubic foot of the kind of wood required drawn in table 3.1

Width	Thickness											
	3⅛	3¼	3⅜	3½	3⅝	3¾	3⅞	4	4⅛	4¼	4⅜	4½
½	0.0109	0.0112	0.0117	0.0122	0.0126	0.0130	0.0135	0.0139	0.0143	0.0148	0.0152	0.0156
⅝	.0136	.0141	.0147	.0152	.0157	.0163	.0168	.0174	.0179	.0185	.0190	.0195
¾	.0163	.0169	.0176	.0182	.0189	.0195	.0202	.0208	.0215	.0221	.0228	.0234
⅞	.0190	.0197	.0205	.0213	.0220	.0228	.0235	.0243	.0251	.0258	.0266	.0273
1	.0217	.0226	.0234	.0243	.0252	.0260	.0268	.0278	.0286	.0295	.0304	.0312
1⅛	.0244	.0254	.0264	.0274	.0283	.0293	.0303	.0313	.0322	.0332	.0342	.0352
1¼	.0271	.0282	.0293	.0304	.0315	.0326	.0336	.0347	.0358	.0369	.0380	.0391
1⅜	.0298	.0310	.0322	.0334	.0346	.0358	.0370	.0382	.0394	.0406	.0418	.0430
1½	.0326	.0339	.0352	.0365	.0378	.0391	.0404	.0417	.0430	.0442	.0456	.0469
1⅝	.0353	.0367	.0381	.0395	.0409	.0423	.0437	.0451	.0466	.0480	.0494	.0508
1¾	.0380	.0395	.0410	.0426	.0440	.0456	.0471	.0486	.0501	.0516	.0532	.0547
1⅞	.0407	.0423	.0440	.0456	.0473	.0488	.0505	.0521	.0437	.0554	.0570	.0586
2	.0434	.0451	.0469	.0485	.0504	.0529	.0538	.0556	.0573	.0590	.0608	.0625
2⅛	.0461	.0480	.0498	.0517	.0535	.0553	.0572	.0590	.0609	.0627	.0646	.0664
2¼	.0468	.0508	.0527	.0547	.0566	.0586	.0605	.0625	.0645	.0664	.0684	.0703
2⅜	.0515	.0536	.0556	.0577	.0598	.0618	.0639	.0660	.0680	.0701	.0722	.0742
2½	.0542	.0564	.0586	.0608	.0629	.0651	.0673	.0694	.0716	.0738	.0759	.0781
2⅝	.0570	.0593	.0615	.0638	.0661	.0684	.0707	.0729	.0752	.0775	.0798	.0821
2¾	.0597	.0621	.0645	.0669	.0692	.0716	.0740	.0764	.0788	.0812	.0836	.0860
2⅞	.0624	.0649	.0674	.0693	.0724	.0749	.0774	.0799	.0824	.0849	.0874	.0899
3	.0651	.0677	.0703	.0729	.0755	.0781	.0807	.0833	.0860	.0886	.0912	.0938
3⅛	.0678	.0705	.0732	.0760	.0787	.0814	.0841	.0868	.0895	.0922	.0949	.0977
3¼	.0705	.0733	.0761	.0790	.0818	.0846	.0874	.0903	.0931	.0959	.0987	.1016
3⅜	.0732	.0761	.0791	.0820	.0850	.0879	.0908	.0937	.0967	.0996	.1025	.1055
3½	.0760	.0790	.0820	.0851	.0881	.0912	.0942	.0972	.1003	.1033	.1064	.1094
3⅝	.0787	.0818	.0850	.0881	.0913	.0944	.0975	.1007	.1039	.1070	.1101	.1133
3¾	.0814	.0846	.0879	.0912	.0944	.0977	.1009	.1042	.1074	.1107	.1139	.1172
3⅞	.0841	.0874	.0908	.0942	.0975	.1009	.1043	.1076	.1110	.1142	.1177	.1211
4	.0868	.0903	.0937	.0972	.1007	.1042	.1076	.1111	.1146	.1181	.1215	.1250
4⅛	.0895	.0931	.0967	.1003	.1039	.1074	.1110	.1146	.1182	.1218	.1253	.1289
4¼	.0922	.0959	.0996	.1033	.1070	.1107	.1144	.1181	.1218	.1255	.1291	.1328
4⅜	.0949	.0987	.1025	.1064	.1101	.1139	.1177	.1215	.1253	.1291	.1329	.1367
4½	.0977	.1016	.1055	.1097	.1133	.1172	.1211	.1250	.1289	.1328	.1367	.1406
4⅝	.1004	.1044	.1084	.1124	.1164	.1204	.1244	.1265	.1325	.1365	.1405	.1445
4¾	.1031	.1072	.1113	.1158	.1196	.1237	.1278	.1319	.1361	.1402	.1443	.1484
4⅞	.1058	.1100	.1143	.1185	.1227	.1270	.1312	.1354	.1397	.1439	.1481	.1526
5	.1085	.1128	.1172	.1216	.1259	.1302	.1346	.1389	.1433	.1456	.1519	.1563
5⅛	.1112	.1157	.1201	.1246	.1290	.1335	.1379	.1424	.1468	.1513	.1557	.1602
5¼	.1139	.1185	.1230	.1276	.1322	.1367	.1413	.1458	.1504	.1550	.1595	.1640
5⅜	1166	.1213	.1259	.1307	.1353	.1400	.1446	.1493	.1540	.1586	.1633	.1680
5½	.1193	.1241	.1289	.1337	.1385	.1432	.1480	.1528	.1576	.1623	.1671	.1719
5⅝	.1221	.1270	.1318	.1367	.1416	.1465	.1513	.1563	.1612	.1660	.1709	.1758
5¾	.1248	.1298	.1347	.1398	.1448	.1498	.1547	.1597	.1647	.1697	.1747	.1797
5⅞	.1275	.1326	.1377	.1428	.1479	.1530	.1581	.1632	.1683	.1734	.1785	.1836
6	.1302	.1354	.1406	.1459	.1511	.1563	.1614	.1667	.1719	.1771	.1823	.1875
6⅛	.1329	.1382	.1435	.1489	.1542	.1595	.1648	.1701	.1755	.1808	.1861	.1914
6¼	.1356	.1410	.1469	.1519	.1573	.1628	.1682	.1736	.1791	.1845	.1899	.1953
6⅜	.1384	.1439	.1494	.1550	.1605	.1660	.1716	.1771	.1827	.1882	.1938	.1992
6½	.1411	.1467	.1523	.1580	.1636	.1693	.1749	.1805	.1862	.1919	.1975	.2031
6⅝	.1438	.1495	.1533	.1610	.1668	.1725	.1783	.1840	.1898	.1956	.2013	.2070
6¾	.1465	.1523	.1582	.1641	.1698	.1758	.1816	.1875	.1934	.1992	.2051	.2109
6⅞	.1492	.1551	.1611	.1671	.1731	.1790	.1850	.1909	.1970	.2029	.2089	.2148
7	.1519	.1580	.1640	.1701	.1762	.1823	.1883	.1941	.2005	.2066	.2127	.2187

The rate at which wood will absorb water when continuously exposed to it varies with the species, whether sapwood or heartwood, the length of the piece, size of the cross section, and other factors. Whereas one species may become waterlogged in a relatively short period of exposure, another may remain under water many years without absorbing so much water that it becomes heavy enough to sink. The heartwood of most species is much more impermeable than the sapwood. Thin boards or short sections absorb water more rapidly and become waterlogged sooner than thick long pieces. Since water diffuses or moves most easily in the longitudinal direction, the wood close to the end surfaces of submerged lumber will normally have a much higher moisture content than wood a foot or more from the ends.

For these reasons, it is extremely difficult to calculate the weight of water that a given ship or boat will take on if continuously in the water over a period of years. The problem of doing so is further complicated by the fact that, in hull planking, for example, drying continues from the inboard face even though the outboard face is in continuous contact with water. Under such conditions, solid planking never becomes saturated, particularly if of an impermeable species, such as the cedars. On the other hand, keels are usually thick enough so that, if made of heartwood of a relatively impermeable species, such as Douglas fir or white oak, they can remain in water for years without becoming waterlogged. Sapwood, however, may become waterlogged in a short time. Plywood, because of exposed end grain and lathe checks, absorbs water more rapidly than lumber.

3.05. SPECIFIC GRAVITY

3.050. General. Specific gravity is the ratio of the weight of a given volume of a substance to the weight of an equal volume of water. The specific gravity of wood tissue is the same among all species, about 1.46. As already explained, however, differences in cell structure account for weight and specific gravity—differences among species.

Since wood exists unaccompanied by water only when ovendried, the specific gravity of wood is sometimes taken as the ratio of the weight of the dry wood substance present in a given volume to that of an equal volume of water. Because the most nearly constant volume of wood is its green volume, which is not subject to the vagaries of shrinkage, specific gravity is often based on ovendry weight and volume when green. Such a value is an index of the weight of dry wood substance present in a certain volume of green wood. Table 3–1 gives the average specific gravity values for various species of wood used in shipbuilding on the basis of their ovendry weight and volume when green.

Although the specific gravity value based on ovendry weight and volume when green is a useful index of the properties of wood, it requires the measurement of the volume when green. For general boat and ship building purposes, the piece can be oven dried and its specific gravity calculated on the basis of ovendry weight and ovendry volume.

The specific gravity of a floating body, such as most woods are, can be determined by ascertaining the ratio of its volume below the waterline to that of the whole piece. The determination of specific gravity by this method is done as follows:

1. Cut a piece of wood from the board to be tested, about 1 by 1 by 12 inches in dimension, the 12-inch dimension being along the grain.

2. Dry the piece for about 48 hours at from 212 to 221 degrees F. This will remove practically all of the moisture.

3. Dress the piece so that the cross section is rectangular and uniform throughout the length, and trim it to exactly 10 inches in length. Mark off and number the length in inches.

4. Place the piece carefully in a tall glass container of water so that it floats in an upright position (figure 3–1). Note the waterline to which it sinks, then quickly remove the piece and mark it. The position of the waterline indicates the specific gravity of the piece on the basis of ovendry weight and ovendry volume. Thus, if the mark is halfway between 4 and 5, the specific gravity is 0.45. The whole operation must be performed quickly, because the ovendry piece will begin to pick up moisture while it is being dressed, cut, and marked and when it is placed in the water.

3.051. Specific Gravity and Strength. Since the specific gravity of a piece of wood is an excellent index of the amount of wood substance the piece contains, it is also an index of the strength properties. The general relation of specific gravity to strength is illustrated by mastic and balsa; tests show that mastic, with an average specific gravity nine times as great as that of balsa, is nine times as high in crushing strength along the grain.

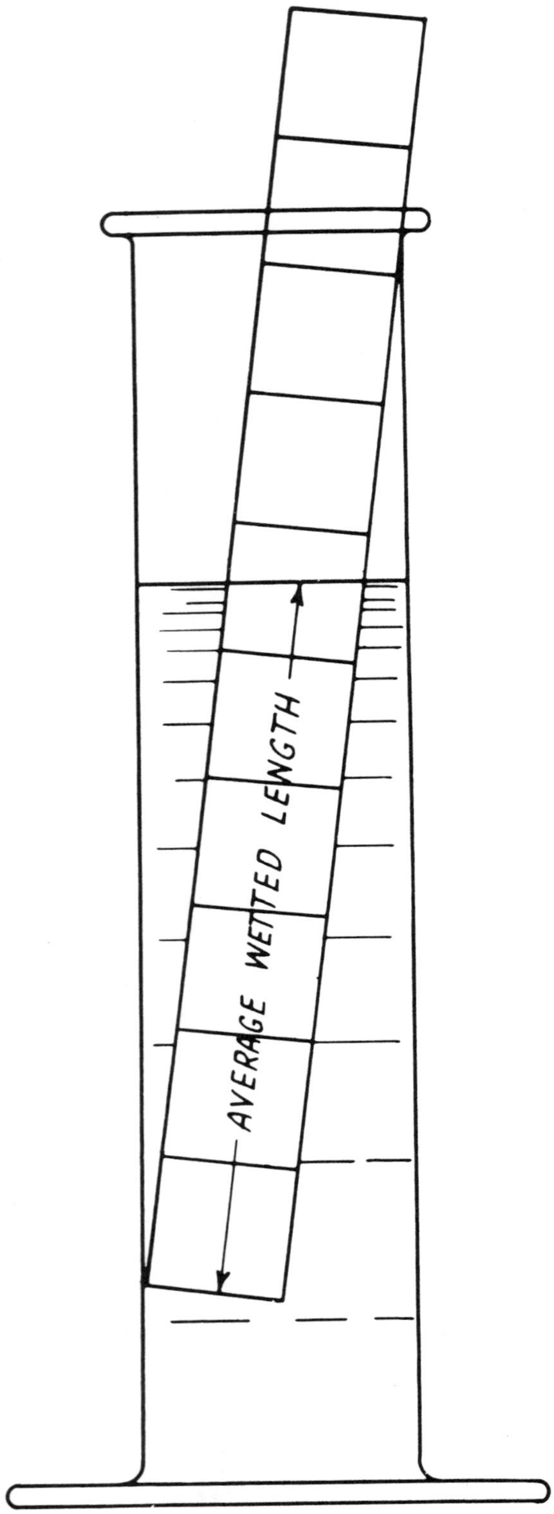

Figure 3–1. Container suitable for making specific gravity tests on wood by the flotation method.

Average specific gravity-strength relations based on 163 species of hardwoods and softwoods show that some properties, such as maximum crushing strength parallel to grain, increase approximately in proportion to the increase in specific gravity, whereas others increase more rapidly. For example, a species twice as high in specific gravity as another has 4¾ times the hardness of the lighter species. It is evident, therefore, that small differences in specific gravity may result in large differences in certain strength properties.

The general relationship between specific gravity and strength does not, however, hold true in all cases. Some species of wood contain relatively large amounts of resins, gums, and other extractives, which add to the weight but do not contribute so much to the strength as would a like amount of wood substance. In addition, species vary in the structural arrangement of their fibers. For these reasons, two species that average the same in specific gravity may exhibit different strength characteristics. For example, although Douglas fir and sweetgum are about equal in specific gravity when dry, Douglas fir averages 39 percent higher in compressive strength, and considerably lower in shock resistance than sweetgum.

This general weight-strength relationship holds for different pieces of wood of the same species, as well as for pieces of different species. Conditions of growth often cause wide variations in weight of a species of wood; some trees grow faster than others of the same species; some are more vigorous than others. Trees with wide annual rings have grown more rapidly than trees with narrow rings. In softwoods, as a rule, exceptionally slow-growth or fast-growth material is lighter in weight and weaker than normal-growth stock. In hardwood, fast growth is generally indicative of good strength properties, although slow growth does not necessarily mean weakness. As a rule, wood of slow growth near the outside of large trees is of light weight and weak.

Locality of growth, within the normal range of growth of a species, has relatively little effect upon quality of lumber. Far greater differences in weight, strength, and other properties are found in trees growing side by side. Douglas fir is an outstanding exception to this general rule among species native to the United States. Coast Douglas fir, as a rule, is of better quality than inland Douglas fir, although, as with all species, exceptions are to be found.

Since weight and strength are so closely related, it is important to set minimum weights for wood to be used in structural parts. This can be done by requiring a minimum weight per cubic foot at a stated moisture content (section 3.06). Weighing wood in this manner, however, is cumbersome as well as open to inaccuracy. Weight determination by means of specific gravity tests on small samples cut from large pieces is preferred (section 3.05).

It is likewise true that some species of wood are equal in some respects to others of higher specific gravity. Coast Douglas fir, although its density is but three-fourths that of commercial white oak is about equal to oak in bending and compressive strength and excels it in stiffness. White oaks, however, averages much higher than Douglas fir in hardness and shock resistance. Specific gravity relationships among species, therefore, represent general trends and not uniform laws. Departure of a species from the general relationship often indicates some exceptional characteristics that may or may not make this species particularly desirable for a certain use.

Specific gravity affords a still better index of strength within a species. The heaviest pieces of any species of wood are generally two to three times as high in specific gravity as the lighest ones of the species, and are correspondingly stronger. The relationship of the pieces within a species is usually represented by a power of specific gravity slightly higher than that representing average values for different species. Furthermore, departures from the average relationship are less marked.

3.06. MOISTURE CONTENT

3.060. General. All wood subjected to normal use conditions contains more or less moisture. The exact quantity depends upon the amount of seasoning the material has received and, more important from the standpoint of long-time use, the atmospheric and other conditions to which it is exposed.

Lumber cut from freshly felled trees contains varying amounts of moisture often called "sap." Sap is composed primarily of water, with varying amounts of other materials in solution. This moisture exists in the wood in two forms: As bulk liquid contained in the cell cavities and known as free water; and as absorbed or "imbibed" moisture held within the cell walls. Sapwood usually contains more moisture than heartwood, the difference being more marked in softwoods than in hardwoods.

Freshly cut or unseasoned lumber is called "green," an indefinite term applied to lumber that has not been seasoned and therefore contains most of its original moisture. Green lumber may contain a weight of water considerably greater than its own dry weight, but usually, the quantity held within it is much less.

As the wood dries, this free water is removed until the cells are virtually empty of it. Only at that stage, called the fiber-saturation point (for most species considered to be about 30 percent moisture content), does the imbibed water within the cell walls begin to leave. As drying continues below the fiber-saturation point, the wood shrinks. Green lumber exposed to the atmosphere dries until it reaches a condition in balance with that of the surrounding air. Its moisture content in this condition is called equilibrium moisture content. It varies with temperature and relative humidity, being affected more by relative humidity than by temperature (fig. 3–2).

Normal daily temperature and relative humidity changes do not have any pronounced effect on the moisture content of wood. Changes over a more extended period of time, such as seasonal changes, cause its moisture content to change until again in equilibrium. Swelling and shrinking of the wood accompanies this change (section 3.07).

The use of figure 3–2 in determining the equilibrium moisture content of wood is illustrated by the following example:

Example: Find the equilibrium moisture content of wood at a temperature of 110 degrees F. and a relative humidity of 70 percent.

First, locate the vertical temperature line for 110 degrees F. on the horizontal scale. Second, locate the intersection of this line with the 70 percent relative humidity curve extending from the right-hand vertical scale. The point on the left-hand scale horizontally opposite this point of intersection gives the reading desired, in this case 12 percent equilibrium moisture content.

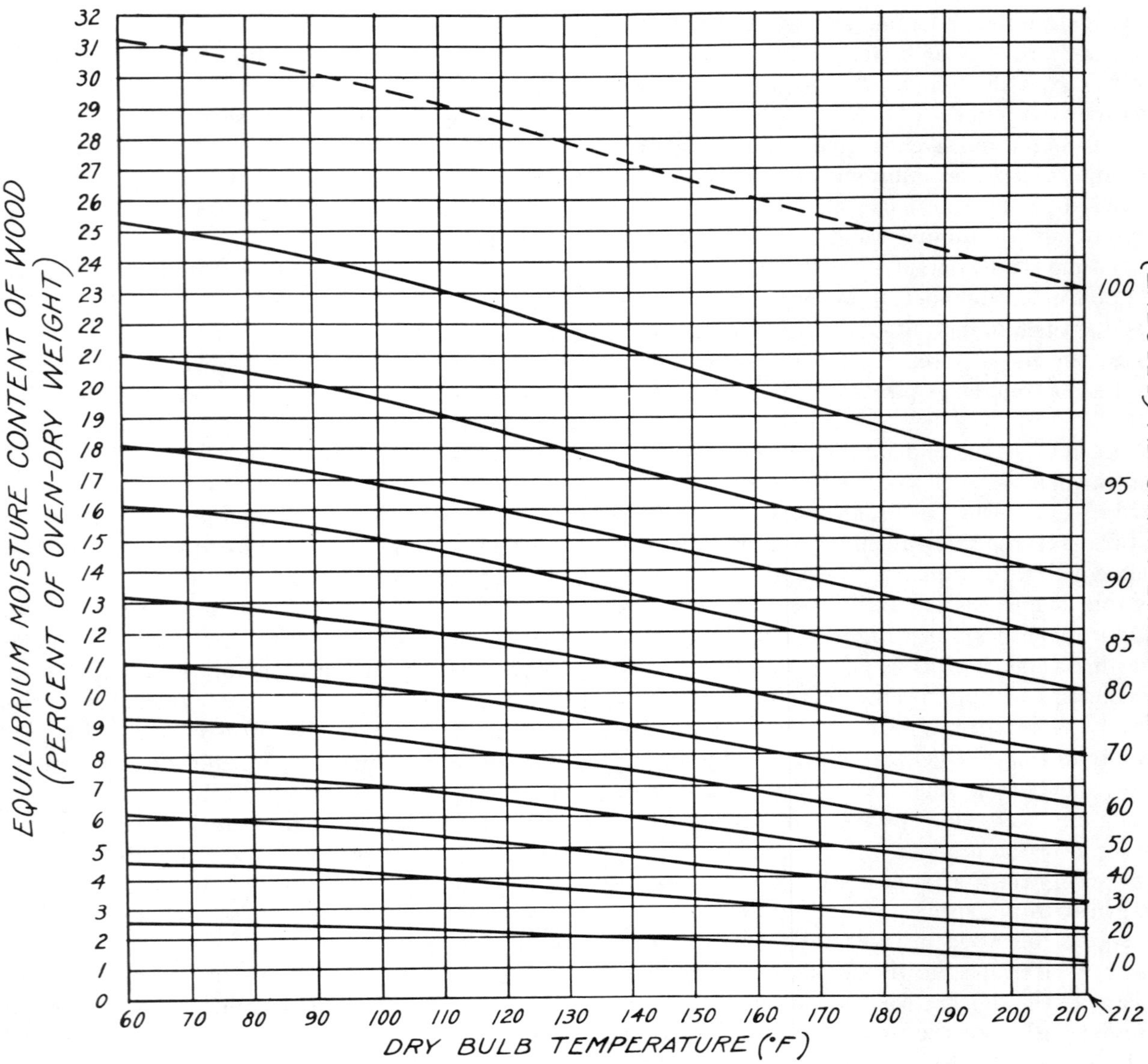

Figure 3–2. Relation of dry-bulb temperature, relative humidity, and equilibrium moisture content of wood.

Moisture content is usually expressed as a percentage of the weight of the ovendry wood. If the moisture content of a board is 10 percent, there are by weight, 10 parts of water to 100 parts of ovendry wood. If the moisture content of a green board is 150 percent, the moisture constitutes three-fifths weight of the board. Moisture content is a relative term and does not give a measure of the absolute amount of water present, unless the density of the wood is taken into consideration. An oak board at 20 percent moisture content, for example, contains more water than a spruce board of the same dimensions at the same moisture content, because the ovendry weight of the oak board is greater than that of the spruce board. Hence, more water is needed to make up 20 percent of its dry weight.

The trade terms "air-dry," "shipping-dry," and "kiln-dried" have no specific meaning with regard to moisture content. Air-dry lumber may have a moisture content ranging from 6 percent in the arid Southwest during the summer months, to 24 percent in parts of the Pacific Northwest during the winter. In general, the average moisture content of thoroughly air-dry lumber is about 12 to 15 percent. Lumber that is partially dried to reduce freight charges is known as "shipping-dry" lumber and may have a moisture content of 30 percent or more. Kiln-dried lumber is commonly reduced to a moisture content of 4 to 12 percent, depending on the requirements for its use, although kiln-dried softwood lumber in common grades is sometimes seasoned only to 15 to 22 percent moisture content.

3.061. Moisture Content and Strength. The strength of wood is intimately related to the amount of moisture it contains (sec. 5.2). As wood dries, most strength properties are increased. On the other hand, there are great differences in the strength of different species of wood at the same moisture content. A relationship also exists between the strength of woods at the same moisture content and their weight, the heavier woods being the stronger. A wood hull cannot be satisfactorily built or repaired unless the moisture content of the lumber used is within established limits. Inasmuch as the lumber used for any wood hull varies in thickness for different components, practical methods of moisture content determination must necessarily be used which will be accurate for all thicknesses.

3.062. Moisture Content Determination. Two methods commonly used to determine the amount of moisture in wood are the oven-drying and electrical methods. The oven-drying method is more exact and is applicable over a wider range of moisture content values, whereas electric moisture meters, although somewhat limited in range and accuracy, offer more rapid means of estimating moisture content without necessitating the cutting of samples from lumber.

3.0620. Ovendrying Method. Five steps are necessary for accurate moisture content determination of wood by the ovendrying method:

1. Select a representative board and cut a cross section about 1 inch long in the direction of the grain.

2. Immediately after sawing, remove all loose splinters and weigh the sample.

3. Put sample in an oven maintained at a temperature of 214 to 221 degrees F. (101 to 105 degrees C.) and dry until constant weight is attained.

4. Reweigh the sample to obtain the ovendry weight.

5. Divide the loss in weight by the ovendry weight and multiply the result by 100 to get the percentage of moisture in the original sample. Thus,

$$\text{Percentage moisture} = \frac{(W-D)}{(D)} \times 100$$

where W is the original weight, as found under 2 above and D is the ovendry weight, as found under 4 above.

First Step. If possible, the sample should be taken at least 2 feet from one end of the piece, because the end may be considerably wetter or drier than the rest of the piece. Short pieces of wood dry out much more rapidly than longer ones. In order to reduce the time required for drying, therefore, the length of the sample in the direction of the grain should be not more than 1 inch. The other dimensions are those of the cross section of the board.

Second Step. The sample should be weighed immediately after it is cut, for the material is subject to moisture changes on exposure to the air. If the sample cannot be weighed immediately, it should be wrapped in foil or waxed paper until weighed.

In order to insure good results, the weights should be correct to within at least 0.50 percent of the dry weight of the sample.

The metric system of weights is convenient in making moisture determinations.

The kind of scales to be used and the size of smallest graduation necessary to insure the specified accuracy will depend on the weight, and consequently the size, of the sample and kind of wood. Small spring postal scales reading to ½ ounce are not suitable for accurate weighing of small moisture samples. A triple-beam balance or torsion balance scale sensitive to 0.01 gram, with auxiliary weights, is satisfactory for small samples.

Third Step. When placed in the oven for drying, the samples should be open-piled to allow free access of air to each piece. The oven should have some ventilation to allow the evaporated moisture to escape. A thermometer is needed on the oven to tell the temperature. Excessive temperatures or excessive periods in the drying oven will cause distillation of the wood, which destroys accuracy of results. Ordinarily, with low-density woods, 12 hours' ovendrying is sufficient, while high-density woods may require 48 hours' ovendrying. Samples containing considerable moisture should remain in the oven longer than those containing less moisture.

Fourth Step. As is the case of the first weight taken, it is essential that the sample be weighed immediately after being removed from the oven.

Fifth Step. A typical example of how to compute the percentage of moisture is:

A 2- by 2- by 1-inch sample of air-dry Sitka spruce weighed 30.8 grams. The sample after ovendrying weighed 27.5 grams. To find the moisture content:

$$\text{Percentage moisture} = \frac{(30.7-27.5)}{27.5} \times 100 = \frac{330}{27.5} = 12$$

3.0621. Electrical Method. Electric moisture meters (figures 3–3, 3–4) are extensively used to determine the moisture content of wood. They are simple to operate, but some judgment and care are required in their use. They are convenient and fast to use requiring only a few seconds to determine the amount of moisture in any piece of wood. Two general types of instruments are available. One evaluates the moisture content by measuring the electrical resistance of the wood, using direct current; the other measuring the electrical capacity or power absorbed by the wood.

The resistance-type electrical moisture meter is the most practical of the several methods that can be used to determine moisture content of ship and boat building lumber.

3.06210. Within a range of 7 to 25 percent, such moisture meters will record the moisture content of lumber with an accuracy of ±1 percent. A moisture meter is calibrated, subject to the correction factors for species, temperature, and similar items, to show the moisture content in percent of the oven dry weight of the lumber being measured.

Wood acts as a resistance element in the electrical circuit of a moisture meter. The current flowing between electrodes will follow the path of least resistance, which is the path with the greatest amount of moisture. In this way, the meter will show the moisture content of the wettest wood contacted by the electrodes at all times.

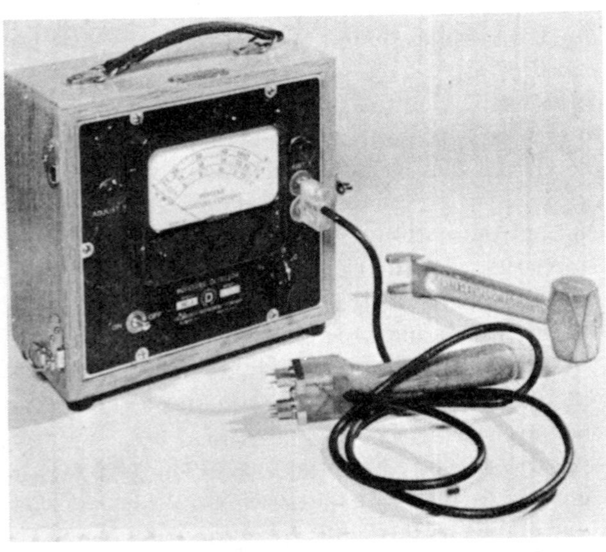

Figure 3–3. Resistance type of electrical moisture meter.

Figure 3–4. High-frequency type of electrical moisture meter.

When a resistance-type moisture meter is used, certain procedures must be followed to make accurate determinations. For lumber exposed to drying conditions, the following procedures are most suitable:

1. For lumber 1⅝ inches (2 inch nominal) and under, use 5/16-inch electrodes and take readings at 1/16- and 5/16-inch depths. A difference of 2 percent or less between the two readings indicates uniform moisture content at the center. If the difference between the two readings is greater than 2 percent, procedure 2 should be used.

2. For lumber more than 1⅝ inches thick, the electrodes or nails should be driven to the center of the piece and a reading taken.

Bright Nails. As indicated in the preceding paragraph, bright nails can be used when regular electrodes are not long enough to reach the required depth. The size of the nail is not important. When nails are used, they should be spaced approximately the same distance apart as the electrodes, and placed in such a manner that the current flow is parallel to the grain of the lumber.

For dense species, it is suggested that a hole about the diameter of the electrodes be drilled to a depth of ¼ inch less than the depth desired to prevent bending the nail in driving. If the nails are touched with the electrodes, or the electrode supports, the moisture content of the wettest wood in contact with the nails can be accurately obtained. Some operators find it more convenient to equip the wire leads to the meter with spring clips to grasp the nail heads.

Electrodes can be obtained which are covered, except for the extreme tip, with a tough insulating resin. This type of electrode is quite useful in determining the moisture content of lumber which has the surface wetted by dew or rain, and for measuring shell and core moisture content separately.

When insulated electrodes are used in very dense wood, it is advisable to drill holes for their placement so that the coating will last much longer. A 3/32- or 1/8-inch drill will usually be of the right diameter. The depth of the hole should be 1/4 inch less than the depth to which the electrode is to be driven. If insulated electrodes are driven from the surface to the center of a board without predrilling, readings can be taken at different depths to obtain a complete picture of the moisture distribution. The electrode responds only to the moisture content of the wood in contact with the tips.

In the case of lumber regaining moisture from surroundings, the surface of the board has a higher moisture content than the interior, so the electrodes pass through a wet region into a drier region. With regular electrodes or nails, the meter indicates the moisture content of the wettest wood in contact with the electrodes, which would be at the surface. Since this is higher than the moisture content of the interior of an otherwise dry board, the reading is actually a false indication.

Insulated Electrodes. Insulated electrodes should be used under these circumstances:

The resin coating insulates the electrode, and allows it to be driven through a moist surface to indicate the actual moisture content within the wood by procedure 2 outlined in a preceding paragraph.

If the surface is covered with free water, such as rain, some water might possibly follow the tips of the electrode into the wood, and again produce a wet low-resistance path between the electrodes. Readings should not be taken under these conditions. If the meter registers a moisture content greater than the limitations specified at a depth of 1/16 inch, the lumber must be dried further before use.

The accuracy of moisture meter readings is influenced by the following factors, which should be taken into account in making readings:

Species. At any given moisture content, the electrical resistance of lumber depends upon the wood species. Moisure meters are calibrated to read directly for one species only. The true value for another common species can be obtained by applying to the meter reading a correction obtained from tables supplied with each instrument.

Temperatures. As the temperature of lumber increases, the electrical resistance decreases, and vice versa. The correction for this can be obtained easily from a chart furnished with the meter. The temperatures referred to in the chart are the temperatures of the lumber upon which the moisture content is being determined. Thus, moisture meter determinations should not be made on lumber which has just come from a kiln unless an actual measurement of the lumber temperature is made.

Contact. The electrodes should be driven into sound wood. If poor contact is made with any type of electrode, too low a moisture content will be indicated. Deduct 1 percent from the reading obtained if nails are used.

Grain Direction. Electrical resistance measured across the grain is slightly greater than that measured along the grain. The electrodes should be driven into the lumber in such manner that the current flows parallel to the grain.

Waterborne Preservative. When lumber is treated with waterborne salts for preservative or fire-retardant purposes, the wood becomes more conductive and indicates a moisture content greater than the correct value. For lumber or plywood pressure treated with waterborne preservative or fire retardant salts, the ovendrying method of moisture content determination should be used.

Glue Lines. Some glue lines affect readings. Therefore, moisture content readings on plywood may be higher than the actual moisture content. To detect this condition, drive the pins through not more than half the thickness of the first ply and read the meter. Then drive the pins so they just pass through the first glue line. If the reading shows no great increase, the glue line may be considered as having no effect, and the reading will be correct.

Oil-borne Preservatives. Pressure preservative treatments with pentachlorophenol have no effect on the accuracy of moisture meter readings below 15 percent. Above this moisture content, 1 percent should be deducted from the indicated reading. Deduct 3 percent from all readings below 25 percent taken on lumber pressure treated with creosote. Corrections for lumber pressure treated with copper naphthenate are not known at this time. However, readings below 15 percent can be relied upon. Brush treating with pentachlorophenol will not cause inaccurate readings. However, copper naphthenate brush treatments might give high readings. A light sanding or scraping of the surface will remove such treatment and accurate readings will be obtained.

Paint Coats. Some paint coatings are conductive. It is best to remove a small portion of the paint on the wood surface and take a reading. Then take a reading through the paint coating adjacent to the first reading. If no great difference is noted in the readings, the paint coating can be assumed to have no effect.

Salt Water. Wood in contact with salt water has a tendency to absorb increasing amounts of salt over long periods of time. The salt will give high readings for which there is no means of correction, since there is no uniform distributon of salt in the wood. The ovendrying method is the only certain means of determination in this case. However, the actual moisture content will never be greater than the meter reading, though wood that has been long exposed to fresh water may give low meter readings when the actual moisture content is near or above 30 percent.

High Relative Humidities. Measurements of lumber with moisture content below 10 percent are not reliable when moisture meters are used in wet weather. The meter surfaces may become damp and provide leakage paths with high resistances comparable to those associated with lumber having a low moisture content.

Number and Location of Measurements. A single measurement should not be depended upon as the true moisture content of a piece of lumber if more measurements can be obtained. Three determinations should be made whenever possible. One measurement per board checked may be enough to obtain an average for the lot, provided enough boards are checked. Measurements should be taken at least 1 foot from the ends of the boards.

Surface Moisture. Surface moisture from rain, dew, or fog forms a layer with low electrical resistance. The wet surface will cause the instrument to record too high an average moisture content. Electrode pins that are covered, except for the extreme tip, by a coating of insulating resin can be used, however, to measure the moisture content, even if the surface is wetter than the interior of the board.

Water Pockets. The presence of water pockets within lumber will cause moisture meters to read too high. Water pockets are sometimes found beneath sticker marks, or where a bit of vertical grain shows on a flat grain surface. Insulated pins will not detect water pockets unless readings are taken at more than one position on the board. When moisture content limitations are specified, any lumber found with pockets is unsuitable for ship or boat building purposes until the moisture content has become uniformly distributed.

Thickness of Material. If 5/16-inch electrodes are used in material thicker than 1⅝ inches, the value obtained will be accurate only when the wood has a uniform moisture distribution; that is, little or no moisture gradient.

Personal Elements. The accuracy of a moisture meter determination often depends upon the care with which the instrument is used. Conditions that require measurements to be made rapidly may lead to inaccuracies due to incorrect balancing, reading, location of points, poor contact, neglect of temperature, and species. As much care as possible should be taken in all measurements to obtain reliable readings.

Meter Maintenance. The maintenance of resistance-type moisture meters consists largely of replacements of exhausted batteries or broken electrodes. Periodic inspection will cut down the probability of failure while in use. Frequency of inspection will depend upon the amount of use the instrument receives.

Portable moisture meters operate on self-contained dry batteries. The batteries should be replaced when they are partially exhausted. The need for replacement is indicated by a low battery voltage measured with the switch on, by the inability to obtain a steady-state balance, or by the meter dial showing a constant drift toward zero. The replacement of batteries can be cut down by keeping the switch in the off position when the instrument is not being used. Vacuum tubes last for as many as several thousand hours of service. They seldom require replacement.

Electrode Contacts. The electrode contacts of moisture meters may become bent or broken in use. A number of spare needles and a wrench should be kept in the moisture meter case. The meter should never be handled roughly, and should be calibrated at intervals. Some moisture meters are furnished with standard resistances for calibration purposes.

If equipment is available for performing oven test, the accuracy of the meter can be checked. The check is made by comparing the meter determination with results obtained by the oven test. The specimen used for the oven drying moisture check should be cut out at the location where the meter reading was taken. The entire cross section of the piece should be used wherever practical, but full thickness of the material should be used in all cases. This check should be conducted on well seasoned lumber.

Properly used, a moisture meter affords a quick means of determining whether or not the lumber being installed is within established limits. Its proper use can prevent much dissatisfaction and many costly maintenance problems.

3.06211. Radio-Frequency Power-Loss Moisture Meters. When the electrodes of a capacity or radio-frequency power-loss meter are brought into contact with wood, the flow of electric current in one branch of the oscillator circuit is affected by the moisture content of the wood adjacent to the electrodes and the moisture content can thus be evaluated. Meters of his type are generally supplied with an arbitrary scale and a table giving the corresponding moisture content readings for a number of species of wood commonly used. In some cases, a calibration for Douglas fir or another species is used. The radio-frequency power-loss method is, in principle, an excellent means of evaluating the quantity of water, by weight, in wood. To convert the evaluation of the weight of water to a percentage moisture content, it is necessary to know the weight or specific gravity of the wood. This property cannot be determined quickly enough by any method now available; therefore, it is present practice to assume that the specific gravity of the individual piece when dry is the same as the average for the species. On this basis, the meters are calibrated, usually for a range of 0 to 25 percent moisture content. Radio-frequency moisture meters make contact with the wood with metal plates on the same side of the board under test.

Factors that may affect the accuracy of the moisture content determinations include:

(*a*) *Variation in Specific Gravity.* Since an assumed specific gravity has to be used for calibrating capacity-type moisture meters, each reading on such a meter carries an error proportional to the actual error in the assumed specific gravity of the piece.

(*b*) *Moisture Distribution.* Moisture distribution may affect the readings of meters using condenser plates on one side only of the piece under test.

(*c*) *Rough Surfaces.* Extremely rough surfaces that do not permit full contact of the condenser plates may affect the accuracy of readings.

3.06212. Limitations of Moisture Meters. Improper use of either resistance or radio-frequency power-loss electrical moisture meters will cause erroneous readings. In most cases, they are designed for certain thicknesses of wood and, if used indiscriminately, incorrect observations are probable. Surfaces that have been treated with paint, varnish, lacquer, dope, and other finish materials do not, as a rule, affect the accuracy of electrical moisture meters. In testing veneer and thin lumber with certain types of meters, care should be exercised to make certain that the readings are not influenced by the character of the material upon which the veneer or lumber is placed during the test. Veneers are easily split by needles, in which case, contact is poor and incorrect readings are possible. Where needles penetrate glue lines, there is some danger that the glue lines may contain electrolytes whose conductivity will obscure the actual moisture content of the wood. Moisture meters cannot be used successfully on wood that has been treated by a chemical to render it fire resistant. Water-borne preservative salts also affect moisture meter accuracy on the high side, especially at higher moisture content values. Preservative oils, on the other hand, affect accuracy on the low side, if at all.

3.07. SHRINKAGE AND SWELLING

3.070. General. Theoretically, shrinkage of wood below the fiber-saturation point is directly proportional to the amount of moisture lost. A curve showing the relationship of shrinkage to average moisture content, however, is rarely a straight line because of the presence of a moisture gradient in the wood (figure 3–5). Shrinkage values for numerous woods used in ship and boat building are given in table 3–2.

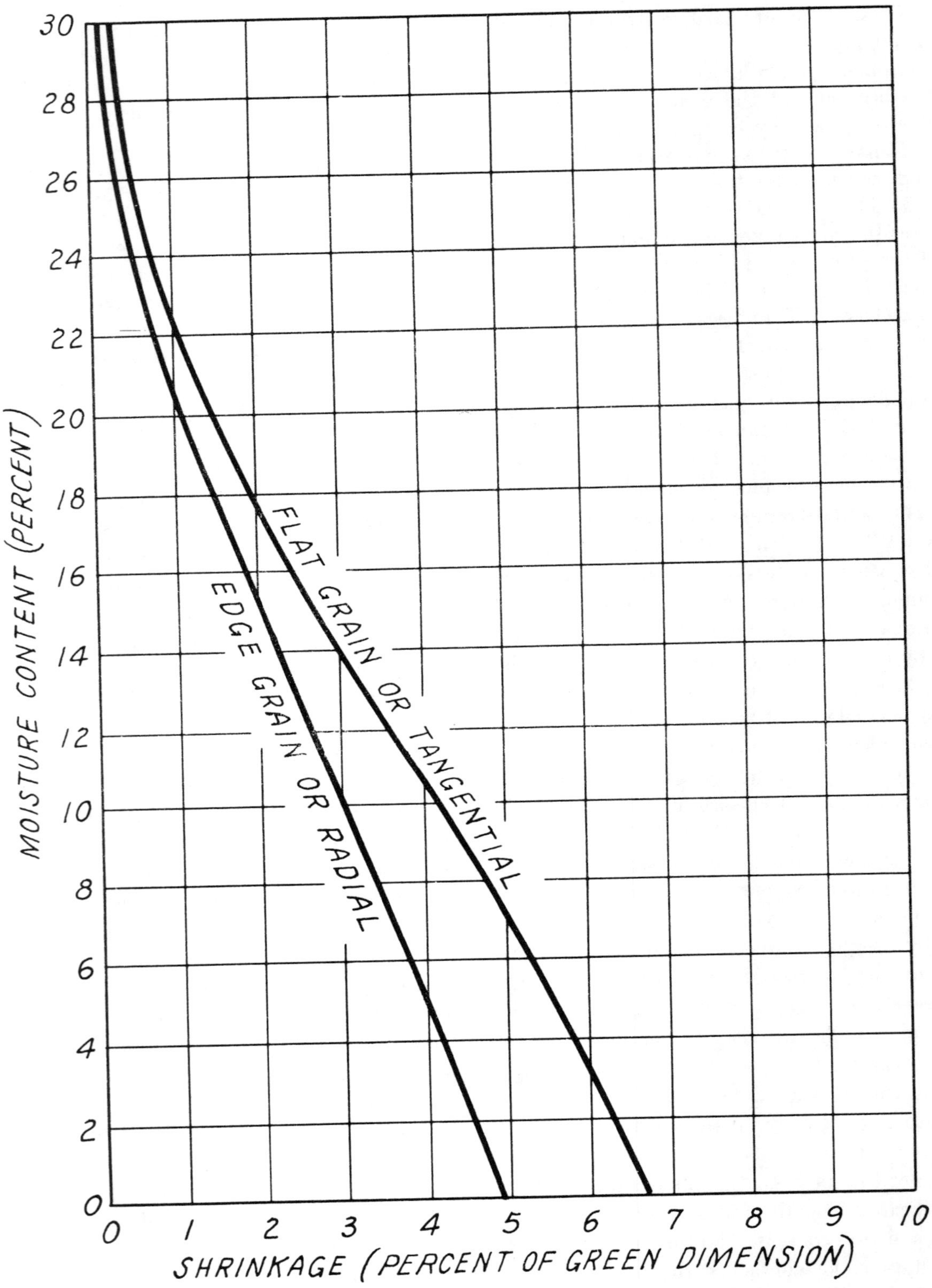

Figure 3–5. Typical moisture content-shrinkage curves for Douglas fir and southern yellow pine. Although the curves are not straight lines, for practical shrinkage calculations they may be considered as such.

TABLE 3-3

SHRINKAGE VALUES OF SHIPBUILDING WOODS BASED ON DIMENSIONS WHEN GREEN

Species	Shrinkage								
	Dried to 20 percent moisture content [1]			Dried to 6 percent moisture content [2]			Dried to 0 percent moisture content		
	Radial	Tangential	Volumetric	Radial	Tangential	Volumetric	Radial	Tangential	Volumetric
SOFTWOODS									
	Percent	*Percent*	*Percent*	*Percent*	*Percent*	*Percent*	*Percent*	*Percent*	*Percent*
Baldcypress	1.3	2.1	3.5	3.0	5.0	8.4	3.8	6.2	10.5
Cedar:									
Alaska	.9	2.0	3.1	2.2	4.8	7.4	2.8	6.0	9.2
Atlantic white	1.0	1.8	2.9	2.3	4.3	7.0	2.9	5.4	8.8
Eastern redcedar	1.0	1.6	2.6	2.5	3.8	6.2	3.1	4.7	7.8
Incense	1.1	1.7	2.5	2.6	4.2	6.1	3.3	5.2	7.6
Northern white	.7	1.6	2.4	1.8	3.9	5.8	2.2	4.9	7.2
Port-Orford	1.5	2.3	3.4	3.7	5.5	8.1	4.6	6.9	10.1
Western redcedar	.8	1.7	2.3	1.9	4.0	5.4	2.4	5.0	6.8
Douglas fir:									
Coast	1.7	2.6	3.9	4.0	6.2	9.4	5.0	7.8	11.8
Intermediate	1.4	2.5	3.6	3.3	6.1	8.7	4.1	7.6	10.9
Rocky Mountain	1.2	2.1	3.5	2.9	5.0	8.5	3.6	6.2	10.6
Larch, western	1.4	2.7	4.4	3.4	6.5	10.6	4.2	8.1	13.2
Pine:									
Eastern white	.8	2.0	2.7	1.8	4.8	6.6	2.3	6.0	8.2
Loblolly	1.6	2.5	4.1	3.8	5.9	9.8	4.8	7.4	12.3
Longleaf	1.7	2.5	4.1	4.1	6.0	9.8	5.1	7.5	12.2
Ponderosa	1.3	2.1	3.2	3.1	5.0	7.7	3.9	6.3	9.6
Shortleaf	1.5	2.6	4.1	3.5	6.2	9.8	4.4	7.7	12.3
Slash	1.8	2.6	4.1	4.4	6.2	9.8	5.5	7.8	12.2
Sugar	1.0	1.9	2.6	2.3	4.5	6.3	2.9	5.6	7.9
Western white	1.4	2.5	3.9	3.3	5.9	9.4	4.1	7.4	11.8
Redwood:									
Old-growth	.9	1.5	2.3	2.1	3.5	5.4	2.6	4.4	6.8
Second-growth	.7	1.6	2.4	1.8	3.9	5.7	2.2	4.9	7.1
Spruce:									
Black	1.4	2.3	3.8	3.3	5.4	9.0	4.1	6.8	11.3
Red	1.3	2.6	3.9	3.0	6.2	9.4	3.8	7.8	11.8
Sitka	1.4	2.5	3.8	3.4	6.0	9.2	4.3	7.5	11.5
White	1.6	2.7	4.6	3.8	6.6	11.0	4.7	8.2	13.7
Tamarack	1.2	2.5	4.5	3.0	5.9	10.9	3.7	7.4	13.6
HARDWOODS									
Angelique	1.5	2.7	4.7	3.7	6.6	11.2	4.6	8.2	14.0
Apitong	1.6	3.9	5.3	3.9	9.3	12.8	4.9	11.6	16.0
Ash:									
Black	1.7	2.6	5.1	4.0	6.2	12.2	5.0	7.8	15.2
Green	1.5	2.4	4.2	3.7	5.7	10.0	4.6	7.1	12.5
Oregon	1.4	2.7	4.4	3.3	6.5	10.6	4.1	8.1	13.2
White	1.6	2.6	4.5	3.8	6.2	10.7	4.8	7.8	13.4
Basswood, American	2.2	3.1	5.3	5.3	7.4	12.6	6.6	9.3	15.8
Beech, American	1.7	3.7	5.4	4.1	8.8	13.0	5.1	11.0	16.3
Birch:									
Paper	2.1	2.9	5.4	5.0	6.9	13.0	6.3	8.6	16.2
Yellow	2.4	3.1	5.6	5.8	7.4	13.4	7.2	9.2	16.7
Cherry, black	1.2	2.4	3.8	3.0	5.7	9.2	3.7	7.1	11.5

Footnotes at end of table.

TABLE 3-3—Continued

SHRINKAGE VALUES OF SHIPBUILDING WOODS BASED ON DIMENSIONS WHEN GREEN—Continued

Species	Shrinkage								
	Dried to 20 percent moisture content [1]			Dried to 6 percent moisture content [2]			Dried to 0 percent moisture content		
	Radial	Tangential	Volumetric	Radial	Tangential	Volumetric	Radial	Tangential	Volumetric
HARDWOODS—continued									
	Percent	*Percent*	*Percent*	*Percent*	*Percent*	*Percent*	*Percent*	*Percent*	*Percent*
Elm:									
Rock	1.6	2.7	4.7	3.8	6.5	11.3	4.8	8.1	14.1
Slippery	1.6	3.0	4.6	3.9	7.1	11.0	4.9	8.9	13.8
Greenheart	1.1	1.4	2.7	2.7	3.4	6.4	3.4	4.2	8.0
Hackberry	1.6	3.0	5.6	3.8	7.1	13.5	4.8	8.9	16.9
Hickory:									
Mockernut	2.6	3.7	6.0	6.2	8.8	14.3	7.8	11.0	17.9
Pignut	2.4	3.8	6.0	5.8	9.2	14.3	7.2	11.5	17.9
Shagbark	2.3	3.3	5.6	5.6	8.0	13.4	7.0	10.0	16.7
Shellbark	2.5	4.2	6.4	6.1	10.1	15.4	7.6	12.6	19.2
Iroko	1.1	1.6	2.8	2.7	3.8	6.8	3.4	4.8	8.5
Ironbark, gray	1.9	2.8	4.7	4.5	6.7	11.2	5.6	8.4	14.0
Khaya ("African mahogany")	1.4	1.9	2.9	3.3	4.6	7.0	4.1	5.8	8.8
Lauan, red	1.1	2.7	3.9	2.6	6.4	9.4	3.3	8.0	11.7
Locust:									
Black	1.5	2.4	3.4	3.7	5.8	8.2	4.6	7.2	10.2
Mahogany	1.2	1.7	2.7	2.9	4.0	6.4	3.6	5.0	8.0
Maple:									
Black	1.6	3.1	4.7	3.8	7.4	11.2	4.8	9.2	14.0
Sugar	1.6	3.2	5.0	3.9	7.6	11.9	4.9	9.5	14.9
Oak, red:									
Black	1.5	3.2	4.7	3.6	7.8	11.4	4.5	9.7	14.2
Northern	1.3	2.7	4.5	3.2	6.6	10.8	4.0	8.2	13.5
Pin	1.4	3.2	4.8	3.4	7.6	11.6	4.3	9.5	14.5
Scarlet	1.5	3.2	4.6	3.7	7.8	11.0	4.6	9.7	13.8
Southern	1.5	2.9	5.4	3.6	7.0	13.0	4.5	8.7	16.3
Oak, white:									
Bur	1.5	2.9	4.2	3.5	7.0	10.2	4.4	8.8	12.7
Chestnut	1.8	3.2	5.6	4.4	7.8	13.4	5.5	9.7	16.7
Live	2.2	3.2	4.9	5.3	7.6	11.8	6.6	9.5	14.7
Post	1.8	3.3	5.4	4.3	7.8	13.0	5.4	9.8	16.2
Water	1.4	3.1	5.5	3.4	7.4	13.1	4.2	9.3	16.4
White	1.8	3.0	5.3	4.2	7.2	12.6	5.3	9.0	15.8
Okoume	------	------	3.6	------	------	8.7	------	------	10.9
Pecan	1.6	3.0	4.5	3.9	7.1	10.9	4.9	8.9	13.6
Peroba do campo	------	------	------	------	------	------	3.4	6.2	9.8
Sweetgum	1.7	3.3	5.0	4.2	7.9	12.0	5.2	9.9	15.0
Tangile	1.4	3.0	4.4	3.4	7.3	10.6	4.3	9.1	13.3
Teak	.8	1.4	2.3	1.8	3.4	5.4	2.3	4.2	6.8
Walnut, black	1.8	2.6	4.3	4.4	6.2	10.2	5.5	7.8	12.8
Willow:									
Black	.9	2.7	4.8	2.1	6.5	11.5	2.6	8.1	14.4
Pacific	1.0	3.0	4.6	2.3	7.2	11.0	2.9	9.0	13.8
Yellow poplar	1.3	2.4	4.1	3.2	5.7	9.8	4.0	7.1	12.3

[1] These shrinkage values have been taken as one-third the shrinkage to the ovendry condition as given in the last three columns of this table.
[2] These shrinkage values have been taken as four-fifths of the shrinkage to the ovendry condition as given in the last three columns of this table.

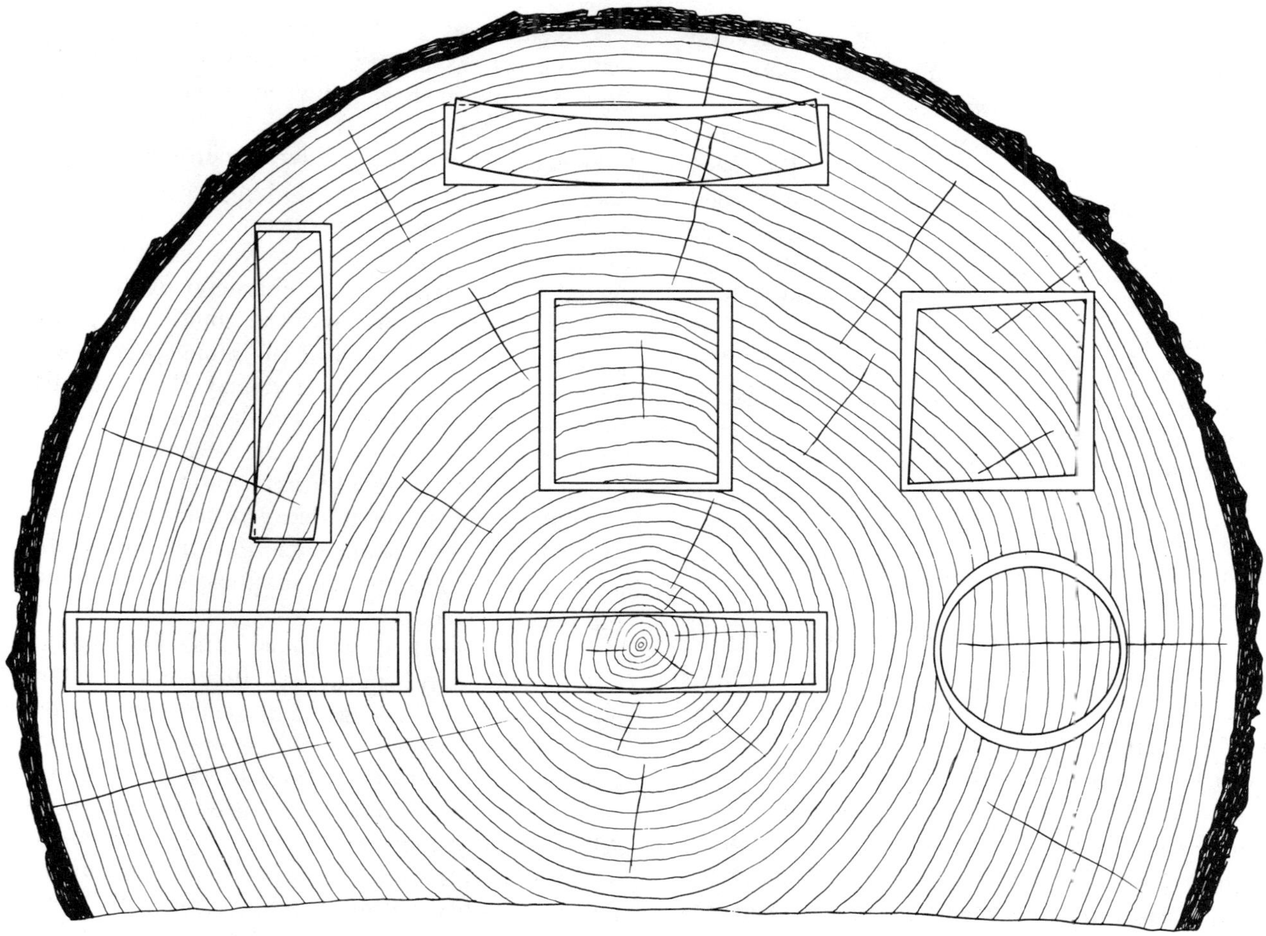

Figure 3–6. Characteristic shrinkage and distortion of flats, squares, and rounds as affected by the direction of the annual rings. Tangential shrinkage is about twice as great as radial.

When wood reaches approximate equilibrium with the temperature and relative humidity of the surrounding atmosphere it may swell or shrink as atmospheric conditions change. This does not mean, however, that the daily temperature and relative humidity changes that occur have any pronounced effect on wood. The rate of exchange of moisture between wood and atmosphere is comparatively slow, and the equilibrium moisture content of the wood is based on the average humidity prevailing over an extended period rather than on changes during very short periods, even though fluctuations may then be through a wide range. Thickness of the wood is also a factor, the moisture content of thin material reacting to atmospheric changes more rapidly than that of thick material.

Since, in seasoning green wood, the surface dries more rapidly than the interior and reaches the fiber-saturation point first, shrinkage may start near the surface of a piece while the average moisture content of the board or timber is still considerably above the fiber-saturation point. Wood shrinks most in the direction of the annual growth rings (tangentially), about ½ to ⅔ as much across these rings (radially), and very little, as a rule, along the grain (longitudinally). The combined effects of radial and tangential shrinkage on the shape of various sections in drying from the green condition are illustrated in figure 3–6.

In general, the heavier species of wood shrink more across the grain than the lighter ones. Heavier wood of the same species also shrinks more in this direction than lighter wood. When freedom from shrinkage is more important than hardness or strength, as, for example, in the planking of small boats, a light-weight species should be chosen. When it is important to combine hardness or strength with low shrinkage, as in the case of treenails, some exceptional species like black locust should be chosen. Table 3–2 gives average radial, tangential, and volumetric shrinkage for various shipbuilding woods.

Plywood bonded with even the highly moisture-resistant synthetic-resin glues will absorb moisture in much the same manner as solid wood, and when the moisture content changes, the plywood will tend to swell or shrink. Since the longitudinal shrinkage of wood is negligible, however, the lateral shrinkage of adjacent plies whose grain is at right angles will be restrained. The shrinkage of plywood in width, is, therefore, reduced but lengthwise is increased. The actual amount of shrinkage varies with the species of wood used, the number of plies, the ratio of the thickness of plies running in one direction to the thickness of plies in the other, and the character of grain.

Tests on a variety of combinations of species and thicknesses of three-ply plywood in drying from the soaked to the ovendry condition showed that the lateral shrinkage of plywood is only about $1/10$ that of solid wood across the grain. Because its shrinkage in the thickness direction is unopposed, plywood will shrink as much as, or slightly more than, normal wood in thickness.

3.08. Resistance of Wood to Natural Hazards. Certain natural hazards of wood must be kept in mind when designing and building boats and ships. Among these are decay fungi, marine borers, land insects, weathering, and fire. By taking advantage of the natural resistance of wood to these hazards and designing, building, using, and maintaining boats and ships intelligently, their damaging effects can be greatly reduced or eliminated entirely. The same holds true, of course, with regard to such minor hazards as chemicals, fumes, and metallic discolorations of various sorts, the effects of which often look much worse than they are.

3.080. Decay. The decay-causing fungus plants that feed on wood (section 2.313) should not be confused with the animal forms of life that attack wood also, notably marine borers and land insects such as termites and certain beetles. The effects of the two types of organisms upon wood are quite different. Decay fungi gradually reduce wood to a punky or crumbly mass. Insects riddle it with holes, but the wood around the holes often remains solid.

The sapwood of all species is low in decay resistance. That is why specifications stringently limit the presence of sapwood in certain grades of hull lumber.

Heartwood varies considerably in decay resistance according to species. Only a few species are rated high in this property. The likelihood of decay in heartwood in service is greatly affected by differences in the character of the wood, the attacking fungus, and the conditions of exposure. Pieces of wood of the same species, even when cut from the same tree, may therefore differ greatly in service life even under apparently similar service conditions. In oak, and in many other woods, the central heartwood, near the pith, is considerably lower in decay resistance than the outer heartwood, being sometimes as susceptible as sapwood. This is one justification for avoiding boxed hearts.

The heartwood of such species as Douglas fir, cedar, mahogany, southern yellow pine, western larch, and white oak is usually classified as moderate in decay resistance and generally gives good service under mild conditions of decay. Among these are our most important shipbuilding woods. All-heartwood material is becoming so difficult to get, however, that all these species except Douglas fir and possibly western larch are generally available only as mixed heartwood and sapwood. In ship and boat hulls, therefore, preservative treatment is considered necessary for these woods if long service life is sought.

Waterproof glues do not improve the decay resistance of boat plywood. Because plywood has end grain on all edges, it absorbs moisture more readily than lumber, so that conditions favorable to decay may be present more often. Preservative treatment, of course, remedies these conditions in proportion to the effectiveness of the preservative and the thoroughness of the treatment.

Service records, laboratory tests, and general experience have made possible the following grouping of native woods according to decay resistance. The woods listed are placed in two groups, one of which consists of species with heartwood high in decay resistance and the other of species moderate to low in decay resistance. Even the woods listed as high in decay resistance may require preservative treatment for important structural uses where decay hazards are high and repairs costly.

Group	Softwoods	Hardwoods
Group 1. Species with heartwood of high resistance to decay.	Baldcypress Cedars Junipers Redwood Yew, Pacific	Catalpa Chestnut Locust, black Mesquite Mulberry, red Osage-orange Walnut, black
Group 2. Species with heartwood of moderate to low resistance to decay.	Douglas fir* Larch, western* Pine, ponderosa Pine, Eastern white* Pine, southern yellow*. Spruces (Tamarack* Eastern larch).	Ashes Beech Birches Cherry, black Elms Hickory Honeylocust* Maples Oaks, red Oaks, white* Sweetgum* Yellow poplar

*Species marked by an asterisk are rated moderate.

Because available data are not so comprehensive or reliable, foreign woods are omitted from the above classification. Information based chiefly on naval experience, however, indicates that some imported species have heartwood that ranks high in decay resistance. These include greenheart, iroko, true mahogany (*Swietenia* spp.), and teak. Plywood of woods such as true mahogany has not shown satisfactory resistance to decay when used under wet conditions in boats and for such uses should also be treated with a preservative.

3.081. Marine Borers. No wood is immune to marine borers and none of the commercially important species of the United States has enough resistance to their attack to justify its use without preservative treatment as hull sheathing, or for other uses where marine borer resistance is a primary requisite. Red oak is preferred for worm sheathing because of its hardness and ease of preservative treatment. Of the foreign woods listed in table 1–1 and described in chapter 2, only greenheart has a wide reputation as possessing considerable resistance to marine borers, and records on this wood do not consistently substantiate this reputation.

3.082. Weathering. Weathering consists of the gradual wearing away of uncoated wood substance by the action of rain, snow, heat, and other components of the weather that cause shrinking, swelling, surface checking, and sometimes warping under normally dry conditions. It is hence not an important consideration with respect to choice of woods for ship and boat hulls, which are generally maintained in a well painted condition.

3.083. Fire and Heat. Wood has considerable fire resistance. Under standard test conditions of time and temperature, wood chars at the rate of about 1½ inches per hour. Laminated wood glued with phenol, resorcinol, and melamine types of resins is at least equal to solid wood in fire resistance. Rate of charring will, of course, vary considerably from that found in tests, depending upon the severity of fire conditions.

The type of wood construction employed greatly affects the fire resistance of any structure. The heavy timber construction of large ships has greater fire resistance than the light construction of small boats. This is primarily due to the size of the framework, planking, and decking used.

Fire resistance can be greatly improved by use of protective materials, including thermal insulation, incombustible materials, and fire-retarding chemicals and coatings. Use of these materials where hazards exist, as in engine rooms and galleys, helps greatly to reduce fire danger aboard ship.

Prolonged high temperatures affect the strength of wood. Maximum heating is seldom prolonged, however, and wood rarely attains the temperature of the hot air surrounding it in actual shipboard service conditions. Good ventilation, use of heat-reflective paints, and other precautions aid greatly in maintaining wood at temperatures that do not impair its strength appreciably, even under difficult conditions.

3.084. Chemicals. Chemicals ordinarily used with wood, such as paints and preservatives, do not impair its strength, although preservative treating or conditioning methods that involve heating at high temperatures for long periods may do so.

The discoloration that often takes place in wood next to iron fastenings, especially in salt water, sometimes results in a deterioration of the joint known as nailsickness. The resistance of wood to decay may also be reduced. While early decay around nails or bolts is probably due to air and water that have penetrated past the plugs, laboratory tests have shown that the presence of iron favors decay in white oak and mahogany, while galanized iron and alloys containing copper had no such effect.

Long continued action by acidic liquids, such as storage battery electrolyte, weakens wood by causing chemical changes due to hydrolysis of the cellulose. Wood has good resistance to mild acid solutions, however. Alkaline solutions are much more harmful, because they tend to dissolve the hemicelluloses and lignin in wood.

3.09. CONDITIONS CONDUCIVE TO DECAY. Where it is practicable, the most economical way to protect wood from decay is to keep it dry. Wood may contain enough moisture internally to decay even if the surface looks and feels dry; but for rapid decay, its moisture content must be more than 30 percent of the weight of the ovendry wood. Decay never takes place if the moisture content of wood remains below 20 percent. There is no such thing as "dry rot!"

On the other hand, too much moisture is unfavorable for most fungi; they must have air as well as water, and the common decay fungi are unable to work in wood that is completely waterlogged or submerged in water. In boats constantly in the water, the keel and other members below the usual bilgewater level would be expected to be too wet for decay and in actual experience are rarely damaged by fungi.

Most decay fungi can make progress at temperatures between 40 and 100 degrees F., with fastest decay in the neighborhood of 75 to 85 degrees F. Although periods of extreme cold halt decay, the fungi are able to resume activity afterwards. High temperatures, such as those to which the wood is subjected in most kiln drying, laminating, exterior plywood manufacture, or steam treatment for bending, kill the fungi. Such treating processes, however, do not prevent reinfection after the wood has been placed in service.

3.090. Factors Affecting Decay in Hulls. It is the universal experience that there is more rapid decay in fresh than in salt water. A survey of boats under repair has substantiated this; more decay, particularly at or below waterline, was found in boats on the Great Lakes than in boats operated most of their time in salt water.

Decay is more rapid when ships are laid up. This is presumably because inactive ships ordinarily are moored in water containing less salt than when in sea service, are subjected to less air motion, and are not so well cared for in respect to keeping seams tight and ventilators in operation. Likewise boats taken out of the water for storage too often show early decay if not given adequate protection (chapter 13).

The factors most obviously associated with important decay in hulls have been the use of decay-susceptible wood, the use of wood in which decay fungi were already active, poor ventilation, and exposure to too much fresh water.

The most important error in choice of wood in the past has probably been in use of red oak where white oak should have been employed. While there have been some very serious decay cases in white oak, the proportion of members decayed always has been higher in red oak where extensive comparison has been possible. Practical methods of distinguishing between red and white oak are available and have come into increasing use at shipyards (section 2.15). Oak is ordinarily used for framing members where decay is especially important and replacements are difficult. Maple, birch, and ash, used in members that are less critical but exposed to the weather, have required many early replacements (figure 3–7). Sweetgum plywood coaming has decayed quickly.

Large timbers, such as are used in the stem, sometimes contain heart rot infections that started in the living tree. These are often impossible to detect (section 2.3122). If decay is detected, the piece should ordinarily be discarded. In oak, active infection may run 4 feet beyond the last recognizable decay, so that attempts sometimes seen in ship construction to dig out the decay and treat the cavity are probably futile.

Excessive fresh water in ship timber comes from three sources. In the heavier solid timbers, too much of the original sap is often still in the wood, and getting it out of a large timber by air seasoning may require years. A more common source of fresh water is rain, supplemented by any fresh water that may be used in washing down. The most harmful leakage is through deck seams, especially at the planksheer or waterways, leading to decay in frameheads, beam ends, and sometimes clamps. The third main source of fresh water in the wood is by condensation. This occurs in cold weather on the inboard face of plankinng, beams, and frames.

3.091. Ventilation. Proper ventilation is essential to prevent condensation of water vapor and remove rain water that gets in through the deck. Forepeaks, lazarettes, steering-gear rooms, and other compartments often have no means of ven-

Figure 3–7. Ash thwart and knees of this motor launch had rotted completely although more durable Douglas fir planking showed decay only at ends butting against thwart.

tilation. Where vents are present, they are too often kept closed. Stems, knightheads, transoms, and after frames too commonly develop decay as a result of inadequate ventilation. Even in mechanically vented crews' quarters, the bays between frames are usually stopped by fillers at the platform deck level. Numerous water-tight bulkheads, linoleum covering of platform decks, and use of double-planking compounds and plywood subdecks, in modern construction, interfere with some of the "breathing" that has occurred in older hulls and increases the need for ventilation. Water that enters the underwater hull through seams, particularly at plank ends, also causes decay in fresh or slightly brackish waters.

3.10. WHAT LUMBER IS. Lumber is broadly defined as the product of the saw and planing mill not further manufactured than by sawing, resawing, and passing lengthwise through a standard planing machine, crosscut to length, and matched. The lumber trade generally recognizes several classes of lumber under this blanket definition, such as "factory and shop lumber," which is lumber intended to be cut up for use in further manufacture; "structural lumber," which is graded by the piece for strength; and "yard lumber," which includes boards, dimension, and strips. Boards are defined as less than 2 inches thick and more than 4 inches wide, dimension as 2 to 4 inches thick and of any width, and strips as less than 2 inches thick and up to 4 inches wide. "Timbers" are recognized as lumber that is 5 inches or larger in its least dimension. All these dimensions are "nominal;" actual dimensions are ¾ to ½ inch smaller.

While these definitions work reasonably well for the average lumber dealer supplying builders of houses, bridges, and other land structures, they are not adapted to shipbuilding. The average landsman's conception of a "plank" for example, is a piece of lumber at least 2 inches thick and 10 inches wide. Ship planking is a term of much more restricted meaning in that it refers exclusively to the planking of the hull exterior; such planking, moreover, may be one-half inch or less in thickness and as narrow as 3 or 4 inches, depending upon the type of boat being built. Ship lumber, therefore, is often procured in accordance with ship terminology and from dealers familiar with its requirements.

Two basic types of lumber are cut at the sawmill. These are known as plain-sawed and quarter-sawed lumber in hardwoods; quarter-

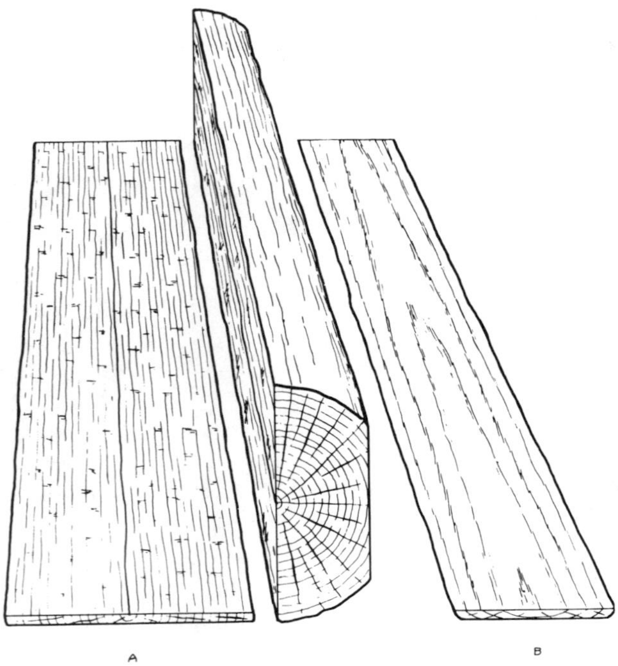

Figure 3–8. ***A*, quarter-sawed and *B*, plain-sawed boards as cut from a log.**

sawed softwood lumber is commonly called edge-grained, vertical-grained, or rift-sawed, while plain-sawed lumber is called flat-grained. Edge-grained lumber is produced by sawing the log along approximately radial lines. Flat-grained lumber is obtained when a log is sawed lengthwise approximately tangent to the annual rings, exposing the tangential surface. Figure 3–8 shows examples of flat-grained and edge-grained lumber with relation to the manner of cutting from the log.

The annual rings often run diagonally across the end of a board, so that it cannot strictly be said to be either plain-sawed or quarter-sawed. Large timbers such as ship beams may be edge-grained on two or more sides while the remaining sides are flat-grained. Or the visible "grain" may not be strictly either type, instead varying over the length of the piece from edge to flat grain. Similarly, planking and ceiling may exhibit various graduations in surface graining. Decking is usually considered to be edge-grained if the slope of the annual rings at the top surface is at an angle of more than 45° with it.

Quarter-sawed lumber shrinks and swells less in width and twists, cups, surface checks, and case-hardens less in seasoning than does plain-sawed lumber. For this reason, edge-grained decking is usually specified because there is less opening of calked seams when it shrinks and less cupping when it swells. On the other hand, plain-sawed lumber is cheaper to produce and does not "collapse" so easily in drying in species subject to this particular defect, such as western redcedar, redwood, and swamp oak. Moreover, the knots it contains will be round or oval instead of long and spike-shaped.

The term "grain" is used rather loosely in connection with lumber. It is used in referring to (*a*) the annual rings, as coarse, fine, even, edge, or flat grain; (*b*) the direction in which the fibers run, as straight, spiral, interlocked, wavy, and curly grain; and (*c*) the relative size of the pores and the fibers, as open grain and close grain. For the sake of clearness, the annual rings should be referred to as such, except with respect to the terms "edge grain," "vertical grain," and "flat grain," the meanings of which are thoroughly established; and (except as just stated) the use of the term "grain" should be confined to the direction of the fibers, as in straight, spiral, wavy, and interlocked grain. The term "texture" should be used to express the relative size of the pores and fibers or the relative amounts of springwood and summerwood, as coarse and fine texture, even and uneven texture. Measurement of slope of grain is discussed in section 2.302.

3.11. PLYWOOD

3.110. Definition. Plywood is an assembly of veneers, or of veneers and lumber, glued together so that the grain of adjoining layers or plies is at right angles. The outside plies are termed "faces" or "face and back;" the center ply is termed the "core;" and the remaining plies are often commonly termed "cross bands" or "inner plies." If lumber is used, it forms the core or center ply and the product is often called a "veneered panel" but the proper designation is "lumber-core plywood." If only veneer is used, the product is properly called "all-veneer plywood" but it is frequently called only by the term "plywood." Plywood for boat and ship construction will usually, if not always, be "all-veneered plywood."

3.111. Unit of measurement. The unit of measurement most commonly used in buying and selling plywood is the square foot, surface measure, and prices are often quoted on a unit of 1,000 square feet of specified thickness, grade, and type. Quantity can be specified in terms of the number of pieces and the thickness, width, and length of each piece in inches.

3.112. Advantages. The chief advantages of plywood, as compared with solid wood, are its more nearly equal strength properties, along the length and width of the panel, greater resistance to deep checking and to splitting (at right angles to the plane of the plies), greatly reduced change in width and length with changes in moisture content, and the fact that it is produced in wider sizes than solid wood. These advantages are obtained by alternating the direction of the grain in successive plies.

The tendency of cross-banded products to warp as the result of stresses that are due to restraint of the normal shrinking and swelling of the wood with changes in moisture content is largely eliminated by balanced construction. This construction consists in arranging the plies in pairs about the core, or center ply, so that for each ply there is a similar parallel ply on the opposite side of the core. Matching the pairs of plies involves consideration of (1) thickness; (2) kind and quality of wood, with particular reference to shrinkage or density; (3) moisture content at the time of gluing; and (4) angle or relative direction of grain.

3.113. Kinds of Plywood. Plywood is produced in a wide variety of thicknesses, sizes, species, and combinations of ply thicknesses. It can be manufactured as flat panels, as curved plywood, or as molded plywood. It may have faces of natural wood veneer, or the facing material may be of resin-impregnated paper (overlaid plywood). The surfaces may be treated with oils or waxes (as for concrete-form plywood) or furnished without surface treatment. Plywood for many naval uses must be treated with preservatives to reduce the hazard of decay under severe conditions of service or with fire-retarding chemicals to reduce flammability. (See section 6.3)

3.12. GLUED LAMINATED WOOD. Laminated wood may be defined as an assembly made by bonding together layers of veneer or lumber with an adhesive so that the grain of all laminations is essentially parallel. A number of advantages make it particularly useful to the shipbuilder.

Members of any desired size can be made by end-joining ordinary lumber lengths, adding laminations, and edge-joining ordinary lumber widths. The use of relatively thin laminations permits the production of curved members of virtually any desired form. Thus, long keels and keelsons of large cross section are feasible and the stem and keel can be fabricated in one piece to increase longitudinal rigidity. Continuous frames of one piece provide increased rigidity and strength with less weight, as compared with sectional sawed or steam-bent solid frames.

Laminated wood is fabricated from seasoned lumber and, therefore, shrinks much less and has fewer seasoning defects, such as checks and splits, than solid lumber, which in large sizes must necessarily be installed at relatively high moisture content.

Adhesives are available which, when properly used, are as durable as the wood. Thus, with proper bonding techniques, every confidence can be placed in the structural integrity of the adhesive bonds.

3.13. MODIFIED WOOD. Various processes have been developed for treating wood to modify and improve certain of its properties. Plywood (section 3.11) and laminated wood (section 3.12) are extensively used for ship and boat building; these are not, however, truly modified woods, as neither their fiber structure nor chemical constitution has been materially affected. Examples of modified wood products, most of which have not yet been adopted for ship and boat use, are briefly described in the following paragraphs.

3.130. Impreg. Impregnated wood, called impreg (sec. 3.2 (6) and (7)), is produced by soaking thin, air-dry veneer, preferably less than 1/8 inch thick, in a phenol-formaldehyde resin-forming mix dissolved in water, or impregnating air-dry veneer with the mix in a cylinder under air pressures of 20 to 75 p. s. i. The veneer is then stored for a day under conditions that prevent drying to permit the resin-forming solution to diffuse into the cell walls of the wood. Next, it is dried in a dry kiln at 150 to 150 degrees F. for about 5 hours before being heated at 200 degrees F. for 24 hours in order to set the resin within the cell-wall structure; or it can be dried in a continuous drier for half an hour at about 180 degees F. and then passed through the drier again at the same speed at 310 degrees F. to set the resin. For most effective results, the weight of resin in the wood should be about 30 percent of the weight of the dry, untreated wood.

Such treatment reduces shrinking and swelling to 25 to 40 percent of that of untreated wood. Where the relative humidity fluctuates so rapidly that the material does not reach equilibrium with atmospheric conditions, swelling and shrinking are limited still further. Face checking under weathering conditions is markedly lower than that of untreated veneer.

Except in hardness and compressive strength, which are appreciably increased, the mechanical properties of wood are not improved by impregnation treatment (see 3.2 (1)). Modulus of rupture and modulus of elasticity are unaffected, while toughness and Izod impact strength are materially reduced.

Decay resistance is increased considerably, apparently because the cell walls of treated wood cannot take up sufficient moisture to support decay. Limited exposure tests indicate that impreg may have reasonably good resistance to marine borers. Acid resistance is high, fire resistance not markedly greater than that of untreated wood. Tests indicate that ammonium phosphate salts in the treating solution increase fire resistance appreciably and are not easily leached out.

Impreg was tested as a facing material for redwood lumber for use in ship decking during World War II (see 3.2.4). Impregnated black gum veneers laminated in a vertical plane to Douglas fir base has also been experimentally used. Due to topside weight limitations this type of decking has not been adopted for carrier decking. Impreg is now finding use in models and molds, where dimensional stability is of great importance.

3.131. Compreg. When wood is treated with phenolic resin-forming mixes, it becomes much more plastic and compressible than untreated wood. This increased compressibility makes possible the production of resin-treated, laminated, compressed wood, called compreg (see 3.2 (6) and (8)). Compreg is much denser than impreg, most of its strength properties being increased about in proportion to the compression. Under pressure of 1,000 p. s. i., most species attain a specific gravity of 1.3 to 1.35, which closely approaches the specific gravity of wood substance.

Compreg is produced by pressing the treated veneers in a heated press, thereby curing the resin within the wood just after compression takes place. Temperatures of 280 to 300 degrees F. are generally used. The material can be molded to slight doble curvatures while being compressed.

Compreg absorbs much less liquid water than does impreg or untreated wood. Highly compressed and stabilized specimens of compreg immersed in water for a year swelled only 5.4 percent in thickness.

A lustrous, varnish-like finish can be attained on compreg compressed between highly polished press platens. Marred finishes can be restored simply by sanding and buffing the damaged face, since the wood is permeated with resin.

Improvement in certain basic strength properties is pronounced (see 3.2 (1) and (3)). Parallel-laminated spruce compreg is 4 times as high as untreated, uncompressed Sitka spruce in maximum crushing strength parallel to grain. Its modulus of rupture and modulus of elasticity in static bending are about 3 times as high. Hardness is increased 15 to 20 times. Shear strength parallel to grain is more than doubled. Impact strength, however, is low compared to that of corresponding untreated wood; the material breaks with a brittle fracture.

Decay resistance is good, regardless of species or whether sapwood or heartwood is used. Marine borers resistance is reasonably good.

Compreg can be molded to rather intricate shapes by a process known as expansion molding (see 3.2 (9)). Dry, uncured, resin-treated veneer is compressed under heat insufficient to set the resin, and then chilled to hold its compressed form. The sheets are cut to templates, laid up in a mold that is then clamped shut, and heated to bend the plies and set the resin. Compreg can be molded to metal or untreated wood cores or sheathing in this way.

Relatively high production costs make compreg a specialty material. It is commercially produced by several firms in the United States for use in handles, metal-forming dies, bearing plates, and textile picker sticks.

3.132. Staypak. By a process developed at the Forest Products Laboratory, normal wood can be compressed to a specific gravity of 1.3 to 1.4 under conditions that relieve internal stresses normally set up when wood is compressed. If not relieved, these stresses would cause it to swell and recover from compression rapidly upon subsequent exposure to moisture. The wood is compressed at pressures of 1,400 to 2,000 p. s. i. and under conditions that cause lignin to flow, thereby relieving these stresses sufficiently to avoid much of the springback they induce upon subsequent swelling. The resulting product is called heat-stabilized compressed wood, or staypak (see 3.2 (5)).

The method of manufacture consists of placing wood preseasoned to a moisture content of 6 to 9 percent in a press and loading the press to the full compression pressure, then applying heat until the center of the piece attains a temperature of about 220 degrees F. At this temperature, the wood can be compressed almost completely, and when full compression has been attained, the the heat is increased to 330 degrees, and then to 340 degrees F. for a length of time necessary to insure stabilization and complete compression to a specific gravity of about 1.3.

Staypak has been produced from veneer, phenolic-resin bonded plywood, and lumber. Species successfully used have been of yellow birch, sugar maple, black walnut, Sitka spruce, yellow poplar, cottonwood, sweetgum, Douglas fir, hickory, white oak, alder, and white fir. Resinous, pitchy, solid wood does not stabilize well.

The strength properties of staypak are considerably higher than those of the normal wood from which it is made (see 3.2 (1)). Tensile strength and static bending strength are about 25 percent higher than for compreg. Staypak is slightly lower in compressive strength but has considerably higher Izod impact strength than compreg. Staypak has lower shear strength at right angles to the direction of compression than compreg.

Although staypak swells less when exposed to moisture than does ordinary compressed wood, it is not comparable to compreg in this respect. Hence, it is less suitable for underwater use than is the latter. It weathers like ordinary wood, the surface becomes gray with many fine hair checks. It should therefore be painted for outdoor use. It has little resistance to marine borers.

3.133. Acetylated Wood. Another means of stabilizing the dimensions of wood without embrittling it results in socalled "acetylated wood" (see 3.2 (10)). Veneer impregnated with small amounts of acetate salts is subjected to the vapors of acetic anhydride. The water-loving hydroxyl groups of the cellulose and lignin in the wood are thereby replaced by acetate groups. The specific gravity and appearance of the wood are but slightly affected. Most strength properties are but slightly affected, but toughness is increased. The dimensional stability is the highest thus far attained by any treatment. Limited decay resistance and marine borer resistance studies indicate the possibilities of high resistance.

3.134. Papreg. By impregnating specially made papers with synthetic thermosetting resins and subsequently pressing sheets of the treated paper together under heat, a special high-strength laminated plastic has been produced that can be molded to single or moderate double curvature. This material is called papreg. Processing more than twice the tensile strength of any paper plastic previously made, as well as other superior strength properties that fit it for use in stressed parts (see 3.2 (2)), papreg has a smooth, hard surface that does not require finishing and is more resistant to scratching and denting than aluminum. It is resistant to penetration by water, is not corroded by salt water, and has good decay resistance. It is somewhat resistant to attack by marine borers.

Papreg has been used for aircraft parts, such as cowlings, gun turret assemblies, ammunition boxes, experimental wing tips and ribs, and for truck floors. It can be produced at relatively low pressures and is easily laminated to veneer for aircraft flooring and other uses. A number of manufacturers have produced it commercially.

3.2. REFERENCES

(1) Erickson, E. C. O.
1947. "Mechanical Properties of Laminated Modified Wood", *Forest Products Laboratory Report No. R1639,* 16 pp. illus.

(2) ———, and Boller, K. H.
1943. "Strength and Related Properties of Forest Products Laboratory Laminated Paper Plastics (Papreg) at Normal Temperature," *Forest Products Laboratory Report No. 1319*, 10 pp., illus.

(3) Findley, W. H., Worley, W. V. and Kacalieff, C. D.
1946. "Effect of Molding Pressure and Resin on Results of Short Time Tests and Fatigue Tests of Compreg," *Transactions American Society Mechanical Engineering*, Vol. 18, No. 317.

(4) Kuenzel, J. G.
1947. "Wood has Big Role on Modern Warships," Wood 2 (5) 14; (6) 26.

(5) Seborg, R. M., Millett, M. A., and Stamm, A. J.
1944. "Heat-Stabilized Compressed Wood (Staypak)," *Forest Products Laboratory Report No. 1580*, 19 pp. illus.

(6) Stamm, A. J.
1948. "Modified Woods," *Modern Plastics Encyclopedia*, 705–707.

(7) Stamm, A. J., and Seborg, R. M.
1950. (Revised) "Forest Products Laboratory Resin-Treated Wood (Impreg)," *Forest Products Laboratory Report No. 1380*, 11 pp.

(8) ———
1951. (Revised) "Forest Products Laboratory Resin-Treated, Laminated, Compressed Wood (Compreg)," *Forest Products Laboratory Report No. 1381*, 13 pp., illus.

(9) ———, and Turner, H. D.
1945. "Methods of Molding," *U. S. Patent 2,391,489.*

(10) Tarkow, Harold, Stamm, A. J., and Erickson, E. C. O.
1950. (Revised) "Acetylated Wood," *Forest Products Laboratory Report No. 1893*, 15 pp., illus.

CHAPTER 4. GLOSSARY OF TERMS RELATING TO WOOD

Air-dried. (See seasoning.)

American lumber standards. American lumber standards embody provisions for softwood lumber dealing with recognized classifications, nomenclature, basic grades, seasoning standards, sizes, uniform workings, description, measurement, tally, shipping provisions, grade marking, tally cards, and inspection of lumber. The primary purpose of these standards is to serve as a guide or basic examples in the preparation or revision of the grading rules of the various lumber manufacturer's associations. However, a purchaser, in order to buy lumber, must use association rules covering the species he wants, because the basic standards are not usable as commercial rules.

Annual growth ring. (See ring, annual growth.)

Bastard sawn. Hardwood lumber in which the annual rings make angles of 30 to 60 degrees with the surface of the piece.

Beams and stringers. Large pieces (nominal dimensions 5 by 8 inches and up) of rectangular cross section graded with respect to their strength in bending when loaded on the narrow face.

Bird's-eye. A small central spot with the wood fibers arranged around it in the form of an ellipse so as to give the appearance of an eye.

Blemish. Anything, not necessarily a defect, marring the appearance of wood.

Blue stain. (See stain, blue.)

Boards. (See lumber.)

Bow. Distortion of a board in which the face is convex or concave longitudinally.

Boxed heart. The term used when the pith falls entirely within the four faces anywhere in the length of a piece.

Brashness. A condition of wood characterized by low resistance to shock and by an abrupt failure across the grain without splintering.

Broad-leaved trees. (See hardwoods.)

Brown stain. (See stain brown.)

Burl. A swirl or twist in the grain of the wood that usually occurs near a knot but does not contain a knot.

Cambium. The layer of tissue just beneath the bark from which the new wood and bark cells of each year's growth develop.

Cant. A thick piece of lumber, with or without squared edges, sawed from a flitch or log and intended for remanufacture into lumber.

Cell. A general term for the minute units of wood structure. It includes fibers, vessel segments, and other elements of divers structure and functions.

Cellulose. The carbohydrate that is the principal constituent of wood and forms the framework of the cells.

Check. A lengthwise separation of the wood, the greater part of which occurs across the rings of annual growth.

Chemical brown stain. (See stain, chemical brown.)

Close-grained wood. (See grain.)

Coarse-grained wood. (See grain.)

Collapse. The flattening of single cells or rows of cells in heartwood during the drying or pressure treatment of wood, characterized externally by a caved-in or corrugated appearance.

Compression failure. Deformation or buckling of the wood fibers resulting from severe stress in compression along the grain. In surface lumber they appear as fine wrinkles across the face of the piece.

Compression wood. Abnormal wood that often forms on the lower side of branches and of leaning trunks of softwood trees. Compression wood is identified by its relatively wide annual rings usually eccentric, and its relatively large amount of summerwood, usually more than 50 percent of the width of the annual rings in which it occurs. Compression wood shrinks excessively lengthwise, as compared with normal wood.

Conifer. (See softwoods.)

Crook. Distortion of a board in which the edge is convex or concave longitudinally.

Crossband. To place the grain of layers of wood at right angles in ordre to minimize shrinking and swelling and consequent warping; also, the layer of veneer at right angles to the face plies.

Cross break. A separation of the wood cells across the grain. Such breaks may be due to internal strains resulting from unequal longitudinal shrinkage or to external forces.

Cross grain. (See grain.)

Cup. Distortion of a board in which the face is convex or concave transversely.

Decay. Disintegration of wood substance through the action of wood-destroying fungi.

Incipient decay. The early stage of decay, in which the disindegration has not proceeded far enough to soften or otherwise impair the hardness of the wood perceptibly.

Typical or advanced decay. The stage of decay in which the disintegration is readily recognized because the wood has become punky, soft and spongy, stringy, pitted, or crumbly.

Defect. Any irregularity occurring in or on wood that may lower its strength.

Density. The mass of a body per unit volume. When expressed in the c. g. s. system, it is numerically equal to the specific gravity of the same substance.

Density rule. Rules for estimating the density of wood based on percentage of summerwood and rate of growth.

Diagonal grain. (See grain.)

Diffuse-porous woods. Hardwoods in which the pores are practically uniform in size throughout each annual ring, or decrease slightly toward the outer border of the ring.

Dimension. (See lumber.)

Dimension stock. Squares or flat stock, usually in pieces under the minimum size admitted in standard lumber grades. They are supplied rough or dressed, green or dry, and cut to the approximate dimensions required for the various products of woodworking factories.

Dote. "Dote," "doze," and "rot" are synonymous with "decay," and are any form of decay that may be evident as either a discoloration or a softening of the wood.

Dry rot. A term loosely applied to many types of decay but especially to that which, when in an advanced stage, permits the wood to be easily crushed to a dry powder. The term is actually a misnomer, since all decay fungi require considerable moisture for growth.

Durability. A general term for permanence or lastingness. Frequently used to refer to the degree of resistance of a species or of an individual piece of wood to attack by wood-destroying fungi under conditions that favor such attack. In this connection, the term "decay resistance" is more specific.

Edge grain. (See grain.)

Empty-cell process. Any process for impregnating wood with preservatives or chemicals in which air is imprisoned in the wood under the pressure of the entering preservative and then expands, when the pressure is released, to drive out part of the injected preservative.

Encased knot. (See knot.)

Extractives. Substance in wood, not an integral part of the cellular structure, that can be dissolved out with hot or cold water, ether, benzene, or other relatively inert solvents.

Equilibrium moisture content. The moisture content at which wood neither gains nor loses moisture when surrounded by air at a given relative humidity and temperature.

Factory and shop lumber. (See lumber.)

Fiber. A wood fiber is a comparatively long (1/25 or less to 1/3 inch), narrow, tapering cell closed at both ends.

Fiber-saturation point. The stage in the drying or in the wetting of wood at which the cell walls are saturated and the cell cavities are free from water.

Figure. The pattern produced in a wood surface by irregular coloration and by annual growth rings, rays, knots, and such deviations from regular grain as interlocked and wavy grain.

Flakes. (See rays, wood.)

Flat grain. (See grain.)

Flitch. A thick piece of lumber with wane (bark) on one or more edges.

Full-cell process. Any process for impregnating wood with preservatives or chemicals in which a vacuum is drawn to remove air from the wood before admitting the preservative.

Grade. The designation of the quality of a manufactured piece of wood.

Grain. The direction, size, arrangement, appearance, or quality of the fibers in wood.

Close-grained wood. Wood with narrow and inconspicuous annual rings. The term is sometimes used to designate wood having small and closely spaced pores, but in this sense, the term "fine textured" is more often used.

Coarse-grained wood. Wood with wide and conspicuous annual rings; that is, rings in which there is considerable difference between springwood and summerwood. The term is sometimes used to designate wood with large pores, such as oak, ash, chestnut, and walnut, but in this sense, the term "coarse textured" is more often used.

Cross grain. Grain not parallel with the axis of a piece. It may be either diagonal or spiral grain or a combination of the two.

Diagonal grain. Annual rings at an angle with the axis of a piece as a result of sawing at an angle with the bark of the tree.

Edge grain. Edge-grained lumber has been sawed parallel with the pith of the log and approximately at right angles to the growth rings; that is, the rings form an angle of 45° or more with the surface of the piece.

Flat grain. Flat-grained lumber has been sawed parallel with the pith of the log and approximately tangent to the growth rings; that is, the rings form and angle of less than 45° with the surface of the piece.

Interlocked-grained wood. Wood in which the fibers are inclined in one direction in a number of rings of annual growth, then gradually reverse and are inclined in an opposite direction in succeeding growth rings, then reverse again.

Open-grained wood. Common classification of painters for woods with large pores, such as oak, ash, chestnut, and walnut. Also known as "coarse textured."

Plain-saved. Another term for flat grain.

Quarter-sawed. Another term for edge grain.

Spiral grain. A type of growth in which the fibers take a spiral coarse about the bole of a tree instead of the normal vertical course. The spiral may extend right-handed or left-handed around the tree trunk.

Vertical grain. Another term for edge grain.

Wavy-grained wood. Wood in which the fibers collectively take the form of waves or undulations.

Green. Unseasoned, wet.

Growth ring. (See ring, annual growth.)

Hardwoods. The botanical group of trees that are broadleaved. The term has no reference to the actual hardness of the wood. Angiosperms is the botanical name for hardwoods.

Heart, Heartwood. The wood, extending from the pith to the sapwood, the cells of which no longer participate in the life processes of the tree. Heartwood may be infiltrated with gums, resins, and other materials which usually make it darker and more decay-resistant than sapwood.

Honeycomb. Checks, often not visible at the surface, that occur in the interior of a piece, usually along the wood rays.

Imperfect manufacture. Includes all defects or blemishes produced in manufacturing, such as chipped grain, loosened grain, raised grain, torn grain, skips in dressing, hit and miss, variation in sawing, miscut lumber, machine burn, machine gouge, mismatching, and insufficient tongue or groove.

Interlocked-grained wood. (See grain.)

Joist and plank. Pieces (nominal dimensions 2 to 4 inches in thickness by 4 inches and wider) of rectangular cross section graded with respect to their strength in bending when loaded either on the narrow face as joists or on the wide face as planks.

Kiln. A heated chamber for drying lumber.

Compartment kiln. A dry kiln designed to keep the same temperature and relative humidity throughout at any given time. In it, the entire charge of lumber is dried as a unit, under drying conditions that increase in severity during the operation.

Progressive kiln. A dry kiln designed to provide dry-conditions that increase in severity from entrance to exit. In it the unit charge is only a part of the total charge of lumber; a unit of perhaps four truck loads is moved through the kiln in a chain of several units, from day to day, with a single unit leaving and another entering at a time.

Kiln brown stain. (See stain, chemical brown.)

Kiln dried. (See seasoning.)

Knot. That portion of a branch or limb has become incorporated in the body of a tree.

Decayed knot. A knot which, due to advanced decay, is not so hard as the surrounding wood.

Encased knot. A knot whose rings of annual growth are not intergrown with those of the surrounding wood.

Intergrown knot. A knot whose rings of annual growth are completely intergrown with those of the surrounding wood.

Round knot. A knot whose sawn section is oval or circular.

Sound knot. A knot which is solid across its face and which is as hard as the surrounding wood.

Spike knot. A knot sawn in a lengthwise direction.

Laminated wood. A wood member built up of laminations that have been joined with glue or with mechanical fastenings. The term is generally applied when the grain of all laminations is parallel.

Lignin. A principal constituent of wood, second in quantity to cellulose. It incrusts the cell walls and cements the cells together.

Lumber. The product of the saw and planing mill not further manufactured than by sawing, resawing, and passing lengthwise through a standard planing machine, crosscutting to length and working.

Factory and shop lumber. Lumber intended to be cut up for use in further manufacture. It is graded on the basis of the percentage of the area which will produce a limited number of cuttings of a specified, or a given minimum, size, and quality.

Yard lumber. Lumber that is less than 5 inches in thickness and is intended for general building purposes.

Boards. Yard lumber less than 2 inches thick, 2 inches or more in width.

Dimensions. All yard lumber except boards and timbers; that is, yard lumber 2 inches and less than 5 inches thick, and of any width.

Millwork. Generally, all building materials made of finished materials made of finished wood and manufactured in millwork plants and planing mills are included under the term "millwork." It includes such items as inside and outside doors, window and door frames, blinds, porch work, mantels, panel work, stairways, moldings, and interior trim. It does not include flooring, ceiling or siding.

Moisture content of wood. Weight of the water contained in the wood, usually expressed as a percentage of the weight of the ovendry wood.

Moisture gradient. A condition of graduated moisture content between the successive layers of a material, such as wood, due to the loss or absorption of moisture. During seasoning, the gradations are between the moisture content of the relatively dry surface layers and the wet layers at the center of the piece.

Open-grained wood. (See grain.)

Peck. Pockets or areas of disintegrated wood caused by advanced stages of localized decay in the living tree. It is usually associated with cypress and incense cedar. There is no further development of peck once the lumber is seasoned.

Pitch pocket. An opening extending parallel to the annual rings of growth usually containing, or which has contained, pitch, either solid or liquid.

Pith. The small soft core occurring in the structural center of a log.

Pith fleck. A narrow streak resembling pith on the surface of a piece, usually brownish and up to several inches in length, it results from the burrowing of larvae in the growing tissue of the tree.

Plywood. A piece of wood made of three or more layers of veneer joined with glue and usually laid with the grain of adjoining plies at right angles. Almost always an odd number of plies are used to secure balanced construction.

Plain-sawed. (See grain.)

Planing mill products. Products worked to pattern, such as flooring, ceiling, and siding.

Pocket rot. Advanced decay, which appears as a hole, pocket, or area of soft rot usually surrounded by apparently sound wood.

Pore. (See vessel.)

Posts and timbers. Pieces of square or approximately square cross section, 5 by 5 inches or larger in nominal dimensions, graded primarily for use as posts or columns but adapted to miscellaneous uses in which strength in bending is not especially important.

Preservative. Any substance that, for a reasonable length of time, will protect wood from the action of wood destroying fungi, borers of various kinds, and similar destructive forms of life when the wood has been properly coated or impregnated with it.

Progressive kiln. (See kiln.)
Quarter-sawed. (See grain.)
Radial. Coincident with a radius from the axis of the tree or log to the circumference.
Rate of growth. The rate at which a tree has laid on wood, measured radially in the trunk or in lumber cut from the trunk. The unit of measure in use is number of annual growth rings per inch.
Rays, wood. Strip of cells extending radially within a tree and varying in height from a few cells in some species to 4 inches or more in oak. The rays serve primarily to store food and transport it horizontally in the tree.
Ring, annual growth. The growth layer put on in a single growth year.
Ring-porous woods. A group of hardwoods in which the pores are comparatively large at the beginning of each annual ring and decrease in size more or less abruptly toward the outer portion of the ring, thus forming a distinct inner zone of pores known as the springwood and an outer zone of smaller pores known as the summerwood.
Rot. (See decay.)
Rotary-cut veneer. (See veneer.)
Sap. All the fluids in a tree. Special secretions and excretions, such as gum, excepted.
Sapwood. The layers of wood next to the bark, usually lighter in color than the heartwood, ½ to 3 or more inches wide. It is actively involved in the life processes of the tree. Under most conditions, sapwood is more susceptible to decay than heartwood; as a rule, it is more permeable to liquids than is heartwood. However, sapwood is not essentially weaker or stronger than heartwood of the same species.
Sawed veneer. (See veneer.)
Seasoning. Removing moisture from green wood in order to improve its serviceability.
Air dried or air seasoned. Dried by exposure to the air, usually in a yard, without artificial heat.
Kiln dried. Dried in a kiln with artificial heat.
Second growth. Timber that has grown after the removal, by any means, of all or a large portion of the previous stand.
Shake. A separation along the grain, the greater part of which occurs between the rings of annual growth.
Shop lumber. (See lumber.)
Side cut. The term used when the pith is not present in a piece.
Sliced veneer. (See veneer.)
Softwoods. The botanical group of trees that have needle or scalelike leaves and are evergreen for the most part, cypress, larch, and tamarack being exceptions. The term has no reference to the actual hardness of the wood. Softwoods are often referred to as conifers, and botanically, they are called gymnosperms.
Specific gravity. The ratio of the weight of a body to the weight of an equal volume of water at some standard temperature.
Spiral grain. (See grain.)
Split. A lengthwise separation of the wood, caused when the wood cells are torn apart.
Springwood. The portion of the annual growth ring that is formed during the early part of the season's growth. It is usually less dense and weaker mechanically than summerwood.
Stain, blue. A bluish or grayish discoloration of the sapwood caused by the growth of certain moldlike fungi on the surface and in the interior of the piece, made possible by the same conditions that favor the growth of other fungi.
Stain, brown. A rich brown to deep chocolate brown discoloration of the sapwood of some pines caused by a fungus that acts similarly to the blue-stain fungus.
Stain, chemical brown. A chemical discoloration of wood which sometimes occurs during the air drying or kiln drying of several species, apparently caused by the oxidation of extractives.
Stain, sap. (See stain, blue.)
Strength. The term in its broader sense embraces collectively all the properties of wood that enable it to resist different forces or loads. In its more restricted sense, strength may apply to any one of the mechanical properties, in which event, the name of the property under consideration should be stated, thus strength in compression parallel to the grain, strength in bending, hardness, etc.
Strips. (See lumber.)
Structural timber. Pieces of wood of relatively large size, the strength of which is the controlling element in their selection and use. Trestle timbers (stringers, caps, posts, sills, bracing, bridge ties, guard rails); car timbers (car framing, including upper framing, car sills); framing for buildings (posts, sills, girders, framing joists); ship timbers, (ship timbers, ship decking); and crossarms for poles are examples of structural timbers.
Summerwood. The portion of the annual growth ring that is formed during the latter part of the yearly growth period. It is usually more dense and stronger mechanically than springwood.
Tangential. Strictly, coincident with a tangent at the circumference of tree or log, or parallel to such a tangent. In practice, however, it often means roughly coincident with a growth ring.
Texture. A term often used interchangeably with grain. In this manual, it refers to the finer structure of the wood (see grain) rather than the annual rings.
Timber, standing. Timber still on the stump.
Timbers. Lumber 5 inches or larger in least dimension.
Timbers, round. Timbers used in the original round form such as poles, piling, and mine timbers.
Tracheid. The elongated cells that constitute the greater part of the structure of the softwoods (frequently referred to as fibers). Also a portion of some hardwoods.
Twists. A distortion caused by the turning or winding of the edges of a board so that the four corners of any face are no longer in the same plane.
Veneer. Thin sheets of wood.
Rotary-cut veneer. Veneer cut in a continuous strip by rotating a log against the edge of a knife in a lathe.
Sawed veneer. Veneer produced by sawing.
Sliced veneer. Veneer that is sliced off by moving a log, bolt, or flitch against a large knife.

Vertical grain. (See grain.)

Vessels. Wood cells of comparatively large diameter which have open ends and are set one above the other to form continuous tubes. The openings of the vessels on the surface of a piece of wood are usually referred to as pores.

Virgin growth. The original growth of mature trees.

Wane. Bark, or lack of wood or bark, on an edge or corner of a piece that is consequently not full size.

Warp. Any variation from a true or plane surface. Warp includes bow, crook, cup, and twist, or any combination thereof.

Wavy-grained wood. (See grain.)

Weathering. The mechanical or chemical disintegration and discoloration of the surface of wood that is caused by exposure to light, the action of dust and sand carried by winds, and the alternate shrinking and swelling of the surface fibers that come with the continual variation in moisture content brought by changes in the weather. Weathering does not include decay.

Wood preservative. (See preservative.)

Workability. The degree of ease and smoothness of cut obtainable with hand or machine tools.

Working of wood. Change in the dimensions of a piece of wood with change in moisture content.

Yard brown stain. (See stain, chemical brown.)

Yard lumber. (See lumber.)

Good constructions result when lumber of the proper moisture content is used.

Result of using lumber with excessive moisture content.

WOOD:
A MANUAL FOR
ITS USE AS
A SHIPBUILDING MATERIAL

Volume II

DEPARTMENT OF THE NAVY

BUREAU OF SHIPS

JANUARY 1958

DEPARTMENT OF THE NAVY
BUREAU OF SHIPS
WASHINGTON 25, D. C.

During World War II and the Korean action, more than 40,000 wooden ships, boats, and landing craft were added to the Fleet. A continuing program of improvement in wood laminating, extreme service adhesives, and development of wood preservatives for protection against marine borers has been largely responsible for the satisfactory performance of these craft. During the past 5 years, many additional wooden ships and boats have been put into service, incorporating the most recent knowledge of wood as an engineering material.

Because of these many new developments in marine uses of wood, a need has arisen for a comprehensive, up-to-date collection of information based on recent investigations and tests. This new publication, "Wood: A Manual for its Use as a Shipbuilding Material," has, therefore, been prepared and should prove most useful to shipyards, naval architects, Supervisors of Shipbuilding, Inspectors of Naval Material, and officers and men concerned with all phases of wooden boat design, construction, or maintenance.

A. G. MUMMA,
Rear Admiral, USN,
Chief, Bureau of Ships.

WOOD: A MANUAL FOR ITS USE AS A SHIPBUILDING MATERIAL

PREPARED BY

BUREAU OF SHIPS

DEPARTMENT OF NAVY

With the cooperation of

FOREST PRODUCTS LABORATORY

FOREST SERVICE

UNITED STATES DEPARTMENT OF AGRICULTURE

ISSUED—JULY 1945

SECOND PRINTING—OCTOBER 1948

EXTENSIVELY REVISED—JANUARY 1958

TABLE OF CONTENTS

ACKNOWLEDGMENTS

In an industry so old as shipbuilding, many men, most of them unknown, have contributed to our present day knowledge and advancement of the art and science. A notable early attempt to organize the improvement of shipbuilding was made by King James I in 1605 when he chartered a Shipwrights Guild with supervision over all shipbuilding in England, and which all English shipwrights were required to join. The Guild made significant progress under the leadership of Master Shipwright Phineas Pett who introduced the use of scale models and mathematically developed plans, making possible the building of two ships alike.

Later, the French took up the banner of progress and carried it well into the 1800's, surpassing the British. This era of shipbuilding witnessed the highest development of the wooden sailing ship in construction, speed, and beauty. During this time, the American forest contributed enormous quantities of shiptimbers, planking, decking, masts, and spars. American shipwrights achieved a reputation for skilled workmanship held to this day.

Shipbuilding began to emerge from an art to a science during the 19th century. Several schools of naval architecture were founded in England, the United States, and Germany, followed by such technical societies as the Institution of Engineers and Shipbuilders in Scotland, the British Institution of Naval Architects, and Institute of Marine Engineers, and in the United States, the Society of Naval Architects and Marine Engineers.

Thus, the modern science of wood shipbuilding was developed over the centuries. Most significant of the achievements during the 20th century has been the adoption of adhesive bonded wood as an engineering material in the construction of wooden hull ships. The bonding together of small pieces of wood into large structural members or panels in either straight or curved form results in a far more efficient use of wood than was formerly possible. This development is the first basic change in wooden shipbuilding in 2,000 years.

Previous editions of this manual have mentioned many individual shipbuilders, researchers, and other contributors who furnished information regarding modern wood ship and boat building practices. The present edition represents the combined efforts of virtually all American technical societies, research agencies, trade association, and universities which foster progress in wood technology, naval architecture, and timber engineering. It is the Bureau's desire to give full recognition to these groups and to the part played by the lumber, laminating, plywood, adhesives, and wood preservation industries in modern wood shipbuilding. Without their cumulative contributions to the advancement of wood as an engineering material, this manual would not be possible.

The basic manuscript for "Wood: A Manual for Its Use as a Shipbuilding Material" was prepared by the Forest Products Laboratory, Forest Service, United States Department of Agriculture, at the request of the

Bureau of Ships, Department of the Navy. A very large share of the credit for the material presented in the manual is due the laboratory, which has been foremost in the field of forest products for almost fifty years. The laboratory's individual staff members which have contributed to the manual are recognized authorities in their individual fields. Thanks are also due the staff of the Division of Forest Disease Research of the United States Forest Service, and the Timber Engineering Co. laboratory for their advice and guidance.

The manuscript was reviewed and edited for the Bureau of Ships by members of its staff. The final arrangement of the material for publication has been made under the direction of Mr. G. I. Dewey, now retired, who until 1948 supervised wood products development for the Bureau of Ships.

WOOD: A MANUAL FOR ITS USE AS A SHIPBUILDING MATERIAL

INTRODUCTION

PURPOSE AND SCOPE. When this publication was first issued in 1945 by the Department of the Navy under the title, "Wood: A Manual for Its Use in Wooden Vessels," its principal purpose was to serve the Nation in the construction of great numbers of wood hulls of all types then being built for World War II service. Hull-building techniques were then in a state of rapid development as new material and methods were introduced. The main objective of the Manual at that time was to guide builders in making intelligent use of lumber, plywood, and allied materials under these rapidly changing conditions. The content of the book was carefully planned with this objective in mind. No attempt was made to present a comprehensive discussion of the many construction techniques involved in building a wood hull. Rather, the properties of wood and allied products were related to the techniques then in use in an effort to demonstrate how these properties affect the intelligent use of wood in hull-building materials.

Since the Manual was first released, developments in both materials and methods of hull construction have continued, if at a slower pace. These developments, notably in the fields of gluing, laminating, preservative treatment, and fabrication, constitute one of the reasons that the Bureau of Ships undertook the complete rewriting of the Manual. This issue, therefore, contains much new information gained from research continuously in progress at the Forest Products Laboratory, Timber Engineering Co., and elsewhere, much of it done in cooperation with the Bureau of Ships.

New information is also presented on hull maintenance, storage, and repair, a field about which little knowledge was available at the time the Manual was originally written. During the postwar years, however, problems of storage and maintenance have become prominent, and considerable research and study have been conducted on them. A summary of the information developed on this subject through research and experience at various Navy activities is presented herein.

The several chapters have been expanded to the extent that a single volume would no longer be convenient to handle. In view of the diverse uses to which the Manual may be put, as well as for convenience in handling, it has been divided into four volumes as follows:

Volume I.—Basic Wood Technology Applicable to Shipbuilding.

Volume II.—Techniques and Practices Applicable to Preservation and Storage.

Volume III. (Confidential)—Technical Data Applicable to Boat and Ship Design.

Volume IV. (Confidential)—Techniques Applicable to Boat and Ship Construction.

The name has been changed to "Wood: A Manual for Its Use as a Shipbuilding Material" to more accurately express the broadened scope.

Throughout recorded history, wood has been a major material for the construction of boats and ships. Its many natural advantages for such use have in recent years been extended by the results of scientific research. This Manual is intended to provide an up-to-date source of basic information about wood materials, the application of which in the construction and maintenance of wood-hulled boats and ships can improve its service under today's conditions.

CHAPTER I. SEASONING AND STORING LUMBER AND PLYWOOD FOR SHIP AND BOAT PARTS

1.0. SEASONING LUMBER.

Moisture content of ship and boat building lumber.

1.00. GENERAL. Lumber is seasoned or dried to make it suitable for use, because green lumber is limited in use to large timbers. Seasoning removes most of the water originally in the living tree and retained in the lumber at the time it is manufactured. With properly controlled drying, seasoning defects are fewer and the moisture content of the lumber is reduced until it is suitable for its end use. Drying is accompanied by shrinkage, and a major reason for seasoning is to shrink the lumber before it is put to use. Seasoning lumber to a moisture content below 20 percent eliminates the hazard of decay and stain unless it becomes wet in use. Seasoning is commonly accomplished by air or kiln drying.

Many remedies have been suggested for correcting shrinkage and swelling damage, but few, if any, actual cures short of replacement are successful. A wooden hull can be satisfactorily built or repaired when close attention is given to the moisture content of the lumber used (fig. 1a).

The shrinkage of lumber is sometimes responsible for the repairs needed on wooden boats and ships. The shrinkage damage can be seen in the excessive opening of plank and deck seams, and by the wide and numerous splits in covering boards, caps, shaft logs, and interior stringers (fig. 1b). Such shrinkage occurs as a result of installation lumber which had too high a moisture content, and results in excessive maintenance and repair.

The moisture content of lumber is expressed as a percentage. That is, the weight of the moisture in the lumber is expressed as a percentage of the weight of the lumber when it is oven-dry. Oven-dry lumber contains zero percent moisture, and lumber at a 100 percent moisture content contains half moisture and half wood substance.

Moisture exists in unseasoned lumber in two forms:

1. Free water within the cell cavities.
2. Bound or absorbed water in the cell walls.

As green lumber dries, the free water in each cell cavity leaves. When all the free water is gone from the cell cavities, the bound water in the cell walls begins to dry. This point where the cell walls begin to dry is known as the fiber-saturation point, which corresponds to a moisture content of approximately 30 percent.

Lumber as it is sawn from the log contains more moisture than it will retain after it is exposed to the air. As a "green" or unseasoned board begins to dry, the surfaces lose moisture first. This means that the moisture content will be relatively low at the surface with a progressively increasing moisture content toward the center of the board. The range of moisture content between the surface and the center is referred to as the moisture gradient. As drying continues, the difference in moisture content between the surface and center of the board becomes less and less, until the moisture is uniformly distributed throughout the cross section and length. If the rate of drying is too rapid, serious defects, such as splits and checks, will develop.

A marked variance between the surface and center moisture content indicates that drying is continuing at a rapid rate and that there is free water in the center of the board. Such lumber is not suitable for most ship and boat building purposes.

Lumber either picks up or loses moisture until it is in equilibrium with the air which surrounds it. The rate at which green lumber dries and the moisture content which is finally reached are dependent on the relative humidity of the surrounding air. When subjected to an atmosphere of given relative humidity for a period ranging from several hours to several months, depending on size, species, and so on, lumber reaches a definite moisture content. This is termed the equilibrium moisture content. Figure 2 shows several relative humidity values with the corresponding equilibrium moisture content which results.

conditions of exposure in the location where the lumber is used. Then the equilibrium moisture content can be closely estimated, and the lumber used for fabrication should be of this estimated moisture content, as shown by figure 2.

Lumber is not like any other material in regard to the cause for dimensional changes. Lumber changes dimensions with a gain or loss of moisture content. This moisture gain or

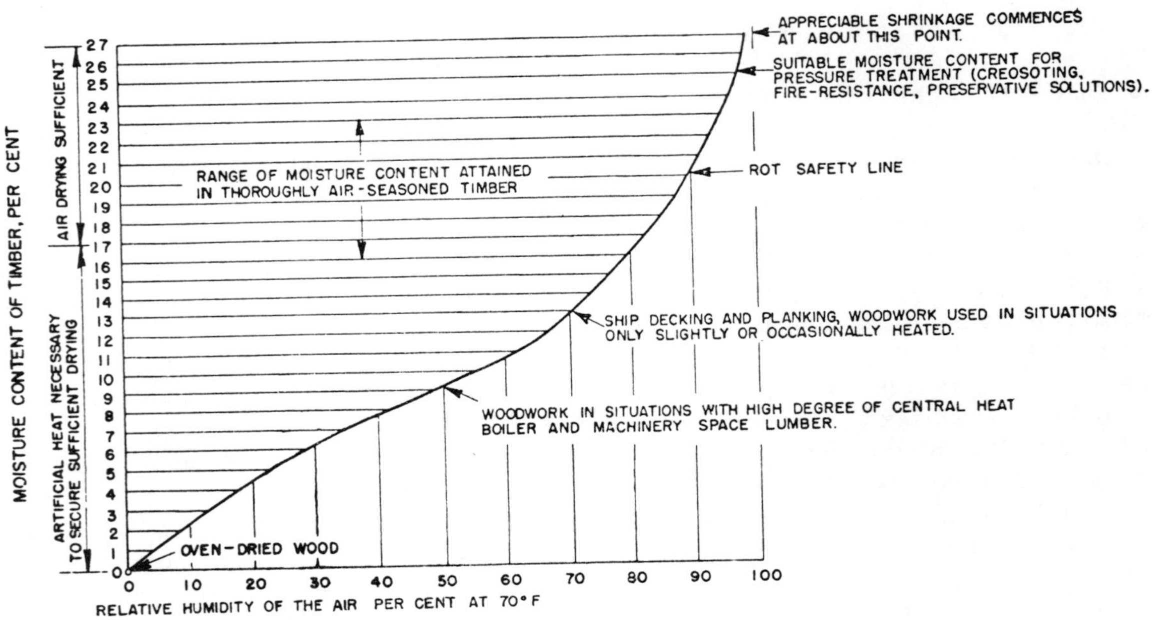

Figure 2. Equilibrium moisture content. Moisture content cited is that considered ideal. For practical purposes a +3 percent tolerance is permissible, preference being given the lower moisture content. B—laminated material is normally around 12 percent in moisture content and is preferable for members of large dimensions. C—waterproof plywood is normally about 5 percent in moisture content after manufacture. Because of its added stability, a 10 percent moisture content with a tolerance of +5 percent is satisfactory for most applications. D—white oak bends best at 25 to 35 percent moisture content. Successful bends can be made with dryer material. Lumber installed with a moisture content in excess of 20 percent should be allowed to dry before painting.

A change in the relative humidity of the air does not result in an immediate change in the moisture content of the lumber. When a change in the relative humidity occurs, the moisture gradient is increased (or decreased) and the moisture content of the lumber slowly comes to equilibrium with the new relative humidity.

The important consideration in fabricating lumber for wooden craft is to determine the average relative humidity which will occur over a period which includes the driest and wettest

loss depends mostly on the relative humidity of the air—not on the temperature, which is responsible for the dimensional changes of most other structural materials. This is the reason why expansion joints are not needed for timber structures and part of the reason why timber structures will withstand extremes of heat without collapse.

Various wood species shrink or swell at different moisture rates, but no wood species shrinks or swells until the moisture content is below 30 percent (fiber saturation point). Below 30 percent, the lumber swells with an increase and shrinks with a decrease in moisture content.

The greatest dimensional change is in the direction parallel to the annual growth rings (tangential). About one-half to two-thirds of this amount occurs in the direction perpendicular to the annual growth rings (radial). No appreciable change occurs in the direction of the grain (longitudinal).

Various widths and thicknesses of lumber of the same species result in different amounts of overall shrinkage or swelling when a change in moisture content occurs. The shrinkage or swelling to be expected for various wood species and sizes can be calculated from the last column of table 3-3, volume I.

The amount of shrinkage or swelling is proportional to the amount of moisture lost or gained below 30 percent. That is, a 1 percent change in moisture content below 30 percent will result in a change amounting to one-thirtieth of the percentage shown in column 3 of the aforementioned table. A formula for this calculation is shown below.

Thus, if a 12-inch vertical grain piece of white oak 2 inches thick is installed as an engine stringer at an average moisture content of 30 percent where the relative humidity is such that the equilibrium moisture content achieved is 10 percent, the calculation would be as follows:

Width: Amount of change from green (30 percent) to oven dry (0 percent) = 5.4 percent.

Thickness: Amount of change from green (30 percent) to oven dry (0 percent) = 9.3 percent.

But, there is only a 20 percent moisture content change, so:

Width $$\frac{.054 \times 20 \times 12}{30} = .43 \text{ inch.}$$

Thickness $$\frac{.093 \times 20 \times 2}{30} = .12 \text{ inch.}$$

$$\frac{\text{Percentage change in dimension (green to oven dry)} \times \text{Change in moisture content (number percentage points)} \times \text{Width or thickness (inches)}}{30} = \text{Change in dimension (inches)}$$

The final dimensions would be 1.88 inches by 11.57 inches. If the piece of lumber was at 10 percent, and somehow acquired moisture to raise it to 30 percent, the calculation would be the same, but the amount would, of course, be added to the dimension existing at 10 percent.

Decay of wood is directly linked with moisture content, and the use of dry lumber for constructing or repairing wood hulls is of vital importance in this respect. The facts to remember are:

1. No decay can occur in any wood species if the moisture content is below 20 percent.
2. No decay can occur in totally submerged lumber.

A large proportion of the decay reported in wooden ships can be directly traced to the use of heavy green timbers with decay infection present before installation. The timbers never dry below the protective level of 20 percent throughout their cross section and length after installation.

During service, the lumber used in underwater portions of a wood hull is subjected to extremes of swelling, whereas, the lumber used in engine rooms and other heated compartments is exposed to extremes of drying. However, the majority of the hull components of wooden boats and ships that are subjected to extremes of wetting and drying are made of wood species with low shrinkage and swelling rates, such as cedar, Douglas fir, and mahogany. When these species swell, they will compress slightly without buckling. All the common wood species used for building wooden boats and ships (mahogany, cedar, Douglas fir, white oak) absorb moisture at a very slow rate when submerged. Paint coats reduce the rate of moisture absorption and evaporation. Most of the underwater components are subjected to drying conditions on their inboard faces—a factor which prevents maximum swelling. Recent tests have shown that underwater hull planking which has the inner face exposed to normal air circulation acquires an average content of 13 percent.

During covered storage ashore, all lumber in a wooden boat will shrink to a degree dependent upon the average relative humidity at the storage site during the driest seasons of the year. However, no continuing climatic condition exists at any boat stocking or repair activity where the average relative humidity is low enough to reduce the moisture content of the lumber below 10 percent, nor raise it above 20 percent; provided, of course, that the boat is properly cradled, drained, and covered.

Solid lumber more than 2½ inches in thickness is difficult to obtain readily in a seasoned condition. This applies more to hardwood species, especially white oak, than to softwoods. Laminated material offers the possibility of obtaining thoroughly seasoned oak and Douglas fir timbers of any dimension and shape.

Laminated members are sterilized by the curing process, and because of their low moisture content, they are much preferred to green timbers, which may have incipient decay infections and which never dry thoroughly. Laminated members can be made to the curvatures required, making it unnecessary to use a forest of lumber and a ton of fasteners to make a large frame or stem. Wood increases in strength rapidly, as the moisture content falls below the fiber saturation point. Because fastenings, hook scarfs, and cross grain are avoided also, a laminated member is much stronger than one of solid material. Many problems and costly repairs can be avoided by using lumber of the proper moisture content for construction or repairing wood hulls.

1.01. AIR SEASONING. Air seasoning is the oldest and simplest method of drying lumber. It consists of exposure of green lumber outdoors or in sheds. Its effectiveness is limited by the vagaries of weather conditions, which in turn limit both the degree, to which the drying rate can be controlled and the final moisture content attainable. As a result, the drying period is relatively long and the final moisture content is dependent on the atmospheric conditions existing shortly preceding the end of the seasoning process (*1–1 and 1–4*) (see note). In general, the moisture content of thoroughly air-seasoned lumber is between 12 and 15 percent during the dry seasons of the year.

Although general weather conditions are beyond control, certain controllable local conditions affect the atmospheric conditions in the yard, and proper spacing and size of piles, as well as methods of piling can speed the rate of air seasoning materially, or, alternatively, retard it to some extent to avoid excessive checking.

1.010. SITE. The lumber yard should be located preferably on high ground that is fairly level, well drained, and not adjacent to water bodies or wind-obstructing objects, such as tall trees or buildings. A yard located in a hollow is sheltered from the full sweep of the winds, and the ground is likely to be relatively damp, promoting decay and stain of lumber. The surface of the yard should be kept free from debris and vegetation. Debris, in the form of pieces of wood such as old stickers and timbers, harbors stain and decay organisms and obstructs the movement of air over the surface and beneath the lumber piles. Vegetation restricts the movement of air in the same manner as debris, and if dry, it becomes a fire hazard. Vegetation can be controlled by applying crude oil, salt, or weedkillers. Covering the ground with cinders, gravel, shells, or crushed stone retards the growth of vegetation. These materials, as well as blacktop and concrete, provide smooth and firm surfaces for the passage of machines used to transport and pile the lumber. If the lumber piles are built by hand, surfacing or paving is of prime importance only on the alleys through which the lumber is transported. If fork-lift trucks are used, surfaced yards are needed.

1.011. LAYOUT. An air-drying yard consists of rows or blocks of lumber piles, with roadways or alleys between, figure 3. The alleys serve as routes for transporting the lumber, as pathways for the movement of air through the yard, and as protection against the spread of fire. The alleys are generally classified as main, cross, and rear (figs. 4 and 5). Spaces between the piles or rows of piles may also form long, straight passageways through the yard. In a yard where the piles are hand stacked, the main alleys are generally 16 to 20 feet wide; but where fork-lift trucks are used, the alleys are commonly 30 feet wide. Cross alleys may be 60 feet wide, rear alleys 8 feet, and spaces between the piles or rows of piles from 1 to 12 feet. Where the lumber is stacked by hand, the piles are in rows along both sides of the main alleys. At the rear of the piles, there is a space designated as the rear alley, which is commonly

NOTE. Italic numbers in parentheses refer to references at the end of this chapter.

about 8 feet wide. Where the lumber piles consist of a number of stickered unit packages, piled by fork-lift trucks, the rows of piles run from one main alley to another. With this method, the number of piles in a row, or the number between main alleys is unlimited; while with hand stacking, the number of piles between main alleys is limited to two.

Figure 3. General view of a well-laid-out lumber storage space at a Naval Shipyard. Fork lift trucks are used for unloading railroad cars and stocking unit packages.

Figure 4. View of a main alley of a well-laid-out yard showing good hand-stacked piles and a clean ground surface.

1.012. FOUNDATIONS. Pile foundations should be mechanically strong and provide a clearance between the lumber and the ground. Wood foundations should be made of heartwood

Figure 5. Main alley in a yard where each pile consists of six unit packages and is built with a fork-lift truck.

of a decay-resistant species or should be pressure treated with a preservative.

Foundation piers are made of concrete, masonry, pressure-creosoted blocks or posts of any species, or untreated heartwood of such decay-resistant species as baldcypress, redwood, and the cedars. Concrete or masonry piers should extend into the ground below the frostline. Wood piers may also be set in the ground or may bear on timbers laid on or slightly below the ground surface. The post, or pier, timber type is preferable to the solid, builtup beam type of foundation because it permits more air movement beneath the piles. Where stringers, running from the front to the rear of the pile, are used, they support the crossbeams (fig. 6); where they are not used, crossbeams are supported directly on the posts or piers (fig. 5). Timbers or steel I–beams or rails make satisfactory stringers. Timbers should be about 6 by 8 inches in cross section and pressure creosoted or given two brush coatings with hot creosote.

Figure 6. Pile foundation showing concrete piers, timber stringers, and beams.

Pile foundations may be level or sloped. Foundations for hand-stacked piles are usually sloped, about 1 inch per foot of length; those for unit-package piles built by machine are usually level. The amount of ground clearance required varies with certain factors. A sloped foundation should have a clearance of about 1 foot at the rear end of the pile, which will give a clearance of about 28 inches at the front end for 16-foot lumber. With level foundations, a clearance of 12 inches may be sufficient. These figures are approximations. Necessary clearance will depend on whether the ground is kept free from debris and vegetation and on climatic conditions, yard layout, pile spacing, and pile width.

The design of any pile foundation should conform to the method of piling—whether by hand, fork-lift truck, or crane—and the construction of the pile itself or the unit packages. Crossbeam spacing should be that of the tiers of stickers in the lumber pile or in the unit packages. Where the lumber piles are built with a fork-lift truck, the foundation must provide operating space for the machine, generally from 8 feet 6 inches to 9 feet (fig. 7). If such a space is not provided, then the foundation members are restricted to a height of not much more than 6 inches, or the foundations are of a temporary rather than a permanent character. With a permanent type of foundation, including the operating space for the fork-lift truck, it is generally necessary to provide one or more auxiliary supports to sustain the lumber over the 8- to 9-foot span.

Figure 7. Foundation for unit-package piles, consisting of four continuous, fixed beams, with a removable central support. A 9-foot space is provided for the passage of the fork lift truck. Notice that the beams, tiers of stickers, and the bolsters or separators between the unit packages, are in alinement.

1.013. PILING. Because different species and thicknesses of lumber dry at different rates, they should be piled separately. Thick stock requires a longer drying period than thin stock and is more likely to check. Moreover, when pieces of different thicknesses are piled in the same layer, the thinner pieces, not being in contact with the stickers, may warp. It is desirable, where practical, to separate stock also on the basis of length and width.

Lumber of random or mixed length, whether built into hand-stacked piles or unit packages, should be box piled. This method of piling avoids overhanging ends, figure 8. The longer boards should be placed in the outer vertical tiers, and if there is a sufficient number of these boards for additional tiers, such tiers should be uniformly distributed across the width of the pile. As far as possible, the boards in any tier should be of the same length. Ends of adjacent tiers of short boards should be ultimately flush with the front and rear of the pile where hand stacking is employed, and should be flush with one end of the pile where the lumber is piled in the form of unit packages, figure 9.

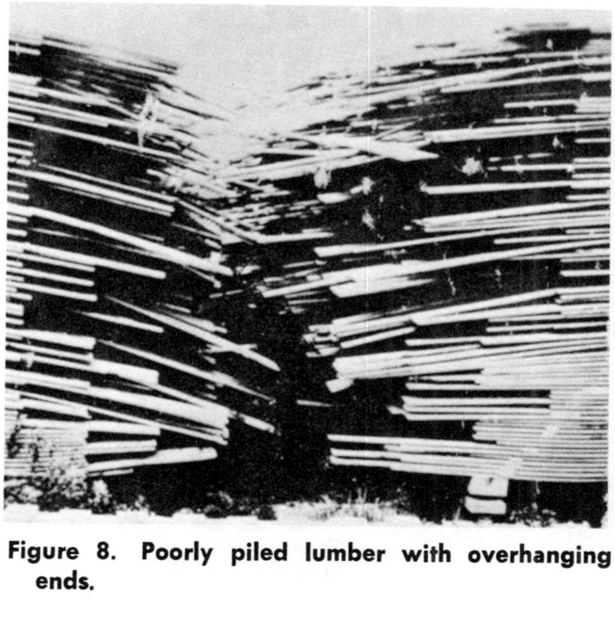

Figure 8. Poorly piled lumber with overhanging ends.

If a lumber pile is hand stacked on a sloped foundation, each successive layer of boards should be placed so that the front ends project slightly beyond those of the layer beneath. This produces the pitch of the pile, 1 inch per foot of height. The resulting overhang excludes rain from the front of the pile. Such overhang is necessary for all sloping piles, because if rain enters the front of such piles, water will flow down the sloping boards and collect where the stickers contact them.

Figure 9. Building a stickered unit package in a stacking rack equipped with sticker guides. The unit package is box-piled.

1.014. STICKERS. Stickers are strips or boards used to separate the layers of lumber in a pile. Stickers should be sound, dry, free from stain and decay, and of uniform thickness. Stickers for 1-inch lumber are usually made from nominal 1-inch stock and should not be more than 4 inches in width. They must be wide enough so that they will not be crushed into the faces of the lumber by the weight of the pile. For softwoods, stickers are frequently 4 inches wide. For hardwoods, stickers as narrow as 1½ inches are often used. For random-length lumber thicker than 1½ inches, stickers more than 1 inch thick are used.

Stickers should be placed in good alinement, either vertical or following the pitch of the front of the pile, each tier resting on a crossbeam (fig. 10). Misalinement of stickers causes warping of the lumber. Stickers are spaced 2 to 5 feet apart, depending upon the thickness of the lumber and its tendency to warp. Thin stock requires closer sticker spacing than thick stock. Closely spaced stickers tend to hold lumber flat. The tiers of stickers at the front of the pile should protrude beyond the ends of the boards to retard end drying. Tiers should follow the pitch of the pile or that of the unit package. If the pile is built on a level foundation, the tiers should form vertical lines.

Figure 10. Good sticker alinement but no forward pitch to pile.

Unit packages should be built in stacking racks or jigs provided with sticker guides, which are needed to space and aline the stickers properly. Proper spacing is necessary to assure that the stickers will be in alinement from top to bottom of the pile, when several unit packages are piled together, figure 7. Also, the unit packages should be placed so that the tiers of stickers are directly above the crossbeams of the pile foundation. Bolsters between the unit packages should also be in alinement with the crossbeams and tiers of stickers.

1.015. BOARD SPACING, FLUES, CHIMNEYS. In wide hand-stacked piles, the principal direction of air movement through a pile is downward, and consequently, ample space should be allowed for this movement. Unobstructed vertical passages should extend from top to bottom of the pile. When lumber is piled in the form of unit packages, the boards in the courses are generally placed edge to edge. With this method, there will be no openings between the edges of the boards, and consequently, no vertical passageways for downward circulation. Since the unit packages are generally relatively narrow, and commonly about 4 feet wide, drying is accomplished by the movement of air horizontally in the pile.

The passage is called a flue or chimney, depending on its width. When it is less than 6 inches wide, it is called a chimney or vent. The chimney may be either straight-sided or tapered (fig. 11) toward the top of the pile. The number, size, character, and arrangement of the flues or chimneys depend on the width of the stock and on the desired speed of drying.

Figure 11. A tapered chimney in a yard pile.

1.016. ROOFS AND SHEDS. It is always desirable to protect lumber from snow, rain, and sunshine. Exposure to alternate precipitation and sunshine causes surface checking, warping, and ultimate disintegration of the wood surface. By heating the surface of wood and the air next to it, direct sunshine dries the surface too rapidly and induces surface checks. Weather can damage seasoned, as well as unseasoned lumber.

Lumber can best be air seasoned in a shed. The shed walls may be tight, slat-sided, or open-sided. Tight sheds must be well ventilated. The roof and walls absorb some solar heat, raising the mean temperature within above that outdoors. If just enough ventilation is provided to get rid of the moisture evaporated from the lumber, the increase in temperature inside will maintain a lower relative humidity than that existing outdoors, thus speeding the drying rate.

If sheds are not available, a temporary roof can be built over the pile to afford protection from direct sunshine, rain, or snow. A tight roof should slope about 1 inch per foot; if it is not tight, it should slope about 1¼ inches per foot in order to provide satisfactory runoff. The roof should project 1 foot beyond the pile at the front and 2½ feet at the rear (fig. 12). The roof must be supported on crossbeams, providing a space of at least 5 inches between the roof and the top layer of lumber. The roof should be fastened to the pile by wire or some other means.

Figure 12. Pile roofs projecting at front end of pile. (The tramway interferes with the air movement around the lower parts of the piles.)

With sloped piles, it has been customary to use loose board roofs. The boards are laid in double-layer, double-length fashion, with the front portion overlapping the rear portion. The pitch of the roof either follows the slope of the pile, or is increased by elevating the front end. It is desirable to provide a space of about 5 inches between the roof and the top layer of lumber. The roof should be fastened to the pile.

Loose board roofs are not efficient on level unit-package piles, because it is impractical to give them enough pitch to make them reasonably watertight. Paneled roofs are more suitable for unit-package piles. The panels can be constructed of boards nailed to light framing members, with battens nailed over the cracks, or covered with roll roofing. The panels should be several feet longer than the pile. If made half length, they should overlap (fig. 13). Since the panels are watertight, little pitch is needed for drainage. By hand or with the fork-lift truck, the roof panels can be placed on the top unit package while it is still on the ground.

Figure 13. Roof panels for unit-package piles.

The essential features of good piling in hand-stacked piles, for rapid air seasoning, are summarized in figure 14. Such piling is intended mainly for low-grade and easily dried lumber. For stock requiring slower drying, the height of foundation, spacing between piles, and width of flues should be decreased, while the width of the pile should be increased and thinner stickers (narrow, dry, special stickers are less conducive to checking than stickers taken from the stock being piled) should be used. The same principles apply to lumber in unit-package piles, except that flues are seldom used.

The two best cold coatings tested are: (1) Phenolic-resin varnish pigmented with aluminum paste or powder; and (2) filled hardened gloss oil. Aluminum paste or powder can be dispersed in a variety of vehicles to form water-resistant paints. Even with the most water-resistant vehicles, such as phenolic-resin tung-oil varnish, two coats are required to form a highly water-resistant end coating; three coats produce the most water-resistant end coating. About 2 to 2½ pounds of aluminum should be used per gallon of vehicle. Aluminum paints should be mixed immediately before use, because the desirable leafing property of the aluminum particles diminishes as the paint stands in the mixed condition. Aluminum coatings do not cause serious abrasion of cutting tools. Coatings containing aluminum are expensive.

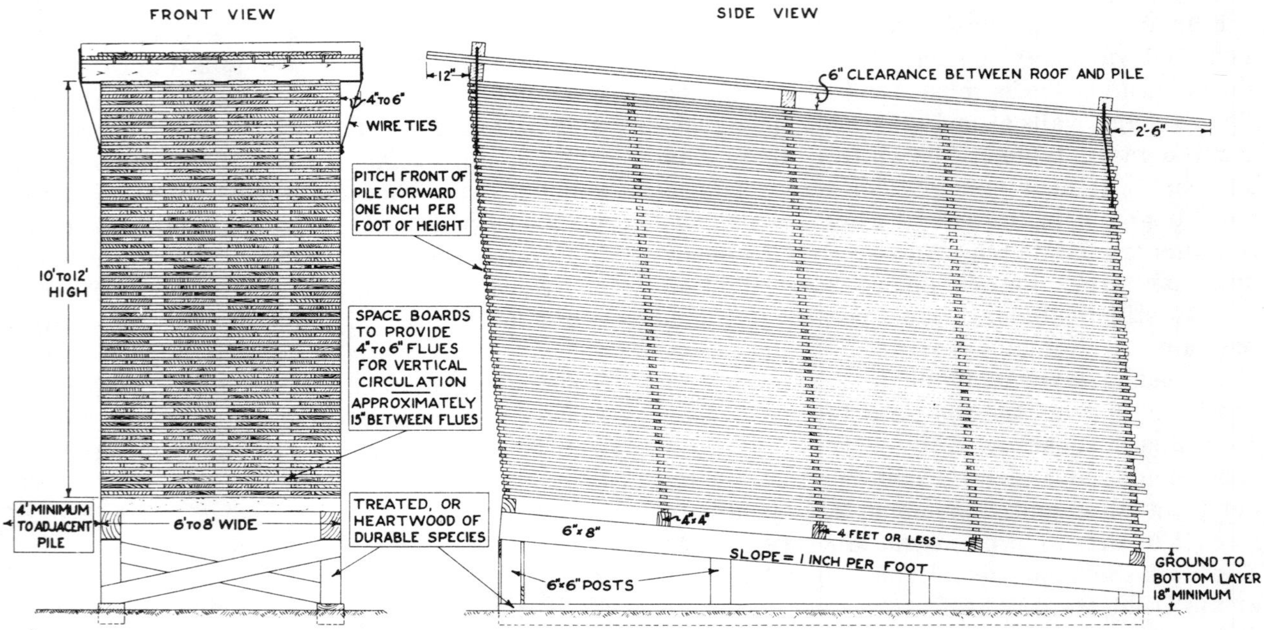

Figure 14. Essential features of piling for rapid and uniform air-seasoning.

1.017. END COATINGS. When lumber dries, areas of end grain should be protected because end checks and end splits are likely to occur. To avoid such defects, end coatings are applied. They are particularly effective if applied as soon as the lumber is received.

To be satisfactory for large-scale use, an end coating should have: (1) Adequate resistance to water movement under all conditions of temperature and relative humidity during the drying of the wood; (2) ease of application; (3) sufficient toughness and adhesion to withstand rough handling and to prevent blistering and cracking during kiln drying; (4) freedom from abrasive substances that may injure saws or cutters during machining of wood; and (5) low cost.

End coatings may be liquid at ordinary temperatures and applied without being heated, or solid at ordinary temperatures and heated for application. Hot coatings are well suited for use on small stock that can be easily handled, but not for large ship timbers.

The filled hardened gloss oil is made by mixing 25 parts (by weight) of barytes and 25 parts of magnesium silicate with 100 parts of a hardened gloss oil made according to the following formula:

Rosin	100 parts by weight
Hydrated lime	6 to 8 parts by weight
Mineral spirits	57.5 parts by weight

The hardened gloss oil should be made up by a varnish manufacturer because of the technical operations involved. Filled hardened gloss oil is moderate in cost. However, the filler contained in this coating is somewhat abrasive to cutting tools. This end coating is intended for preventing checking in large timbers and in logs.

A heavy paste containing a pigment, as white lead in a vehicle such as linseed oil, makes a simple coating that is moderately effective when applied in a single coat. The paste should be prepared with only enough oil to allow its application with a trowel or stiff brush. This type of coating is generally available and admirably suited for shipyard use in drying small lots of material. Ordinary paints and varnishes are too thin to be effective as end coatings, although they are helpful if several coats are applied. A paint consisting of red iron oxide mixed with a vehicle consisting of a mixture of 75 percent linseed oil, 15 percent nonreactive spar varnish, and 10 percent combined drier and thinner is in common use. There are also numerous proprietary end coatings on the market.

The effectiveness of an end coating depends as much on its thickness as on the material of which it consists. For all coatings, with the exception of the pastes, it is necessary for best results to apply two or three coats, allowing each to dry before the application of the next.

1.02. KILN DRYING. In kiln drying, the same principles are employed as in air-seasoning (*1–2*). Kiln drying is accomplished at higher temperatures than is air seasoning, but in modern, efficient dry kilns, the drying conditions are at all times under control, so that they may be suited to the requirements of the lumber. The temperatures used in kiln drying are generally sufficiently high to kill any fungus that may be present. It is possible to hasten or retard drying and to obtain a predetermined final moisture content. In general, lumber up to 2 or 3 inches in thickness can be dried in a shorter period and with fewer drying defects in a kiln than by air seasoning. Somewhat thicker lumber can be kiln dried by the use of special schedules, but, in general, the kiln drying of heavy material is impractical because of the development of defects and the length of time required.

In the past, there has been some prejudice among shipbuilders against kiln-dried lumber. This prejudice has probably arisen because of unfortunate experience with lumber that was weakened by unnecessarily severe drying schedules, or that was dried to a moisture content too low for general ship use. Excessively dry wood is stiff and, since many boat parts have to be sprung into position, wetter and more flexible wood is preferred. In addition, an extremely low moisture content is undesirable for those parts that are to be submerged in water because they swell excessively when submerged. Properly kiln-dried lumber is just as strong and fully as decay-resistant as air-seasoned lumber of equivalent moisture content.

1.020. DRY KILNS. A dry kiln is essentially an enclosure or building provided with suitable apparatus for the control of temperature and humidity, and to provide air circulation, within which lumber can be piled and dried to the desired moisture content. Lumber is piled on trucks and wheeled into and out of the kiln. In some of the newer kilns, the lumber is placed directly on the floor of the kiln by fork-lift trucks, and after drying is removed in like fashion.

Heat is usually provided by piping live or exhaust steam through coils or radiators, but it may be provided by furnace or smoke pipe. The air within many kilns is humidified by steam sprays. Air is circulated through the kiln and the lumber piles by gravity or by mechanical means, such as internal fans or external blowers. The temperature and relative humidity may be controlled by manipulating the heating and steam spray valves and vents, or by instruments that automatically control temperature and relative humidity.

1.0200. TYPES OF KILNS. Dry kilns for wood are of many forms and types. On the basis of construction and operation, they can be divided into two groups, progressive and compartment (sometimes called "box" or "charge" kilns). Both types are suited to the drying of wood for ship construction.

In a progressive kiln, a number of truckloads of lumber at different stages of drying extend through the kiln in an unbroken line. Truckloads of green stock are periodically fed into the kiln, and truckloads of dried stock are removed at the same time from the other end of the line. The stock thus moves progressively through the kiln until it emerges, supposedly dry, after a specified time of exposure to the various conditions along the length of the kiln. The temperature and relative humidity at any point are intended to remain constant, but the kiln is hotter and drier at the dry end than at the green end.

The compartment kiln is loaded fully at one time, the entire charge remaining in place throughout the drying period. The temperature and relative humidity are as nearly uniform as possible throughout the kiln at any given time.

The usefulness of a progressive kiln is limited, whereas a compartment kiln can be used for any kind of drying. A progressive kiln must be supplied continually with lumber of a single kind and thickness to function properly and, therefore, is not satisfactory except where a constant supply of such stock is available. Further, because it lacks flexible control of temperature and relative humidity, it is not adapted to drying that requires great accuracy of control. Progressive kilns are used to dry easily dried softwoods, such as southern yellow pine and Douglas fir, or air-seasoned hardwoods.

Kilns may also be classified on the basis of the means used to obtain circulation. Natural-circulation kilns are equipped with inlets and vents. Hot air tends to rise through the chimney from the kiln, causing fresh air to enter the kiln through leakage or through the inlet ducts. Moist air is thus removed and replaced by drier air, reducing the relative humidity within the kiln. The drying of the lumber is accomplished chiefly by the recirculation of air past the steam coils and through the lumber. This type of circulation is slow, but it is well distributed throughout the lumber pile if suitable chimneys and flues are built into the pile and the air is correctly baffled. Natural-circulation kilns are adapted to the drying of most lumber except the green lumber of most hardwoods. Steam sprays or jets are used in natural-circulation systems, being placed so that the impact of the spray helps the natural air movement.

Forced-circulation kilns usually produce air movement with internal fans or external blowers. The internal fans force the air into a space from which it must pass laterally through the lumber pile in order to return to the fans. The external blower forces air into a duct, supplied with openings, which extends the length of the kiln. After passing out of the openings, the air travels through the lumber and back to the blower through return ducts. Forced-circulation kilns accomplish more rapid drying and maintain more nearly uniform conditions throughout the kiln than do natural-circulation kilns. They are, therefore, suited to dry all lumber for shipbuilding, with the exception of heavy timbers.

1.021. PILING OF LUMBER. Lumber is piled in a dry kiln in such a manner that surfaces are exposed to the currents of moving air. This requires that the boards be separated by stickers, as in air seasoning, and that the stickers be parallel to the direction of the air currents. In natural-circulation kilns, where vertical movement of the air is strongly relied on, vertical flues at least 4 inches wide and spaced 12 to 14 inches apart should be provided. In forced-circulation kilns, with lateral circulation, the boards should be piled edge to edge. In some designs, a central chimney is utilized. When material is nearly square in cross section, spaces are usually left between the edges to obtain some edge drying.

1.022. DRYING SCHEDULES. In kiln drying, wood is exposed gradually to higher temperatures and lower relative humidities until the desired moisture content is attained. The series of temperature and relative humidity changes used to dry a kiln charge of lumber is termed a drying schedule. Woods differing in species, thickness, character of grain, and moisture content require different drying conditions or schedules; thus, 3-inch Douglas fir decking will require a different schedule than 1-inch Atlantic white-cedar planking. The Forest Products Laboratory has prepared kiln-drying schedules for most native United States species. These schedules, which are applicable to wood for ship construction, are not based on time but on the moisture content of the lumber and therefore require the use of kiln samples for proper operation. The conditions of the first step in the appropriate schedule should be established as quickly as possible. The steam sprays should be used to help in heating the kiln and in maintaining a high relative humidity during the time when the kiln and lumber are being heated, but in no case is a long initial steaming period recommended. A long initial steaming treatment is particularly disadvantageous when the charge consists of air-seasoned lumber, which is likely to contain surface checks.

Kiln-drying schedules suitable for boat and ship lumber of various species and thicknesses are listed in table 1. The schedules themselves are presented in table 2, and a sample combined schedule derived from table 2 is given in the table 3 for 1-inch green red oak.

There are 14 temperature schedules, ranging from the mildest (T1) to the most severe (T14). For instance, T1 has been found suitable for hickory handle stock to prevent pinking, and T14 is commonly used to accelerate the drying of such species as southern pine lumber. The initial temperature is maintained until the moisture content of the stock dries to 30 percent. The final temperature is reached in 5 steps during the time the moisture content changes from 30 to 15 percent.

The humidity schedules are given in terms of the wet-bulb depression and consist of 8 sets of values, identified by numbers from 1 to 8. Each schedule, or set, has a different initial value, but the same final value of 50° F. In each, the final value is reached in six steps, at progressively greater rates that have been found safe and desirable. Another phase of the relative humidity schedule is the moisture content at which the various wet-bulb depressions are used. As mentioned before, it has been found that the first change in relative humidity can be made at a moisture content that bears some relationship to the original green moisture content of the wood. For that reason, six classes of moisture content steps have been made, identified by the letters A to F, any one of which can be used with any one of the eight wet-bulb-depression sets to make up a humidity schedule that is best suited to the wood being dried. Forty-eight combinations are possible.

1.03. CONTROL OF CHECKING WITH CHEMICALS. Hygroscopic chemicals can be applied to green wood (*1–3*) to reduce the amount of checking and splitting that occurs during drying. The outer zones of the wood are impregnated with the chemical solution, which changes the hygroscopic properties of the wood so that, during drying, a higher moisture content is

TABLE 1.

INDEX OF SCHEDULES FOR KILN DRYING SHIP AND BOAT LUMBER 1 TO 2 INCHES THICK

Species	Schedule designation for lumber			
	4/4, 5/4, 6/4		8/4	
	Temperature	Relative humidity	Temperature	Relative humidity
HARDWOODS				
Ash, White	T8	B4	T5	B3
Basswood	T12	E7	T10	E6
Beech	T8	C2	T5	C1
Birch, yellow	T8	C4	T5	C3
Cherry, black	T8	B4	T5	B3
Hickory	T8	D3	T6	D1
Locust, black	T6	A3	T3	A1
Mahogany (*Swietenia* spp.)	T6	C4	T4	C3
Maple, hard	T8	C3	T5	C2
Oak, red	T4	D2	T3	D1
Oak, white	T4	C2	T3	C1
Sweetgum (sap gum)	T12	F5	T11	D4
Sweetgum (red gum)	T8	C4	T5	C3
Tupelo, black	T12	E5	T11	D3
Tupelo, water	T12	E5	T11	D3
Walnut, black	T6	D4	T3	D3
Yellow poplar	T11	D4	T10	D3
SOFTWOODS[1]				
Baldcypress	T12	E4	T11	E3
Cedar:				
Alaska	T12	A4	T11	A3
Atlantic white (South. white)	T12	A5	T11	A4
Eastern red	T10	A4	T8	A3
Incense	T12	E5	T11	E4
Northern white	T12	B5	T11	B4
Port Orford	T12	B4	T11	B3
Western red (light)	T12	B5	T11	B4
(sinker)	T8	F5	T6	F4
Douglas fir	T13	A4	T12	A3
Larch, eastern (Tamarack)	T12	B4	T11	B3
Western	T12	B4	T11	B3
Pine, eastern White	T11	C5	T10	C4
Ponderosa	T10	C6	T8	C5
Southern yellow	T14	B5	T12	B4
Sugar	T10	E6	T8	E5
Western white (Idaho)	T10	B5	T8	B4
Redwood (light)	T6	D5	T6	D4
(sinker)	T6	F5	T6	F4
Spruce, eastern	T14	B5	T12	B4
Sitka	T14	B5	T12	B4

[1] The schedules given are based mainly on the moisture content of the heartwood and are intended for average stock and for all grades and items that do not have special drying requirements. One modification would be in the drying of commons to prevent the development of loose knots. In this case, the temperature schedule should be T7 and the conditions given in step 3 of both temperature and relative-humidity schedules should be maintained as final conditions until the desired moisture content is reached. As trouble of this kind becomes somewhat less with increases in thickness, step 4 instead of step 3 in the relative-humidity schedule can be used for final conditions, if desired, to accelerate the drying of 8/4 stock.

TABLE 2.

Temperature and relative humidity kiln schedules for ship and boat lumber 1 to 2 inches thick

Temperature schedules (dry-bulb temperatures)

Step number	Moisture content		Temperature schedule													
	From	To	T1	T2	T3	T4	T5	T6	T7	T8	T9	T10	T11	T12	T13	T14
	Percent	*Percent*	°*F.*	°*F.*	°*F.*	°*F.*	°*F.*	°*F.*	°*F.*	°*F.*	°*F.*	°*F.*	°*F.*	°*F.*	°*F.*	°*F.*
1	Initial	30	100	100	110	110	120	120	130	130	140	140	150	160	170	180
2	30	25	105	110	120	120	130	130	140	140	150	150	160	170	180	190
3	25	20	105	120	130	130	140	140	150	150	160	160	160	170	180	190
4	20	15	115	130	140	140	150	150	160	160	160	170	180	180	190	200
5	15	Final	120	150	160	180	160	180	160	180	160	180	180	180	190	200

Relative humidity schedules (wet-bulb depressions)

"A" schedules

Step number	Moisture content		Relative humidity schedule							
	From	To	A1	A2	A3	A4	A5	A6	A7	A8
	Percent	*Percent*	°*F.*	°*F.*	°*F.*	°*F.*	°*F.*	°*F.*	°*F.*	°*F.*
1	Initial	30	3	4	5	7	10	15	20	25
2	30	25	4	5	7	10	14	20	30	35
3	25	20	6	8	11	15	20	30	40	50
4	20	15	10	14	19	25	35	50	50	50
5	15	10	25	30	35	40	50	50	50	50
6	10	Final	50	50	50	50	50	50	50	50

"D" schedules

Step number	Moisture content		Relative humidity schedule							
	From	To	D1	D2	D3	D4	D5	D6	D7	D8
	Percent	*Percent*	°*F.*	°*F.*	°*F.*	°*F.*	°*F.*	°*F.*	°*F.*	°*F.*
1	Initial	50	3	4	5	7	10	15	20	25
2	50	40	4	5	7	10	14	20	30	35
3	40	35	6	8	11	15	20	30	40	50
4	35	30	10	14	19	25	35	50	50	50
5	30	25	25	30	35	40	50	50	50	50
6	25	Final	50	50	50	50	50	50	50	50

"B" schedules

Step number	Moisture content		Relative humidity schedule							
	From	To	B1	B2	B3	B4	B5	B6	B7	B8
	Percent	*Percent*	°*F.*	°*F.*	°*F.*	°*F.*	°*F.*	°*F.*	°*F.*	°*F.*
1	Initial	35	3	4	5	7	10	15	20	25
2	35	30	4	5	7	10	14	20	30	35
3	30	25	6	8	11	15	20	30	40	50
4	25	20	10	14	19	25	35	50	50	50
5	20	15	25	30	35	40	50	50	50	50
6	15	Final	50	50	50	50	50	50	50	50

"E" schedules

Step number	Moisture content		Relative humidity schedule							
	From	To	E1	E2	E3	E4	E5	E6	E7	E8
	Percent	*Percent*	°*F.*	°*F.*	°*F.*	°*F.*	°*F.*	°*F.*	°*F.*	°*F.*
1	Initial	60	3	4	5	7	10	15	20	25
2	60	50	4	5	7	10	14	20	30	35
3	50	40	6	8	11	15	20	30	40	50
4	40	35	10	14	19	25	35	50	50	50
5	35	30	25	30	35	40	50	50	50	50
6	30	Final	50	50	50	50	50	50	50	50

"C" schedules

Step number	Moisture content		Relative humidity schedule							
	From	To	C1	C2	C3	C4	C5	C6	C7	C8
	Percent	*Percent*	°*F.*	°*F.*	°*F.*	°*F.*	°*F.*	°*F.*	°*F.*	°*F.*
1	Initial	40	3	4	5	7	10	15	20	25
2	40	35	4	5	7	10	14	20	30	35
3	35	30	6	8	11	15	20	30	40	50
4	30	25	10	14	19	25	35	50	50	50
5	25	20	25	30	35	40	50	50	50	50
6	20	Final	50	50	50	50	50	50	50	50

"F" schedules

Step number	Moisture content		Relative humidity schedule							
	From	To	F1	F2	F3	F4	F5	F6	F7	F8
	Percent	*Percent*	°*F.*	°*F.*	°*F.*	°*F.*	°*F.*	°*F.*	°*F.*	°*F.*
1	Initial	70	3	4	5	7	10	15	20	25
2	70	60	4	5	7	10	14	20	30	35
3	60	50	6	8	11	15	20	30	40	50
4	50	40	10	14	19	25	35	50	50	50
5	40	35	25	30	35	40	50	50	50	50
6	35	Final	50	50	50	50	50	50	50	50

TABLE 3.

COMBINED TEMPERATURE AND RELATIVE HUMIDITY SCHEDULES FOR KILN DRYING GREEN 4/4 RED OAK

Temperature step number	Relative humidity step number	Moisture content Class D		Temperature schedule T4	Relative humidity schedule D2	Wet Bulb	Relative humidity	Equilibrium moisture content
		Moisture content						
		From	To	Dry Bulb	Wet-bulb depression			
		Percent	*Percent*	°*F.*	°*F.*	°*F.*	*Percent*	*Percent*
1	1	Initial	50	110	4	106	87	17.5
1	2	50	40	110	5	105	84	16.2
1	3	40	35	110	8	102	75	13.3
1	4	35	30	110	14	96	60	9.9
2	5	30	25	120	30	90	31	5.4
3	6	25	20	130	50	80	10	2.0
4	6	20	15	140	50	90	14	2.6
5	6	15	Final	180	50	130	26	3.3

maintained on the surface than on that of untreated wood. Wood develops surface checks because the surfaces dry and shrink more rapidly than the interior. The tendency to check is reduced when a higher surface moisture content is maintained. The presence of the chemical does not seriously interfere with the drying as a whole. The moisture from the interior parts passes through the treated surface zones and is evaporated from the surfaces, Some chemicals not only alter the hygroscopic properties of the treated wood but also exert an antishrink effect. This is an added advantage in reducing the tendency of the wood to surface check.

It is advantageous to use chemicals on timbers of large dimension, such as heavy oak flitches and material for keels, frames, stems, stern posts, and softwood timbers for keels, shelves, clamps, and deck beams, that are difficult to season by ordinary methods without developing serious drying defects. The presence of the chemical in the seasoned wood is seldom advantageous and may be disadvantageous. Chemicals may discolor the wood, corrode metals, dull woodworking machines, and cause moisture to condense on the surfaces of the wood. Because wood is usually dressed or cut up after seasoning and before use, thereby removing most of the chemical in the surface layers, some of these disadvantages can be eliminated.

Numerous chemicals have been used, among them common salt, urea, various low-grade sugars, diethylene glycol, and urea-aldehyde. Common salt is an effective reagent but corrodes metals, and lumber treated with it should not be kiln dried because kiln equipment will be corroded. Common salt has been used in the past by lumbermen to close up surface checks in lumber and by shipbuilders to keep parts damp during construction and to provide some decay resistance. Urea is moderate in cost, available, noncorrosive to ferrous metals, slightly corrosive to copper, and variable in its effectiveness, but it has a tendency to discolor wood. Urea has been used with marked success in the seasoning of heavy Douglas fir timbers, but is not so effective when applied to 3- and 4-inch oak, although the surface checking during kiln drying is sometimes reduced by one-half. Low-grade cane and corn sugars are moderate in cost, in ordinary times available, and when used in conjunction with a hygroscopic chemical, highly effective; but they tend to discolor wood. Diethylene glycol is somewhat more expensive but appears to be a promising reagent, although its properties have not been extensively studied. The urealdehyde solutions are among the most effective of the reagents, but are expensive.

Chemicals may be applied to green wood by soaking, dipping, spraying, brushing, and dry spreading. The soaking method consists of submerging the wood in a saturated or nearly saturated aqueous solution of the chemical for about 3 days per inch of thickness. A large investment in chemical and such equipment as mixing tanks, pumps, and pipelines is necessary for efficient operation. Dipping, spraying, and brushing methods have been developed to eliminate the need for tanks and other equipment and to lower the investment in the chemical. A thickening reagent, such as starch, can be used in the chemical solution in order to get a coating thick enough on the surface of the lumber. After the coating is applied, the lumber is either piled directly on stickers for air seasoning or on thin wood strips, about the thickness of the coating, for about 3 days per inch of thickness to allow the surface layers to absorb some of the chemical before piling it for seasoning.

The dry-spreading method is the simplest. A solid pile of wood is built up, from 60 to 100 pounds of chemical per 1,000 board feet being spread over each layer of wood. When the dry chemical has disappeared, the lumber is piled on stickers for seasoning. Unless the ends of the lumber or timbers have already been end coated, it is necessary to apply the chemical, in the form of a saturated solution, to the ends about once a day during the treating period. Unless this is done, the ends of the material are likely to check. Although the various methods of application appear to accomplish the same result as to chemical impregnation, the soaking method is generally more effective in reducing defects during drying.

1.04. SEASONING DEFECTS. Shrinkage causes most of the defects that develop during seasoning.

1.040. SURFACE CHECKS. A surface check usually originates at a point where there is a natural break in the continuity of the wood fibers, such as a knot, a resin duct in some softwoods, or a ray in hardwoods. It is a tensile failure perpendicular to the grain, caused usually by drying and shrinking of the surface zones while the interior is still at a relatively high moisture content and unshrunken. Surface checks are confined to flat-grain faces and are usually more common on the side of the piece that was nearest the bark when in the tree. Wide stock surface checks more readily than narrow stock, and thick stock more readily than thin stock. Boxed-heart timbers, which have four flat-sawed faces, develop surface checks on all faces. These checks may penetrate deeply into the timber, even to the pith.

1.041. END CHECKS. End checks belong in the same category as surface checks, but they originate on end-grain surfaces. They start as separations of the wood fibers extending across the rings of annual growth, and soon turn into a lengthwise separation of the wood as they progress inward. Because of rapid drying from the end grain and the exposure of the ends of the wood fibers, end checks form more quickly than do surface checks. End checks are responsible for much waste of material because, for many purposes, checked ends must be trimmed off.

1.042. SPLITS. A seasoning split results when an end check and a surface check meet. A split is usually more serious than an end check or a surface check, beause it is likely to extend farther along the face and deeper into the piece. In boards, splits often extend from face to face. In some species, splits develop from shakes, particularly at or near the end of a board.

1.043. WARP. When a board or timber becomes distorted so that its faces are no longer true planes, it is said to be warped. Warp includes, bow, crook, cup, and twist. Bow distorts a board so that its faces are convex or concave longitudinally; crook distorts the edges of a board so that they are convex or concave longitudinally; cup distorts a board so that its faces are convex or concave transversely; and twist distorts a board so that the four corners of any face are no longer in the same plane.

Warp is more common in stock of small dimension than in heavy stock. It is caused by natural shrinkage differences in the different directions of grain. Such differences may be present in the two faces of a flat-sawed board, and are marked in boards containing distorted grain such as is caused by knots and cross grain. Warp is also caused by differences in shrinkage between normal wood and such abnormalities as compression wood. Uneven drying also causes boards to warp. Projecting board ends, insufficient stickering, and poor vertical sticker alinement contribute to warp. Most warping takes place during seasoning, but a board may warp after it has been seasoned if it is alternately wetted and dried.

1.044. HONEYCOMB. Honeycomb checks occur in the interior of a piece, usually along the rays. They may be the result of an extension of an end or surface check. Like end checks and surface checks, they are the result of failure of the wood in tension across the grain. They occur during the later stages of drying, when the interior is losing moisture and shrinking under restraint. Honeycomb is more prevalent in thick stock, hardwoods, and kiln-dried stock than in thin lumber, softwoods, and air-seasoned lumber. Excessively high kiln temperatures contribute to the formation of honeycomb checks. Honeycombing often can be detected by the appearance on the surfaces of grooves running along the length of the piece.

1.045. CASEHARDENING. The term "casehardening," as applied to lumber, is often misunderstood and confused with the meaning as applied to metals. Casehardening as applied to wood does not mean that the surface is hardened, but denotes a stressed condition present in dry lumber; namely, compression on the surfaces and tension in the interior. This condition is brought about by the fact that the various zones of the wood have dried under different degrees of restraint and have become "set.". The differences in set between the various zones cause the stressed condition in the dry lumber.

Although casehardening exists when the wood is dry and the moisture is uniformly distributed throughout the piece, those factors which cause the final condition are at work throughout the entire drying process. Because of the manner in which wood dries, some degree of set is unavoidable. This set, however, can be minimized in kiln drying by the use of suitable drying schedules and can be removed near the end of the drying period by conditioning treatments. It is possible, therefore, to kiln-dry lumber without serious casehardening.

1.046. COLLAPSE. Collapse is the flattening of single cells or rows of cells, usually in the heartwood, during drying of wood or pressure treatment with preservatives. The cell walls come together, practically eliminating the cell cavity. Collapsed portions have abnormally large shrinkage, which is manifested by a corrugated appearance of the faces of a board. Collapse is common in the heartwood of those woods through which moisture moves slowly and whose moisture content when green is usually high, such as lowland oak, redwood, and western red cedar. The exact causes of collapse are not known, but it may be caused by liquid tensile forces exerted in cells completely full of water or by compressive stresses developed during drying. Collapse is often associated with honeycombing and, as in honeycombing, high drying temperatures contribute to its formation.

1.1. HANDLING PRACTICES FOR LUMBER.

1.10. GENERAL. Handling includes all unloading, loading, stacking, and transportation of lumber (*1–5*). Even if machines are employed to the fullest extent, there still may remain considerable manual handling of individual boards. The shape of lumber lends itself to packaging for machine handling. The packages, of course, are usually made up and broken down by hand.

1.100. HANDLING LOOSE LUMBER. Equipment to handle single boards is limited. Dead rolls and bars are used in unloading boards from a railroad box car. The roll or bar supports the piece while it is shoved or pulled laterally and holds it at a certain elevation, so that it may be picked up and piled. Inclines equipped with dead rolls are also used to unload lumber piece by piece. Conveyor belts and chains are other devices used to move boards, dimensions, and timbers in unloading and other plant operations (fig. 15).

Loads of boards are transported by motor truck and by trailers, dollies, or wagons. The beds, platforms, or bunks of these conveyances are often fitted with rollers to facilitate loading. Trailers, dollies, and wagons hauled by other vehicles can be loaded piece by piece without much loss of efficiency. When motor trucks are used, however, loading an entire load as a unit avoids tying up the truck during loading. The load may be made up on rollers and hauled by cable and winch over the rollers onto the truck, or it may be built on temporary supports that are knocked down when the truck backs against them.

Figure 15. Conveyor belt used to carry boards away from railroad car.

Loads of loose lumber carried on trailers, dollies, or wagons are generally unloaded by hand. Dump trucks are sometimes used, and the loads are dumped on the ground. Loads resting on rollers may be dumped by suddenly starting the vehicle and literally running it out from beneath the load. Dumping of whole loads, however, is likely to result in breakage and other damage, as, for example, driving grit into some boards.

1.101. HANDLING PACKAGED LUMBER. Lumber is often received in the form of packages that are solid piled and generally strapped. Strapped packages are preferred for all types of package shipment because strapping prevents movement of boards. Sometimes, unstrapped packages shift so much in transit that they cannot be unloaded by machine (fig. 16). Even well strapped packages sometimes shift and are damaged during shipment.

Lumber procured through regular Department of Defense channels is packaged according to specifications. Strapped solid packages of thoroughly seasoned lumber can be left intact throughout handling, transporting, and storing operations until they reach the shop. Strapped solid packages of green or inadequately seasoned lumber should be opened and the lumber piled with stickers between courses for air drying in a storage yard or open shed.

Cranes and fork-lift trucks are used to unload, pile, and unpile lumber and fork-lift or straddle trucks to transport solid or stickered packages of lumber (figs. 17 and 18). The versatility of fork-lift trucks, however, often justifies their exclusive use, particularly at small storage yards. If a large volume of lumber is handled, a combination of fork-lift trucks for unloading, piling, and unpiling with straddle trucks for transporting is efficient.

Figure 16. A fork-lift truck spilling part of an unstrapped solid package of lumber. Strapped packages are preferred for machine handling and for all types of package shipment because strapping prevents movement of boards.

Figure 17. Unloading strapped solid packages from a flat car by tractor crane. Fork-lift trucks are also used in unloading railroad cars.

Figure 18. Straddle truck carrying a solid package of lumber. When large volumes of lumber are handled, the use of straddle trucks for transporting and fork-lift trucks for piling and unpiling is efficient.

1.2. STORING LUMBER AND PLYWOOD.

1.20. GENERAL. Lumber for boat and ship construction should be stored under conditions that will prevent its deterioration and, where necessary, improve its moisture-content condition (fig. 19).

The methods used to store lumber depend to a large extent on its state of seasoning, the use to which it is to be put, and the species and dimensions of the stock. Green lumber requires different storage methods than does air-seasoned or kiln-dried stock. Large timbers require different methods than does 1-inch lumber. All lumber, however, should be protected from direct sunshine and wetting.

1.21. GREEN LUMBER. A large proportion of ship lumber, particularly for the building of large ships, arrives at the shipyard green or only partially seasoned. For underwater parts and bending stock, green material is satisfactory or preferred if it can be kept without serious deterioration until the ship is launched or the part is bent. Green lumber intended for other parts may require seasoning before it is suitable for use. Deterioration of such material can be prevented by various methods during the storage period.

1.210. WATER STORAGE. Submersion in water is a simple and effective means of preserving wood in the green condition. This can be done either in fresh or in salt water. If

Figure 19. Various types of faulty piling. A, poor, with overhanging ends and vegetation on ground providing a decay hazard; B, lumber carelessly handled and piled too closely to the ground; C, thick stock solid piled carelessly and exposed to direct sun and rain; D, reasonably well piled, but without a roof.

Figure 19b.

Figure 19c.

Figure 19d.

timbers are merely rafted or floated on water, some checking and splitting may originate in the parts above the waterline. Complete submersion is better, but in most instances, the material must be weighted down. Completely submerged timbers will not dry or decay so long as they are under water. Since they do not dry, they will not check or split.

It is Navy practice to store lignum vitae logs (used for fabricating propeller shaft bearings) under fresh water. Live-oak logs, cut when this species was preferred for sawn frames and knees, have been stored for many years in fresh or slightly brackish ponds at Atlantic and Gulf Coast Naval installations. However, wood held in some waters for long periods is subject to slight surface deterioration and to attack by such organisms as marine borers.

1.22. PILING. Methods of piling that facilitate drying are discussed in section 1.01. Other piling methods are available that prevent or retard drying. Close piling restricts air movement and retards drying. Solid piling permits material to dry only at the exposed ends, edges, and top of the pile. Solid piling favors the development of mold, stain, and decay. By dipping the lumber in an antistain chemical solution (sec. 2.3133, vol. I) the action of the organisms is retarded. The lumber should be piled edge to edge on stickers, as for air seasoning, but no attempt should be made to build vertical flues or chimneys into the pile. This method permits some surface drying but, in the main, considerably retards drying. As for air seasoning, a good roof should be provided for protection from the weather. Adequate pile foundations are also necessary. The best place to pile lumber is in unheated but ventilated sheds.

1.23. COATINGS. Moisture-resistant end and side coatings retard drying and help to maintain a high moisture content in the lumber. End coatings are discussed in section 1.017; the points mentioned apply as well to lumber held in storage. Coatings may also be applied to the faces and edges of lumber.

A coating composed of 2 pounds of secondary aluminum in 1 gallon of phenol-aldehyde synthetic resin vehicle has been found an effective treatment for reducing side and face checking of bending oak in tests. This coating is also considered suitable for preventing excessive checking on oak ship timbers intended for shaft logs and keels.

An application of aluminum paint to all surfaces of white oak bending stock at the sawmill within 24 hours after cutting the pieces from the logs, followed by a second coat within 24 hours of the first, is useful for protecting high-grade oak ship timbers and bending oak from excessive degrade. Laminated members stored in the weather or shipped in open cars are also protected in this manner.

The application of 2 coats of hull paint to thick white oak prevented the development of surface checks when the material was stored in a stickered pile for 171 days. The same coating reduced the surface checking to about one-sixth that of uncoated material, where the test material was exposed to sunshine and rain.

Two coats of wood sealer or two of water-repellent preservative reduced surface checking of white oak to about one-half that in the untreated wood. They interfere with gluing, caulking, and painting, however. Common salt applied to white oak stored in a stickered and protected pile reduced surface checking by two-thirds.

The effectiveness of face coatings in reducing surface checking is governed by their efficiency in retarding drying. Where side or face coatings are used, the lumber should be piled on stickers with openings to permit some circulation of air. Side or face coatings should be of such consistency that they will not be objectionable from the standpoint of handling and should not contain abrasive materials that will dull tools.

Shaping and installation of ship parts during construction exposes fresh green surfaces of frames, bulwark stanchions, stems, and other parts not only to the drying atmosphere, but to the full effect of sunshine and precipitation. Protective coatings are the only means of shielding parts at this stage. Most shipbuilders apply a coat of paint or a combined preservative and moisture repellent to the parts soon after placing them in position. A single coat of oil paint or an application of a moisture repellent has some slight protective value, depending upon the amount that is applied; a thick coating is more efficient than a thin one.

1.24. HYGROSCOPIC CHEMICALS. The use of hygroscopic chemicals in the seasoning of lumber is discussed in section 1.03. The principle involved is applicable to the storing of green material.

Dry spreading is the most practical means of applying the chemical for this purpose. The treated lumber is solid piled and allowed to stand from 1 week to 10 days, until the chemical has been absorbed by the wood. From 40 to 100 pounds of chemical per 1,000 board feet, which usually form a layer about one-fourth inch thick, are used. During this time, if the ends of the lumber have not already been end coated, it is advisable to apply an additional amount of the chemical, in the form of a saturated solution, to the ends about once each day.

When the dry chemical has been absorbed, the lumber can be piled for storage.

Urea or common salt may be used. Common salt is cheap and effective, but is corrosive to ferrous fastenings. There is a product on the market that consists of common salt plus a small amount of other chemicals that apparently reduce the corrosiveness of the common salt.

Urea is somewhat more expensive, effective with some species (notably Douglas fir), and in general, noncorrosive to ferrous metals. Common salt, in general, is somewhat more effective than urea when applied dry.

The chemicals can be used not only on the rough green lumber and timbers but on the shaped parts after they have been installed in the ship. The method of application must be modified to suit the new circumstances. Ships formerly were "salted" during construction by placing dry salt on shelves built in the frame bays. The shelves were octagonal in shape, the corners being cut off to allow air to circulate through the bays. Absorbing moisture from the air, the salt tended to spread over nearby wood surfaces, thus helping to keep the wood moist. The chemical can also be applied to ship parts by brushing or swabbing a saturated solution of it over the wood. It is difficult to apply much chemical by this method unless several applications are made. The protection from surface and end checking afforded by this method is not great.

1.25. SEASONED LUMBER.

1.250. GENERAL. Such parts of large boats as decking, planking, superstructure, and joinery require seasoned lumber for best results. Seasoned lumber is, as a rule, used for all parts of small boats. Seasoned material should be stored in such a manner as to maintain or, where necessary, improve its moisture content condition.

1.251. AIR-SEASONED LUMBER. Air-seasoned lumber should be stored so as to attain and maintain a moisture content of about 15 percent. Lumber at this moisture content may be solid piled clear of the ground on good foundations, either outdoors or in an unheated shed (fig. 20). If it is piled outdoors, it should be protected by a good roof, as for air-seasoning (sec. 1.013). If the lumber is not thoroughly air-seasoned, it should be piled on stickers, as for air seasoning, either outdoors or in a shed.

Figure 20. Open storage shed suitable for the storage of air-dry lumber.

1.252. KILN-DRIED LUMBER. Lumber kiln-dried to a moisture content of between 6 and 10 percent should be piled inside heated sheds or shops. It is preferable to pile with stickers, but solid piling is permissible, especially if the lumber is well seasoned. Kiln-dried lumber thus stored will absorb some moisture under humid conditions, but the amount will not be great. If the lumber is solid piled outdoors under a protecting roof, or in unheated sheds, the ends and exposed edges and faces of the boards will absorb moisture.

If lumber intended for use in such parts as planking and decking is kiln-dried to an excessively low moisture content, it should be conditioned to a moisture content of from 12 to 15 percent by piling it on stickers, as for air-seasoning, either outdoors or in an unheated shed.

1.253. STORING AND HANDLING PLYWOOD. The moisture content of plywood for boat and ship construction should be within the following percentage limits when finished ready for shipment:

Class	*Moisture Content Percent*
Preservative treated	8 to 15
Fire-retardant treated	10 to 17
Untreated	5 to 12

Upon receipt at the yard, plywood should be stored under conditions that will maintain the moisture content at levels approximating those to be expected in service. Storage under roof in dry, unheated buildings or sheds is suggested as an economical means of maintaining the plywood in a condition favorable for prompt use. The plywood may be piled solid (fig. 21), if uniformly conditioned as required by specifications, but the top of each pile should be covered with unserviceable plywood or boards to prevent warping of the top panels of the pile. Weighting the piles also helps in reducing the warping tendencies of the top panels. If suitable racks are available, plywood may be stored on edge, but exposed face surfaces should be protected from rapid moisture changes by waste panels to reduce warping.

Figure 21. Solid piles of plywood panels in a closed but unheated shed. Long-time storage of plywood panels under these conditions is not recommended because the stickers are too far apart and the tops are not weighted.

If it is necessary to store plywood temporarily in the open, the piles should be well off the ground and firmly supported by cross pieces no more than 18 inches apart. The piles should be covered in such a manner that the ends and sides, as well as the top, are protected from sun and rain by a paulin, or other effective covering. It is not advisable to store plywood outdoors for long periods lest conditions develop under the paulin or cover that favor development of mold and eventual decay. Careless storage can lead to increased difficulties with warping, face checking, and discoloration.

Since plywood is usually in the form of comparatively large, thin sheets with some end grain showing on all edges, it is rather easily damaged by careless mechanical handling. It is sometimes possible to protect the corners of the piles by suitable bracing or supports, but continuous care is required in handling if the plywood is to reach the point of installation free from mechanical damage to edges and ends.

1.26. MANUFACTURING CONDITIONS. Boats and ships are built under various conditions, from closed, heated shops to open ways. Preferred practice is to build in open sheds or outdoors. The conditions of manufacture affect the moisture content of the wood and the development of defects. A boat in service will never be exposed to atmospheric conditions so dry as the air in a heated building; it is therefore disadvantageous to expose any wood parts except interior parts, cabin finish, or cabinet work to the dry air of a heated shop. Hulls under construction or repair should never be exposed to a direct source of heat, such as a space heater or steam pipes. From the standpoint of best woodworking practice, it is preferable to make and assemble wood boat and ship parts in an unheated enclosure. Such enclosures should be heated during cold weather only to the minimum temperature consistent with the comfort of the workmen.

1.3. REFERENCES.

(1–1) FOREST PRODUCTS LABORATORY

1951. "Properties of Wood Related to Drying," *Report No. 1900–1,* 10 pp., illus.

(1–2) ———

1952. "Auxiliary Kiln Equipment," *Report No. R1900–3,* 15 pp., illus.

(1–3) LOUGHBOROUGH, W. KARL

1948. "Chemical Seasoning: Its Effectiveness and Present Status," *Report No. D1721,* 23 pp., illus.

(1–4) PECK, E. C.

1950. "Air Drying of Lumber," *Forest Products Laboratory Report No. D1657,* 93 pp., illus.

1952. "Storage and Handling of Lumber," *Forest Products Laboratory Report No. D1919,* 31 pp., illus.

CHAPTER 2. WOOD TREATING PROCESSES FOR PROTECTION AND SERVICE

2.0. GENERAL.

It is now generally recognized in the boat and shipbuilding industry that preservative treatment is of considerable value in protecting wood from decay and marine borers, particularly if the ships are not stored undercover when laid up. Preservatives are widely employed for the protection of wood exposed to infection or borer attack unless the heartwood of highly resistant species such as teak, green heart white oak, or the cedars are involved. Preservative treatment is essential when such species as red oak, the Philippine or African mahoganies, laminated oak, cedar planking, or plywoods, in which sapwood is present.

Military and Navy Specifications define the quality of various kinds of preservatives and prescribe when and how they are to be used in ships and boats. The purpose of this chapter, therefore, is to describe the kinds of preservatives that are available and discuss the merits and shortcomings of common wood treating methods.

While preservative treatment prolongs the life of wood, much of the benefit is dependent on proper processing. Haphazard treating methods or careless handling of treated wood can defeat the best intentions of the builder. Neither the perfect preservative nor the perfect treating method has yet been found. It is necessary that builders use the available preservatives and treating processes with understanding and effectiveness.

2.1. PRESERVATIVE TREATMENT OF WOOD.

The life of wood used in many parts of ships and boats can be considerably extended if it has been treated with suitable preservatives. The degree of protection given it, however, depends upon the retention (quantity in pounds per cubic foot of wood) of the preservative that is put into the wood and the thoroughness of its distribution in the wood. Poor distribution (penetration) and insufficient retention of preservative can be the reason why wood is not protected adequately when exposed to conditions highly favorable to growth of decay fungi and marine borers (sec. 3.08, vol. I).

2.10. PRESERVATIVE CHEMICALS. For protection from fungi, insects, and marine borers, preservatives must either be toxic to or inhibit the growth of the organism. The preservative must also be able to remain in the wood for long periods in order to provide protection when it is needed. Preservatives should not, however, be a health hazard to those who apply them or to those who handle and use the treated wood. For special uses, preservatives must also meet other requirements, such as cleanness, compatibility with paint and other wood finishes, freedom from objectionable odor, and lack of fire hazard.

Some preservatives are more effective than others. All possess certain disadvantages that limit their use, as well as advantages that make them especially suitable for specific purposes.

Wood preservatives fall into two general classes. They are: (1) Oils that are relatively insoluble in water; and (2) inorganic salts that are injected into the wood in the form of water solutions.

2.11. PRESERVATIVES BORNE IN OILS. Preservative chemicals borne in oils are resistant to leaching and are suitable for outdoor uses, particularly where the treated wood is in contact with water. Oils do not cause the wood to swell. The heavier oils tend to protect the wood from weathering Residual oils remaining in the wood after treatment bleed through and interfere with paint coatings. The properties of the treated wood with regard to color, odor, cleanness, and combustibility are influenced by the presence of residual oils.

2.110. COAL-TAR CREOSOTE. Coal-tar creosote is a black or brownish oil made by distilling coal tar. The character of the tar used, the method of distillation, and the temperature range in which the creosote fraction is collected all influence the character of the creosote oil. The character of the various coal-tar creosotes available, therefore, may vary to a considerable extent. Small differences in character, however, do not prevent creosotes from giving good service, and satisfactory results in preventing decay may be expected, as a general rule, from any coal-tar creosote that complies with the requirements of Federal Specification TT-W-556.

Advantages of coal-tar creosote are: (1) Its high toxicity to wood-destroying organisms; (2) its relative insolubility in water and its low volatility, which impart to it a great degree of permanence under the most varied use conditions; (3) its ease of application; (4) the ease with which its depth of penetration can be determined; (5) it is not corrosive to metals; (6) its general availability and relatively low cost when purchased in wholesale quantities; and (7) its long record of satisfactory use.

The color and odor of creosote and the fact that wood treated with it usually cannot be painted satisfactorily make it unsuitable for most boat uses. Creosote is subject to marine fouling. Studies are now underway on procedures to improve the paintability of creosoted wood. Creosote is used for the treatment of some wood barges, also in floating dry docks constructed of wood.

Crystal-free coal-tar creosote is creosote from which some of the crystal-forming materials have been removed, so that the oil will flow freely at ordinary temperatures and will not deposit crystals in its container. Oils of this type are more convenient to handle than ordinary creosotes, but their general properties and effectiveness are similar to those of the ordinary coal-tar creosote. They are intended primarily for brush and spray applications. American Wood-Preservers' Association Standard P7 and Federal Specification TT–W–560 cover crystal-free creosotes.

2.111. CREOSOTE SOLUTIONS. For many years, either coal tar or petroleum oil has been mixed with coal-tar creosote, in various proportions, as a means of lowering preservative costs. These creosote solutions have a satisfactory record of performance and are covered by Federal Specification TT–W–566, "Creosote-Coal-Tar Solution," and Federal Specification TT–W–568, "Creosote-Petroleum Solution."

Creosote solutions not only cost less than straight creosote but also tend to reduce weathering and checking of the treated wood. The solutions may have a greater tendency to accumulate on the surface of the treated wood (bleed) and may penetrate the wood with greater difficulty, particularly since they generally are more viscous than straight creosote. Higher temperatures and pressures during treatment, when they can safely be used, will often improve penetration of solutions of high viscosity.

Creosote solutions, while important wood preservatives, have some of the objectionable features of creosote and are, therefore, unsuited for most boat uses.

2.112. PENTACHLOROPHENOL SOLUTIONS AND COPPER NAPHTHENATE SOLUTIONS. The use of petroleum oils fortified with chlorinated phenols, principally pentachlorophenol, or with copper napthenate as wood preservatives, became large-scale during and following World War II. These preservatives were first used primarily for surface applications. The chlorinated phenols in volatile light-colored solvents, such as mineral spirits, were first used for window sash and millwork that required a clean, nonswelling, and paintable treatment. Copper naphthenate and pentachlorophenol solutions are specified for the brush, dip, or soak treatment of wood boat parts that do not require pressure treatment. They are also used on surfaces exposed by machining of treated wood.

The petroleum oil customarily used in these solutions for pressure treatment varies from the diesel-oil type to the heavier types ordinarily used in creosote-petroleum solutions. Wood pressure treated with solutions containing these oils is not paintable unless suitable conditioning methods are employed after treatment to remove the residual oil. To meet boat and ship requirements for cleanliness, paintability, and solvent drying of the treated wood, lighter petroleum solvents are used.

Federal Specification TT-W-571 covers the pressure treatment of wood with solutions of pentachlorophenol; Federal Specification TT–W–570 covers pentachlorophenol; and American Wood-Preservers' Association Standard P8 covers both pentachlorophenol and copper naphthenate.

Pentachlorophenol solutions generally contain 5 percent of this chemical; copper naphthenate solutions have a copper metal concentration varying from 0.5 to 3.0 percent for different use requirements. The performance of these preservatives is influenced by the properties of the petroleum oil in the treating solution; best protection is obtained with fairly heavy oils of low volatility. Petroleum oils that remain in the wood in substantial quantities after treatment, however, are likely to interfere with painting.

Petroleum oils containing 5 percent of pentachlorophenol or copper naphthenate equivalent to 0.5 percent or more of copper metal provide a high degree of protection from decay fungi and termites when the wood is pressure treated. Wood well impregnated with copper naphthenate solution resists marine borer attack only for limited periods. Southern yellow pine heavily impregnated with petroleum oil containing 5 percent pentachlorophenol is severely riddled by marine borers within 1 year.

Pentachlorophenol does not appreciably alter the natural color of the wood, but copper naphthenate gives it a green color. The properties of the wood treated with petroleum oils containing these preservatives, with respect to cleanness, paintability, color, odor, and combustibility, depend largely upon the properties of the petroleum oil used. Pentachlorophenol in both the dry concentrate form and in solutions irritates the skin of workers, but careful handling and protective clothing help to avoid harmful effects. The use of "bloom" preventives, such as ester gum, is required in pentachlorophenol solutions with volatile solvents in order to prevent the formation of objectionable crystals of the preservative on the surface of the wood after treatment. This loss of pentachlorophenol through blooming reduces the effectiveness of the treatment.

2.113. WATER-REPELLENT PRESERVATIVES. Preservative oils containing water-repellent components are sold under various trade names, principally for the treatment of window sash and other millwork. Such preservatives generally consist of volatile solvents, such as mineral spirits, that do not cause appreciable swelling of the wood and not less than 10 percent or more than 25 percent of nonvolatile matter, including the preservative and water-repellent components. The preservative chemicals may be not less than 5 percent of pentachlorophenol or not less than 2 percent of metallic copper in the form of copper naphthenate.

Effective water-repellent preservatives retard moisture changes in the wood but do not prevent them. They thus help reduce dimensional changes in wood exposed to rainwater or dampness for short periods, but not when it is in contact with the ground or other sources of prolonged dampness. As with any wood preservative, their effectiveness in protecting wood from decay depends upon the thoroughness of the method of application.

Preservatives having large amounts of wax as a water repellent have been found to interfere with adhesion of marine paints, glues, and some calking compounds, and their use for boat and ship construction is therefore limited.

2.12. WATERBORNE PRESERVATIVES. Standard wood preservatives used in water solution include chromated zinc chloride, copperized chromated zinc chloride, Wolman salts (Tanalith), acid copper chromate (Celcure), zinc meta arsenite, ammoniacal copper arsenite (Chemonite), chromated zinc arsenate (Boliden salt), and chromated copper arsenate (Greensalt or Erdalith). These preservatives are used principally for the pressure treatment of wood that will not be in contact with the ground or water or that will be painted.

As a general rule, waterborne preservatives are less resistant to leaching and do not perform so satisfactorily as the preservative oils under conditions favorable to leaching. The leaching resistance of some of these preservatives has been developed to the extent that good performance can be expected in ground contact or in other wet installations. They are not considered equal in effectiveness to creosote, however, under such conditions. On the other hand, waterborne preservatives can give indefinitely long life where not subject to leaching. Paint retards leaching considerably. The treated wood, when seasoned, is only slightly heavier than untreated wood. This is an advantage in watercraft where weight is often a critical factor.

A major consideration favoring waterborne

preservatives for ship and boat use is that they do not add to the flaming hazard, while some preservative oils do. Tests indicate that waterborne inorganic salts cause wood to ignite more slowly, slow down the rate of flame spread, and reduce the intensity of the fire. Waterborne preservatives containing copper or chromium, however, tend to prolong the glowing of wood after flaming has stopped. Preservative oils, on the other hand, definitely increase flaming time of wood and the rate of flame spread, although tests have shown that interior wood temperatures remain low, and the oil treated members are less readily consumed in a fire.

Waterborne preservatives leave the wood surface comparatively clean, paintable, and free from objectionable odor. With several exceptions, they must be used at low treating temperatures (100° to 140° F.) because of their instability at the higher temperatures. This may involve some difficulty when higher temperatures are needed to obtain good penetration in such woods as Douglas fir. Since a large quantity of water is added during treatment with waterborne preservatives, the wood must be dried after treatment to the moisture content required for use. Recommended retentions of waterborne preservatives in lumber are shown in table 4.

TABLE 4.—RECOMMENDED RETENTIONS OF WATERBORNE PRESERVATIVES FOR VARIOUS USES

Product and service condition	Recommended minimum net retention (pounds per cubic foot)						
	Celcure (acid cupric chromate)	Chemonite (ammoniacal copper arsenite)	Chromated zinc chloride	Tanalith	Boliden salt (chromated zinc arsenate)	Copperized chromated zinc chloride	Greensalt
Lumber and structural timber:							
For use under moderate leaching conditions	0.75	0.50	1.15	0.55	1.00	1.15	0.75
For use not in contact with ground or water	.50	.30	.75	.35	.50	.75	.35
For boat & ship members (special) (marine borer protection)	1.5	1.5	----	----	----	----	1.5

2.120. ZINC CHLORIDE PRESERVATIVES. Chromated zinc chloride is covered in Federal Specification TT–W–551 and American Wood-Preservers' Association Standard P5. Copperized chromated zinc chloride is included in American Wood-Preservers' Association Standard P5 and chromated zinc chloride (FR) in P10.

Chromated zinc chloride is sometimes used as a fire retardant for lumber, but at retentions higher than those used only to preserve wood from decay. Waterborne preservatives with the retentions generally used tend to increase the resistance of the wood to flaming and glowing.

Copperized chromated zinc chloride has shown better permanence in laboratory and field tests than chromated zinc chloride. Some degree of protection against marine borers is indicated, and the two preservatives compare favorably in other properties. Chromated zinc chloride (FR) in retentions of 1½ to 3 pounds per cubic foot of wood provides combined protection from fire, decay, and insect attack.

2.121. WOLMAN SALTS (TANALITH). Wolman salt (Tanalith) is covered by U. S. Patent No. 1,957,873 (1934); and by Federal Specification TT–W–573 and American Wood-Preservers' Association Standard P5. Service records on various types of wood structures treated with this preservative, and not used in contact with the ground show good performance.

2.122. ACID COPPER CHROMATE (CELCURE). Acid copper chromate (Celcure) is covered by U. S. Patents Nos. 1,684,222 (1926) and 2,041,655 (1936) and Federal Specification TT–W–546. Tests on stakes and posts exposed to decay and termite attack indicate that wood well impregnated with Celcure gives good service. Wood thoroughly impregnated with at least 1.5 pounds or more of Celcure per cubic foot has some resistance to marine borer attack.

2.123. AMMONIACAL COPPER ARSENITE (CHEMONITE). Ammoniacal copper arsenite (Chemonite) is covered by U. S. Patent No. 2,149,284 (1939) and Federal Specification TT–W–549. It has been in commercial use for about 15 years.

Service records for land structures show that this preservative provides good protection against decay and termites. Chemonite adds somewhat to the life of wood exposed to attack by marine borers.

2.124. CHROMATED COPPER ARSENATE (GREENSALT OR ERDALITH). Chromated copper arsenate (Greensalt or Erdalith) is covered by U. S. Patent No. 2,106,978 (1938) and Standard P5

of the American Wood-Preservers' Association. Service data on poles, posts, and stakes treated with this preservative show good protection from decay and termites over a period of 12 years or less. Limited tests show that Greensalt provides some protection from marine borers.

2.125. CHROMATED ZINC ARSENATE (BOLIDEN SALT). Chromated zinc arsenate (Boliden Salt) is covered by U. S. Patent No. 2,139,747 (1938) and Standard P5 of the American Wood-Preservers' Association. Test data on stakes show that chromated zinc arsenate, in retentions of 0.33 pound per cubic foot or higher, provides good protection against decay and termites.

2.13. METHODS OF APPLYING PRESERVATIVES. The degree of protection given by any good wood preservative depends upon the method of application and the resulting preservative retention and penetration obtained. In descending order of effectiveness, methods of treatment are: Pressure treatment; hot and cold bath; cold soaking; vacuum; dipping; and brushing.

Pressure treatment is not always practicable for boat members, particularly when considerable cutting and boring of the wood is done during assembly. Less thorough nonpressure methods can often be used to advantage in such cases.

With any process, it is important to see that the treatment is well done. This means careful and efficient inspection (sec. 2.17) to assure: (1) That the wood is in a satisfactory and treatable condition; (2) that the preservative is of standard quality and thoroughly applied; (3) that the time, temperatures, and pressures used during treatment comply with specifications and are not harmful to the wood; (4) that retentions and penetrations meet the minimum requirements of the specifications; (5) that the wood after treatment is undamaged and suitable for its required use; and (6) that in the case of waterborne preservative treatments, the subsequent kiln drying schedule is such that a minimum of seasoning degrade occurs in the treated lumber.

2.130. PREPARING TIMBER FOR PRESERVATIVE TREATMENT. For satisfactory results with any treating process, the timber must be sound and in proper condition. For best results with most processes, wood should be well seasoned. The surface application of preservatives to cold or wet surfaces is to be discouraged. Since preservatives will not make weak timber strong nor restore the strength of timber that has been partially destroyed by decay, only sound timber should be used. Except in small material, the preservative and the heat of the treating process cannot be expected to kill all fungus growth in an infected stick unless care is taken to attain complete sterilization. Any live fungus left in the wood may continue to grow after treatment, perhaps destroying completely the unprotected interior of the timber.

All cutting, framing, and boring of holes should be done before treatment, if practicable. In shipbuilding, this requirement often determines the type of treatment given to the wood. Where pieces can be cut, shaped, and bored on a production scale, pressure treatments can well be utilized if equipment is available; if it is not, the more convenient nonpressure treatments can be resorted to, although the results will be less satisfactory. Where, however, it is necessary to saw, adze, bore, or otherwise fit large wood sections, such as floor timbers and frames on the ship, the effectiveness of chemical pretreatments is largely lost by the exposure of untreated interior wood that results unless the surfaces exposed by cutting are again treated.

In the case of sawn ends, holes for fastenings, and other faired surfaces, the application of brush, dip, spray, or oil can preservatives, such as those covered in section 2.112 will help extend maintenance free service life.

Consideration should be given so far as practicable, by both designer and builder, in the development of plans for the ship to prevent loss of effectiveness of the pretreatment. In the event cutting away of pretreated wood becomes necessary for faying surfaces, protection can be restored by application of fungicide fortified bedding compound.

2.131. CONDITIONING GREEN TIMBER FOR PRESSURE TREATMENT. When green timber is to be treated under pressure, either of two commonly used methods recognized in AWPA and Federal specifications for conditioning may be selected. One is the steaming-and-vacuum process, which is employed mainly for southern yellow pine treated with oils and waterborne preservatives and Douglas fir treated with waterborne preservatives. The other is the

Boulton or boiling-under-vacuum process, which is used for Douglas fir and to some extent for hardwoods.

In the steaming process, the green timber is steamed in the treating cylinder for several hours, usually at a maximum temperature of 259° F. for pine and 240° F. for Douglas fir, and when the steaming is completed, a vacuum is immediately applied. During the steaming period, the outer part of the wood is heated to a temperature approaching that of the steam, and the subsequent vacuum lowers the boiling point so much that part of the water is evaporated or is forced out of the wood by the steam produced when the vacuum is applied. The steaming and vacuum periods employed depend upon the size, species, and moisture content of the wood. The steaming method usually reduces the moisture content of green wood somewhat, and the heating assists greatly in getting the preservative to penetrate, but it does not dry the wood to the "seasoned" condition. If the steaming period is sufficient, it will also sterilize the wood.

In the Boulton or boiling-under-vacuum method of partial seasoning, the timber is heated under vacuum in the preservative oil, such as creosote or pentachlorophenol solution, usually at temperatures of about 180° to 220° F. This temperature range, lower than that of the steaming process, is of considerable advantage in treating woods that are especially susceptible to injury by high temperatures. The Boulton method, however, usually removes only a very limited amount of moisture from the heartwood.

2.132. PRESSURE PROCESS. In commercial practice, wood is most often treated by surrounding it with preservative in high pressure cylinders (fig. 22) and applying pressure to the preservative. A number of pressure processes differ from one another in a few details, but the general principle is the same in all. The wood is run on steel cars into a long steel cylinder, which is then closed and filled with preservative. Pressure forces preservative into the wood. A considerable amount of preservative is absorbed which results in a relatively deep penetration. Two principal types of pressure treatment, the full-cell (Bethell) process and the empty-cell processes, are in common use.

Figure 22. A typical cylinder for pressure preservative treatment.

The full-cell process involves an initial vacuum followed by impregnation under pressure. It is used when the retention of a maximum quantity of preservative is desired. For timbers to be treated with creosote and exposed to marine borers, this process is standard. Waterborne preservatives are generally applied by this process, and control over preservative retention is obtained by regulating the concentration of the treating solution.

The object of empty-cell processes is to obtain deep penetration for a specified, and usually low, net retention of preservative. For treatment with oil preservatives, the empty-cell process should always be used if it will result in the desired retention. Two empty-cell treatments, the Rueping and the Lowry processes, are commonly employed, both of which use the expansive force of compressed air to drive out part of the preservative absorbed during the pressure period.

The Rueping process involves preliminary air pressure, while the Lowry process does not. The two empty-cell processes are practically the same, except for this difference. Both are usually followed by a final vacuum to help the air trapped in the wood force out part of the preservative absorbed during the pressure period. The choice of an empty-cell process should be left to the treating engineer, and the selection would depend on the species, size, and condition of the wood as well as on the retention requirements in the specifications.

2.133. NONPRESSURE TREATMENTS. The numerous nonpressure processes differ widely in the retentions and penetrations of preservative they give and consequently in the degree of protection that they provide to the treated wood. When similar retentions and penetrations are achieved, wood treated by a nonpressure method should have a service life comparable to that of wood treated by pressure. The results of nonpressure treatments, especially the superficial treatments, are not so generally satisfactory as the results of pressure treatment. Even the less effective, superficial nonpressure processes, however, serve a useful purpose when more thorough treatments are impractical.

Nonpressure methods for application to boat members in general consist of: (1) Superficial applications of preservative oils by spraying, brushing, or brief dipping; (2) Cold soaking in preservative oils; (3) Various adaptations of the hot-and-cold-bath process; and (4) Vacuum treatment.

The simplest treatment is the brush or spray application of a preservative to the wood. It is also the least effective The oil should be flooded over the wood, rather than merely painted upon it, and every check and depression in the wood should be thoroughly filled with the preservative, since exposed untreated wood is an open door to fungi. Rough lumber may require as much as 10 gallons of oil to 1,000 square feet of surface, but surfaced lumber requires considerably less. The transverse penetrations obtained will usually be less than one-sixteenth inch. Brush and spray treatments should be used only when more effective treatments cannot be employed. Preservative oils are suitable for brushing and spraying.

Dipping for a few seconds to several minutes in a preservative gives greater assurance than brushing or spraying that all surfaces and checks are thoroughly coated with the oil, and the oil usually penetrates the wood a little deeper. Preservative penetrates into end surfaces of ponderosa pine sapwood as much as 1 to 3 inches. End penetration is much less, however, in such woods as southern yellow pine and Douglas fir, particularly in the heartwood. Transverse penetration is very shallow, but usually greater than that obtained in spray or brush applications. Since the exposed end surfaces at joints are the most vulnerable to decay, good end penetration is especially advantageous. Dip applications provide very limited protection to wood used in contact with the ground or under very moist conditions. They do have value, however, for exterior or interior wood members that are painted.

Cold soaking of well seasoned wood for several hours or days in low-viscosity preservative oils has been tried with varying success on fence posts, lumber, and timbers. Pine sapwood treated by cold soaking in a solution of 5 percent pentachlorophenol in No. 2 fuel oil for 24 to 48 hours or longer, is reasonably well penetrated with the preservative. Retentions are generally below those required in pressure treatment. Most species are not treated so satisfactorily as the pines by cold soaking. Soaking periods of several days have shown better results than those lasting only a few hours. In 2- by 8-inch lumber of 12 commercial species cold soaked for 24 hours, preservative oil retentions varied from 0.5 pound to 5.4 pounds per cubic foot for different woods. The cold soaking method has had some use for treating boat members of irregular form which could not be treated in pressure cylinders (fig. 23).

Figure 23. Tank used for the cold soaking treatment of laminated frames for minesweepers.

Cold soaking with copper naphthenate has been used at Naval Shipyards on parts to be painted. Preservative bleeding through the paint has been traced to too high concentration of copper naphthenate in the vat, due to solvent evaporation. The bleeding through has been avoided by placing covers on the dipping tank to maintain the proper preservative-solvent ratio.

The hot- and cold-bath treatment with coal-tar creosote and other oil preservatives is among the more effective of the nonpressure processes, and the thoroughness of the treatment obtainable may, under favorable conditions, approach that of the pressure processes. The treatment consists of heating the wood in the preservative in an open tank for several hours, then quickly submerging it in cold preservative for several hours. Two methods may be used: (1) Transferring the wood at the proper time from a hot-bath tank to a cold-bath tank; or (2) draining the hot preservative from a single tank and quickly refilling the tank with cold preservative. The same result can also be accomplished, although more slowly, by shutting off the heat at the proper time and allowing the wood and the hot preservative to cool together.

During the hot bath, the heating causes the air in the wood to expand, forcing some out. Heating the wood also improves the penetration of the preservative. When the cooling takes place, whether it is sudden or slow, the air in the wood contracts. A partial vacuum is thus created and the liquid is forced into the wood by the pressure of the atmosphere. Some preservative is absorbed during the hot bath, but most of it is taken up during the cold bath. Coal-tar creosote and pentachlorophenol solutions are the preservatives ordinarily used in the conventional hot- and cold-bath process, but water solutions of preservatives such as zinc chloride solution may also be employed if the solution is kept at uniform strength and the water-soluble preservative is not adversely affected by heating. For the preservatives that cannot safely be heated, the process must be modified.

Certain preservative chemicals, such as chromium salts in water solutions, cannot safely be heated to high temperatures because at such temperatures, some of the chemical may precipitate out of the solution or the solution may be otherwise impaired. Such solutions are not practical for use in the hot bath. This difficulty has been solved in some open-tank plants by heating the wood with steam instead of in the preservative. The tank is covered after the wood is in place, and steam is applied at atmospheric pressure for a few hours. At the conclusion of the steaming period, cold preservative solution is pumped into the tank and absorption takes place as the wood cools (see U. S. Patent No. 2,235,822 (1941)).

Preservatives consisting of toxic chemicals in volatile organic solvents cannot safely be heated in open tanks because of the loss of solvent by evaporation and the considerable danger of fire or explosion. The obstacle is sometimes overcome by heating the wood in a dry kiln or hot box and then immersing it quickly in cold preservative. A kiln-heating process for this purpose, in which the relative humidity in the kiln is controlled during the heating process, is covered by U. S. Patent 1,991,811 (1935).

2.14. TREATING SPECIFICATIONS. Specifications for the treatment of various wood products by pressure processes have been developed by the American Wood-Preservers' Association. These specifications limit pressures, temperatures, and time during conditioning and treatment, in order to avoid conditions that will seriously injure the wood. They also contain minimum requirements as to preservative retentions and penetrations and recommendations for handling wood after treatment, as means of providing a quality product.

The specifications are rather broad in some respects, allowing the purchaser some latitude in specifying the details of his individual requirements. The purchaser should, however, avoid limiting the operator of the treating plant so closely that he cannot do a good treating job, and not require treating conditions so severe that they will damage the wood. Federal Specification TT–W–571 lists recommended treatments for pressure-treated wood products. Military Specifications MIL–P–19550 covers the preservative treatment of plywood for boat and ship use, and MIL–W–17445 describes the preservative treatment and subsequent laminating of red oak and of white oak with unlimited sapwood for ship and boat use.

Oak lumber that is to be pressure treated before it is glued for ship and boat use is air dried or kiln dried and uniformly surfaced on each face to a thickness slightly more than that required for gluing. For protection from decay, pressure treatment with 0.75 pound of Tanalith per cubic foot is required. Complete penetration of the sapwood is required in both red oak and white oak. There is no heartwood penetration requirement for white oak, but 80 percent of the annual rings of red oak heartwood must be penetrated.

In surfacing laminations after treatment,

the amount of wood removed from each face should be as small as practicable in order to retain the effectiveness of the preservative treatment. The final surfacing must be accomplished not more than 24 hours before gluing is started.

2.140. PRESSURE TREATMENT OF LAMINATED MEMBERS. Research on the pressure treatment of laminated timbers, after gluing, has shown that the preservative retentions required for conventional timbers can usually be obtained in laminated timbers, although the penetration of preservative is frequently blocked by the glue lines (fig. 24).

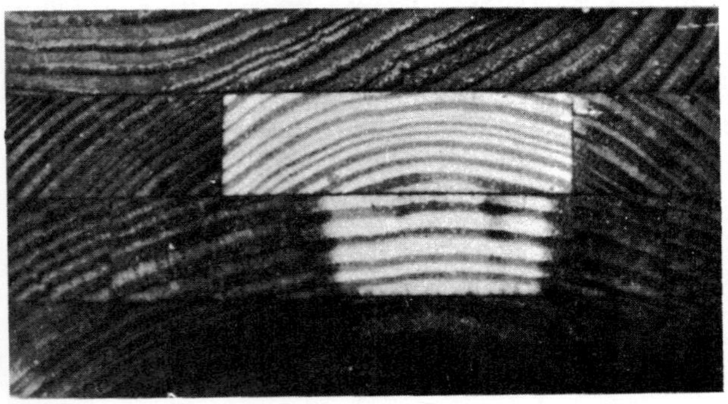

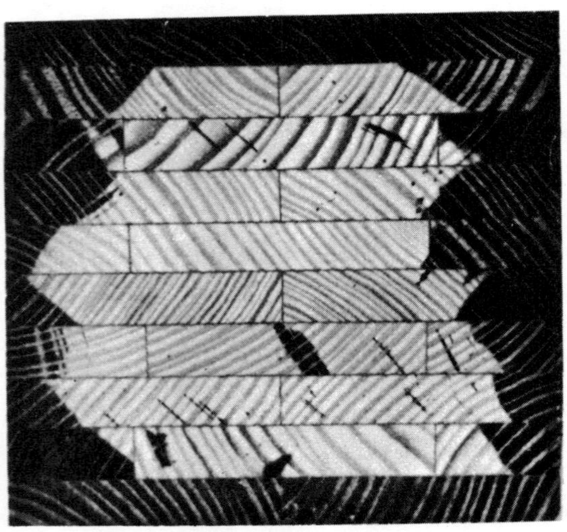

Figure 24. Cross sections of pressure-treated laminated southern yellow pine timbers show blocking effect of glue lines in penetration of preservative.

2.141. PRESSURE TREATMENT OF PLYWOOD. Preservative treatment of ship and boat plywood is, generally by full-cell process with one of the waterborne preservatives, listed in table 5. To insure adequate penetration, the retention, as measured by gage, should be no less than that specified in table 5, and the retention of treating solution following treatment should not be less than 20 pounds per cubic foot.

To avoid impairing the properties of the plywood, the treating temperature should not exceed 110° F. (sec. 2.18). Pressure treatment of plywood is covered by Military Specification MIL–P–19550.

TABLE 5.
WATERBORNE PRESERVATIVES
FULL CELL PRESSURE TREATMENT FOR PLYWOOD

Preservative	Minimum retention
	Lb. per cu. ft.
Chromated zinc chloride	1.15
Tanalith	.55
Acid cupric chromate (Celcure)	.75
Ammoniacal copper arsenite (Chemonite)	.50
Chromated copper arsenite (Greensalt erdalith)	.75
Chromated zinc arsenate (Boliden salt)	1.00

2.142. REQUIREMENTS FOR WOOD PRESERVATIVE TREATMENT IN WOOD HULL SHIPS. As typical of the preservative treatment afforded various wood members to be used in wood hull ships, the following examples are given:

Red oak or southern pine exterior hull sheathing, keel shoe, also red oak bilge keels and similar members requiring marine borer protection must be surfaced four sides and given pressure preservative treatment with one of the following water borne preservatives:

Schedule of Pressure Treatment for Marine Borer and Decay Protection

Preservative	Retention (Lbs. cu. ft.)	Material specification	Procedure specification
		AWPA Stds.	
Erdalith (Greensalt)	1.5	P-5-57	TT-W-571c
Acid Cupric Chromate (Clecure)	1.5	TT-W-546	TT-W-571c
Chemonite	1.5	TT-W-549	TT-W-571c

Treated red oak or southern pine material

described above is also used for chafing pieces, and chain locker lining to protect against decay in these members.

When the wood is of a species requiring pressure preservative treatment, both courses of diagonal inner planking are surfaced four sides and pressure treated as follows:

Schedule of Pressure Treatment for Decay Protection of Nondurable Woods Used as Diagonal Planking

Preservative	Retention (Lbs. cu. ft.)	Material specifications	Procedure specifications
Tanalith (Wolman salts)	0.55	TT-W-573	TT-W-571c
Celcure (acid-cupric-chromate)	.75	TT-W-546	TT-W-571c
Chromated zinc chloride	1.15	TT-W-551	TT-W-571c
Chemonite (ammonical-copper-arsenite	.45	TT-W-549	TT-W-571c
Copper napthenate 0.75 metallic copper by weight	10.00	American Wood Preservers Association Standards	TT-W-571c
Pentachlorophenol (5% solution)	10.00	TT-W-570	TT-W-571c

Lumber or plywood used for doors, cupboards, tables, or components receiving a bright finish require no preservative treatment. Plywood other than that noted above must be given a full-cell pressure preservative treatment with waterborne preservatives and minimum retentions of Military Specification MIL–P–19550, as shown in table 5. Tanalith (Wolman salts) is one plywood treatment for which considerable service experience exists, and also laboratory testing which shows that reduction in plywood strength is less severe than with several other waterborne preservative treatments used for treating structural plywood.

In the case of plywood treated with waterborne preservative, kiln drying after drying is necessary in order to remove the water used as the vehicle to carry the preservative salt into the wood. Moisture content should not exceed that outlined for untreated material in detail hull specification.

All solid and laminated wood and plywood members not specifically exempted from treatment, and for which no other preservative treatment is specified are given preservative treatments as follows:

Schedule of Preservative Soak Treatment (30 Minutes) for Decay Protection

Preservative	Applicable MIL specification
Copper napthenate	MIL-W-18142 (type A)
Pentachlorophenol	MIL-W-18142 (type B)

Where treatment is in accordance with any of the above schedules, all surfacing must be accomplished prior to treatment, as far as practicable. Any surface cut, bored, faired, or sanded after any scheduled preservative treatments must be thoroughly brushed or dipped in a preservative conforming to Military Specification MIL–W–18142. End grain surfaces are particularly decay susceptible and thus, require special attention. Treat surfaces which are to be varnished with type B, clear preservatives. Brush or other application of Military Specification MIL–W–18142 preservative is not required for bored holes in planking provided planking fasteners are dipped in sealing compound (wood bedding) conforming to Military Specification MIL–S–19653, type I, class I, before being driven.

Preservatives applied by brushing or soaking must be allowed a minimum drying time of 72 hours prior to the application of any paint. Allow longer time, if necessary, to assure good paintability characteristics.

Wooden plugs, dowels, and stop waters are given a preservative soak treatment. Decking is similarly treated, and outgaged after treatment. After sanding, decks do not require retreatment. The inner and outer planking above the light load water line is given a preservative soak treatment.

A thorough brush treatment with the same preservatives used in the schedule for soak treatment is applied to all surfaces of the frames and stem. The inner and outer planking below the light load water line, as well as the bilges and chain locker must not be painted, but must receive a thorough brush treatment with a Military Specification MIL–W–18142 wood preservative. Make periodic checks of dipping vats containing Military Specification MIL–W–18142 preservative to assure that solvent evaporation has not resulted in a concentrated solution which could cause painting difficulty.

For the oil-borne preservatives, copper napthenate, and pentachlorophenol, the solvent vehicle should be removed by air seasoning. Kiln drying is not recommended for this purpose because of the danger involved in handling volatile solvents in the presence of treated air.

Sealing Compound (Wood Bedding) fortified, conforming to Military Specification MIL–S–19653, type I, class I, must be used beneath butt blocks, gunwhale guards, and superstructure sills and coamings; on the outboard faces of transverse frames and transom stiffeners, the frame ends, the faying surfaces of deck and deck beams and on the faying surfaces of mechanical joints where an adhesive is not specified; and in general, on all faying surfaces exposed to moisture. Sealing compound (Wood Bedding) fortified between the layers of hull planking must conform to Military Specification MIL–S–19653, type I, class I. Sealing compound (Wood Bedding), fortified is not required by frame courses.

2.15. EFFECT OF TREATMENT ON STRENGTH. Wood preservatives are not regarded as harmful in themselves; however, the treatment necessary to inject them into the wood by pressure, if unusually severe or if not properly carried out, may result in considerable loss in strength of the wood. Preservative treated oak lumber for marine laminating requires that retention, penetration, rough surfacing prior to treatment with a waterborne preservative, and subsequent kiln drying must all be accomplished with great care so that maximum decay resistance and structural strength is not impaired.

Lumber for underwater sheathing, treated for marine borer protection with such waterborne chemicals as Erdilith, Chemonite, or Celcure is generally specified to be dried to a moisture content of 15 ± 3 percent, in order that copper bottom paints will adhere. Here again emphasis is placed on the need for most careful control of kiln drying. Treated plywood also requires a mild drying schedule in order that toughness and static bending quality is not reduced and checking is avoided.

The placing of orders for pressure preservative treated lumber and plywood should be accomplished with sufficient lead time to meet ship production schedules, and allow treating and drying periods which will result in a minimum of degrade.

Results of tests of pressure-treated timbers and lumber show reductions in strength sufficient to affect working stresses. Such reductions range up to 25 percent, or occasionally more, for stress in extreme fiber in bending and in compression perpendicular to grain. Strength in compression parallel to grain ordinarily is affected less, and modulus of elasticity very little. The effect on resistance to horizontal shear can be estimated by inspection for shakes and checks after treatment.

Strength losses may be minimized by restricting temperatures, heating periods, and pressures as much as is consistent with proper retention and penetration of preservative. Treating conditions specified by the American Wood-Preservers' Association and any governing Military Specification should not be exceeded if strength losses in excess of those mentioned are to be avoided.

2.16. EFFECTS OF PRESERVATIVE AND TREATMENT ON GLUE JOINTS. The results of treating and 6-year exposure tests on plywood and laminated timbers indicate that well made glue joints of waterproof phenolic and resorcinol resins are not significantly affected by standard pressure treatment with wood preservatives.

2.17. INSPECTION OF TREATED WOOD. Timber should be inspected for quality and grade before treatment. Grade-marked lumber and timber can often be obtained. Such material usually does not need to be inspected. When inspection before treatment is impractical, the purchaser can usually inspect for quality and grade after treatment; if this is to be done by Navy contractors, however, it should be made clear in the purchase order.

Currently, there are no generally accepted procedures for accurately determining retention of various preservatives, other than measurement of quantity of preservative used in treatment. Some specifications provide for a determination of minimum retentions by chemical analysis of borings from the treated wood, but such sampling and analysis have not been used to determine the overall retention of a charge, as it is commonly specified. Furthermore, the nature and quantity of the preservative in the wood change as the period of service increases, so that samples of treated wood taken sometime after treatment may not be representative of those taken when the treatment was completed. The purchaser, therefore, must either accept the statements of affidavit of the treating-plant operator or have an inspector at the treating plant to observe the treatments and insure compliance with the specifications.

Commercial inspection and consulting service is available for fee to purchasers not using enough treated timber to justify employing inspectors of their own. Care should be taken to choose experienced, competent, and reliable inspectors, since premature failures of treated structures can often be attributed to faulty inspection. Penetration measurements should be made at the treating plant if inspection service is provided, but can be made by the purchaser at any time after the timber has been treated. They give about the best single measure of the thoroughness of treatment. Rejection of treated timber for insufficient penetration is hardly enforceable, however, unless there has been a previous agreement with the plant operator on minimum penetration requirements.

The depth of penetration of dark-colored preservatives can be determined directly by observing a boring removed by an increment borer. The boring should usually be taken at about midlength of the piece, or at least several feet from the end of the piece, in order to avoid the unrepresentative end portion that is sometimes completely treated by end penetration. Since preservative oils tend to creep over cut surfaces, the observation should be made promptly after the boring is taken. Holes made for penetration measurements should be tightly filled with a thoroughly treated wood plug.

The penetration of preservatives that are practically colorless must be determined by chemical dips or sprays that show the penetration by color reactions.

American Wood-Preservers' Association Standard M2 contains instructions for the inspection of preservative treatment of wood.

2.18. HANDLING AND SEASONING AFTER TREATMENT. Treated wood should be handled with care, so as to avoid breaking through the treated portion. The use of cant hooks, picks, tongs, or other pointed tools that dig deeply into the wood should be prohibited. Handling rope or cable slips around heavy loads of treated lumber or sawed timber may crush the corners or edges of the outside pieces. Breakage or deep abrasions may also result from throwing the lumber or dropping it any considerable distance. Damaged places should be retreated as thoroughly as conditions permit. Treated wood that must be stored before use should be protected in order to avoid checking or mechanical injury.

Cutting of wood after treatment cannot always be avoided. When cutting is necessary, the exposed, untreated, or partially treated wood for land or fresh-water use may be protected by thoroughly brushing the cut surfaces with pentachlorophenol or copper naphthenate solution. Brush-coating of cut surfaces, however, gives little protection from marine borers. A special device is available for pressure treatment of bolt holes bored after treatment.

Timber treated with preservative oils may be used immediately after treatment, if desired, but at least 2 weeks of seasoning of open-piled lumber in a warm, well ventilated area help dry the oily surfaces and make the timber less unpleasant to handle. The removal of petroleum-oil solvents is particularly important if wood requires painting after treatment with oil-borne preservatives, such as pentachlorophenol. Vapor cleaning and steaming are used to hasten removal of surplus oils, but these conditioning procedures also remove some of the preservative.

With waterborne preservatives, seasoning after treatment is important for timbers that are to be used where shrinkage after placement in the structure is undesirable. Injection of waterborne preservatives puts large amounts of water into the wood, and considerable shrinkage is to be expected as subsequent seasoning takes place. For best results, the wood should be dried to approximately the moisture content it will ultimately reach in service (sec. 1.00). Air drying or kiln drying at temperatures less than 130° F. is advantageous, so that a minimum of seasoning degrade and strength loss occur in the treated lumber (sec. 2.15). With some waterborne preservatives, seasoning of the wood after treatment is recommended in order to complete the chemical reactions that are intended to take place within the wood, which increase the resistance of the preservative to leaching by water.

Plywood pressure-treated with waterborne chemicals should be kiln dried under a mild drying schedule to avoid excessive warping and face checking. Warping during drying can be controlled by proper stickering between panels, adequate support, and weights or spring takeup devices that hold down the pile. A level base should be used under each pile. Supports under each tier of stickers should be uniform in thickness and preferably square in cross section or of a width equal to 1½ times that thickness. A thickness of three-fourths inch has been found generally satisfactory. Stickers should be spaced not more than 12 inches apart. Sticker guides should be used to aline stickers one above another (fig. 25).

It is difficult to dry treated plywood without surface checking, and completely reliable methods for all woods are not known. Kiln schedules should be mild if a quality product is to be obtained. The time at which changes in kiln conditions can be made will depend upon the thickness and composition of the plywood. The following kiln schedule was derived from one used commercially to dry Douglas fir plywood treated with waterborne preservatives:

Portion of total drying time	Dry-bulb temperature	Wet-bulb temperature	Relative humidity
	°F.	*°F.*	*Percent*
First 1/6	130	121	76
Second 1/6	140	123	60
Third 1/6	150	123	45
Last 1/2	165	123	30

An equalization treatment at 45 percent relative humidity can be used when the driest material in the kiln charge gets down to 7 percent moisture content. If a surface reconditioning treatment is used, a relative humidity of 70 percent should be used if the average moisture content sought is 11 percent or below, and 75 percent if an average of 12 percent or higher is desired.

Figure 25. Plywood treated with waterborne preservatives piled for air seasoning. Note spacing and alinement of stickers between plywood panels, and use of heavy timbers to weigh down the pile. (Photo courtesy of American Lumber and Treating Co.)

2.2. MARINE BORERS.

Damage by marine boring organisms to ships in salt or brackish waters is practically worldwide and has called forth all the protective measures known to man from ancient times to the present. The seriousness of the menace varies, however, in different waters, the Pacific, Gulf, and South Atlantic waters surrounding the United States being more heavily infested than the North Atlantic Coast.

No wood is immune to marine-borer attack. A few tropical woods appear to be resistant and usually are attacked, but slowly. No commercially important wood of the United States has sufficient marine-borer resistance to guarantee its freedom from attack over any prolonged period of use in heavily infested waters unless heavily treated with a good preservative such as coal-tar creosote or creosote-coal-tar solutions.

The principal marine borers found in waters surrounding the United States are shipworms, gribble, *Martesia,* and *Sphaeroma.*

Shipworms include several species of *Teredo* and several species of *Bankia,* all of which are similar in several respects. In early life they are minute, free-swimming organisms. Lodging on wood, they quickly develop into a new form and bury themselves in the wood. A pair of boring shells on the head grows rapidly in size as the boring progresses, while the tail part or siphon remains at the original entrance. Thus the animal grows in length and diameter within the wood but remains a prisoner in its burrow, which it lines with a shell-like deposit. It lives upon the wood borings and upon the organic matter extracted from the sea water that is continuously being pumped through its system. The entrance holes never grow large, and the interior of a timber may be completely honeycombed and ruined while the surface shows only slight perforations. When present in great numbers, borers grow only a few inches before the wood is so completely occupied that their growth is stopped, but when not crowded they can grow to lengths of 1 to 4 feet according to species.

Limnoria or gribble are small crustaceans about one-eighth to one-sixth inch long that bore small burrows in the surface of wood. They can move freely from place to place, but usually continue to bore in one place. When great numbers attack wood, their burrows are separated by very thin walls of wood easily washed away by the ship's movement through water, causing the *Limnoria* to burrow continually deeper.

Martesia are wood-boring mollusks that resemble clams in general appearance. Like shipworms, they enter the wood when very small, leaving a small entrance hole, but grow larger as they burrow into the wood. They generally do not exceed 2½ inches in length and 1 inch in diameter, but are capable of doing considerable damage. Their activities in United States waters appear to be confined mainly to the Gulf of Mexico.

Sphaeroma are somewhat similar to *Limnoria* but larger, sometimes reaching a length of one-half inch and a width of one-fourth inch. They are widely distributed, but not so plentiful as *Limnoria.* Occasionally, they have been found working in fresh water.

Chelura are similar to *Limnoria,* but slightly larger, and apparently work always with the latter in marine structures. They appear to be active chiefly in tropical or subtropical waters, although one species is reported as a wood destroyer along the Atlantic Coast of North America, causing at least as much damage as the native *Limnoria.*

2.20. PROTECTION OF WOOD FROM MARINE BORERS. From the viewpoint of the ship and boat builder, the conception of a satisfactory preservative for underwater protection from marine borers is quite different from that of wharf and pier constructors. If a ship can be protected from attack during its period of operation between drydockings, the preservative has done its part. Thus, if preservative treated wood will resist all attacks of marine borers for 10 months without any paint coating, a ship could conceivably operate without worm damage to planking for 7 or 8 months without docking. While hauled out, the hull may be repaired and repainted with Navy copper bottom paint to protect the hull until the next drydocking, provided this interval is not over 6 or 7 months. Since paint retards the leaching of preservative from the timber, it is quite likely that the preservative-treated, painted timber would last considerably longer than the 10 months postulated above. Ordinarily, the docking interval for wooden boats and ships in severely infested waters should not be longer than 6 months. Many unforeseen hazards exist, such as coral beds or sand bars, that may gouge the planking, or some pieces of planking may not have received a complete penetration of preservative and untreated wood may be exposed when paint and surface wood is rubbed off by hauling lines or other ships' hulls at moorings.

Preservatives that have been investigated by extensive tests and found satisfactory for treatment of hull planking are Celcure, Chemonite, and Greensalt. Unpainted panels of southern yellow pine and Douglas fir pressure treated with 1.3 pounds of dry salt per cubic foot of Celcure has withstood attack of marine borers for 3 years. Figure 26 illustrates the interior of test panels which were 1 inch thick, after being submerged at Daytona Beach, Fla., for periods up to 18 months; the differences in the efficiency of the preservatives are evident.

tive, has disadvantages that render it unsuitable for use as planking in a wooden ship or boat. Creosoted wood is subject to marine fouling. It bleeds through and interferes with the adhesion of copper bottom paint and renders less effective antifouling properties of the paint. In single-planked ships, especially, it gives off odors objectionable to personnel.

In the selection of lumber for planking and sheathing, sapwood is preferred to heartwood where this material is to be pressure treated with a preservative. Sapwood is far more easily treated in many species of wood than is heartwood, and maximum absorption and penetration are the goals strived for in pressure treating wood with preservatives.

Unless planking or sheathing is thoroughly impregnated with preservatives, cutting or

Figure 26. Panels of pine pressure treated with various preservatives and exposed at Daytona Beach, Florida, to activity of marine borers.

The effect is shown on marine borer attack on 3 groups of panels exposed for periods of 3 months, 1 year, and 1½ years, respectively.

The 4 panels in each group are as follows:

From left to right: 1. Untreated. 2. Chlorinated phenol. 3. Celcure. 4. Creosote.

Note that at 1½ years exposure, the untreated panel had disintegrated and less than 25 percent of the chlorinated phenol treated panel remained.

From a standpoint of resistance to marine borers, creosote has been found to be the best of the preservatives tested. When injected into southern pine under pressure, in retentions of 20 pounds per cubic foot of wood, protection from marine borers has been afforded to piling for as much as 20 or 25 years in infested waters. Creosote, although a proved preserva-

fairing the planking before installation below the water line exposes untreated wood to the attacks of marine borers. Any such untreated surface should be given a heavy application of preservative when the planks are fixed in place on the hull.

Various pitch, tar, and bituminous coatings have been held with favor by some boat operators for prevention of marine borer attack of ships' bottoms. On most Naval craft they cannot be used without antifouling paint applied over them and in tests, various pitches have not added to the effectiveness of 16X paint, when used alone in protection against borer attack. It would seem, then, that these pitches have little to offer in the matter of protection of ships' bottoms.

2.3. SHIP BARNACLES.

The common ship or goose barnacle is a form of marine life that attaches itself to ship bottoms by means of its antennules, which are equipped with glands that produce an adhesive secretion. They do not, however, destroy the wood, but feed on bits of material floating in the water which they capture by means of six pairs of long, curled appendages used like a net.

The most effective protection developed against them is antifouling paint. Such paints contain a proportion of poisonous material, such as cuprous oxide or mercuric oxide. These poisonous materials gradually dissolve, presumably forming toxic solutions that retard growths from forming on the ship's bottom. Bureau of Ships specifications provide approved formulations for antifouling paints.

2.4. WEATHERING OF WOOD.

Wood exposed to the weather without a protective coating soon becomes weathered. Weathering is the mechanical or chemical disintegration and discoloration of the surface of wood that is caused by exposure to light, freezing, and thawing, the action of dust and sand carried by winds, and the alternate shrinking and swelling of the surface fibers that come with the continual variation in moisture content brought by changes in the weather.

Besides change in color, weathering may involve roughening and checking of the surface and, if one side of the piece only is fully exposed, cupping and tearing loose of relatively thin lumber from fastenings. Weathering does not, however, include decay.

With all species, edge-grain boards check less conspicuously and cup less than flat-grain boards of the same species. Twisting, although less common, is another effect of weathering. Caused primarily by uneven shrinkage resulting from spiral and interlocked grain, twisting is more pronounced in plain-sawed than in quarter-sawed stock.

The principal function of exterior paints, aside from appearance, is to protect wood from weathering. Fundamentally, of course, prevention of weathering is itself largely a matter of appearance. Paints and other finishes applied to wood ships and boats are discussed elsewhere.

2.5. FIRE RESISTANCE OF WOOD.

The tendency of wood to ignite when exposed to fire or high temperature depends on several factors. One of these is size. The greater the volume in relation to the surface area of the piece, the less likely it is to ignite. Thick ceiling, decking, deck beams, and framing members do not present the same fire hazard as do lighter members used in lockers, cabin partitions, bulkheads, and other fittings. Other factors, such as moisture content, contained chemicals, sap, and pitch, also have an appreciable effect. The likelihood of setting fire to wood can be minimized by keeping its surfaces free from such highly combustible materials as oils, tar, pitch, grease, gasoline, other inflammable solvents, and thick paint films.

The two general methods of imparting fire resistance to wood are by: (1) Impregnation treatments; and (2) surface coatings. Fire-retardant treatments, either impregnations or coatings, are of value in delaying ignition and checking flame spread. Well impregnated, fire-retardant treated wood will not ignite or sustain and spread flame unless an external source of fire is directed upon it for a considerable period of time.

2.50. FIRE-RETARDANT COATINGS. The customary finishing materials, such as ordinary oil paints, stains, varnishes, and lacquers, are of no particular value in protecting wood against fire. The fire-retardant constituents of the most effective paints are water-soluble chemicals. The use of such paints on ships is, therefore, limited to those interior locations not subjected to washing. Many paints of this type have as bases: (1) Ammonium phosphate in a vehicle of starch and urea resin; (2) borax in linseed oil; (3) finely ground monoammonium phosphate dissolved and suspended in an aqueous polyvinyl latex; or (4) antimony oxide and calcium carbonate in a vehicle with chlorinated rubber or chlorinated paraffin. All of these paints contain other ingredients to give them improved hiding power, brushability or film forming properties, and many are formulated to swell and bubble at fire temperature to form a frothy insulation to protect the face material from the flame.

With any fire-retardant paint, a thick, continuous film is necessary. The best of these coatings are themselves noninflammable and offer a protective film over the wood.

2.51. FIRE-RETARDANT IMPREGNATIONS. Uniform and thorough impregnation with chemicals of demonstrated effectiveness is superior to surface coatings. Wood is impregnated with fire-retardant chemicals by methods similar to those in use for the injection of preservatives (sec. 2.13). When heat is applied to fire-retardant treated wood, the injected chemicals do not prevent thermal decomposition of the wood, but they do form products that inhibit combustion. Checking, charring, and slow destruction of the locally heated wood will take place by virtue of dry distillation, but with relatively little flaming and spread of flame. Effective fire-retardant compounds in treated wood increase formation of charcoal at high temperatures but decrease the formation of flammable tars and gases.

Many chemicals have a fire-retardant effect on wood. No one chemical combines all the desirable properties. For this reason, chemicals are used in various combinations to obtain desirable characteristics, including high fire resistance, decay resistance, permanence, noncorrosiveness, paintability, workability, little effect on strength, and low cost. Among the most effective chemicals of those in commercial use are monoammonium phosphate, diammonium phosphate, ammonium sulphate, and a mixture of borax and boric acid.

There are combinations of chemicals which have been in commercial use over sufficient periods of time to demonstrate satisfactory stability and permanence under ordinary conditions of exposure. However, they are subject to leaching when exposed to water or weather. Leaching causes some of the combustibility characteristic of untreated wood to return. It has been demonstrated that sealing coats prevent leaching to a substantial degree.

Impregnation treatments may be complete, with penetration throughout the cross section, or they may be partial, in which case only an outside zone is impregnated. Lumber selected for ease of treatment and in thicknesses of 2 inches or less may be completed impregnated. Fully impregnated material can be machined, sawn, or otherwise manufactured without exposing untreated areas.

Partial impregnation is usually considered adequate for fire-retardant treatment of most structural lumber and timber, because protection in the outer zone is sufficient to delay the spread of fire, to make small fires self-extinguishable, and to permit larger fires to be more easily controlled. Complete impregnation is usually impossible and not economical in timbers of large cross section. Generally, the retention of fire-retardant chemicals varies, heavier retentions being required in thin material than in structural sizes. Also, retention may vary somewhat with the density of the wood, to achieve equal fire resistance in different species.

Fire-retardant treatments are most appropriate to wood parts of small sizes, in which combustion is rapid and any delay in ignition and flame spread may be of critical importance.

Plywood that is to be treated with fire-retardants must be bonded with water-proof types of glue, usually of the phenolic-resin group, which prevent delamination of the plies during treatment. Carefully handled, plywood can be treated and dried without affecting its original shape or flatness. Every ply should be fully impregnated so that panels may be cut without exposing untreated areas. Fire-retardant plywood is required for some types of ships and boats as decking, floors, bulkheads, superstructure, storage boxes, chests, and lockers, and many other parts.

All fire-retardant-treated wood, regardless of species, type, or dimensions, must be kiln dried or thoroughly air seasoned before use (sec. 2.18). This is because the chemicals are injected in waterborne solutions and the wood comes from the treating tank very wet. Kiln-drying schedules are somewhat slower than on untreated material, so as to disturb as little as possible the position of the salts which have been impregnated into the wood. Fast schedules tend to pull both moisture and chemicals to the surface quickly, before the chemicals can be precipitated throughout the wood.

Research on the effects of some preservative treatments on the flammability of oak wood has shown that: (1) The time required to set the wood afire is increased slightly by the inorganic salts of waterborne preservatives; (2) it is practically unaffected by the amounts of organic pentachlorophenol, ester gum, and copper naphthenate compounds retained after the solvent carrier has evaporated; and (3) it is decreased a little by No. 2 fuel oil. The rate at which flame spreads over the wood surface is increased by the presence of flammable oils, but decreased by waterborne salts, particularly the fire-retardant compounds in moderate and high retentions. Copper and chromium compounds in preservatives stimulate and prolong glowing of carbonized wood.

In general, preservative organic compounds in the retentions used for preservatives have little or no important fire effect after all volatile solvent is gone, but the oil solvents, while present, have unfavorable fire influence. Inorganic waterborne wood preservatives reduce the fire hazard somewhat, but not so much as moderate to heavy retentions of the fire-retardant inorganic chemicals. The overall conclusion drawn from research is that treatment of oak wood with standard waterborne preservatives decreases the danger of ignition and burning. Preservatives dissolved in oils not subsequently removed from the wood increase the danger of ignition and rapid spread of flame.

Several fire tests have been developed that are applicable to fire-retardant treated wood. In addition to flame spread and glowing, limiting requirements may be put on leaching, metal corrosion, effects on strength, finishing properties, and appearance of the treated wood.

Military Specification MIL–F–19140 covers the fire-retardant treatment of lumber and plywood.

CHAPTER 3. STORAGE AND REPAIR OF WOOD BOATS AND SHIPS

3.0. GENERAL.

When wood-hulled ships are improperly stored, certain hazards are accentuated. Chief among these is the danger of decay (sec. 3.09, vol. I). Others include weathering, marine borers, rodents, termites, ice, and fire.

Good storage conditions help safeguard the craft from these hazards. To provide such conditions, three factors must be taken into account: (1) A favorable site; (2) adequate preparation of the craft for storage; and (3) proper maintenance during storage.

No practical system of storage will guarantee indefinite freedom from deterioration. Wood-hulled craft can be kept in reasonably good condition for extended periods of time at costs that, considering the replacement cost of the craft, appear well within reason. It has been demonstrated that lack of proper storage can be enormously expensive. Repair bills for poorly stored boats have frequently exceeded one-half the price of a new hull. For these reasons, directives providing for efficient storage have been issued (see section 3.30).

3.1. STORAGE HAZARDS.

The hazards to which a wood hull is exposed in storage depend to a considerable extent upon the kind of storage possible for the particular type. A ship must of necessity be stored afloat, while most boats can be taken ashore—preferably for storage under roof.

3.10. FIRE AND THEFT. Ships and boats are rarely exposed to fire danger when in storage. Smoking by workmen and intruders is the most common cause, and occasionally, grass fires spread into a storage area. The necessary precautions are obvious. Likewise, steps should be taken to prevent looting of fittings by fencing in the area and policing it.

3.11. MECHANICAL DAMAGE. Damage incurred while inactivated boats are being handled is generally minor. Superstructure of inactive ships is sometimes damaged when they are towed or even at dock moorings in high winds. Hogging, sagging, or other deformation sometimes occurs when boats are suspended in davits or improperly stored on keel blocks and cradles—especially when the supports decay or sink unevenly because of poor footings.

3.12. WEATHERING. Damage caused by weathering is usually limited to surface coatings and more or less superficial checking and cracking. Paint and natural finishes are the most frequent victims. Long-time shrinkage and swelling of the surface fibers of bare wood causes gradual but rarely consequential loss of fiber (sec. 3.08, vol. I). Douglas fir plywood appears to sustain more pronounced face checking than plywood of other species. On the whole, however, weathering is chiefly harmful to appearance.

3.13. DRYING DAMAGE. Boats stored out of the water commonly open somewhat at the seams. As a rule, these seams close tightly when the craft is put back into the water. More serious damage may occur, however, in boats stored in a very dry climate or in buildings that are heated or poorly ventilated. Large members, such as keels, knees, and stems, may split if drying conditions are severe enough (fig. 27).

3.14. CORROSION. Rusting of iron or steel fastenings and fittings sometimes damages not only the metal parts but the wood around them. With the usual galvanized fasteners, there is little difficulty. Storage in or near seawater or brackish water adds the threat of electrochemical action, the products of which can damage the wood around fastenings as well as the fastenings themselves where dissimilar metals are used. Electrochemical action can take place even with fresh water that has taken on acidic extractives from wood, thereby creating electrolytic conditions.

3.15. DECAY. The most destructive hazard

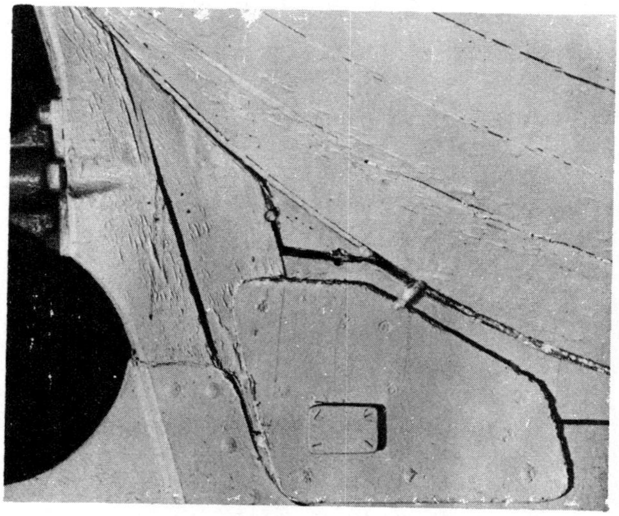

Figure 27. Open seams and split shaft block of boat stored in an unventilated building with much glass in walls that admitted direct sunlight and excessive heating and drying.

of stored boats and ships—and for the most part needlessly so—has been decay (sec. 2.313 and 3.08, vol. I). Open storage, with little or no protection given the hull from rain, snow, and soil moisture, has been chiefly responsible. Contributory causes have been due to more or less unavoidable conditions of storage, such as berthing of large ships in fresh water, the difficulty of protecting large hulls from rain, and failure to provide adequate ventilation below decks. Even ships moored in salt water, which suffer little or no decay of planking and other exposed parts below waterline, can, without proper care, develop much decay near the weather-deck level, as a result of rainwater leaks or condensation of vapor from the air on cold surfaces.

3.16. INSECTS, RODENTS. To craft stored ashore, termites may be a serious threat; colonies can riddle structural members usually without leaving any evidence. Other insects that occasionally damage small craft berthed ashore—or even ships in long-time water moorings—include carpenter ants and powderpost beetles which push shredded or powderlike borings from openings in the wood.

Ships moored in brackish water may be attacked by marine borers (sec. 3.08, vol. I). Bottoms can also be fouled by barnacles in salt water moorings.

Rodent damage is seldom consequential as far as hull structure is concerned. Chief damage may be done to upholstery, mattresses, and similar equipment that has been left aboard.

3.2. STORAGE LOCATION.

The properties and enemies of wood should be given full consideration when a storage site is to be selected. If the site has already been chosen, the same considerations should be weighed to determine how and to what extent unfavorable conditions can be modified to make the site more suitable for storage of wood-hulled craft.

Properties of wood that should be considered are: (1) Its tendency to shrink, swell, and develop checks or splits with extreme fluctuations of wet and dry weather; and (2) its susceptibility to decay if kept at too high a moisture content under warm conditions. Wood's principal enemies are, of course, decay fungi and wood-consuming insects, with fire and weathering relatively much less important.

For ships, fresh-water berths in localities where winters are relatively long is preferred. Localities with warm winters increase the decay hazard to the underwater hull beyond the point at which dehumidification can be positively relied on to prevent decay.

Obviously, there are few available storage areas that provide ideal conditions for wood boats. The site should provide good water drainage and wind movement. Slightly sloping firm ground away from wind-obstructing buildings and wooded areas is to be preferred. The ground should preferably be covered with crushed stone, gravel, or other material that facilitates drainage and discourages plant growth.

3.3. STORAGE OF BOATS.

For the purpose of this discussion, all wood-hulled craft stored ashore are classed as boats. Craft of a size such that they must be stored in water moorings are classed as ships (sec. 3.4).

3.30. STANDARDS FOR STORAGE OF BOATS. Boats and boat components stored in the open are subject to rapid deterioration from the effects of sun and moisture. Such deterioration can be prevented at a great saving in cost by applying simple preservative methods prior to and during inactivation.

A Navy program was inaugurated in 1954 to

provide portable boat covers at all stocking activities. Long range plans are in process for more permanent pole frame barn type structures which will substantially reduce boat handling costs, as well as prevent deterioration of the hull structure. It is the intent that all new, repaired, and economically repairable wooden boats be placed in properly covered storage.

To further insure the proper storage of boats, a set of standards has been prepared in the form of a checkoff list. This checkoff list should be used for each boat placed in storage and for conducting routine, technical, and preservation inspections. It is intended that this checkoff list be used to ascertain whether a boat and the boat components are properly stored, which fact should be entered on the boat record card. The checkoff list itself need not be made a permanent record. It is suggested that the checkoff list be made up in such a form that it can be attached to the boat for information purposes.

The following items should be checked for all boats placed in storage:

a. All fresh water has been removed from piping, fresh water tanks, and bilges. All drain plugs have been removed and attached in a packet to the steering wheel or tiller. The boat is properly leveled to facilitate drainage. Check_____

b. The bilges have been cleaned and limber holes and scuppers are unobstructed. Check_____

c. The boat is rigidly supported by keel blocking timbers (6″x8″) placed beneath the keel at not greater than 6-foot intervals, or by a cradle designed for the boat. Check_____

d. Side bunks are preventing racking and are not bearing weight as evidenced by absence of local hull deformation. Check_____

e. All canvas, dunnage, life jackets, and rope have been removed, dried, and stored separately in accordance with Military Specifications MIL–P–3131, MIL–P–3454 or PPP–P–51, as applicable. Check_____

f. The engine cooling system has been drained, flushed with fresh water, and the engine and accessory equipment has been preserved in accordance with Bureau of Ships Manual, chapter 9 and the detailed requirements of chapter 41, section 2, part 6. Check_____

g. Batteries have been removed, packaged, and packed in accordance with Military Specifications MIL-P-207, MIL-P-208, and placed in separate storage. Check_____

h. Fuel and fresh water tanks have been drained and preserved in accordance with chapter 9, article 9-164 of Bureau of Ships. Check_____

i. All chrome fittings, piping, shafting, propellers, and bearings have been preserved in accordance with Type III, Military Specification MIL-P-18147. Check_____

j. An adequate cover has been provided with sufficient side protection to prevent the access of driven rain or snow. There are no leaks in the cover as evidenced by the absence of fresh water accumulation on the decks or in the bilges. Check_____

k. All ventilators, doors, and hatches have been secured open for ventilation. Check_____

l. The shelter permits adequate circulation of fresh air. Check_____

m. There is no source of artificial heat within the shelter. Check_____

n. The shelter has no large expanses of glass surfaces. Check_____

o. Drain holes have been drilled at low points in compartments not capable of being drained by leveling the boat, existing drain plugs or portable pump. Check_____

p. Deformation of the hull by overhang at the transom end is prevented by rigid supporting timbers and adequate padding to prevent damage to the planking. Check_____

q. Frost heaving or settling of the ground has not caused racking of the hull. Check_____

r. No equipment is stored in the forepeaks or transoms which interferes with thorough ventilation. Check____

s. Exterior paint coats are maintained. Flaking paint has been removed by scraping on plywood hulls or by scraping or careful sanding on planked hulls. Bare spots have been touched-up in accordance with the following:

(1) Above waterline: Use one coat of aluminum paint prepared by mixing one gallon of Formula 80 varnish (MIL-V-1174) with two pounds of Aluminum paste, Fed. Spec. TT-A-468, Type II, Class B.

(2) Below waterline: Use one pre-treatment coat of Formula 117 (MIL-C-15328) followed by one coat of Formula 119 (MIL-P-15929) vinyl red lead primer. Precaution: Vinyl paints are fire-hazardous until dry. Check____

3.31. STOWAGE. Handling and stowage can be done in various ways, depending upon the type of boat to be stored and the equipment available. The simplest method is to lift the boat out of the water with slings and a crane, load it on a truck or trailer, and haul it to the place of storage, where it is placed on keel chocks. Some types of boats, however, are too large to be handled in this manner. A more convenient method is to haul them out of the water on a marine railway that leads either to a loading dock or directly into a storage building. Sometimes boats are lifted out of the water with cranes and carried thus to the place of storage. This method, however, is inconvenient when storage is to be under roof; it requires either a removable roof or rehandling with trucks or dollies to get the craft under roof. Crane handling may also be dangerous when a strong wind is blowing. Overhead bridge cranes are convenient in large buildings for moving and stacking boats.

Whether stored outdoors or under roof, a boat should be supported full length on its keel when possible. (See sec. 3.30, item C.) Wood timbers are suitable for this purpose, provided they are preservative treated or bear on concrete footings that extend several inches above ground. Figure 28 shows fitted cradles that provide satisfactory support. To prevent water from collecting at the points where planking rests on supports, water drips or guards should be placed just above the contact points to deflect the water.

Figure 28. Fitted cradles support this hull at three points. Note concrete footings to keep wood timbers from contact with ground.

Figure 29. Interior view of long term single level storage.

The single level type of storage, as shown in figure 29, permits handling with the minimum of difficulty. However, indoor storage usually calls for effective use of available space, and boats are often stowed in tiers.

Figure 30a shows a simple post-and-beam type of support for stowing small boats. The posts and beams are notched where they join and bolted securely together; diagonal bracing contributes stiffness to this simple structure. Three-point cradle support is provided under the keels. Weight of upper hulls is kept off lower ones by such a structure, thereby avoiding the possibility of distorting the craft underneath. Figure 30b shows a post and platform support of a more permanent type.

Figure 30a. Three-deck post-and-beam mount for stowage of small boats.

Figure 30b. Interior view of long term storage with boats stacked.

Some types of boats, chiefly the smaller ones without cabins or engines, can be stowed upside down. This is common procedure with punts and wherries. Chocks or other mounting timbers will need to be placed fairly close together to support the hull along the gunwales, and should be high enough to keep projecting parts, such as coaming, well off the ground. Some kinds of V-bottom craft can probably also be stored upside down, but care must be taken to place supporting chocks principally at bulkhead frames to avoid undue stress at the chine joints of the frames. Crossties may also be necessary at the heads of certain frames to prevent spreading and relieve stress on frame joints, especially at keel and chine. Obviously, all heavy machinery and other equipment should be removed, since their weight suspended from the keel and floors would contribute greatly to damaging stresses in an inverted hull.

The purpose of inverted stowage is to give a hull some protection from rain and snow when stored outdoors. Direct sunshine on bottom planking, keels, and other exposed members of a boat so stowed can cause damaging shrinkage when the hull is in this position for some time. A rough board roof or at least a lattice shade frame will help to avoid opened seams and checked planking. Some open seams and checks are to be expected, of course, in practically any type of storage, though they usually close when the hull is returned to the water; however, when excessive, they must be repaired. Aluminum or asphalt paint applied to the bottom may help retard drying under such conditions; or, a reflective paint such as aluminum will keep the surface temperature down if combined with some type of shade cover (sec. 3.30, item 5). On the whole, however, it is questionable whether outdoor upside-down storage would prove much more economical — even from a first-cost standpoint and ignoring depreciation rates — than indoor storage in open sheds when all costs are included.

3.32. OPEN STORAGE. For periods longer than a few months, storage of boats in the open, with no protection from rain and sun, is by far the most expensive type because of the early and rapid depreciation it causes. Outdoor storage leaves a boat exposed to the worst enemies of wood — decay fungi, insects, weathering, and the harmful effects of successive severe changes in moisture content — but decay most of all. A hull stowed outdoors in its normal position, keel down, is a natural receptacle for moisture, which inevitably finds cracks and joints into which to penetrate and collect. Ideal conditions are thus quickly set up for the growth of decay fungi (sec. 3.09, vol. I). Even when inverted, a hull is not immune to the penetration of water, especially after a few months' exposure. Expensive repairs are the rule, and hulls eventually become unsalvageable.

3.33. STORAGE SHELTERS. Because outdoor storage without shelter can prove so damaging, it should not be concluded that a tightly closed, heated building is the best possible storage place. Again, the properties of wood must be taken into consideration and related to the condition of the boat when it leaves the water. Clearly, the least possible damage will occur to a boat that is stored under uniform moisture conditions approximating those of its service, except that it is protected from fresh water and from excessive shrinkage. Boats may vary considerably in moisture content of their wood parts when they leave the water, depending on the type of service; those constantly in water will, of course, have more moisture in their keels, underwater planking, and other such parts than boats intermittently afloat during service. However, the above-water parts of any boat can be assumed to have a relatively high moisture content, in the neighborhood of 15 percent.

While 15 percent is well below the danger point for decay, it is still high enough to permit excessive shrinkage in heated buildings. Wood readily dries to around 6 or 8 percent moisture content in a heated building during the winter. This can cause serious shrinkage and seam opening, as well as checks and possible splitting. Underwater parts of considerably higher moisture content will be much more drastically shrunken — which is the cause of split keels, stems and sternposts, shaftlogs, and other large parts of boats stored in heated buildings.

Obviously, then, a heated building is not an ideal place to store boats. Even a closed unheated building, especially if it has much glass wall or roof area, can be the cause of serious damage (fig. 27).

The best type of storage shelter is one that allows free circulation of outdoor air but keeps sun, rain, and dew off the hulls stored within it. A shelter that protects the hull from rain but allows free circulation of air will prevent further decay after boats have had time to dry out, for as long a period as may be desired and irrespective of the kind of wood, construction details, or shortcomings in preservative treatment. Wood under cover does not pick up enough moisture from free circulating air to enable decay fungi to work. Infections already established will not continue active while the wood is kept dry.

Moisture tends to condense out of the air on cold surfaces of uncovered boats, just as dew collects on grass, during clear nights. Most of this evaporates during the day. Some, however, may run down the sides and gradually build up to dangerous amounts in joints. An open shed reduces the heat loss from boats to such an extent that condensation is negligible even on nights when uncovered equipment is streaming with water. Some condensation may collect on the sides of boats under roofed sheds without walls, but much less than collects on the tops of uncovered boats. Moisture meter tests have shown that the moisture content of covered boats is generally low and unaffected by condensation.

3.330. SINGLE-UNIT SHELTERS. A variety of shelters have been tried for single boats, ranging from canvas paulins to gableroofed wood structures. The principal difficulty in all cases is to get a cover that is effective, permanent, and reasonably low in cost.

Canvas paulins have been found to be effective for only 2 or 3 years. By that time, they usually tear, sag, and generally deteriorate enough to lose their effectiveness in keeping the boat dry. For long-time storage, they represent problems of maintenance and renewal that bring total costs to a high level.

Figure 31. Portable wooden covers with canvas skirts showing outside view of construction similar to that used in Figure 32.

Individual boat covers consisting of gabled roofs are relatively high in first cost and are not conveniently moved from boat to boat as needed. For longtime storage, the latter may not be an important consideration. Nevertheless, their cost is such that they should be reusable as conditions require.

A composite shelter utilizing plastic naval gun covers has been successfully used. The gun covers are cut in half and set over the bow and stern. Canvas is then laid over pipes connecting the two halves, to cover the midships section.

One of the most practical types of single-boat shelters tried has been the shed-roof type shown in figures 31 and 32. These shelters were developed primarily for LCVPs. They are equipped with grab rings for easy handling with a crane. The depth of the longitudinals, as shown in figure 32, could be increased in order to stiffen the structure and eliminate any possible need for supports at the midpoint. On LCVPs this roof rests directly on the hull at both ends. Canvas skirting around sides and stern is intended to keep wind-driven rain out of the boat.

Figure 32. Shed-roof type of reusable single-boat shelter developed for LCVP landing craft. Roof slopes from ramp, on which it rests, toward stern for water runoff and has good overhang at all sides, plus canvas skirting on sides and stern to exclude wind-driven rain, and provide good ventilation.

Semipermanent roofs over boats have been constructed of a variety of materials. They may be framed with steel, or new or salvaged lumber. The cover may be corrugated aluminum, galvanized iron, or wood sheathing covered with asphalt roll roofing or canvas. "Tar paper" applied to the frames directly without sheathing boards has been unsatisfactory, but an inexpensive and serviceable shelter can be obtained with roll roofing of standard quality laid over sheathing boards. Canvas, even on tightly laid sheathing, cannot be expected to last as long. Single-layer roll roofing should be an asphalt-saturated and coated felt, of at least 50 pounds weight per roll of 108 square feet, and preferably mineral faced. If it is to last 15 years or more without periodic treatment, the wide-selvage type, which has 50 percent overlap that is cemented down in use, is advisable. This roofing can be obtained with a reflective upper surface.

Such roofs should have a slope of not less than an inch to the foot. Metal roofing can be supported on properly spaced frames without sheathing. For roof frames and sheathing, preservative treatment is not needed.

Roofs should be so mounted as to leave at least 4 inches of clearance between gunwales and shelter for free air movement. All parts of the boat should be covered. A roof that covers only the cockpit is insufficient; by discharging a large amount of water on the cover board and transom, it may cause more decay than it prevents. The roofs should, in fact, have at least a 4-inch overhang, or 8-inches, if the eaves are far above the deck. Excessive overhang, however, can cause stowage difficulty where storage space is limited and boats must be parked close together.

3.331. MULTIUNIT SHELTERS. Permanent buildings have been used for storage of boats. Occasionally, warehouse space is available for such storage, but results have not always been successful. The chief problems, as already noted, are heat and poor ventilation. Where unheated, well ventilated warehouses are available, they prove satisfactory provided direct sunlight does not fall upon stored boats due to large expanses of glass windows.

Recent experiments with low cost types of buildings especially designed to overcome boat-storage problems appear to promise a considerable degree of success. These are low cost buildings of the pole type, with at least one side entirely open. Figure 33 shows one such type of building, sheathed with corrugated metal on three sides and the roof. The roof is panelized for convenient production and erection. The wood poles are set in the ground deeply enough to make a foundation unnecessary, and are preservative treated for decay resistance. If it is necessary to lock the buildings, lower walls can be partly of lattice construction to permit ventilation.

Figure 33. Pole-type structure for storage of boats. Roof is panelized, with corrugated metal covering that overlaps at adjoining edges of panels to prevent rain leakage. Note let-in diagonal bracing of roof panels.

Figure 33a. Shed similar to Figure 33 with boats stowed.

A somewhat similar structure, designed for two-tier storage of boats, is shown in figure 34. This structure has lumber sheathing on walls, diagonally laid for greater stiffness, and a panelized roof. It is built on concrete sleepers originally used for lumber storage. Diagonals are used to stiffen the columns between each storage bay.

3.4. CARE OF INACTIVE SHIPS.

The larger wood ships are berthed in water during periods of inactivation. In salt water, frequent dry docking, replacement of damaged members, and periodic renewal of antifouling paint are needed to safeguard against barnacles and marine borers. Docking is needed at therefore ordinarily in fresh or nearly fresh

Figure 33b. Group of LCVP boats with covers installed.

12-month intervals. Long-time inactivation is water, where docking is needed only at 5-year intervals, but where decay at or below waterline is much more likely than in salt water.

Figure 34. Post and beam type of structure for two-tier storage of boats. Columns bear on concrete sleepers. Lumber sheathing is laid diagonally for greater stiffness.

In fresh water in the North, it may be necessary to cut ice away immediately around the ship to avoid damage to plank and calking. Unfortunately, escape from ice by locating berthing areas in the warm South aggravates the much more serious hazard of decay.

Outside paint before or during storage has value in hindering weathering, but little in preventing decay unless it is so unusually well maintained as to avoid cracking at joints. If water gets into joints, as is too often the case, paint on the surfaces may actually increase the decay hazard by hindering evaporation. Additional paint coats on interior surfaces of plank and frames below waterline probably increase the decay hazard somewhat during inactivation.

Weathering of decks can be somewhat diminished and water absorption decreased without material interference with evaporation, by use of a water-repellent preservative. Repeat treatment at 2-year intervals are desirable. There is little opportunity for effective application of preservatives elsewhere in ships after the original fabrication and painting are completed. Salt packing can be considered, particularly in ceiled portions of hulls, but would have little bearing in ships which are entirely dehumidified. Various fungicides could be added to the bilge water with probable advantage to hulls in fresh-water moorings, if bilges are not pumped too often or dried up as a result of dehumidification. In addition to common salt, dry pentachlorophenol and orthophenylphenol appear promising, but there is little experience bearing on this type of application.

Watertightness of the hull is important for decay prevention, and calls for attention in preliminary overhaul. Weathertightness of the above-water structure is also an important objective in preparation of ships for inactivation. Scuppers should be cleared, and additional ones cut, if necessary, to keep water from standing on any part of the deck that may have poor drainage. Seams that leak water lead to decay, and those that admit air interfere with dehumidification. Deck seams can be reinforced by beads of seam compound or calking, which could not be employed in active ships; these can also be used on exterior joints at the deckhouse sills. Light asphalt membranes or steel plate used to exclude water from decks have resulted in increased wetting, since they often admit liquid water at cracks, joints, or fastenings, but greatly hinder evaporation. Heavier and better reinforced membranes and better attention to the fitting of the joints where horizontal and vertical surfaces come together may help.

On ships held for exhibition or long-term reserve-storage purposes, roofs have frequently been erected. These have generally been deficient in overhang, dumping on the region of the planksheer and transom the water that would normally be distributed over the whole ship. This, and their interference with sun and air movement, have made them liabilities rather than assets from the standpoint of decay prevention; but with sufficient overhang, they can be very helpful.

During inactivation, ships suffer particularly from lack of ventilation. Mechanical ventilation ceases, and hatches and manholes are too often closed. Some water inevitably leaks in, especially in the way of the planksheer, beam ends, and frameheads. Reduction in the decay hazard can be accomplished through proper ventilation. Ventilation based on a 10-minute rate of change of air, using either a mechanical supply and natural exhaust, or a mechanical exhaust and natural supply, is suitable in hulls which are decked over. Where mechanical ventilation is impractical, the space should be left open to the extent that all spaces will have natural ventilation as a minimum. The transome and the forepeak areas require particular attention, since decay is commonly encountered in these areas.

In cold weather, water vapor from the bilge region tends to condense on the inner surface of the skin above waterline.

Considerable movement of air is needed to counteract such accumulations. The movement of air in the hull and its interchange with the air outside is often insufficient to remove the moisture from these combined sources. The dynamic dehumidification that has become general practice in steel ships that are inactivated is the best prospect for protecting wood hulls. For success, the hulls must be tightly sealed. It is not considered probable that wood hulls will tolerate a relative humidity below 45 percent without troublesome opening of seams.

In small spaces not adapted to dynamic dehumidification, static dehumidification can be accomplished with dessicants that are occasionally renewed. All of the larger spaces in wood hulls should be included in the dynamic dehumidification system as a safeguard against structural decay, as well as against deterioration of equipment; or they should be given effective natural ventilation.

3.5. HULL REPAIRS.

3.50. GENERAL. With reasonable care, wood-hulled craft can give many years of economical service. Ordinary wear seldom necessitates more than superficial maintenance during the life of a boat or ship—barnacle removal, replacement of worn decking and sheathing, and the like. Accidents do happen, however, that require more extensive repair. Likewise, fresh-water leakage and inadequate ventilation are among the major causes of decay. Such damage sometimes requires extensive repairs.

Hull repairs can be conveniently classified for purposes of analysis into two kinds: Those which can be made by the ship's force, and those for which the more extensive facilities and materials available at a central repair sta-

Figure 35. Boats under repair at a large facility.

tion are required (fig. 35). Fleet repairs are often temporary or minor in character, such as patching a small stove-in area of planking or splicing an occasional broken frame or floor. For extensive permanent repairs, ships are sent to a boat repair shop or to drydock.

NOTE—Repainting, calking, of opened seams, and similar repairs are not considered here because they comprise routine maintenance operations guided by well-established rules.

In general, the same standards and procedures that apply to hull construction hold good for repair work. The basic objective of all repair is to return the ship to service in seaworthy condition.

Requirements for an adequate repair job are:

1. The use of material generally comparable in properties and quality to those replaced.
2. Repairs extensive enough to assure that the hull is essentially as strong and serviceable as originally.
3. Careful planning of repairs, so that replacement parts, including fastenings, are adequately designed and equal in size and number to those replaced.
4. Good workmanship.

Neglect of any one of these requirements will result in an inferior repair job.

The following discussion of repair work is intended to illustrate these four requirements. No attempt is made to catalog the many individual repair operations or to set forth specific techniques for doing them. Procedures and methods mentioned are presented only for illustrative purposes; it is recognized that other methods may be equally or more effective.

3.51. TYPES OF HULL DAMAGE. Mechanical breakage is the most common type of damage that happens to hulls in use. A boat dropped from its slings even a few feet can sustain broken frames—most likely at bilge or chine, where curvature is sharpest or gusseted joints most severely stressed. Stove-in planking and stem or transom collision damage are other common causes of mechanical breakage.

A type of damage less obvious than mechanical breakage, but fully as destructive in the long run, is that caused by decay fungi (sec. 3.08, vol. I). Decay is no accident, and in wood craft, it usually results from either fresh-water leakage or inadequate ventilation. These conditions, outlined in section 3.09, volume I, should be avoided wherever possible, both during use and when the craft is in storage. The responsibility for prevention of decay rests only partly with the builder; no hull carries a built-in guarantee against it. Reasonable maintenance and adequate storage assure a long, troublefree life.

Rainwater leakage through decks is often responsible for decay. Especially in boats stored without cover, it is often found along the planksheer and around gunwale stanchions; under fenders, armorplate, and sheathing; and in frames, especially at the chine or keel. Lack of ventilation is also frequently responsible, permitting moisture to build up in certain parts, such as stems, forefeet, transom frames, and sternposts when forepeaks, lazarets, and similar compartments are kept tightly closed. Hulls with full ceiling are especially likely to develop decay in frames and the adjacent faying surfaces of planking, because ventilation of the bays is blocked. Condensation may set up decay conditions in the way of refrigerated compartments, galleys, laundry facilities, or wherever moist, warm air contacts cold surfaces.

Causes of wood deterioration sometimes associated with decay are chemical disintegration and ordinary weathering. Chemical disintegration is uncommon but sometimes does enough damage to require extensive repair in the way of fastenings. A common cause is electrochemical action due to the use of dissimilar metals for fastenings. Weathering (sec. 3.082, vol. I) is a very slow process even on bare wood and is virtually halted by proper paint maintenance. Damage from marine borers occurs most often in waterline planking, keels, and garboards. If tightly abutted, preservative-treated sheathing protects planking in waters heavily infested with these pests. However, due to the difficulty of maintaining tight seams between the strakes of sheathing, a good quality tarred felt is used as a further barrier between the sheathing and the planking.

3.52. REPAIR MATERIALS. The rule that repair materials must be equal in size and quality to the damaged material removed is basically sound. Like most rules, however, it has its exceptions. If experience indicates that the original materials were inadequate, this should be taken into account when repair materials are selected. Likewise, if improved materials are available and suitable, they should be considered for use, especially if extensive repair is necessary and calls for removal of most or all of the original material.

An example of replacement of material with stock of better quality would be the replacement of a decayed red oak frame with one of white oak heartwood. The assumption is, of course, that the adjacent parts are sound and promise long service.

Another example of the use of improved materials would be the replacement of solid untreated stem with a laminated stem adequately treated with a good preservative.

In matching repair materials with those being replaced, the following factors need to be considered:

1. Species.
2. Heartwood and sapwood limitations.
3. Sizes.
4. Quality, as indicated by grade, type, moisture content, and other elements.

The information can generally be obtained from the original specifications and drawings for the class of ship being repaired, if they are available. If not, species identification can usually be made from the part to be repaired, or from comparable sound parts elsewhere in the hull. Likewise, the usual heartwood requirements for the part in question should be adhered to. Size is important because of strength and weight considerations and to provide adequate space for fastenings. The fastenings themselves should be at least equal, in both size and number, to those used in the original joint; with fewer or smaller fastenings, the joint will be weaker. Slightly larger fastenings may be needed to fasten a new member to an old member than were originally used.

The quality of the replacement part should conform to the usual specifications for such parts. For lumber, this requires that the appropriate grade and moisture content be used. Laminated wood suitable for hull repairs must be manufactured to meet special requirements. Lumber should have a moisture content at the time of installation which is approximately the same as that which will prevail in service. This is generaly about 13±3 percent. Underwater portions are normally installed at this moisture content also. This assures that they will become tighter as they take on moisture in service. Requirements for plywood express the quality of the veneers and the glue bonding them together. Plywood for boat and ship construction, Military Specification MIL–P–18066, should be restricted exclusively for repair and new construction on boats and ships due to its special characteristics and high cost.

3.520. FABRICATION OF REPAIR PARTS. Essentially, there is no difference between the fabrication of parts for repair and of those for new construction. Consequently, the discussion of new parts production is applicable to production of repair parts, as well as those for new hulls.

Where the volume of repair work justifies it, repair parts can be prefabricated in quantity to reduce the unit cost. In general, such production methods are economical where numerous craft of the same or similar classes are frequently repaired and where the material can be produced with conventional woodworking machinery to known patterns. Parts most frequently prefabricated under such conditions include sawed-frame futtocks or bent-frame blanks, gussets, stems, transom framing, knees, guards, and floors for the repair of small craft, such as launches, landing craft, utility boats, picket boats, and whaleboats. A shrinkage and fitting allowance should be established for all prefabricated parts.

3.53. PREPARING HULLS FOR REPAIR. Bilges should be drained and cleaned and, wherever possible, hulls should be opened up by removal of hatches, floor boards, bilge ceiling, and other covering to permit wet parts to dry before repair is undertaken. This is generally necessary whether the damaged craft has been in use or in storage. There are several reasons. Wet parts will probably dry anyway before the craft is again put to use, so that shrinkage stresses will develop in the joints between old and new parts, with possible checking, splitting, and other damage. Also, parts that are wet because of leakage, resulting from the broken or decayed parts being replaced, may very well contain incipient decay and, if still moist, are likely to infect the new parts, causing early failure. And thirdly, decayed and otherwise damaged parts are more likely to be detected when dry than when water-soaked.

Preliminary drying is, of course, more practical in small craft than in ships framed with heavy timbers. Many parts below waterline in a ship are necessarily wet and are so large that they cannot be dried within a reasonable length of time. Even with small craft, drying may take several months during the winter in damp coastal areas. If building space is available, it is recommended that hulls be opened up and stored in a ventilated building for a few weeks before repair is started. In any case, hulls so exposed should be protected from rain and snow, at least with paulins supported a few inches above the hull to permit ventilation; a shed with open sides is better.

Drying will be further hastened if obviously damaged parts are removed first. Thereby, wood that will be in contact with new parts will have a better chance to dry first. And, of course, a thorough housecleaning is in order, with removal of all water and dirt from the bilge, cleaning of drains and limbers, and scraping of barnacles from the bottom (fig. 36).

Figure 36. Scraping barnacles from bottom of an LCVP in for repairs.

3.54. INSPECTION. Inspection is needed to determine: (1) The repair work to be done; and (2) the adequacy of the repairs. A three-step inspection procedure is often followed, consisting of: (1) Initial inspection; (2) prefinal inspection; and (3) final inspection for acceptance.

3.540. INITIAL INSPECTION. Unless damage is purely superficial, the nature and extent of necessary repairs can be determined only by careful inspection. This is more important when decay is present than when the repairs are necessitated by mechanical damage, because decayed parts are often difficult to find. But mechanical damage also is likely to be more extensive than first appearances indicate. Collision shocks that are severe enough to break some parts, for example, undoubtedly will also cause lesser damage to joints between other parts before the stresses imposed have been dissipated Even unbroken parts may thus be racked out of alinement, so that the joints need to be rebuilt or, in extreme cases, the parts need to be replaced.

Initial inspection, then, calls for close scrutiny of all parts in the way of the obviously damaged ones. Even small cracks around joints should be carefully noted, for they may have serious weakening effects and at least expose untreated wood and allow moisture to gain access to critical points in the hull, thus raising the threat of decay. The fact that they can be hidden with paint means nothing; paint films are easily broken, especially over such weakened parts.

Inspection for decay is more detailed. Not only must the immediate vicinity of the obvious damage be carefully examined; other critical parts of the hull must also be inspected. Any sort of thorough inspection, in fact, often times cannot be made until some parts, such as floor boards, ceiling, sheathing, and armor, are removed. Probing for soft spots with an oyster knife, ice pick, or screwdriver is the usual method of seeking out decay (fig. 37). Mere visual inspection is not enough; discolorations can be very misleading, since stains caused by water, chemicals, metal fastenings and fittings, certain staining fungi, and wood extractives may look like those of decay, but seldom impair strength materially. Softened or crumbly wood, on the other hand, is weak regardless of the cause, and is a warning sign that adjacent wood not obviously damaged usually contains incipient decay which has caused some weakening and may continue to spread. For a high degree of safety, it is essential to mark wood for removal at least 2 feet each way along the grain and 2 inches across the grain of each member showing an area of obvious decay. (See also sec. 1.550 regarding application of wood preservative to such areas.)

Figure 37. Examining frames in contact with decay planking (removed) for presence of rot.

Adequate inspection should make possible a definite job order specifying in detail what repairs are to be made. For example, it should specify the precise strakes of planking or ceiling and the exact frames, beams, clamps, knees, and other parts to be replaced.

An important part of this inspection should be the determination of the cause of the trouble, especially if decay is responsible, and corrective measures should be taken to prevent recurrence of the same trouble. Corrective measures may include ventilators in closed compartments, alteration of flooring and ceiling to permit air circulation, and correction of any feature that permits leakage or prevents complete water runoff.

3.541. PREFINAL INSPECTION. The purpose of a prefinal inspection is to determine whether repair work has been adequately done. Of necessity, this inspection must be made before the hull has been completely finished and painted. In actual practice, resident inspectors are usually available for spot inspections as work proceeds, so that costs will not be unduly increased and the work delayed because partly finished work must be redone. Where such continuous inspection is maintained, a formal prefinal inspection is readily accomplished.

3.542. INSPECTION FOR ACCEPTANCE. Inspection for acceptance involves not only a visual inspection of the repaired craft in finished conditions, but may also include test runs for performance checks. Test runs are made if hull repairs were of the overhaul type or involved replacement of structural parts critical to the performance of the craft, such as shaftlogs, rudders, power plants, and other operating gear. Such inspection is intended to certify that all repairs have been satisfactorily completed and the craft is seaworthy.

3.55. MINOR REPAIRS. Any damage that can be repaired without removing the structural parts of a hull is classified in this Manual as a minor repair job. Examples are a small punc-

Figure 38. Plywood patch of a planking puncture, backed up with a butt plate.

Figure 39. Splicing a broken bent frame with a sister frame.

ture in planking that can be patched by one of several means (fig. 38) or a broken frame that can be spliced with a sister frame (fig. 39). Many such repairs are temporary; others can make the hull as serviceable as a new one.

3.550. SUPERFICIAL DAMAGE. Among the most frequently damaged parts of a hull are fenders and guards, coaming, and the like. The purpose of such parts as fenders is, of course, to protect the hull from unavoidable rubbing and bumping against docks and other craft. Damage is therefore inevitable and the protective members must be replaced at intervals to keep them serviceable. Likewise, on small craft particularly, deck rails, coaming, and similar parts are exposed to frequent mechanical damage by handling of cargo and stores and of the boat itself during lowering, launching, and landing operations.

Usual practice on small craft is to replace the entire damaged part. The reason is that this can be done much more quickly than a splicing operation, which, to be mechanically strong, must be done skillfully. For such parts as fenders, the practice has an added advantage, since removal of the entire part permits inspection of the wood underneath it for evidence of decay or other damage. Experience has shown that planking underneath such parts is particularly vulnerable to decay attack because moisture has a better than ordinary chance to collect there. It is good practice, therefore, to examine carefully planking thus exposed by removal of fenders, and to replace damaged strakes. The same is true of coaming, rails, deck fittings and the parts next to them. Exposed parts in sound condition should be brush-treated liberally with a preservative solution, such as pentachlorophenol or copper naphthenate.

On large hulls, it is practical to replace only damaged portions of fenders, guards, rails, and the like if they were installed in sections. Again, it is advisable to replace the entire section. Planking, decking, frameheads, and other parts adjacent to those removed should be examined for evidence of decay.

3.551. SOFT PATCHES. Emergency repairs of small planking holes in boats are often made with so-called soft patches. Sheet lead is commonly used, being tacked over the puncture in a bed of white lead paste to seal the edges of the sheet against entry of water. Such patches are ordinarily made by the ship's force to keep the craft in service until more extensive damage warrants thorough-going repair, or until the pressing need for the boat is past.

3.552. BUTT-BLOCK PATCHES. The simplest permanent repair for small holes in plywood planking and decking is the butt-block patch (figure 38). To make this patch, a hole large enough to remove all splintered and cracked wood is sawed out; for convenience in fitting the patch, the hole generally is square or rectangular. A butt block is then screwed to the solid planking around the opening on the inboard side between frames, leaving a space for water runoff and ventilation between the ends of the block and the frames. A patch is cut to fit the opening in the planking as snugly as possible on all edges and screwed in place. Resorcinol glue will help hold both butt block and patch more firmly in place than screws alone. If glue is used, the gluing surfaces should first be sanded. The grain of the face and back plies of the plywood patch should run in the same direction as the grain of the planking, and the plywood patch should be of the same thickness and number of plies as the planking. Coating the edges of the patch with white lead paste before it is set in place will help seal out water. Plywood patches are not outgaged for calking. If treated plywood is used, the edges should be brush-treated with preservative; untreated plywood should be soaked 2 hours in preservative after it is cut to shape.

3.553. LUMBER PLANKING REPAIR. When one or two adjacent strakes of lumber planking are stove in, without damage to frames, repair can follow the steps shown in figure 40. Each strake of damaged planking must be removed for several frames fore and aft of the damaged portion. This is necessary because short repair pieces cannot be bent to the curvature of the hull; a piece spanning two frames, for example, would be perfectly flat, and a plank extending over a number of frames is needed in order to get a fair curve. Moreover, the longer repair section is stronger and stiffer than a short section.

In making measurements for the repair sections, or "spiling," a template is lightly tacked in place while points are laid off on it at each

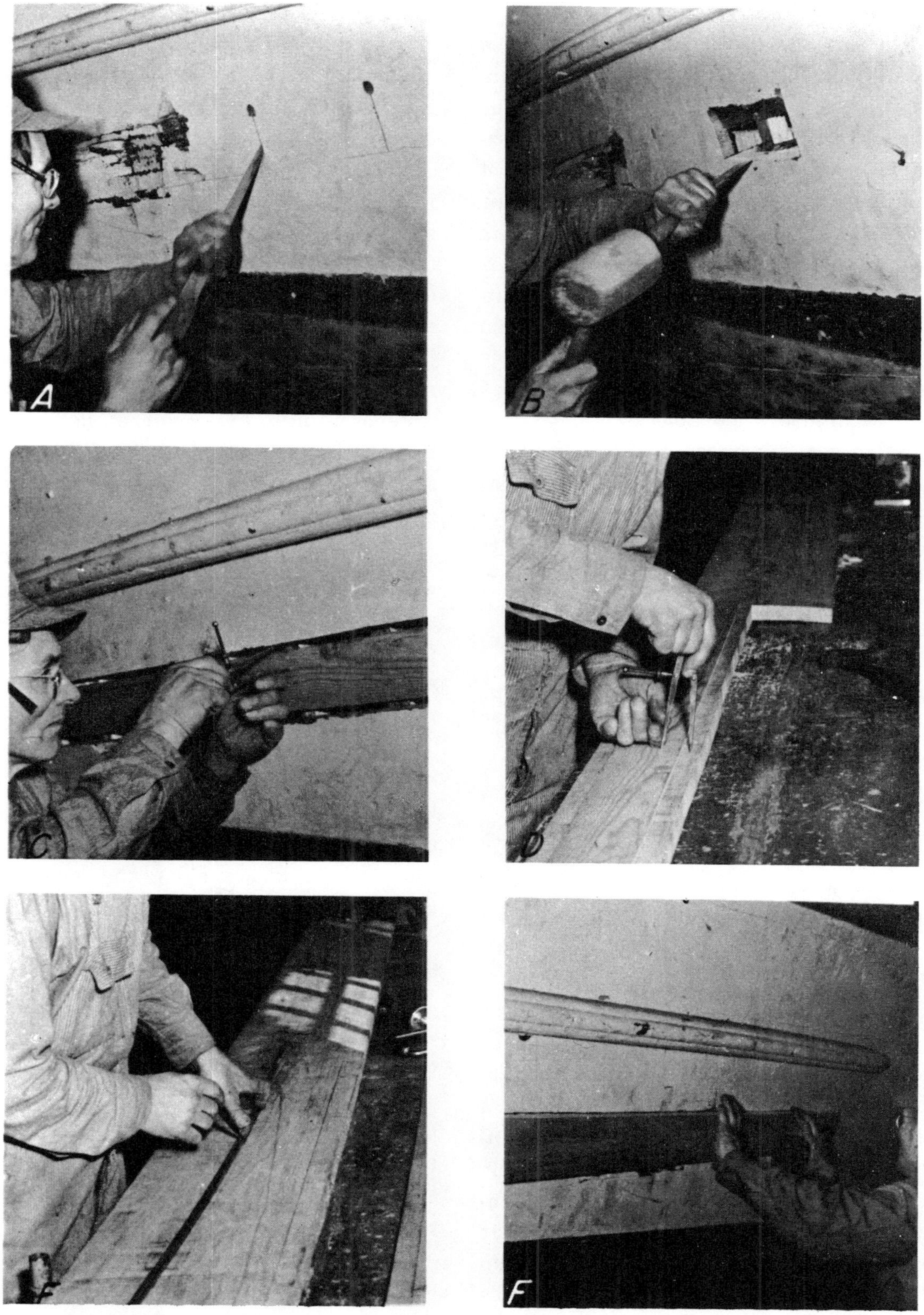

Figure 40. Steps in repairing stove-in planking. A, Sawing sections of damaged plank; B, removing plank with mallet and chisel; C, spiling with dividers to get measurements for repair section; D, laying out repair section with template and dividers; E, connecting divider points with flexible wood strip; F, installing repair section.

frame (fig. 40, C.). In this way, any slight variations in width of the strake can be noted and laid off on the repair material. The template is marked "fore" and "aft" to avoid mistakes in laying out the repair section. A flexible strip of wood is a convenient means of connecting the layout points on the repair material with a solid line for convenience in sawing out the repair section.

The sawed-out repair section can be outgaged with a hand plane for calking and should then be treated with preservative—preferably soaked 2 hours—before it is installed. Plank ends are secured with butt blocks.

The same general procedure is followed for repairing planking of heavy hulls, except as technique needs to be modified for the larger machines used. Repair planking may need to be hollowed to fit the curve of the frames, particularly at the turn of the bilge. And steaming may be needed where sharp curvatures occur. Brush treatment with preservative should include exposed framing, as well as the planking.

3.554. MINOR FRAME REPAIRS. Occasional damage to one, or even a very few frames of a small craft can be repaired well enough to keep the hull seaworthy by splicing the broken parts with sister frames. As a rule, this amounts to placing a full-length frame beside the damaged one. This should be done wherever possible. Installation of sister frames is poor practice when decay is present. Sometimes, however, other parts of the hull, such as longitudinals, engine mounts, and the like, prevent installation of a full-length frame. In such cases, a partial sister can be installed (fig. 39) for temporary use until the hull can be more adequately repaired. The partial sister should be through-bolted to the damaged frame and extend as far as possible beyond the damaged area both up and down. The use of partial sisters should be recognized as a purely temporary expedient when the hull cannot be spared for the time necessary to make more adequate repairs.

3.555. SPLICES FOR LONG MEMBERS. In repairing hull parts of considerable length, it is often necessary to splice sections together, either to salvage part of a broken member or to make replacements. Shipbuilders over the centuries have developed various ingenious joint designs for such parts. Bolts and other mechanical devices are the principal means of fastening such parts together. The availability of waterproof glues, however, has made possible the production, with shipyard tools, of joints considerably superior to joints with mechanical fastenings alone. Such joints have wide usefulness for repair of hulls—notably, for longitudinal members of small or big cross section, provided they are dry enough for gluing.

The plain scarf joint is the best all-around glue joint for repair purposes, because it is simple to make and stronger than more complicated types. Its simplicity, however, can be misleading; it must be well fitted. For repair purposes, a slope of 1 in 12 is recommended. That is, the cut should slant through the length of the piece a distance of 12 inches for every inch of depth or width. Thus, a cut through a 6-inch beam would extend a horizontal distance of 72 inches along the length of the member. The cut can be made through the depth or the width of the member, whichever is more advantageous. The diagonal cut can be shorter when made across the narrower face (fig. 41). Just before they are to be glued, the two surfaces should be carefully machined smooth and flat. For pieces too large to machine in a planer or scarf cutter, a portable router can be made to produce acceptably smooth surfaces. Such surfaces should be checked for trueness with a steel straightedge and feeler gage. Surfaces should not vary more than 0.003 inch from perfect flatness—a tolerance that has been shown to be practical in work at a shipyard. A hand plane can be used to smooth small irregularities (fig. 42).

Figure 41. Cutting a vertical scarf through the width of a laminated beam on a band saw.

Figure 42. Smoothing a scarf cut with a hand plane.

A resorcinol or phenol-resorcinol glue that cures at moderate temperatures should be applied and cured according to the manufacturer's directions. This is a critical gluing operation, hence the optimum (not the maximum or minimum) assembly time for a particular glue should be used. Spreads somewhat heavier than normal are required, because glue penetrates more readily into end grain than into side grain. Cauls and rockerhead-type clamps are convenient for shipyard use in applying pressure while the glue cures. Pressures of 150 to 250 p.s.i. are applied, depending on the species of wood and the size of the members.

3.56. MAJOR REPAIRS. Any damage that requires replacement of major structural parts of a hull or more than a few scattered strakes of planking or decking is considered a major repair job.

In an emergency, large-scale repairs may have to be undertaken by the ship's force with the materials and equipment at hand. In these instances, the methods will depend largely upon the ingenuity of the repair crew. Normally, however, repairs of a permanent nature will be accomplished after careful planning by a well-equipped repair shop. Under these conditions, a thorough job can be done using the best techniques.

During major repairs, partial dismantling of a hull is often required. This presents an opportunity to examine parts, especially in the bow or stern, that are normally hidden. Consequently, every opportunity should be taken by the inspector on the job to examine exposed parts adjacent to obviously damaged ones.

If probing and boring at the highest risk locations give any indication of decay, the suspected areas should be uncovered to adequately determine the extent of such damage. Such opportunities are especially useful for inspection of older hulls and of ships with full ceiling, such as sawn-frame harbor tubs and bent-frame minesweepers and submarine chasers, because ceiling frequently hides serious decay in frames. This is particularly true of sawn frames with their many joints between futtocks where moisture can gather. It is essential to brush-treat sound exposed parts including neighboring members with preservative before repair parts are installed and to apply a fungicide fortified bedding compound to all faying surfaces.

3.560. PLANKING REPAIRS. Major repairs to planking generally involve replacement of a number of adjacent strakes for some distance fore and aft of the damaged portion. As a rule, the greater the area of planking damaged, the more necessary it becomes to remove extensive portions of undamaged planking above and below, as well as fore and aft, of the damaged area. The reason has already been pointed out but bears repeating: Repair pieces must be long enough to assure adequate strength and retention of the hull's lines, both of which are essential in preserving the craft's seaworthiness. In major damage, moreover, framing is usually affected more or less extensively.

Methods of removing damaged planking vary, of course, with the type of planking—single-thickness, fore-and-aft; double-thickness, fore-and-aft; diagonal inner and fore-and-aft outer planking; both inner and outer diagonally laid; plywood single or double skin; or a combination of plywood inner skin and strip-plank outer skin. Methods also vary with respect to the comparatively light planking of boats and the heavy single or double planking on ships.

Only two methods of plank removal and repair are discussed herein, single-skin fore-and-aft strip and single-skin plywood. These methods are, in general, adaptable for use with hulls planked by other systems.

3.5600. Repair of Strip Planking. Essentially, the method described for repair of planking in section 3.553 and figure 40 can be adapted for repair of strip planking where damage is extensive enough to be classed as major. The principal precaution to be observed in replacing a number of adjacent strakes of plank is to stagger the butts. On ships, it is generally advisable to remove full planks, and on any hull, no plank should be partially removed if this will leave only a short length remaining from the point where the plank was cut off to its butt joint with the next plank in the strake. To leave a length of plank spanning only four or five frames means that excessive stresses will be put on the fastenings, especially at points of the hull where curvature is appreciable, to hold the plank ends to the frames or butt blocks. Short lengths of plank may readily "work" enough in heavy seas to open up the seams. With diagonally laid planking, only full-length planks are generally used (fig. 43).

Figure 43. Diagonal planking installed as inner skin repair.

Repair of damaged frames is discussed in sections 3.5621 and 3.5622. Even undamaged frames, however, should be plugged to fill all fastening holes before new planking is installed. Wood plugs of a diameter one thirty-second inch or so larger than bolt holes give a suitable tight fit, and plugs for screw holes should be similarly snug. Plugs should be of heartwood of the same species as the frame. If the frame is of treated wood, the plugs should also be treated; before they are inserted, preservative should be squirted into the holes.

Unless otherwise specified, planking should be of the same species and sizes as that which it replaces, and should be fastened with bolts, screws, or rivets of the same kind, size, and metal used in the rest of the hull. Calking and finishing should adhere to Navy standards (fig. 44); finishes should not be applied to freshly preservative-treated surfaces for 72 hours after treatment.

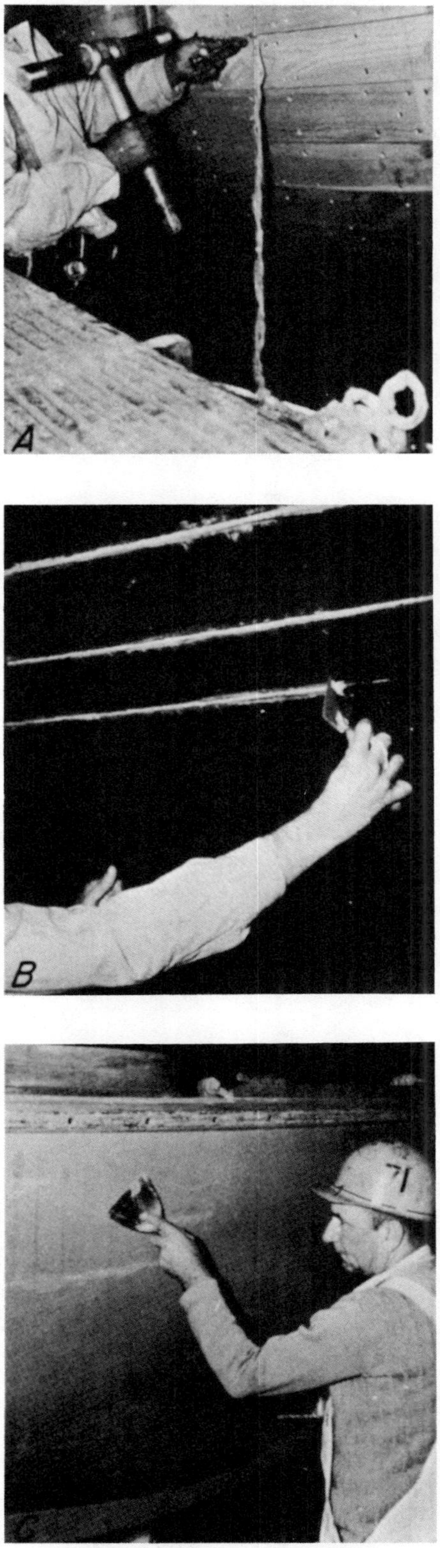

Figure 44. A, Calking strip planking of a repaired motor launch; B, Applying seam sealer to calked seams; C, Glazing compound applied to provide smooth surface for paint.

3.5601. Repair of Plywood Planking. Major damage to plywood single-skin planking necessitates removal of the entire sheet originally installed, even if it extends the length of the hull (fig. 45). Only if the plywood serves as inner skin covered by a mastic or other moisture barrier and an outer skin of plywood or strip planking, is it feasible to cut on the damaged portion and splice a repair sheet to the undamaged part of the original skin with a butt plate glued and screwed over the joint.

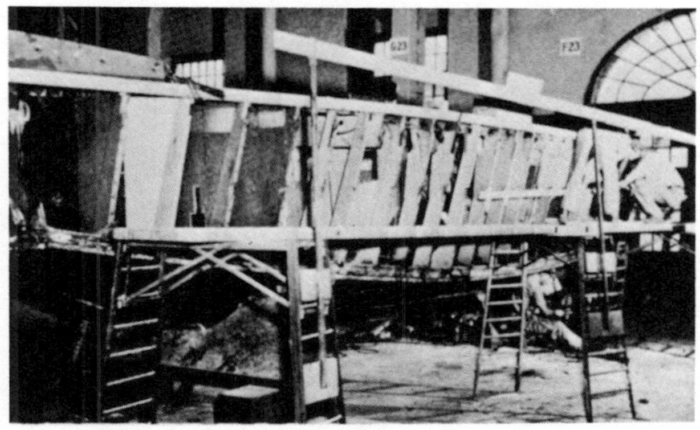

Figure 45. LCVP with plywood side removed for repair of hull.

A rapid method of removing plywood planking is shown in figure 46. Fastenings are located and a groove is cut through the plywood around each with a two-wing circular groove cutter. The sheet is then pried off and the round plugs left with each fastening knocked off; after which, the fastenings are withdrawn. Fastening holes in undamaged frames are filled with wood plugs.

Only boat and ship plywood, Military Specification MIL–P–18066, is used for major boat hull repairs. The replacement sheet of plywood is carefully scribed to fit the area to be covered, and sawed out. Before it is lifted into place, the perimeter of the side to which it is to be fastened is coated with bedding compound. The new plywood is then set in place, trimmed for secure fit, and fastened. Figure 47 shows the steps in this process.

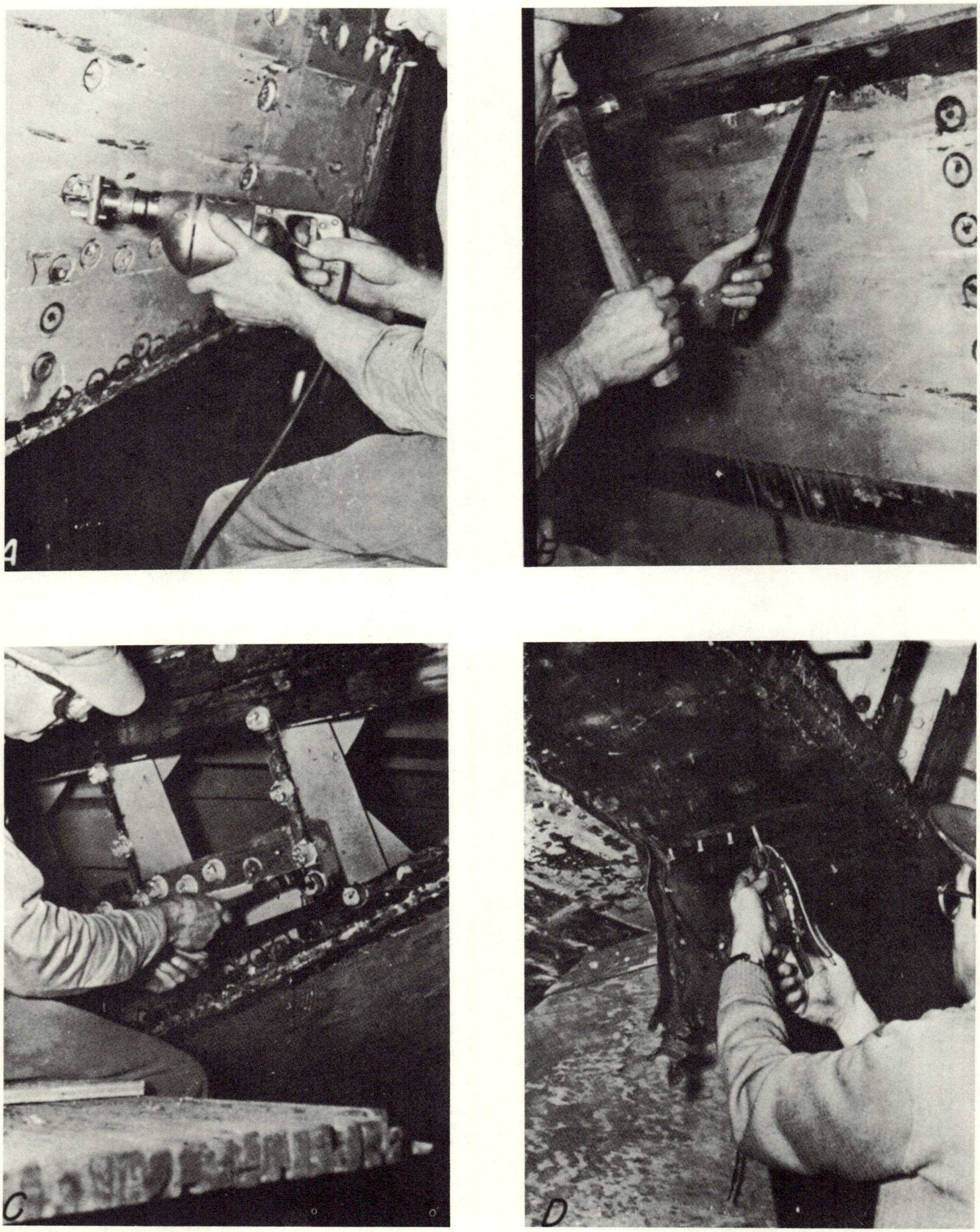

Figure 46. Removing plywood side from LCVP. A, Cutting grooves around fastenings with two-wing circular groove cutter. B, Prying plywood off. C, Knocking off wood plugs at fastenings. D, Removing fastenings.

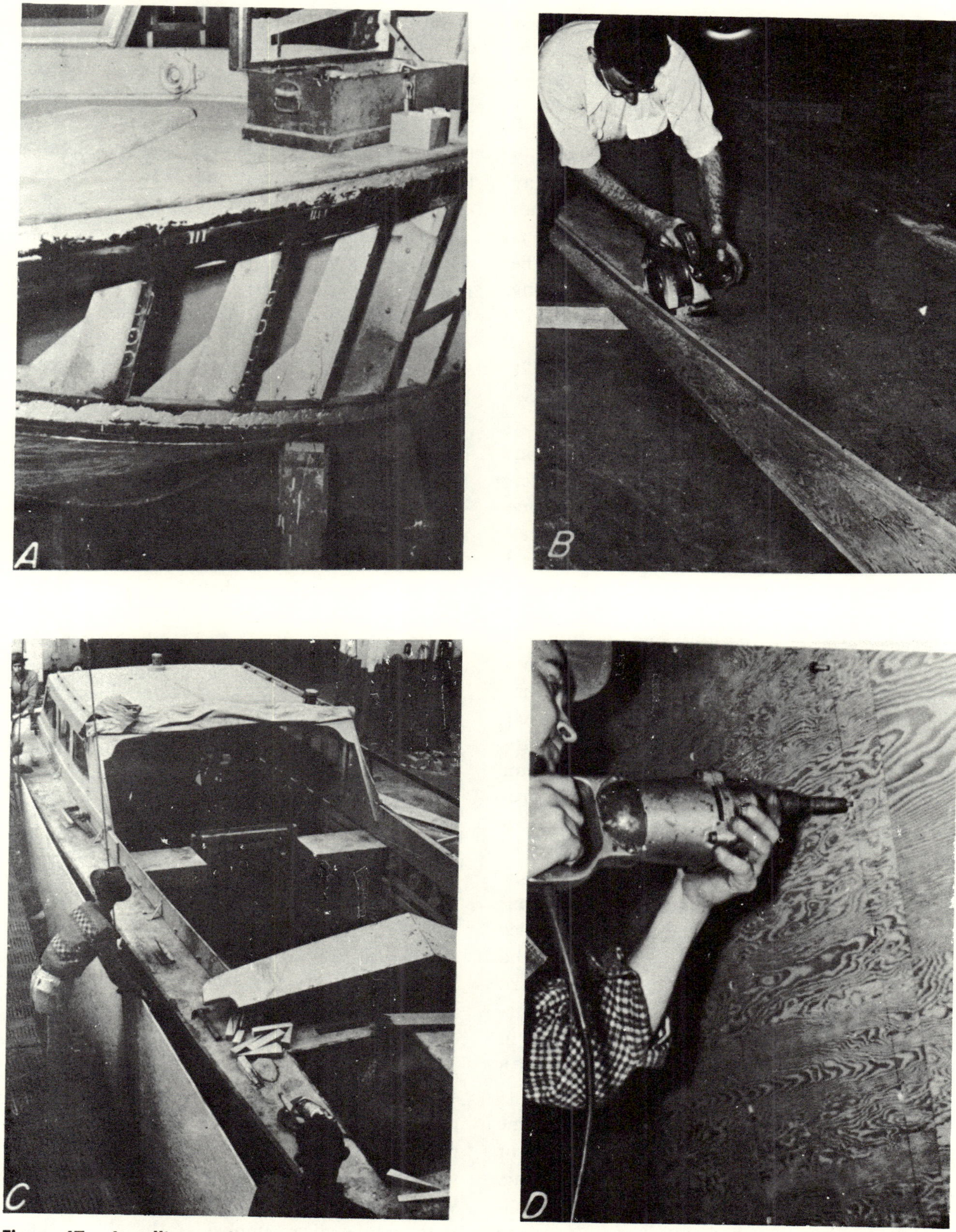

Figure 47. Installing replacement sheet of plywood on side of motor launch. A, Chine, sheer clamp, and bow frames coated with bedding compound. B, Cutting replacement plywood to shape. C, Lifting plywood in place with overhead crane. D, Fastening plywood to frames.

If the plywood has not been pressure-treated with preservative, it should be given at least a 2-hour soak; or, lacking soaking tanks, several liberal brush coats should be applied before it is installed. Even with pressure-treated plywood, it pays to coat freshly cut edges liberally with preservative, since untreated wood may be exposed. This can best be done by suspending the plywood edgewise in a trough of preservative for several minutes or longer. The exposed end grain of some of the plies along each edge will soak up preservative freely. Finishing is done in accordance with prescribed Navy standards, with a drying period of 72 hours allowed for freshly preservative-treated surfaces.

3.561. DECKING REPAIRS. Although occasionally a deck sustains major damage from accidents, most deck repairs are necessitated by one thing—fresh (rain) water leaks that let decay develop, especially in margin pieces, around deckhouses, and sometimes in the deckbeams, sheer clamps, shelves, and other supporting members. Decay in decking and its supports may also be caused by condensation of vapors from galleys, baths, and even refrigerated compartments or the cold undersurface of weather decks. If the cause is determined during inspection of the extent of the damage, steps can often be taken during repair operations to prevent or minimize future trouble. Better maintenance of calking and seam compounds, use of reflective insulation, installation of metal flashing and vapor barriers, and use of bedding compound on all faying surfaces, are examples of things that can be done to cut down the frequency of such troubles. Particular care should be taken to avoid low spots in deck surfaces that would allow rainwater to collect in pools. Sanding should be carefully done with this in mind.

3.5610. Repair of Lumber Decking. The same rules and precautions that apply to repair of strip planking also apply to lumber decking. While it is not so critically important that full-length members be installed to replace damaged pieces, it is usually more convenient to remove an entire deck plank than to cut it off centered on a beam. Deck plank butts should be staggered and repair pieces should span at least three deck beams for adequate rigidity and to lessen the stress on end fastenings under heavy loads. Even though such stresses do not cause bending deflection, they may affect tightness of calked seams.

Repair material should conform to Navy requirements for edge grain and other quality criteria important for decking. Width and thickness must match those of the original decking; this is essential for tight calking as well as for adequate strength, since seams must line up with those of the undamaged decking.

3.5611. Repair of Plywood Decking. Plywood decking is generally used only on boats and high-speed craft. It is mounted in various sizes from small pieces to deck-length sections scarf-glued from standard 4-by-8-foot sheets. To retain the design strength and rigidity of the hull, repair sheets should, as a rule, be as large as those damaged. Only when this is manifestly impractical for lack of scarfing facilities should smaller pieces be used to replace large scarf-joined deck sections. By so doing, the number of seams is increased and these must be carefully sealed. The use of preservative on all edges of plywood is recommended; all joints should be saturated with it before they are sealed, unless the plywood has been treated thoroughly beforehand. Freshly treated surfaces should dry 72 hours before they are painted.

3.562. REPAIRS TO STRUCTURAL FRAMEWORK. Major damage to planking and decking generally also involves damage to supporting framework. Regardless of the type of damage, whether due to mechanical causes, decay, or something else, it is usually impractical to repair the part in place. Generally, the damaged part must be removed and repaired, or replaced with a new part (fig. 48). Exceptions are V-type frames and similar parts of small craft that can be disassembled at the chine, sheer, and keel for removal of the broken component.

Figure 48. Replacing a solid stem with a laminated one.

3.5620. Repair of Long Straight Members. If straight, or even mildly curved, long members such as sheer clamps, shelves, and chines sustain localized damage, the undamaged part can sometimes be spliced with a plain scarf joint. The method used is described in section 3.555. The plain scarf joint is not only simpler to make, but tests have shown that it is stronger than more complex joints, such as hooked and fingered scarfs.

Long members, such as keels and clamps, that consist of several parts bolted together at hook scarfs can, of course, be disassembled for replacement of only the damaged parts. Figure 49 shows a keel being replaced in a boat and figure 50, a keel shoe being installed on a V-bottom hull. Timber connectors are useful sometimes for strengthening bolted joints.

Figure 49. Replacing keel under skeg of bent-frame boat.

Figure 50. Replacing keel shoe on V-bottom hull. Braces hold shoe in place while it is screwed to keel.

3.5621. Repair of Bent Frames. The practicability of repairing bent frames depends upon the type and extent of damage done to them and the size and soundness of the frames. For boats, it is generally more practical to replace the frame with a new one, either entirely removing the old frame or splicing the new one to it, as described in section 3.554.

New techniques in the experimental stage of development indicate that it may be feasible to repair large bent frames, such as the one-piece laminated U-frames of minesweepers that extend from sheer to sheer, by a scarf splicing technique. This method has been found in experiment under shipyard conditions to be practical for splicing minesweeper frames shipped in sections for assembly by scarf gluing at the shipyard, with the further requirement that the scarf must be cut into an existing damaged frame that has been removed from the hull, and a mating scarf cut on the frame section to be spliced to it. If sections to be spliced are of uniform moisture content, and proper techniques are used, spliced frames may be expected to be about as strong as one-piece frames.

The method of cutting a scarf joint is illustrated in figure 51. A two-man combination crosscut-rip saw or other tool can be used to rough-cut the scarf to the approximate slope. Then the rough surfaces must be machined smooth, for example with a portable router and guide boards. The main requirements of this method of splicing are: (1) The two surfaces to be joined must be of the same size and shape; and (2) each scarfed surface must be perfectly flat and smooth.

Guidelines are marked off on all four faces of the piece to be cut. These guidelines should be positioned so that the scarf joint is not near scarf joints in either the top or bottom lamination of the member. The scarf lines should also avoid visible knots in inner laminations. A slope of 1 in 12 is the steepest slope recommended. A preliminary guide cut can be made with a portable circular saw, as shown in figure 51.

3.5622. Repair of Sawn Frames. Damaged sawn frames of V-bottom craft can often be repaired without removing the entire frame from the hull. Only the broken member need be removed (fig. 52), by taking out the fastenings that secure it to longitudinals, such as keel, chine, or sheerclamp, and taking off gusset plates and other joint reinforcements. A replacement member is then installed (fig. 53).

Sawn frames of ships can sometimes be repaired in much the same way by replacing only the damaged futtocks. For example,

Figure 51. Cutting a scarf joint in a laminated minesweeper frame for repair splicing. A, Guide cut is made along scarf guideline on both sides. B, Through cut is made with combination crosscut-rip saw. C, Surface is machined true and smooth with router guided on edge strips.

Figure 52. Removing broken side member of frame from LCVP.

seriously decayed floor futtocks can be replaced one at a time with new ones. It is good practice, however, to replace also the futtocks immediately adjacent to those showing obvious decay, since they too are probably seriously infected, although still apparently sound.

3.57. OVERHAUL REPAIRS. Overhaul repairs are generally made to small craft that have: (1) Been in use long enough so that all parts, including motor, propeller gear, and other mechanical adjuncts need complete examination along with the hull; or (2) have been in storage under favorable conditions for a prolonged period (see chapter II).

Overhaul repair techniques differ from those for major repairs only in extent. Usually, the hull is dismantled and stripped of planking, and each framework member is inspected for mechanical breakage, decay, corrosion damage, and other effects of long, hard use or unfavorable conditions. The objective is to put the complete craft, engine and equipment, as well as hull, in as nearly new condition as possible, so that it can be assigned to a ship as a new boat.

Figure 53. Installing new side member of frame of LCVP.

3.58. SUMMARY. Storage conditions and repair practices are the factors which usually determine the serviceable life of a wood ship or boat. If storage and repairs are carried out with full recognition of the principles stated in this chapter, practically unlimited economical service can be expected. On the other hand, if these principles are ignored or handled in a slip-shod way, a well designed and constructed craft can become practically worthless in a few short years.

There are no set formulas or rules which can be applied to insure satisfactory results in storage and repair. Good storage and repair are primarily matters of planning and a working knowledge of the factors which lead to deterioration.

Minehunter MHC-43, first U.S. Navy ship framed with laminated pressure-treated red oak.

WOOD:

A MANUAL FOR
ITS USE AS
A SHIPBUILDING MATERIAL

VOLUME III

TECHNICAL DATA APPLICABLE TO
BOAT AND SHIP DESIGN

PREPARED BY
BUREAU OF SHIPS
DEPARTMENT OF THE NAVY

WITH THE COOPERATION OF

FOREST PRODUCTS LABORATORY
FOREST SERVICE
UNITED STATES DEPARTMENT OF AGRICULTURE

BUREAU OF SHIPS · NAVY DEPARTMENT · WASHINGTON 25, D.C.

JULY 1962

DEPARTMENT OF THE NAVY
BUREAU OF SHIPS
WASHINGTON 25, D. C.

IN REPLY REFER TO

During World War II and the Korean action, more than 40,000 wooden ships, boats and landing craft were added to the Fleet. In the ensuing years, many additional wooden ships and boats have been put into service, incorporating the most recent knowledge of wood as an engineering material. A continuing program of improvement in wood laminating, extreme service adhesives and development of wood preservatives for protection against marine borers has been largely responsible for the satisfactory performance of these craft.

Because of the many new developments in marine uses of wood, a need has arisen for a comprehensive, up-to-date collection of information based on recent investigations and tests. This new publication, Volume III of "Wood: A Manual For Its Use As A Shipbuilding Material," has therefore been prepared and should prove most useful to shipyards, naval architects, Supervisors of Shipbuilding, Inspectors of Naval Materials, and officers and men concerned with all phases of wooden ship or boat design, construction or maintenance.

ACKNOWLEDGMENTS

In an industry so old as shipbuilding, many men, most of them unknown, have contributed to our present day knowledge and advancement of the art and science. A notable early attempt to organize the improvement of shipbuilding was made by King James I in 1605 when he chartered a Shipwrights Guild with supervision over all shipbuilding in England, and which all English shipwrights were required to join. The Guild made significant progress under the leadership of Master Shipwright Phineas Pett who introduced the use of scale models and mathematically developed plans, making possible the building of two ships alike.

Later, the French took up the banner of progress and carried it well into the 1800's, surpassing the British. This era of shipbuilding witnessed the highest development of the wooden sailing ship in construction, speed, and beauty. During this time, the American forest contributed enormous quantities of shiptimbers, planking, decking, masts, and spars. American shipwrights achieved a reputation for skilled workmanship held to this day.

Shipbuilding began to emerge from an art to a science during the 19th century. Several schools of naval architecture were founded in England, the United States, and Germany, followed by such technical societies as the Institution of Engineers and Shipbuilders in Scotland, the British Institution of Naval Architects, and Institute of Marine Engineers, and in the United States, the Society of Naval Architects and Marine Engineers.

Thus, the modern science of wood shipbuilding has developed over the centuries. Most significant of the achievements during the 20th century has been the adopting of adhesive bonded wood as an engineering material in the construction of wooden hull ships. The bonding together of small pieces of wood into large structural members or panels in either straight or curved form results in a far more efficient use of wood than was formerly possible. This development is the first basic change in wooden shipbuilding in 2,000 years.

Previous editions of this manual have mentioned many individual shipbuilders, researchers, and other contributors who furnished information regarding modern wood ship and boat building practices. The present edition represents the combined efforts of virtually all American technical societies, research agencies, trade associations, and universities which foster progress in wood technology, naval architecture, and timber engineering. It is the Bureau's desire to give full recognition to these groups and to the part played by the lumber, laminating, plywood, adhesives, and wood preservation industries in modern wood shipbuilding. Without their cumulative contributions to the advancement of wood as an engineering material, this manual would not be possible.

The basic manuscript for "Wood: A Manual for Its Use as a Shipbuilding Material" was prepared by the Forest Products Laboratory, Forest Service, United States Department of Agriculture, at the request of the Bureau of Ships, Department of the Navy. A very large share of the credit for the material presented in the manual is due the Laboratory, which has been foremost in the field of forest products for almost 50 years. The Laboratory's individual staff members who have contributed to the manual are recognized authorities in their individual fields. Thanks are also due the staff of the Division of Forest Disease Research of the United States Forest Service, and the Timber Engineering Company Laboratory for their advice and guidance.

The manuscript was reviewed and edited for the Bureau of Ships by members of its staff. The final arrangement of the material for publication was made under the direction of the late Mr. G.I. Dewey, who supervised wood products development for the Bureau of Ships until 1948, and Mr. J. G. Kuenzel, in charge of wood products development until 1962.

TABLE OF CONTENTS

WOOD: A MANUAL FOR ITS USE AS A SHIPBUILDING MATERIAL
INTRODUCTION TO VOLUME III

Purpose and Scope. The six chapters which make up Volume III contain basic technical data for use primarily by designers of wood hull boats and ships. Included are data related to design with solid or laminated wood, plywood composite members, and their attendant joints and fastenings.

The complete manual is composed of four volumes, as follows:

Volume I Basic Wood Technology Applicable to Boat and Shipbuilding.
Volume II Techniques and Practices Applicable to Preservation and Storage.
Volume III Technical Data Applicable to Boat and Ship Design.
Volume IV Techniques Applicable to Boat and Ship Construction.

CHAPTER 1. BASIC STRENGTH OF WOOD

1.1 GENERAL.--Strength is a general term for a number of interrelated mechanical properties. The strength properties of wood of principal importance to the shipbuilder are bending strength, as in a beam; stiffness; shearing strength; compressive strength parallel to grain, as in a post; compressive strength perpendicular to grain, as in the bearing of a beam on a post; hardness, a measure of resistance to indentation and, to some degree, workability; and toughness, or impact strength, which is a measure of the ability of wood to absorb shocks. A set of standard tests is used to evaluate these and other strength properties. Some of the strength properties of the principal woods used in boatbuilding, as determined by the standard American Society for Testing Materials (ASTM) mechanical tests on green wood and wood at 12 percent moisture content, are shown in table 1-1. "Green wood" is defined as unseasoned wood with a moisture content at or above 25 to 30 percent. Wood at 12 percent has the approximate moisture content it attains when protected from precipitation but otherwise exposed to the weather.

The values of table 1-1 are average values of the strength of clear wood as determined in laboratory tests. These require adjustment from the conditions of laboratory test to those of structural use, and the application of a factor of safety. Among the more important considerations in this adjustment are the variability of the material and the difference in strength between short-time and long-time loading. This adjustment and the application of a factor of safety lead to basic stresses that are allowable design stresses for clear wood. Basic stresses derived in this manner are given in Chapter 2. Basic stresses are used to derive allowable working stresses for structural design with lumber of a particular grade in boats and ships.

CHAPTER 2. EFFECT OF MOISTURE CONTENT AND TEMPERATURE ON STRENGTH OF WOOD

2.1 EFFECT OF MOISTURE CONTENT ON STRENGTH.--Most of the strength properties of wood improve as it dries. The increase in strength varies with the property. Values for compression parallel to grain of small clear pieces increase about twice and those for stress in extreme fiber about two-thirds in drying from the green condition to 12 percent moisture content. Impact and work values increase only slightly, or in some species may even decrease as the wood dries from a green condition to 12 percent moisture content. Twelve percent is the approximate average moisture content reached by wood protected from precipitation but otherwise exposed to the weather.

During drying, timbers of large size that contain defects do not increase in strength in the same proportion as small clear pieces, because these defects cause injurious effects that increase with drying, partially or wholly offsetting the increase in the strength of the fibers.

Table 1-1.--Strength properties of various native and imported woods

(Results of tests on small, clear specimens in the green and air-dry condition[1])

Species	Moisture content	Specific gravity[2]	Static bending: Fiber stress at proportional limit	Static bending: Modulus of-- Rupture	Static bending: Modulus of-- Elasticity	Static bending: Work to-- Proportional limit	Static bending: Work to-- Maximum load	Impact bending--height of drop causing complete failure (50-pound hammer)	Compression parallel to grain: Fiber stress at proportional limit	Compression parallel to grain: Maximum crushing strength	Compression perpendicular to grain--fiber stress at proportional limit	Shear parallel to grain--maximum shearing strength	Tension perpendicular to grain--maximum tensile strength	Hardness: Load required to embed a 0.444-inch ball to 1/2 its diameter: End	Hardness: Side
1	2	3	4	5	6	7	8	9	10	11	12	13	14	15	16
	Percent		P.s.i.	P.s.i.	1,000 p.s.i.	In.-lb. per cu. in.	In.-lb. per cu. in.	In.	P.s.i.	P.s.i.	P.s.i.	P.s.i.	P.s.i.	Lb.	Lb.
HARDWOODS															
Angelique	79	0.60	...	11,400	1,840	...	12.0	...	...	5,590	1,000	1,340	...	1,100	1,100
	12	...	...	17,400	2,190	...	15.2	...	...	8,770	1,280	1,660	...	1,700	1,290
Apitong	39	.58	...	9,400	1,940	...	7.7	...	...	4,500	680	1,140	...	870	780
	12	.63	...	12,300	1,940	...	14.8	...	...	6,200	990	1,330	...	910	880
Ash:															
Black	85	.45	2,600	6,000	1,040	0.41	12.1	33	1,690	2,300	430	860	490	590	520
	12	.49	7,200	12,600	1,600	1.57	14.9	35	4,520	5,970	940	1,570	700	1,150	850
Commercial white[3]	43	.54	5,300	9,500	1,410	1.14	14.7	37	3,360	4,060	860	1,350	580	1,010	940
	12	.58	8,800	14,600	1,680	2.68	15.6	40	5,580	7,280	1,510	1,920	850	1,680	1,260
Balsa[4]	12	.072	...	870	180	...	...	...	...	420	50	120	...	60	30
	12	.129	...	2,100	420	...	...	...	...	1,240	90	240	...	170	75
	12	.200	...	3,550	730	...	...	...	...	2,250	140	380	...	300	130
Basswood, American	105	.32	2,700	5,000	1,040	.40	5.3	16	1,690	2,220	210	600	280	290	250
	12	.37	5,900	8,700	1,460	1.37	7.2	16	3,800	4,730	450	990	350	520	410
Beech	54	.56	4,300	8,600	1,380	.85	11.9	43	2,550	3,550	670	1,290	720	970	850
	12	.64	8,700	14,900	1,720	2.63	15.1	41	4,880	7,300	1,250	2,010	1,010	1,590	1,300
Birch, sweet and yellow[5]	62	.57	4,400	8,700	1,560	.79	16.0	48	2,640	3,510	550	1,160	430	910	850
	12	.63	10,100	16,700	2,070	2.83	19.8	52	6,200	8,310	1,250	2,020	930	1,660	1,340
Cherry, black	55	.47	4,200	8,000	1,310	.80	12.8	33	2,940	3,540	440	1,130	570	750	660
	12	.50	9,000	12,300	1,490	3.11	11.4	29	5,960	7,110	850	1,700	560	1,470	950
Rock	48	.57	4,600	9,500	1,190	1.05	19.8	54	2,970	3,780	750	1,270	...	980	940
	12	.63	8,000	14,800	1,540	2.45	19.2	56	4,700	7,050	1,520	1,920	...	1,510	1,320
Slippery	85	.48	4,000	8,000	1,230	.82	15.4	47	2,790	3,320	510	1,110	640	750	660
	12	.53	7,700	13,000	1,490	2.35	16.9	45	4,760	6,360	1,010	1,630	530	1,120	860

(Sheet 1 of 5)

Table 1-1.--Strength properties of various native and imported woods--Continued

(Results of tests on small, clear specimens in the green and air-dry condition[1])

Species	Moisture content	Specific gravity[2]	Static bending: Fiber stress at proportional limit	Static bending: Modulus of-- Rupture	Static bending: Modulus of-- Elasticity	Static bending: Work to-- Proportional limit	Static bending: Work to-- Maximum load	Impact bending--height of drop causing complete failure (50-pound hammer)	Compression parallel to grain: Fiber stress at proportional limit	Compression parallel to grain: Maximum crushing strength	Compression perpendicular to grain--fiber stress at proportional limit	Shear parallel to grain--maximum shearing strength	Tension perpendicular to grain--maximum tensile strength	Hardness: Load required to embed a 0.444-inch ball to 1/2 its diameter: End	Hardness: Side
1	2	3	4	5	6	7	8	9	10	11	12	13	14	15	16
	Percent		P.s.i.	P.s.i.	1,000 p.s.i.	In.-lb. per cu. in	In.-lb. per cu. in.	In.	P.s.i.	P.s.i.	P.s.i.	P.s.i.	P.s.i.	Lb.	Lb.
Greenheart	41	.81		19,300	2,970					10,500		1,320		2,160	2,110
	12	.89		30,500	3,400					14,940		2,830			2,650
Hickory:															
Pecan[6]	68	.59	5,300	10,000	1,380	1.18	19.3	60	3,810	4,320	980	1,260	680	1,270	1,310
	12	.65	9,100	16,300	1,780	2.61	18.8	58	5,240	8,280	2,040	2,080		1,930	1,820
True[7]	57	.65	6,100	11,300	1,570	1.33	28.9	88	3,650	4,570	1,080	1,360			
	12	.73	10,900	19,700	2,100	3.07	27.2	75		8,970	2,310	2,130			
Khaya	55	.43		7,600	1,180		8.5			3,670	650	1,270		570	510
	12	.45		10,700	1,480		7.8			5,680	980	1,340		1,080	790
Lauans:															
Red	48	.40		7,300	1,280		11.0			3,680	480			480	510
	12	.52		11,600	1,820		11.0			6,500	980			790	710
Tangile	37	.53		8,600	1,480		8.2			4,240	600	1,000		670	680
	12	.56		11,000	1,730		10.6			5,450	850	1,120		670	640
White	12	.51		11,700	1,810		12.4			6,050	850	1,200		950	710
Lignumvitae	12	1.09								11,400				3,600	4,500
Locust, black	40	.66	8,800	13,800	1,850	2.36	15.4	44	6,120	6,800	1,430	1,760	770	1,640	1,570
	12	.69	12,800	19,400	2,050	4.62	18.4	57	6,800	10,180	2,260	2,480	640	1,580	1,700
Mahogany	58	.45		9,200	1,290		10.2			4,540	710	1,310		750	650
	12	.46		11,100	1,430		6.8			6,430	1,210	1,050		880	760
Maple:															
Sugar	58	.56	5,100	9,400	1,550	1.03	13.3	40	2,850	4,020	800	1,460		1,070	970
	12	.63	9,500	15,800	1,830	2.76	16.5	39	5,390	7,830	1,810	2,330		1,840	1,450
Maria (Santa Maria)	Green	.50		9,900	1,540					5,160	640	1,200		990	860
	12	.51		12,600	1,700					6,670	1,210	1,480		1,370	870

(Sheet 2 of 5)

Table 1-1.--Strength properties of various native and imported woods--Continued

(Results of tests on small, clear specimens in the green and air-dry condition[1])

Species	Moisture content	Specific gravity[2]	Static bending: Fiber stress at proportional limit	Static bending: Modulus of-- Rupture	Static bending: Modulus of-- Elasticity	Static bending: Work to-- Proportional limit	Static bending: Work to-- Maximum load	Impact bending--height of drop causing complete failure (50-pound hammer)	Compression parallel to grain: Fiber stress at proportional limit	Compression parallel to grain: Maximum crushing strength	Compression perpendicular to grain--fiber stress at proportional limit	Shear parallel to grain--maximum shearing strength	Tension perpendicular to grain--maximum tensile strength	Hardness: Load required to embed a 0.444-inch ball to 1/2 its diameter: End	Hardness: Side
1	2	3	4	5	6	7	8	9	10	11	12	13	14	15	16
	Percent		P.s.i.	P.s.i.	1,000 p.s.i.	In.-lb. per cu. in.	In.-lb. per cu. in.	In.	P.s.i.	P.s.i.	P.s.i.	P.s.i.	P.s.i.	Lb.	Lb.
Oak:															
Red[8]	80	0.56	4,400	8,500	1,360	0.85	12.6	43	2,590	3,520	800	1,220	740	1,050	1,030
	12	.63	8,400	14,400	1,810	2.30	15.0	43	4,610	6,920	1,260	1,830	820	1,490	1,300
White[9]	70	.59	4,700	8,100	1,200	1.08	11.3	42	2,940	3,520	850	1,270	760	1,110	1,070
	12	.67	7,900	13,900	1,620	2.31	13.3	39	4,350	7,040	1,410	1,890	770	1,420	1,330
Okoume	12	.37		7,600	1,300		4.9			4,710	430	1,230		800	450
Peroba do Campo	Green			14,100	1,500					6,520					
	15	.73		16,900						7,840					
Sweetgum	81	.44	3,700	6,800	1,150	.81	9.4	33	2,230	2,840	460	1,070	510	630	520
	12	.49	8,100	11,900	1,490	2.57	11.3	32	4,700	5,800	860	1,610	800	950	690
Teak	52	.58		11,400	1,580		10.0			5,490	1,040	1,300		900	980
	12			13,800	1,670		9.3			7,520	1,190	1,360		1,010	1,100
Walnut, black	81	.51	5,400	9,500	1,420	1.16	14.6	37	3,520	4,300	600	1,220	570	960	900
	12	.55	10,500	14,600	1,680	3.70	10.7	34	5,780	7,580	1,250	1,370	690	1,050	1,010
Yellow-poplar	64	.38	3,400	5,400	1,090	.62	5.4	18	1,930	2,420	330	740	450	390	340
	12	.40	6,100	9,200	1,500	1.43	6.8	20	3,550	5,290	580	1,100	520	560	450
SOFTWOODS															
Baldcypress	91	.42	4,200	6,600	1,180	.91	6.6	25	3,100	3,580	500	810	300	440	390
	12	.46	7,200	10,600	1,440	2.15	8.2	24	4,740	6,360	900	1,000	270	660	510

(Sheet 3 of 5)

Table 1-1.--Strength properties of various native and imported woods--Continued

(Results of tests on small, clear specimens in the green and air-dry condition[1])

Species	Moisture content	Specific gravity[2]	Static bending: Fiber stress at proportional limit	Static bending: Modulus of-- Rupture	Static bending: Modulus of-- Elasticity	Static bending: Work to-- Proportional limit	Static bending: Work to-- Maximum load	Impact bending--height of drop causing complete failure (50-pound hammer)	Compression parallel to grain: Fiber stress at proportional limit	Compression parallel to grain: Maximum crushing strength	Compression perpendicular to grain--fiber stress at proportional limit	Shear parallel to grain--maximum shearing strength	Tension perpendicular to grain--maximum tensile strength	Hardness: Load required to embed a 0.444-inch ball to 1/2 its diameter: End	Hardness: Side
1	2	3	4	5	6	7	8	9	10	11	12	13	14	15	16
	Percent		P.s.i.	P.s.i.	1,000 p.s.i.	In.-lb. per cu. in.	In.-lb. per cu. in.	In.	P.s.i.	P.s.i.	P.s.i.	P.s.i.	P.s.i.	Lb.	Lb.
Cedar:															
Alaska-...............	38	0.42	3,800	6,400	1,140	0.77	9.2	27	2,500	3,050	430	840	330	540	440
	12	.44	7,100	11,100	1,420	2.06	10.4	29	5,210	6,310	770	1,130	360	790	580
Atlantic white-.......	35	.31	2,500	4,700	750	.51	5.9	18	1,660	2,390	300	690	180	400	290
	12	.32	4,800	6,800	930	1.46	4.1	13	2,740	4,700	500	800	220	520	350
Eastern red-..........	35	.44	3,400	7,000	650	1.08	15.0	35	2,540	3,570	860	1,010	330	760	650
	12	.47	3,800	8,800	880	1.01	8.3	22	...	6,020	1,140	...	...	900	900
Incense-..............	108	.35	3,900	6,200	840	.94	6.4	17	2,940	3,150	460	830	280	570	390
	12	.37	5,900	8,000	1,040	1.67	5.4	17	4,760	5,200	730	880	270	830	470
Northern white-.......	55	.29	2,600	4,200	640	.60	5.7	15	1,490	1,990	290	620	240	320	230
	12	.31	4,900	6,500	800	1.72	4.8	12	2,630	3,960	380	850	240	450	320
Port-Orford-..........	43	.40	4,000	6,200	1,420	.65	7.4	22	2,770	3,130	350	830	180	460	400
	12	.42	7,700	11,300	1,730	1.97	9.1	28	5,890	6,470	760	1,080	400	730	560
Western red-..........	37	.31	3,200	5,100	920	.63	5.0	17	2,470	2,750	340	710	230	430	270
	12	.33	5,300	7,700	1,120	1.44	5.8	17	4,360	5,020	610	860	220	660	350
Douglas-fir, Coast type.	38	.45	4,500	7,600	1,570	.75	7.6	26	3,130	3,860	440	930	300	570	500
	12	.48	7,800	12,200	1,950	1.77	9.8	31	5,850	7,430	870	1,160	340	900	710
Larch, western..........	58	.51	4,600	8,200	1,530	.81	10.3	29	3,010	3,990	420	900	330	580	510
	12	.55	8,300	13,900	1,960	1.99	12.6	35	5,620	8,110	980	1,410	430	1,120	830
Pine:															
Eastern white.........	73	.34	3,000	4,900	990	.53	5.2	17	2,040	2,440	260	680	250	300	290
	12	.35	5,700	8,600	1,240	1.51	6.8	18	3,670	4,800	510	900	310	480	380
Ponderosa.............	91	.38	3,100	5,000	970	.59	5.1	20	2,070	2,400	360	680	290	300	310
	12	.40	6,300	9,200	1,260	1.85	6.6	17	4,060	5,270	740	1,160	400	550	450

(Sheet 4 of 5)

Table 1-1.--Strength properties of various native and imported woods--Continued

(Results of tests on small, clear specimens in the green and air-dry condition[1])

Species	Moisture content	Specific gravity[2]	Static bending: Fiber stress at proportional limit	Static bending: Modulus of-- Rupture	Static bending: Modulus of-- Elasticity	Static bending: Work to-- Proportional limit	Static bending: Work to-- Maximum load	Impact bending--height of drop causing complete failure (50-pound hammer)	Compression parallel to grain: Fiber stress at proportional limit	Compression parallel to grain: Maximum load	Compression perpendicular to grain--fiber stress at proportional limit	Shear parallel to grain--maximum shearing strength	Tension perpendicular to grain--maximum tensile strength	Hardness: Load required to embed a 0.444-inch ball to 1/2 its diameter: End	Hardness: Side
1	2	3	4	5	6	7	8	9	10	11	12	13	14	15	16
	Percent		P.s.i.	P.s.i.	1,000 p.s.i.	In.-lb. per cu. in.	In.-lb. per cu. in.	In.	P.s.i.	P.s.i.	P.s.i.	P.s.i.	P.s.i.	Lb.	Lb.
Pines: (continued)															
Southern yellow:															
Loblolly	81	0.47	4,100	7,300	1,410	0.68	8.2	30	2,550	3,490	480	850	260	420	450
	12	.51	7,800	12,800	1,800	1.92	10.4	30	4,820	7,080	980	1,370	470	750	690
Longleaf	63	.54	5,200	8,700	1,600	.95	8.9	35	3,430	4,300	590	1,040	330	550	590
	12	.58	9,300	14,700	1,990	2.44	11.8	34	6,150	8,440	1,190	1,500	470	920	870
Shortleaf	81	.46	3,900	7,300	1,390	.63	8.2	30	2,500	3,430	440	850	320	410	440
	12	.51	7,700	12,800	1,760	1.93	11.0	33	5,090	7,070	1,000	1,310	470	750	690
Slash	66	.56	5,100	8,900	1,580	1.02	9.5	36	3,040	4,340	680	1,000	400	600	630
	12	.61	9,800	15,900	2,060	2.76	12.6	36	6,280	9,100	1,390	1,730	570	1,080	1,010
Virginia	88	.45	4,000	7,300	1,220	.75	10.9	34	2,500	3,420	390	890	400	590	540
	12	.48	7,100	13,000	1,520	1.86	13.7	32	3,820	6,710	910	1,350	380	980	740
Sugar	137	.35	3,400	5,100	940	.70	5.4	17	2,330	2,530	350	680	270	320	310
	12	.36	5,700	8,000	1,200	1.53	5.5	18	4,140	4,770	590	1,050	350	530	380
Western white	54	.36	3,400	5,200	1,170	.56	5.0	19	2,430	2,650	290	640	260	310	310
	12	.38	6,200	9,500	1,510	1.47	8.8	23	4,480	5,620	540	850	....	440	370
Redwood (old growth)	112	.38	4,800	7,500	1,180	1.18	7.4	21	3,700	4,200	520	800	260	570	410
	12	.40	6,900	10,000	1,340	2.04	6.9	19	4,560	6,150	860	940	240	790	480
Spruce:															
Eastern[10]	45	.38	3,300	5,600	1,120	.56	6.6	21	2,120	2,600	290	710	200	390	340
	12	.40	6,500	10,100	1,450	1.67	8.5	23	4,240	5,620	590	1,070	350	640	490
Sitka	42	.37	3,300	5,700	1,230	.53	6.3	24	2,240	2,670	340	760	250	430	350
	12	.40	6,700	10,200	1,570	1.62	9.4	25	4,780	5,610	710	1,150	370	760	510

[1]The values in the first line for each species are from tests of green material; those in the second line are from tests of seasoned material adjusted to an average air-dry condition of 12 percent moisture content.

[2]Based on weight when oven-dry and volume when green or at 12 percent moisture content.

[3]Average of blue ash (F. quadrangulata), green ash (F. pennsylvanica), and white ash (F. americana).

[4]Because of the wide range in density of balsa estimated values are shown for balsa for densities of 5, 9, and 14 pounds per cubic foot, in that order. It is not known what proportion of any particular shipment will fall in these classes.

[5]Average of sweet birch (B. lenta) and yellow birch (B. alleghaniensis).

[6]Average of bitternut hickory (C. cordiformis), nutmeg hickory (C. myristicaeformis), water hickory (C. aquatica), and pecan (C. illinoënsis).

[7]Average of shellbark hickory (C. lacinosa), mockernut hickory (C. tomentosa), pignut hickory (C. glabra), and shagbark hickory (C. ovata).

[8]Average of black oak (Q. velutina), laurel oak (Q. laurifolia), pin oak (Q. palustris), northern red oak (Q. rubra), scarlet oak (Q. coccinea), southern red oak (Q. falcata), water oak (Q. nigra), and willow oak (Q. phellos).

[9]Average of bur oak (Q. macrocarpa), chestnut oak (Q. prinus), post oak (Q. stellata), swamp chestnut oak (Q. michauxii), swamp white oak (Q. bicolor), and white oak (Q. alba).

[10]Average of black spruce (P. mariana), red spruce (P. rubens), and white spruce (P. glauca).

(Sheet 5 of 5)

In ship and boat use, wood may be subjected to a variety of conditions affecting its moisture content, ranging from the extreme dryness associated with artificial heating to saturated wet conditions. Based on much experience, however, moisture conditions existing in various ship and boat members may generally be classified in the following categories:

a. Wet condition of use.--The wood in service generally maintains an equilibrium moisture content greater than 15 percent.

b. Intermittent wetting and drying conditions of use.--

   (1) Wood in service is usually at an equilibrium moisture of 15 percent or less, but at times is subjected to prolonged wetting conditions which will raise the moisture content about 15 percent.

   (2) Wood in service is generally at a moisture greater than 15 percent, but dries out at times.

c. Dry conditions of use.--The wood is installed at a moisture content of 15 percent or less, and this moisture content remains approximately the same in use.

Under wet conditions or intermittent wetting and drying conditions, the basic stress values for wet wood of table 2-1 are considered applicable. The basic working values in table 2-1 are for clear material and, besides the adjustments mentioned, contain a factor of safety. The value of this safety factor depends upon the variability of a particular strength property within a species. It varies among the individual timbers in proportion to the amount the actual strength value exceeds the basic stress rating and is influenced by the conditions under which the timbers are used. The most probable value for the true factor of safety for the basic stresses in table 2-1 is estimated at about 2-1/2. This is based on statistical studies of the combination of all the factors affecting the structural safety of timbers. About 95 percent of the timbers have a factor of safety of at least 1-1/4. An exceptionally strong timber used in an especially favorable environment can have a factor of safety as high as 5.

For dry solid members over 2 inches in the minimum dimension, the stress values of table 2-1 for wet wood are considered applicable. As indicated previously, the increase in strength resulting from lower moisture content is generally offset by resulting drying defects.

In drying lumber 2 inches or less in the minimum dimension, the drying defects are not so numerous nor so severe. Therefore, for solid dry lumber 2 inches or less in thickness and for glued laminated members fabricated from thin lumber and maintained in the dry condition, the basic dry condition stresses of table 2-2 are applicable.

2.2 EFFECT OF TEMPERATURE ON STRENGTH.

2.2.1 Immediate Effect.--In general, when the temperature of wood is raised, most strength properties decline. At lower temperatures, wood becomes stronger. This effect on strength is immediate. Its magnitude depends upon the moisture content of the wood and, when the temperature is elevated, upon the length of exposure. For most structural uses under ordinary atmospheric conditions, wood exposed to moderately high temperatures for a short time recovers essentially its original strength when the temperature is reduced to normal. The stress values given in tables 2-1 and 2-2 may be considered applicable to wood at a normal temperature of about 70°F.

The immediate effect of temperature on most static strength properties of dry wood (12 percent moisture content) within the range of 0° to 150° F. can be approximated as an increase or a decrease in the strength at 70° F. of about 0.3 to 0.5 percent for each 1° F. decrease or increase in temperature. The change will be greater if the moisture content is high, and less if the moisture content is low.

Tests of wood conducted at about -300° F. show that the important strength properties of dry wood in bending and in compression, including stiffness and shock resistance, are much higher at the extremely low temperature than at normal temperature.

2.2.2 Permanent Effect.--When wood is exposed to temperatures of 150° F. or more for extended periods of time, it is permanently weakened even though subsequently used at normal temperatures. The permanent strength loss will depend upon a number of factors, including the moisture content and the maximum temperature reached by the wood, the heating medium and length of exposure, and, to some ex-

Table 2-1--Basic stresses for clear lumber under long-time service at full design load under wet conditions.[1]

Species[2]	Extreme fiber in bending or tension parallel to grain	Maximum horizontal shear	Compression perpendicular to grain	Compression parallel to grain L/d = 11 or less	Modulus of elasticity in bending
1	2	3	4	5	6
	P.s.i.	P.s.i.	P.s.i.	P.s.i.	1,000 p.s.i.
SOFTWOODS					
Baldcypress	1,900	150	220	1,450	1,200
Cedar:					
Alaska-	1,600	130	185	1,050	1,200
Atlantic and northern white-	1,100	100	130	750	800
Port-Orford-	1,600	130	185	1,200	1,500
Western red-	1,300	120	145	950	1,000
Douglas-fir, coast type, medium-grain	2,200	130	235	1,450	1,600
Douglas-fir, coast type, close-grain	2,350	130	250	1,550	1,600
Douglas-fir, all regions, dense	2,550	130	275	1,700	1,600
Larch, western	2,200	130	235	1,450	1,500
Pine:					
Eastern white, ponderosa, sugar, and western white	1,300	120	185	1,000	1,000
Southern yellow	2,200	160	235	1,450	1,600
Southern yellow, dense	2,550	160	275	1,700	1,600
Redwood	1,750	100	185	1,350	1,200
Redwood, close-grained	1,900	100	195	1,450	1,200
Spruce, red, white, and Sitka	1,600	120	185	1,050	1,200

(Sheet 1 of 2)

Table 2-1.--Basic stresses for clear lumber under long-time service at full design load under wet conditions.--Continued [1]

Species[2]	Extreme fiber in bending or tension parallel to grain	Maximum horizontal shear	Compression perpendicular to grain	Compression parallel to grain L/d = 11 or less	Modulus of elasticity in bending
1	2	3	4	5	6
	P.s.i.	P.s.i.	P.s.i.	P.s.i.	1,000 p.s.i.
HARDWOODS					
Ash, black	1,450	130	220	850	1,100
Ash, commercial white	2,050	185	365	1,450	1,500
Beech, American	2,200	185	365	1,600	1,600
Birch, sweet and yellow	2,200	185	365	1,600	1,600
Elm, American and slippery	1,600	150	185	1,050	1,200
Elm, rock	2,200	185	365	1,600	1,300
Hickory, true and pecan	2,800	205	440	2,000	1,800
Khaya	2,000	175	430	1,300	1,300
Mahogany	2,200	175	450	1,500	1,300
Maple, black and sugar	2,200	185	365	1,600	1,600
Oak	2,050	185	365	1,350	1,500
Sweetgum	1,600	150	220	1,050	1,200
Tangile	2,000	175	430	1,500	1,300
Teak	2,750	200	600	1,900	1,600

[1]These stresses are applicable with adjustments described in Chapters 2 and 3 to lumber of any degree of seasoning, or lumber used under any condition of duration of load, for use in determining working stresses

[2]Species names from approved check list, U.S. Forest Service, 1953.

(Sheet 2 of 2)

Table 2-2.--Basic stresses for 1- and 2-inch nominal clear lumber under long-time service at full design load under dry conditions[1]

Species[2]	Extreme fiber in bending or tension parallel to grain	Maximum longitudinal shear	Compression perpendicular to grain	Compression parallel to grain	Modulus of elasticity in bending
1	2	3	4	5	6
	P.s.i.	P.s.i.	P.s.i.	P.s.i.	1,000 p.s.i.
SOFTWOODS					
Baldcypress	2,400	170	330	2,000	1,300
Cedar:					
Alaska-	2,000	150	275	1,450	1,300
Atlantic and northern white-	1,400	115	195	1,050	900
Port-Orford-	2,000	150	275	1,650	1,600
Western red-	1,600	135	220	1,300	1,100
Douglas-fir, coast type, medium-grain	2,750	150	350	2,000	1,800
Douglas-fir, coast type, close-grain	2,950	150	375	2,150	1,800
Douglas-fir, all regions, dense	3,200	150	410	2,350	1,800
Larch, western	2,750	150	350	2,000	1,600
Pine:					
Eastern white, ponderosa, sugar, and western white	1,600	135	275	1,400	1,100
Southern yellow	2,750	180	350	2,000	1,800
Southern yellow, dense	3,200	180	410	2,350	1,800
Redwood	2,200	115	275	1,850	1,300
Redwood, close-grained	2,400	115	295	2,000	1,300
Spruce, red, white, and Sitka	2,000	135	275	1,450	1,300
Tamarack	2,200	160	330	1,850	1,400

(Sheet 1 of 2)

Table 2-2.--Basic stresses for 1- and 2-inch nominal clear lumber under long-time service at full design load under dry conditions--Continued [1]

Species[2]	Extreme fiber in bending or tension parallel to grain	Maximum longitudinal shear	Compression perpendicular to grain	Compression parallel to grain	Modulus of elasticity in bending
1	2	3	4	5	6
	P.s.i.	P.s.i.	P.s.i.	P.s.i.	1,000 p.s.i.
HARDWOODS					
Ash, black	1,800	150	330	1,150	1,200
Ash, commercial white	2,550	210	550	2,000	1,600
Beech, American	2,750	210	550	2,200	1,800
Birch, sweet and yellow	2,750	210	550	2,200	1,800
Elm, rock	2,750	210	550	2,200	1,400
Hickory, true and pecan	3,500	235	660	2,750	2,000
Khaya	2,500	200	645	1,800	1,400
Mahogany	2,750	200	675	2,050	1,400
Maple, black and sugar	2,750	210	550	2,200	1,800
Oak	2,550	210	550	1,850	1,600
Sweetgum	2,000	170	330	1,450	1,300
Tangile	2,500	200	645	2,050	1,400
Teak	3,450	230	900	2,600	1,800
Tupelo, black	2,000	170	330	1,450	1,300

[1]These stresses are applicable with adjustments described in Chapters 2 and 3 to lumber used under any condition of duration of load, for use in determining working stresses.

[2]Species names from approved check list, U.S. Forest Service, 1953.

(Sheet 2 of 2)

tent, the species and size of the piece. Strength losses will become increasingly greater for specific periods of time as the temperature is increased. The shock resistance of wood (work to maximum load) is first affected; modulus of elasticity, a measure of stiffness, is affected least. Dry wood is much less affected by heat than green or wet wood.

The effect of high temperatures on various species differs, but, in general, hardwoods are affected much more than softwoods.

The source of heat has a considerable bearing on the amount of reduction of strength that results from exposure of wood at a particular temperature for various periods of time. At all temperatures, hot water causes somewhat less strength loss in a given period of time than does steam; and hot, dry air has the least effect.

Exposure to high temperatures has a cumulative effect on the strength of wood. For example, if wood at a particular moisture content is exposed at six different times to a temperature of 180° F. for 1 month each, the overall effect would be approximately the same as that for a single exposure of 6 months.

CHAPTER 3. INFLUENCE OF DEFECTS AND OTHER FACTORS ON STRENGTH OF WOOD

3.1 GENERAL.--The major defects which materially affect the strength of structural timbers and must therefore be limited by grading rules are decay, knots, cross grain, holes, checks, shakes, and splits. Other defects also affect the strength. In large structural timbers, they normally may be ignored because their effect is less than that of the permitted defects. They must, however, be controlled in small members of boats because of their relation to the size of the member and because such members usually should have none or only small defects of the type permitted in structural timbers. Thus, a pitch pocket may materially reduce the strength of a 1 - by 2-inch member. In a member 4 by 8 inches in cross section, however, it would be relatively unimportant, since the percentage loss in strength it would cause would be much less than that caused by a permitted knot, for which due allowance has been made in establishing the working stresses. The defects that are unimportant in large timbers but must be controlled in small timbers used for structural purposes include pith fleck, pitch pockets, pitch streak, and bark pockets.

Other defects or blemishes occur which do not affect the strength or produce an effect so small as to be of no importance in members of the size used even in small boats. They include mineral streaks, indented rings, bird's-eye, occasional bird pecks, red streak in spruce, and black streak in other conifers.

3.1.1 General Definition of Strength Ratio.--Basic stress is a generalized working strength value for the clear wood of a species. Since the several defect characteristics that exist in structural members affect their strength, it is necessary to employ a method of assigning strength properties either to individual pieces containing these characteristics or to a structural grade. Strength ratios are used for this purpose. The strength ratio of a piece of structural lumber or of a structural grade represents the strength remaining after allowance is made for knots, cross grain, or other strength-reducing characteristics present. Strength ratios are applied to stresses in transverse bending, tension parallel to grain, compression parallel to grain, and horizontal shear. Modulus of elasticity and compression perpendicular to grain are little affected by most of these factors, and strength ratios of 100 percent are assumed for all grades.

Strength reductions for knots and cross grain generally are not cumulative; only the lower strength ratio is applied. The strength ratio for horizontal shear, in the case of beams, may or may not be critical.

The basic stresses for a species, when multiplied by the strength ratio for a piece or a grade, become the working stresses for the structural member of the grade.

If the structural lumber is seasoned and is to remain continuously dry in use (as in most covered structures), the strength ratios in bending, tension, and compression may be increased by one-half of their excess over 50 percent; thus, material with a strength ratio of 60 percent when green is given a strength ratio of 65 percent when dry.

The working stress in compression is subject to an additional 10 percent increase for drying.

3.2 INFLUENCE OF DEFECTS ON LUMBER

3.2.1 Cross Grain.--The strength of wood differs greatly in the two principal directions, along and across the grain. Strength parallel to grain is 10 or more times greater than that perpendicular to grain. Any deviation of the fibers from a direction parallel to the long edges of the piece constitutes cross grain and causes part of the stress to act in the weak direction, or perpendicular to the grain. When this part of the stress exceeds the strength of the wood perpendicular to grain, failure occurs at a lower average stress than would have been the case had all the stress been directed parallel to grain, or parallel to the strong direction of the wood. On the basis of actual tests, relations between strength properties and the measured combined slope of grain determined for several species of wood are given in table 3-1.

Table 3-1.--Strength of wood members with various combined slopes of grain[1]

Maximum combined slope of grain in member	Static bending[2]		Impact bending[2] --height of drop causing complete failure (50-lb hammer)	Compression parallel to grain[2] (maximum crushing strength)
	Modulus of rupture	Modulus of elasticity		
	Percent	Percent	Percent	Percent
Straight grained	100	100	100	100
1 in 25	96	97	95	100
1 in 20	93	96	90	100
1 in 15	89	94	81	100
1 in 10	81	89	62	99
1 in 5	55	67	36	93

[1]Wood members containing both spiral and diagonal grain must consider the combined slopes rather than the greater of the two slopes. Combined slope is determined by taking the square root of the sum of the squares of the slopes of the two types of cross grain. As an example, if these slopes are 1 in 18 and 1 in 12, the combined slope is.

$$\sqrt{\left(\frac{1}{18}\right)^2 + \left(\frac{1}{12}\right)^2} = \frac{1}{10} \text{ or a slope of 1 in 10.}$$

[2]Expressed as percentages of the strength of straight-grained members.

Because of practical considerations in stress grading, slope-of-grain limitations in structural members are generally measured on the face with the greatest (steepest) slope of grain. This fact is taken into account in the reduction factors applied for slope of grain in establishing working stresses for various grades. Table 3-2 shows the strength ratios for stress in extreme fiber in bending, tension, and compression parallel to grain for slopes of grain from 1 in 6 to 1 in 24.

Table 3-2.--Strength ratios for wood in bending and in compression parallel to grain as related to slope of grain

Slope of grain[1]	Maximum strength ratio[2]	
	For stress in extreme fiber in bending	For stress in compression parallel to grain
	Percent	Percent
1 in 20	90	100
1 in 18	85	100
1 in 16	80	100
1 in 15	76	90
1 in 14	74	87
1 in 12	69	82
1 in 10	61	74
1 in 8	53	66
1 in 6	45	56

[1]Measured on the surface with maximum slope.

[2]Expressed as percentages of the strength of straight-grained wood.

3.2.2 Knots.--Knots reduce beam strength because of the localized cross grain of steep slope that is usually present. This condition of cross grain around a knot is at an angle to the fiber direction of the surrounding wood, so that stresses acting parallel to the grain of the member set up stresses in the weak direction, or along the cross grain of the knot. Cross-grained wood surrounding a knot generally checks severely in seasoning. Knots are more injurious to bending strength than to compressive strength, but they have little effect on shear and stiffness.

3.2.2.1 Effect of Knots on Bending Strength.--In a beam of any particular span length and cross section, it is possible to estimate fairly accurately the effect of knots on the strength properties of the beam by their size and position within the cross section of the beam and span length.

3.2.2.1.1 Simply Supported Beam

3.2.2.1.1.1 Large Cross Section.--In a simply supported beam, knots are most injurious when they are on or near the top and bottom face of the beam within the center portion of the span length. In that portion of the beam, a knot reduces the bending strength in a ratio proportional to the ratio of the knot diameter to the width of the top or bottom face. (Figure 3-1.) Thus, a 1-inch knot on the bottom face, located at the center one-third of the span, will reduce the bending strength of a 4-inch-thick beam approximately 25 percent.

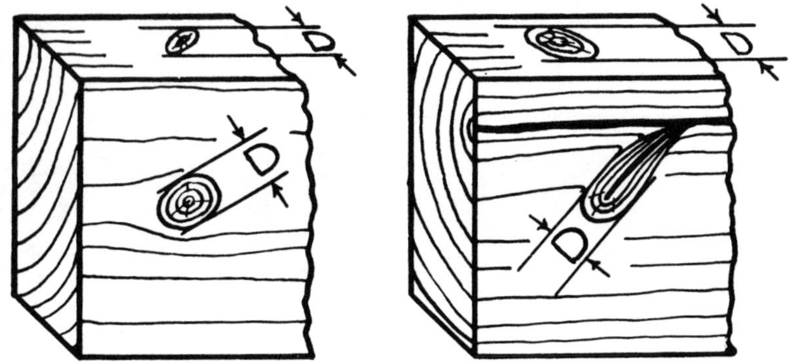

Figure 3-1.--Measurement of knots in beams and stringers, where D is the diameter of the knot.

(M 22545 F)

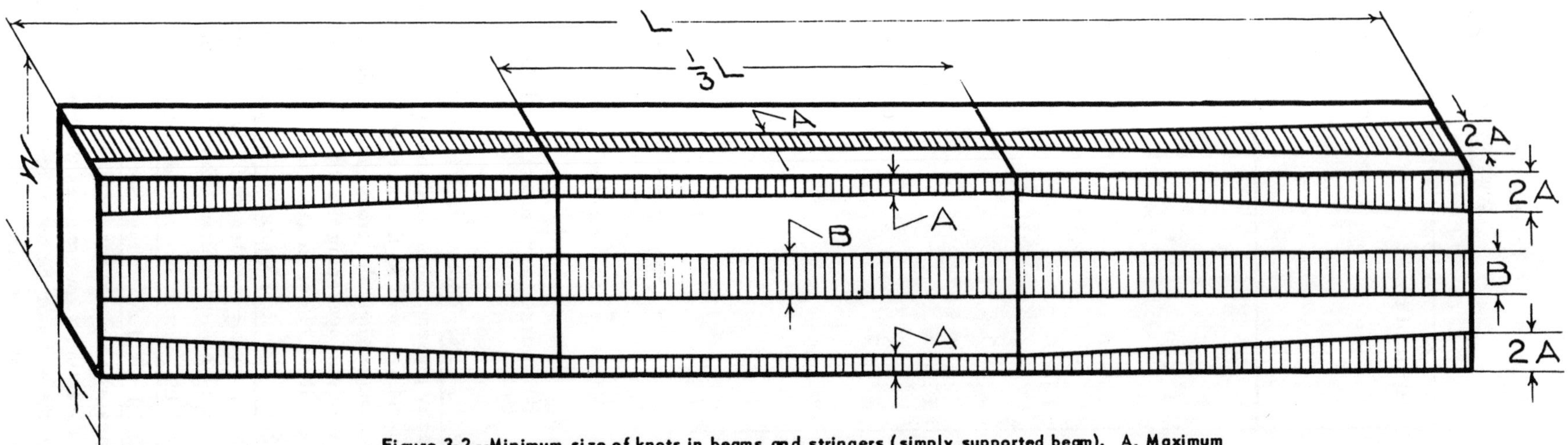

Figure 3-2.--Minimum size of knots in beams and stringers (simply supported beam). A, Maximum size on narrow face or edge of wide face, middle third of length, with gradual increase to 2A, at ends of piece; B, maximum size at centerline of wide face; L, length; W, width; T, thickness.

(M 22546 F)

The effect of a knot on the strength of a beam is proportionately lower with increasing distance of the knot from the narrow face or from the center of the span length. At the center of the vertical face of the beam, the effect is only approximately half that of a knot on a narrow face or the edge of a wide face. When a knot is located at or near either or both ends of the span length of a beam, it has little effect on the strength; regardless of where it may be located in the cross section (fig. 3-2).

Strength ratios that correspond to various combinations of knot size and width of face are given in tables 3-3, 3-4, and 3-5 for stress in extreme fiber in bending for simply supported beams, and tension parallel to grain. The following criteria are used in connection with measuring knots.

Measurements of knots in joists and planks (2 inches and greater in nominal thickness, 4 inches and greater in nominal width)--

a) The size of a knot on the narrow face of a joist or plank is taken as its width between lines enclosing the knot and parallel to the edges of the piece. A knot that appears on a narrow and a wide face of a side-cut-joist, but does not contain the intersection of those faces, is measured and graded only on the wide face.

b) On the wide faces of joists and planks, the average diameter of a knot is taken as its size. In such thin and relatively wide material used either on edge or flat, this is a safe measure of the influence of knots on strength, and it has the commercial advantage of being directly applicable to yard grades of lumber. This method of measurement will exclude damaging spike knots, and may be applied to them as well as to round or oval knots.

Measurements of knots in beams and stringers (nominal cross-sectional dimensions 5 by 8 inches and up)--

c) On the narrow faces of a beam or stringer member to be used for stress in bending, the size of a knot is taken as its width between lines enclosing the knot and parallel to the edges of the piece, except that when a knot on a narrow face of a side-cut piece extends into the adjacent one-fourth of the width of a wide face, its least dimension is taken as its size.

d) On the vertical faces of a beam or stringer to be used for stress in bending, the smallest diameter of knots shall be used. A knot at the edge of a wide face is limited to the same size as a knot on a narrow face but is measured by the smallest diameter.

e) Intergrown or live knots offer some resistance to stress, while encased knots or knot-holes are of little or no value. On the other hand, the distortion of fibers is greater around a live or intergrown knot than around a dead or encased knot of the same size. Effects on strength are thus roughly equalized, and no distinction is made in strength grading between live knots, dead knots, and knotholes.

f) The distribution and sum of the diameters of knots are important, as well as the maximum size of a single knot. In beams and stringers up to 20 feet long, the sum of the sizes of all knots within the middle half of the length of a face is limited to four times the size of the largest knot allowed on that face. Limits in longer beams and stringers are proportionate. In joists and planks, or in structural timber to be stressed in axial compression or tension, the sum of the sizes of all knots in any 6 inches of length of the piece is limited to twice the size of the largest permitted single knot. Two or more knots of maximum or near maximum size are not allowed in the same 6 inches of length on any face.

g) Because large seasoning defects generally develop in boxed heart, boxed heart is usually not permitted in members under bending stress in ships and boats. Therefore, special knot measurements for boxed heart are not discussed.

Knot limitations are established on the assumption that the piece will be used in its full size. If such a beam is altered by cutting into smaller sections, the original assigned grade is no longer valid for any of the resulting sections. In such event, each section would require regrading to assure that it has the required stress rating.

Table 3-3.--Strength ratios[1] for various knot sizes occurring on the narrowest face of beams within middle third of length of piece; cross section nominal thickness of beam 2 inches or greater, and cross section nominal width of beam 4 inches or greater (beams and stringer or joists and planks)[2,3]

Size of knot	Percentage strength ratio when nominal width of narrow face in inches is:									
	2	3	4	5	6	8	10	12	14	16
Inches										
1/4	90	93	95	96	96	97	97	97	98	98
3/8	83	89	92	93	94	95	96	96	96	97
1/2	77	85	88	91	92	93	94	95	95	95
5/8	71	81	85	88	90	92	92	93	94	94
3/4	65	76	82	86	88	90	91	92	92	93
7/8	58	72	79	83	86	88	89	90	91	91
1	52	68	76	81	84	86	88	89	89	90
1-1/8	46	64	73	78	82	84	86	87	88	89
1-1/4	...	60	70	76	80	83	84	86	87	88
1-3/8	...	56	67	73	78	81	83	84	85	86
1-1/2	...	51	63	71	76	79	81	83	84	85
1-5/8	...	47	60	68	74	77	80	81	83	84
1-3/4	...	35	57	66	71	75	78	80	81	83
1-7/8	...	...	54	63	69	73	76	78	80	81
2	...	...	51	61	67	72	75	77	79	80
2-1/8	...	...	48	58	65	70	73	75	77	79
2-1/4	...	...	45	56	63	68	71	74	76	77
2-3/8	...	...	36	53	61	66	70	72	74	76
2-1/2	...	...	32	51	59	64	68	71	73	75
2-5/8	...	...	29	48	57	63	67	70	72	74
2-3/4	...	...	...	46	55	61	65	68	70	72
2-7/8	...	...	...	...	53	59	63	67	69	71
3	...	...	...	...	51	57	62	65	68	70
3-1/8	...	...	...	...	...	55	60	64	66	68
3-1/4	...	...	...	...	...	54	59	62	65	67

(Sheet 1 of 2)

Table 3-3.--Strength ratios[1] for various knot sizes occurring on the narrowest face of beams within middle third of length of piece; cross section nominal thickness of beam 2 inches or greater, and cross section nominal width of beam 4 inches or greater (beams and stringer or joists and planks)[2,3]--Continued

Size of knot	Percentage strength ratio when nominal width of narrow face in inches is:									
	2	3	4	5	6	8	10	12	14	16
Inches										
3-3/8						52	57	61	64	66
3-1/2						50	55	59	62	65
3-5/8							54	58	61	63
3-3/4							52	56	59	62
3-7/8							50	55	58	61
4								53	57	60
4-1/8								52	55	58
4-1/4								50	54	57
4-3/8									53	56
4-1/2									51	54
4-5/8									50	53
4-3/4										52
4-7/8										51

[1]Strength ratios are for stress in extreme fiber in bending or tension parallel to grain.

[2]Strength ratios of 45 percent or higher are based on the relation of the size of knot to the nominal width of face; lower ratios are based on the relation of the size of knot to the actual standard dressed width of face. In tabulating the sizes of knots, the rule used in approximating decimals by common fractions with 1/8-inch intervals has the effect of subtracting 1/24 inch from the size of each knot. For example, the strength ratios listed for a 3-1/2-inch knot are those which actually would obtain for a size of 3-11/24 inch. In view of the allowance thus introduced and the fact that nominal instead of actual widths of face are used, the strength ratios given should be taken as maximum values.

[3]Sizes of knots permitted for a given strength ratio in narrow faces wider than 6 inches in beams and stringers are increased only in proportion to the square root of the increase of width of the face. Below these limits, the sizes permitted are proportioned to the width of the face.

(Sheet 2 of 2)

Table 3-4.--Effect on strength of knots occurring along the centerline of the wide face of beam; cross section nominal thickness of beam 2 inches or greater; cross section nominal width 4 inches or greater; (beams and stringers, joists and planks) also applies for 1-inch boards with knots occurring anywhere in face[1,2]

Size of knot[3]	Percentage strength ratio when nominal width of wide face in inches is:											
	3	4	5	6	8	10	12	14	16	18	20	22
Inches												
1/4	......	95	96	96	97	98	98	98	98	99	99	99
1/2	......	88	91	92	94	95	96	96	97	97	97	97
3/4	......	82	86	88	91	93	94	94	95	95	95	96
1	......	76	81	84	88	90	92	93	93	93	94	94
1-1/4	......	70	76	80	85	88	90	91	91	92	92	93
1-1/2	......	63	71	76	82	85	88	89	89	90	91	91
1-3/4	......	57	66	71	79	83	86	87	88	88	89	89
2	......	51	61	67	75	80	84	85	86	87	87	88
2-1/4	......	45	56	63	72	78	82	83	84	85	86	86
2-1/2	......	......	51	59	69	75	79	81	82	83	84	85
2-3/4	......	......	46	55	66	73	77	79	80	82	82	83
3	......	......	36	51	63	70	75	77	79	80	81	82
3-1/4	......	......	......	47	60	68	73	75	77	78	79	80
3-1/2	......	......	......	37	57	65	71	73	75	76	78	79
3-3/4	......	......	......	......	54	63	69	71	73	75	76	77
4	......	......	......	......	50	60	67	69	71	73	74	76
4-1/4	......	......	......	......	47	58	65	67	70	71	73	74
4-1/2	......	......	......	......	41	55	63	66	68	70	71	73
4-3/4	......	......	......	......	37	53	61	64	66	68	70	71
5	......	......	......	......	......	50	59	62	64	66	68	70
5-1/4	......	......	......	......	......	48	57	60	62	65	66	68

(Sheet 1 of 2)

Table 3-4.--Effect on strength of knots occurring along the centerline of the wide face of beam; cross section nominal thickness of beam 2 inches or greater; cross section nominal width 4 inches or greater; (beams and stringers, joists and planks) also applies for 1-inch boards with knots occurring anywhere in face[1,2]--Continued

Size of knot[3]	Percentage strength ratio when nominal width of wide face in inches is:										
	3	4	5	6	8	10	12	14	16	18	20	22
Inches												
5-1/2						45	54	58	61	63	65	66
5-3/4						40	52	56	59	61	63	65
6						37	50	54	57	59	61	63
6-1/4						35	48	52	55	58	60	62
6-1/2							46	50	53	56	58	60
6-3/4							42	48	52	54	57	59
7							39	46	50	53	55	57
7-1/4							37	42	48	51	53	56
7-1/2							35	40	46	49	52	54
7-3/4							33	38	42	48	50	53
8							31	36	40	46	49	51
8-1/4							29	34	38	42	47	49
8-1/2							26	32	37	40	45	48

[1]See footnote 2 of Table 3-3.

[2]Sizes of knots permitted for a given strength ratio in wide faces wider than 12 inches in beams and stringers and joists and planks, or faces wider than 12 inches in posts and timbers, are increased only in proportion to the square root of the increase of width of the face. Below these limits, the sizes permitted are proportioned to the width of the face.

[3]Ratios corresponding to other sizes of knot can be found by interpolation.

(Sheet 2 of 2)

Table 3-5.--Effect on strength of knots occurring at edge of wide face of beam; cross section nominal thickness from 2 to 4 inches, and cross section nominal width 4 inches or greater (joists and planks).[1,2]

Size of knot	Percentage strength ratio when nominal width of wide face in inches is--												
	2	3	4	5	6	8	10	12	14	16	18	20	22
Inches													
1/4	80	86	90	92	93	95	96	97	97	97	97	97	97
3/8	70	79	84	87	89	92	93	94	95	95	95	96	96
1/2	60	72	78	82	85	89	91	92	93	93	94	94	94
5/8	50	65	73	78	81	86	89	90	91	92	92	93	93
3/4	32	58	68	74	78	83	86	88	89	90	91	91	91
7/8	......	52	63	69	74	80	84	87	88	88	89	89	90
1	......	46	58	65	71	77	82	85	86	87	87	88	89
1-1/8	......	34	53	61	67	75	79	83	84	85	86	86	87
1-1/4	......	......	49	57	64	72	77	81	82	83	84	85	86
1-3/8	......	......	40	54	60	69	75	79	80	82	83	83	84
1-1/2	......	......	36	50	57	67	73	77	79	80	81	82	83
1-5/8	......	......	......	47	54	64	71	75	77	78	80	81	81
1-3/4	......	......	......	40	51	62	69	74	75	77	78	79	80
1-7/8	......	......	......	36	48	59	67	72	74	75	77	78	79
2	......	......	......	......	45	57	65	70	72	74	75	76	77
2-1/8	......	......	......	......	40	55	63	68	70	72	74	75	76
2-1/4	......	......	......	......	37	52	61	67	69	71	72	73	75
2-3/8	......	......	......	......	34	50	59	65	67	69	71	72	73
2-1/2	......	......	......	......	......	48	57	63	66	68	69	71	72
2-5/8	......	......	......	......	......	46	55	62	64	66	68	69	71
2-3/4	......	......	......	......	......	41	53	60	63	65	66	68	69
2-7/8	......	......	......	......	......	39	51	58	61	63	65	67	68

(Sheet 1 of 2)

Table 3-5.--Effect on strength of knots occurring at edge of wide face of beam; cross section nominal thickness from 2 to 4 inches, and cross section nominal width 4 inches or greater (joists and planks).[1,2]--Continued

Size of knot	Percentage strength ratio when nominal width of wide face in inches is--											
	2	3	4	5	6	8	10	12	14	16	18	20	22
Inches													
3						37	50	57	60	62	64	65	67
3-1/8						35	48	55	58	60	62	64	66
3-1/4							46	54	57	59	61	63	64
3-3/8							42	52	55	58	60	62	63
3-1/2							40	51	54	56	58	60	62
3-5/8							39	49	52	55	57	59	61
3-3/4							37	48	51	54	56	58	60
3-7/8							36	46	50	52	55	57	58
4							34	45	48	51	53	55	57
4-1/8							33	42	47	50	52	54	56
4-1/4							31	40	46	48	51	53	55
4-3/8							29	39	43	47	50	52	54
4-1/2							28	37	41	46	48	51	53
4-5/8							27	36	40	45	47	50	51
4-3/4							25	35	39	42	46	48	50
4-7/8								34	37	41	45	47	49
5								32	36	40	42	46	48

[1]See footnote 2 of Table 3-3.

[2]Sizes of knots permitted for a given strength ratio at the edges of wide faces wider than 12 inches in joists and planks are increased only in proportion to the square root of the increase of width of the face. Below these limits, sizes are proportioned to the square of the remaining width of the face after the width of the knot is subtracted.

(Sheet 2 of 2)

3.2.2.1.1.2 Members of Square Cross Section Used as Joists, Planks or Beams.--With members of 4-by 4-inch nominal size, each of the four faces should be graded as a narrow face under joist and plank limitations. In larger members, each face should be graded as a wide face under either joist and plank or beam and stringer limitations, whichever applies (that is, based on nominal dimensions).

3.2.2.1.1.3 Members of Small Cross Section.--In small pieces of lumber, the distortion of grain caused by a knot may occupy a large part of the cross section, though the knot itself is partly or entirely absent. In like manner, a knot at the center of a wide face may cause distortion extending to the edge of that face. For this reason, grading for strength of pieces 3 by 3 inches or smaller is more difficult than grading of larger pieces, and ordinarily it is not done. Where such small pieces are graded for strength, all knots should be considered and limited as if at the edge of the face where they appear.

In addition, severe cross grain resulting from large knots that may be partly or entirely absent will impair the strength drastically. Therefore, in specifying and grading members of small dimension for use in ships and boats, maximum limitations on local slope of grain need to be specified.

3.2.2.1.2 Continuous Beams.--Since, in a continuous beam, stresses are distributed differently than in a simple beam, the areas of knot limitation for bending strength need to be modified. In general, this is accomplished by enlarging the maximum stress concentration area to conform to the condition of the beam. If a member is to be used as a continuous beam, on three supports, knot-strength ratios given in tables 3-3, 3-4, and 3-5 apply to knots measured in the center two-thirds of the length of the piece rather than the center one-third. If a member is to be used as a continuous beam, on four or more supports, knot-strength ratios given in tables 3-3, 3-4, and 3-5 apply to knots measured along the full length of the piece rather than the center one-third.

3.2.2.2 Effect of Knots on Compression Strength Parallel to Grain.--Knots in posts reduce compressive strength in proportion to the percentage of the cross section they occupy, regardless of their position in the post. The diameter of a knot in posts and timbers is taken as the diameter of a round knot, the lesser of the two diameters of an oval knot, and the greatest diameter perpendicular to the length of a spike knot. Unless more than one knot is in the same cross section, the effect is not cumulative. To be acceptable for posts, lumber should therefore have the knots so dispersed that no two knots of maximum size will occur in any 6 inches of the length.

Figure 3-3 illustrates the measurement used to determine the diameter of knots in posts and timbers (nominal dimensions 5 by 5 inches and larger) and in other compression members. Strength ratios for compression parallel to grain that correspond to knots of various sizes and various face widths are given in table 3-6.

3.2.2.3 Knot Clusters.--Knot clusters (three or more knots in a compact circular group with the grain between them highly contorted) are prohibited in timbers where strength is of prime importance, because neither the size of the individual knots nor the size of the cluster is a good measure of the amount of grain distortion they cause and consequently does not indicate their effect on strength.

3.2.3 Shakes and Checks.

3.2.3.1 Effect on Strength of Beams.--Shakes and checks reduce the strength of beams by reducing the area available to resist horizontal shear. The size of a shake is normally determined by measuring, at the end of a timber, the distance between lines that enclose the shake and are parallel to the wide faces of the piece (fig. 3-4). In actual practice, however, it is very difficult to discern the full depth of a shake; thus, it is not always possible to determine full extent of its effect on shear strength.

Checks are measured by estimating, from the faces of the piece, the amount of reduction in shear-resisting area.

The effect of shakes or checks measured as discussed above on the strength of simply supported beams are given in table 3-7, assuming no hidden shake is present. If a beam is used as a continuous member over more than one span, restrictions for visible checks are applied to the middle half of the height of the piece and for a distance equal to three times the height of the piece (height equals width of wide face) anywhere throughout its length.

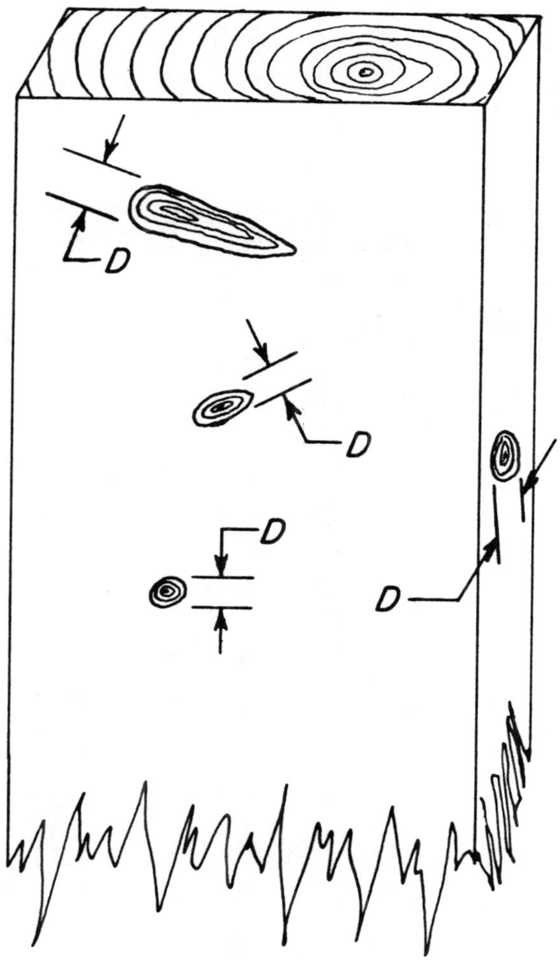

Figure 3-3.--Measurement of knots in posts and timbers or other compression members.

(M 91872 F)

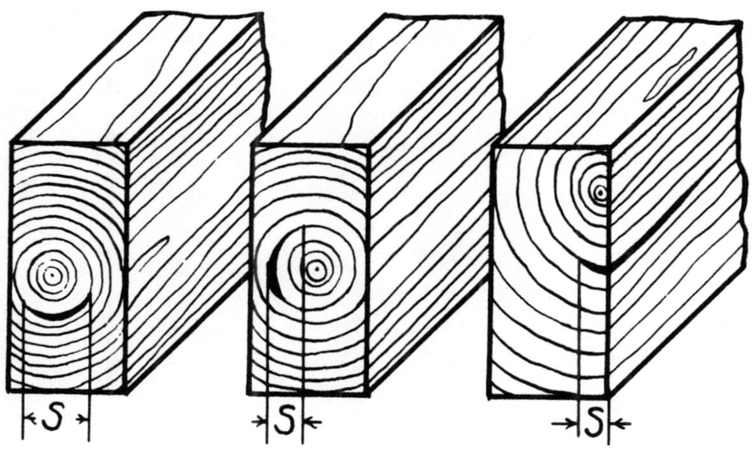

Figure 3-4.--Measurement of shakes in beams and stringers or joists and plank. S is the measured size.

(M 22547 F)

Table 3-6.--Strength ratios for compression parallel to grain for pieces with knots occurring at any point on any face[1,2]

Size of knot	Percentage strength ratio when nominal width of wide face in inches is										
	3	4	5	6	8	10	12	14	16	18	20	22
Inches												
1/4	......	95	96	96	97	98	98	98	98	99	99	99
1/2	......	88	91	92	94	95	96	96	97	97	97	97
3/4	......	82	86	88	91	93	94	94	95	95	95	96
1	......	76	81	84	88	90	92	93	93	93	94	94
1-1/4	......	70	76	80	85	88	90	91	91	92	92	93
1-1/2	......	63	71	76	82	85	88	89	89	90	91	91
1-3/4	......	57	66	71	79	83	86	87	88	88	89	89
2	......	51	61	67	75	80	84	85	86	87	87	88
2-1/4	......	45	56	63	72	78	82	83	84	85	86	86
2-1/2	......	32	51	59	69	75	79	81	82	83	84	85
2-3/4	......	25	46	55	66	73	77	79	80	82	82	83
3	......	......	36	51	63	70	75	77	79	80	81	82
3-1/4	......	......	31	47	60	68	73	75	77	78	79	80
3-1/2	......	......	25	37	57	65	71	73	75	76	78	79
3-3/4	......	......	......	33	54	63	69	71	73	75	76	77
4	......	......	......	28	50	60	67	69	71	73	74	76
4-1/4	......	......	......	......	47	58	65	67	70	71	73	74
4-1/2	......	......	......	......	41	55	63	66	68	70	71	73
4-3/4	......	......	......	......	37	53	61	64	66	68	70	71

(Sheet 1 of 2)

Table 3-6.--Strength ratios for compression parallel to grain for pieces with knots occurring at any point on any face[1,2]--Continued

Size of knot[3]	Percentage strength ratio when nominal width of wide face in inches is										
	3	4	5	6	8	10	12	14	16	18	20	22
Inches												
5					34	50	59	62	64	66	68	70
5-1/4					31	48	57	60	62	65	66	68
5-1/2					27	45	54	58	61	63	65	66
5-3/4						40	52	56	59	61	63	65
6						37	50	54	57	59	61	63
6-1/4						35	48	52	55	58	60	62
6-1/2						32	46	50	53	56	58	60
6-3/4						29	42	48	52	54	57	59
7						27	39	46	50	53	55	57
7-1/4							37	42	48	51	53	56
7-1/2							35	40	46	49	52	54
7-3/4							33	38	42	48	50	53
8							31	36	40	46	49	51
8-1/4							29	34	38	42	47	49
8-1/2							26	32	37	40	45	48

[1]See footnote 2 of Table 3-3.

[2]Sizes of knots permitted for a given strength ratio in wide faces wider than 12 inches in beams and stringers and joists and planks, or faces wider than 12 inches in posts and timbers, are increased only in proportion to the square root of the increase of width of the face. Below these limits, the sizes permitted are proportioned to the width of the face.

[3]Ratios corresponding to other sizes of knot can be found by interpolation.

(Sheet 2 of 2)

Table 3-7.--Strength ratios in horizontal shear corresponding to various combinations of size of shake or check and thickness at end of piece--beams and stringers or joists and planks--shakes or checks in middle one-half of height[1]

Size of shake or check	Green lumber									Seasoned lumber[2]								
	Percentage strength ratio when nominal end thickness of piece in inches is--									Percentage strength ratio when nominal end thickness of piece in inches is--								
	2	3	4	5	6	8	10	12	14	2	3	4	5	6	8	10	12	14
Inches																		
1/4	90	93	95	96	96	97	98	98	98	100	100	100	100	100	100	100	100	100
3/8	83	89	92	93	94	96	97	97	98	94	100	100	100	100	100	100	100	100
1/2	77	85	88	91	92	94	95	96	97	87	95	100	100	100	100	100	100	100
5/8	71	81	85	88	90	93	94	95	96	80	91	96	99	100	100	100	100	100
3/4	65	76	82	86	88	91	93	94	95	73	86	93	97	99	100	100	100	100
7/8	58	72	79	83	86	90	92	93	94	66	81	89	94	97	100	100	100	100
1	52	68	76	81	84	88	90	92	93	59	77	85	91	94	99	100	100	100
1-1/8		64	73	78	82	86	89	91	92	52	72	82	88	92	97	100	100	100
1-1/4		60	70	76	80	85	88	90	91		67	78	85	90	95	99	100	100
1-3/8		56	67	73	78	83	87	89	90		62	75	82	87	94	97	100	100
1-1/2		51	63	71	76	82	85	88	90		58	71	80	85	92	96	99	100
1-5/8			60	68	74	80	84	87	89		53	68	77	83	90	95	98	100
1-3/4			57	66	71	79	83	86	88		48	64	74	80	88	93	96	99
1-7/8			54	63	69	77	82	85	87			61	71	78	87	92	95	98
2			51	61	67	75	80	84	86			57	68	76	85	90	94	97
2-1/8			48	58	65	74	79	83	85			54	66	73	83	89	93	96
2-1/4				56	63	72	78	82	84			50	63	71	81	88	92	95
2-3/8				53	61	71	77	81	83				60	69	80	86	91	94
2-1/2				51	59	69	75	79	82				57	66	78	85	89	93
2-5/8				48	57	68	74	78	81				54	64	76	83	88	92
2-3/4					55	66	73	77	81				52	62	74	82	87	91
2-7/8					53	65	72	76	.80				49	59	73	81	86	90

(Sheet 1 of 2)

Table 3-7.--Strength ratios in horizontal shear corresponding to various combinations of size or shake of check and thickness at end of piece--beams and stringers of joists and planks--shakes or checks in middle one-half of height[1]--Continued

Size of shake or check	Green lumber: Percentage strength ratio when nominal end thickness of piece in inches is--									Seasoned lumber[2]: Percentage strength ratio when nominal end thickness of piece in inches is--								
	2	3	4	5	6	8	10	12	14	2	3	4	5	6	8	10	12	14
Inches																		
3[3]					51	63	70	75	79					57	71	79	85	89
3-1/4						60	68	73	77					52	67	76	82	87
3-1/2						57	65	71	75					48	64	74	80	85
3-3/4						54	63	69	73						60	71	78	83
4						50	60	67	72						57	68	75	81
4-1/4							58	65	70						53	65	73	79
4-1/2							55	63	68						50	62	71	77
4-3/4							53	61	66							59	68	75
5							50	59	65							57	63	73
5-1/4								57	63							54	64	71
5-1/2								54	61							51	61	69
5-3/4								52	59								59	67
6								50	57								57	65
6-1/4									56								54	63
6-1/2									54								52	61
6-3/4									52								50	59
7									50									57
7-1/4																		55
7-1/2																		53
7-3/4																		51
8																		

[1]In tabulating the sizes of shakes or checks, the rule used in approximating decimals by common fractions with 1/8-inch intervals has the effect of subtracting 1/24 inch from the size of each shake or check. For example, the strength ratios listed for a 3-1/2-inch shake or check are those that actually would obtain for a size of 3-11/24 inches. In view of the allowance thus introduced and the face that nominal instead of actual widths of face are used, the strength ratios in this table should be taken as maximum values.

[2]Moisture content of 15 percent or less.

[3]Ratios for sizes of shake or check other than those listed between 3 and 8 inches can be found by interpolation.

(Sheet 2 of 2)

28

3.2.3.2 Posts and Columns.--Shakes and checks, as a rule, have little influence on the strength of members to be loaded in pure compression, provided the piece can act as a unit.

3.2.3.3 Other Considerations.--In addition to the difficulty of discerning the full extent of shake in lumber in service, the following considerations make lumber with shakes, checks, and splits undesirable for ship and boat use:

a) When exposed to wetting, shakes, checks, and splits retain water and create conditions conducive to decay.

b) In direct contact with salt water, shakes, checks, and splits provide an access for marine borer entrance and attack.

c) Shakes, checks, and splits can cause leaks that are difficult to stop.

For the reasons listed above, wood with shakes and splits should not be used in ship and boat applications. Wood with checks may be used, depending on their number and size and where the member is to be used.

3.2.4 Holes.--Holes in timber may be caused by knots that fall out or by worms, insects, or careless use of tools. Knotholes generally impair strength of timber less than tight knots of the same size, because there is usually less distortion of grain around them. The effect of holes from other causes is also less than the effect of a knot of the same size, because no distortion of grain occurs around them. Holes other than knotholes seldom affect strength of timbers as much as other permitted defects. In small wood members, however, holes may become important because of the high percentage of the cross section they occupy and the fibers that are cut.

3. 2.5 Compression Wood.--Bending and tensile strength, stiffness, and shock resistance of compression wood are low for its weight and also erratic. Its shrinkage behavior is also erratic. Pronounced compression wood should be eliminated from all lumber going into ships. Fortunately, the percentage of lumber containing pronounced compression wood is small and because of the tendency of such lumber to warp, it usually is culled if the lumber is seasoned before it reaches a shipyard.

3.2.6 Compression Failures.--Lumber containing compression failures has only a portion of its strength left because it has already been broken. Any lumber or timber containing them should be rejected. Failures that are not readily detectable with the unaided eye are not likely to impair seriously the strength of large timbers. But those that are almost invisible, however, seriously reduce the bending strength and shock resistance of small members. On the tension side of a beam, they usually cause an abrupt and complete fracture across the grain under somewhat lower loads than those normally supported. Only a small percentage of lumber contains serious compression failures.

3.2.7. Minor Defects.--Pitch pockets, bark pockets, pitch streaks, pith flecks, and burls are comparatively small and, with the exception of burls, do not distort the fibers to any great extent. The amount of damage from such causes as compared to other defects which are generally permitted, such as knots, is so small in large structural timbers that these defects may be ignored insofar as their effect on strength is concerned. However, pitch pockets and bark pockets, if large in number and extent, indicate a general fiber separation in the wood, and such a piece should be carefully inspected for shake.

Because they create openings and thus reduce the effective cross section, pitch and bark pockets can appreciably weaken small members, especially in small boats. Their effect on such members may be judged, as is that of holes, by their size in relation to that of the member in which they occur. Burls, by distorting the grain surrounding them, weaken thin members and should be treated similarly to knots with respect to their position in the member under stress.

3.2.8 Decay.--Decay in any form should be prohibited in all types of timbers to be used in ships and boats, because no satisfactory method of estimating its extent or its damaging effect on strength has been devised. In addition, the chance is great that decay will spread and weaken the wood further.

3.3 EFFECT OF DEFECTS AND OTHER FACTORS ON STRENGTH OF GLUED WOOD LAMINATES.

3.3.1 General.--The strength of a laminated timber is affected by more factors than that of a solid timber. It is dependent not only on species, quality, and grade of lumber, as is a solid timber, but also on whether the laminations are bolted, screwed, nailed, or glued together; the kind of glue used; the orientation of the laminations with respect to the load; the type, quality, and location of end joints in individual laminations; and the quality of the workmanship. Bolted, screwed, and nailed members that are horizontally laminated (those in which the load acts perpendicular to the width of the lamination) are not so strong as solid members. Their strength is somewhere between that of a solid member and the total strength of the individual laminations acting independently. Within those limits, the strength depends on the position of the laminations relative to the direction of load and on the efficiency and number of fastenings. Glued laminated members are as strong as solid members of equal quality, provided the glue joints are as strong as the wood and are permanent. Glues are now available that will produce such joints. The sections that follow deal only with glued laminated members.

3.3.2 Cross Grain.--In laminated beams or other bending members, the cross grain requirements may be permitted to vary through the cross section in accordance with stress requirements, as indicated by tables 3-1 and 3-2. Data indicate, however, that the strength of a laminate with laminations containing cross grain of a specified slope may be improved by requiring that outer laminations be made from straighter grained material.

3.3.3 Knots.--The effect of knots on the strength of laminated members is, of course, similar to their effect on the strength of solid timbers (sec. 3.2.2). A knot visible on the surface of a solid timber is known to extend some distance into the interior. In a laminated timber, however, a knot extends through only one lamination; the probability that another knot of the same or similar size will occur in the adjacent lamination at the same location is small. Thus, the dispersion of knots may be expected to result in somewhat higher strength for laminated members than for solid members of the same grade.

Tests have shown that the bending strength of a horizontally laminated member is dependent on both the size of the knot and its position with respect to the neutral axis.

The direct use of this information to establish strength ratios and design stresses requires that both the size and position of knots be known. The size of knots presents little difficulty, since knot limitation may be incorporated in various grades. Knowledge of location of knots, however, would require preassembly of all members and adjustment of the various laminations until the specific limitations were met. This is not in agreement with general laminating practice, in which laminations are assembled randomly, and would add considerably to the cost.

Test data generally show, however, that the cumulative effect of knots of a certain size is less damaging to the strength of a beam when the knots are located near the center of the depth than when they are near the top or bottom. This may permit some economy in the selection and layup of lamination grades, provided that a careful placement is made of lower grade material and laminated members are not beveled or shaped.

In laminated members subjected to tension and compression stresses, the location of knots in the cross section was found to have no effect on strength reduction as in solid beams. The random dispersion of knots in laminated members, however, has the effect of improving strength properties over those for comparable size solid members.

3.3.4 End Joints.--In many laminated members, end joints are required to produce laminations of the necessary length. The type of end joints used, the quality of the workmanship in their fabrication, and their location affect the strength of the completed member.

Butt joints, formed by abutting the squared ends of two pieces, are never glued because experience has indicated that it is not possible to obtain strong dependable butt joints. Butt joints cannot transmit stress and are therefore ineffective, so that at the cross section where they occur, the lamination containing them is essentially wasted. Furthermore, they have been found to be a serious source of stress concentration, both of longitudinal stress in the adjacent laminations and of shear stress in the glued joints. They are seldom, if ever, used in commercial production of important structural members and are not permitted in wood laminates used in wood hulls.

Scarf joints, formed by gluing the feathered end-scarfs of boards, can be made to carry considerable proportions of the stress that could be carried by a continuous piece. Well-made scarf joints with a slope of 1 in 12 or flatter develop 90 percent of the strength of unscarfed laminations. Scarf joints of steeper slope have less strength and are not permitted.

Scarf joints in adjacent laminations should be separated from each other by a distance of 24 times the lamination thickness. If it is impractical to maintain this spacing, the scarf joints of 1 in 12 or flatter should be calculated, as developing 85 percent of the strength of unscarfed laminations. A conscious effort should be made to prevent concentration of joints in adjacent laminations.

Other types of end joints have been proposed and sometimes used, but, in general, those that were used did not give consistently acceptable results.

3.3.5 Curvature of Laminations.--When laminations are bent to produce members of curved form, bending stresses result. If the curvature is too sharp, the laminations may break in bending. Minimum permissible bending radii for various thicknesses are given in table 3-8.

Even if laminations do not fail in bending, a stress of some magnitude will exist, depending upon the relative values of thickness and radius. While tests have demonstrated that much of this stress will be relieved by plastic flow, such stress as remains must be considered in design. The effect of lamination curvature is given by the factor

$$F_c = 1 - 2{,}000\,(t/R)^2$$

where t is the thickness of the lamination in inches and R is the radius of curvature of the lamination in inches.

3.3.6 Vertically Laminated Beams.--Tests have shown that the strength ratio for vertically laminated beams is equal to the average of the strength ratios of the component laminations. Thus, if a beam is made up of laminations of a single stress grade, the design stress for that grade will be applicable also to the laminated beam.

End joints in such beams have not been studied experimentally, but the information on end joints presented in section 3.3.4 is considered applicable.

3.4 EFFECT OF DEFECTS AND OTHER FACTORS ON STRENGTH OF PLYWOOD.--The chief advantage of plywood as compared with solid wood is its more nearly equal strength along the length and width of the panel. The characteristic is obtained by alternating the direction of the grain of successive plies. Thus, the parallel-to-grain direction (the strong direction) of some plies is parallel to the length of the panel and that of the others is parallel to the width. With this construction, "strong" wood is both parallel and perpendicular to the face-grain direction. A piece of plywood in which the grain of the face plies is parallel to its span is not so strong as a piece of solid wood of the same size and quality with its grain parallel to its span. On the other hand, plywood is stronger in the direction across the grain. As a result of its more nearly balanced construction, plywood tends to split, check, shrink, and swell less than solid wood. Because it offers greater resistance to splitting, plywood can hold fastenings closer to the edges than can solid wood.

The strength of plywood is dependent not only on the strength and size of the wood used, but also on the number of plies, their relative thickness, and the grain direction of adjacent plies. In addition, veneers of different species are sometimes combined. The determination of the strength of a particular piece of plywood is therefore complicated, and tables that show the strength properties of all possible combinations of species and ply thicknesses are impractical. The design of plywood parts in ships seldom requires the use of such detailed data. Approximate methods of calculating the various strength properties of plywood are given in table 3-9. These methods, used in combination with the strength data on solid wood shown in tables 2-1 and 2-2, will generally suffice for the degree of

Table 3-8.--Minimum permissible bending radius of various thicknesses of clear, straight-grained lumber of white oak and Douglas-fir

Lamination thickness	Minimum permissible radius of curvature	
	White oak	Douglas-fir
Inches	Inches	Inches
1/4	18	43
5/16	24	56
3/8	31	69
7/16	38	82
1/2	45	95
9/16	53	108
5/8	61	121
11/16	70	134
3/4	81	147
[1]13/16	87	160
7/8		173
15/16		187
[2]1		200

[1]Maximum permitted for white oak to avoid failure in bending or stresses of a magnitude that fail in service.

[2]Maximum permitted for Douglas-fir to avoid failure in bending or stresses of a magnitude that fail in service.

Table 3-9.--Approximate methods for calculating the properties of plywood

Property	Direction of stress with respect to direction of face grain	Area to be considered	Unit stress to be used
Tension	Parallel or perpendicular	Parallel plies[2] only	Unit stress for extreme fiber in bending
	± 45°	Full cross-sectional area	1/6 unit stress for extreme fiber in bending
Compression	Parallel or perpendicular	Parallel plies[2] only	Unit stress in compression parallel to grain
	± 45°	Full cross-sectional area	1/3 unit stress in compression parallel to grain
Bearing at right angles to plane of plywood		Loaded area	Unit stress in compression perpendicular to grain
Load in bending	Parallel or perpendicular	Bending moment M = KSI/c where S = unit stress for extreme fiber in bending; I = moment of inertia computed on basis of parallel plies only; c = distance from neutral axis to outer fiber of outermost ply having its grain in the direction of the span; K = 1.50 for three-ply plywood having the grain of the outer plies perpendicular to the span; K = 0.85 for all other plywood	Unit stress for extreme fiber in bending
Deflection in bending	Parallel or perpendicular	Deflection may be calculated by the usual formulas, taking as the moment of inertia that of the parallel plies plus one-twentieth that of the perpendicular plies. (When face plies are parallel, the calculation may be simplified, with but little error, by taking the moment of inertia as that of the parallel plies only.)	Unit value for modulus of elasticity

(Sheet 1 of 3)

Table 3-9.--Approximate methods for calculating the properties of plywood--Continued

Property	Direction of stress with respect to direction of face grain	Area to be considered	Unit stress to be used
Deformation in tension or compression	Parallel or perpendicular	Parallel plies[2] only	Unit value for modulus of elasticity
Shear through thickness	Parallel or perpendicular	Full cross-sectional area	Double unit stress in horizontal shear
	± 45°	Full cross-sectional area	4 times unit stress in horizontal shear
Shear in plane of plies	Parallel or perpendicular	Full shear area. Plywood beams. Horizontal shear: 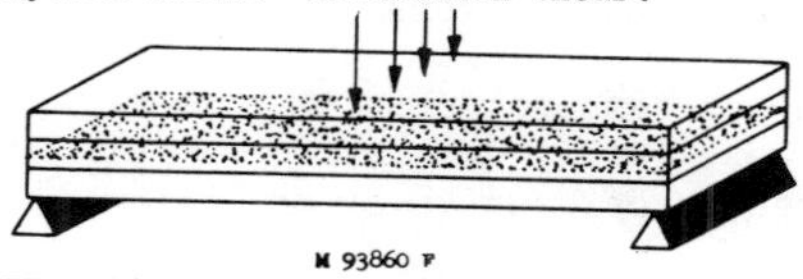	3/4 unit stress in horizontal shear
		Area of contact between plywood and flange or framing member. I- or box-beams with plywood webs. Shear between plies of web or between web and flange: 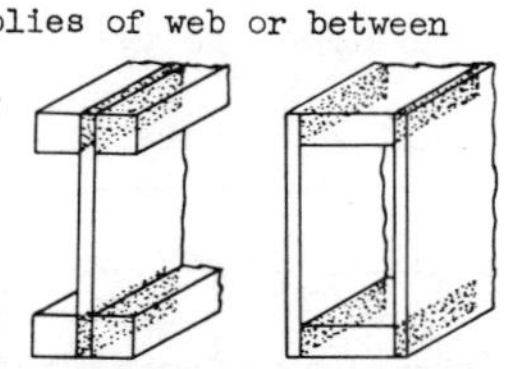 Panels having plywood covers stressed in compression or tension, or both. Shear between plies or between cover and framing members when depth of member exceeds twice its width and end headers are used or when depth is not more than twice the width and no headers are used.	3/8 unit stress in horizontal shear
		Interior framing members: 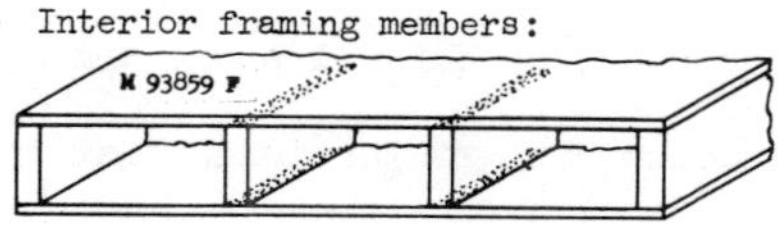	3/4 unit stress in horizontal shear
		Framing members at edge of panel: 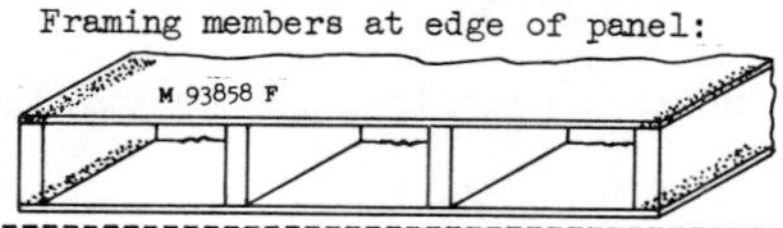	3/8 unit stress in horizontal shear

(Sheet 2 of 3)

Table 3-9.--Approximate methods for calculating the properties of plywood--Continued

Property	Direction of stress with respect to direction of face grain	Area to be considered	Unit stress to be used
Shear in plane of plies	± 45°	Area of contact between plywood and flange or framing member. I- or box-beams with plywood webs. Shear between plies of web or between web and flange: M 93861 F 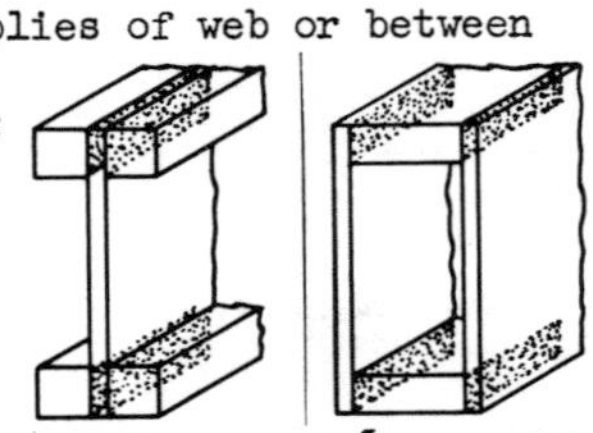	1/2 unit stress in horizontal shear
		Panels having plywood covers stressed in compression, or tension, or both. Shear between plies or between cover and framing members when depth of member exceeds twice its width and end headers are used or when depth is not more than twice the width and no headers are used.	
		Interior framing members: 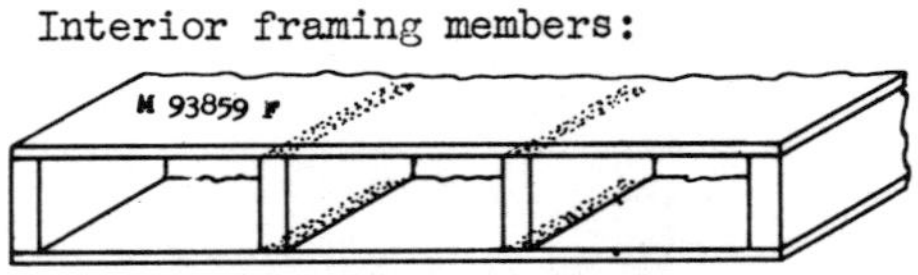	Unit stress in horizontal shear
		Framing members at edge of panel: 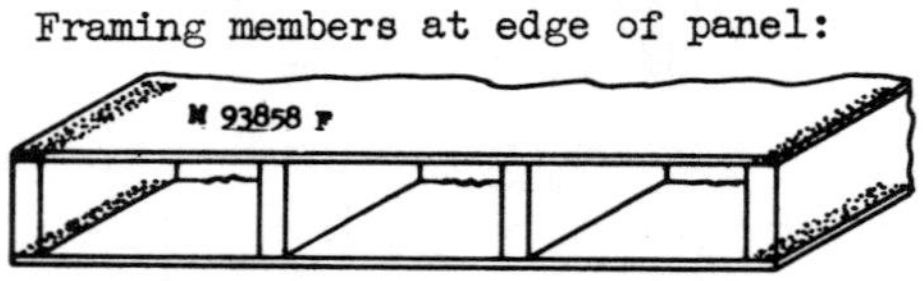	1/2 unit stress in horizontal shear

[1]The suggested simplified methods of calculation apply reasonably well with usual plywood types under ordinary conditions of service. It is recognized, however, that they are not entirely valid for all types of plywood and plywood construction, or for all spans and span-depth ratios. Also the methods given are not applicable to structures so proportioned that the plywood is in the buckling range, in which event the results will be too high.

[2]By "parallel plies" is meant those plies whose grain direction is parallel to the direction of principal stress.

(Sheet 3 of 3)

accuracy required in the design of most of the parts of boats for which plywood is used. No general rules for estimating the effect of defects on the strength of plywood have been developed. For plywood containing large defects, some reduction in strength properties should be assumed.

3.5 REFERENCES

(3-1) AMERICAN SOCIETY FOR TESTING MATERIALS
1958. Methods for Establishing Structural Grades of Lumber (Tentative). ASTM Standards, Part 6, D 245-57T.

(3-2) FREAS, A.D., and Selbo, M. L.
1953. Fabrication and Design of Glued Laminated Wood Structural Members. U.S. Dept. Agr. Tech. Bull. No. 1069.

(3-3) FOREST PRODUCTS LABORATORY
1955. Wood Handbook, U.S. Department of Agriculture (No. 72) (Revised).

CHAPTER 4. REQUIREMENTS FOR WOOD USED IN SHIPS AND BOATS

4.1 GENERAL.--The structural members of ships and boats require a combination of wood properties that is found in only a limited number of species. The principal properties required in most structural members of ships are decay resistance, ability to hold fastenings, bending strength, shock resistance, and stiffness. Ease of bending to curved shapes is an additional requirement of many parts. A minimum of warping, swelling, shrinking, and water absorption is also generally desirable as well as ease of working, especially in woods used for planking and ceiling. For small boats, woods that combine light weight with adequate strength, decay resistance, and freedom from shrinkage, swelling, warping, and excessive water absorption are sometimes sought.

4.2 Woods Used in Ship and Boat Construction.--Both hardwoods and softwoods are used in ship and boat construction. Of the hardwoods, white oak has been used predominantly, while Douglas-fir has been the most extensively used softwood, followed by southern yellow pine.

Typical of the amounts of lumber used for large- and medium-size wood ships are the following, taken from bills of materials submitted by builders and designers:

AM-421 Class Minesweeper: (Now designated MSO-421) 328,202 board feet of lumber, 26,706 square feet of Douglas-fir plywood. Species and amounts (board feet): Douglas-fir, 229,095; southern pine, 5,088; red oak (treated), 13,585; white oak, 80,374; sugar pine, 51; maple, 9 board feet.

AMS 60 Class Minesweeper: (Now designated MSC 60) 143,154 board feet of lumber, 36,392 square feet of Douglas-fir plywood. Species and amounts (board feet): Douglas-fir 81,759; Sitka spruce (treated), 1,634; southern pine, 21,173; white pine, 230; white oak, 11,468; white bending oak, 17,880; red oak (treated), 8,520; white ash, 290; mahogany, 280 board feet.

MSB 5 Class Minesweeper: 12,575 board feet of lumber, 7,005 square feet of Douglas-fir plywood. Species and amounts (board feet): Douglas-fir, 8,786; white cedar, 935; white oak, 2,485; mahogany, 2,425 board feet.

4.2.1 Hardwoods

4.2.1.1 White Oak.--White oak has a better combination of properties which meet the principal requirements of ship and boat parts than any other native wood. The heartwood of white oak ranks moderately high in decay resistance, holds fastenings well, and is one of the stronger and stiffer native woods. It is an excellent bending wood and can be satisfactorily bent to curvatures required in boatbuilding. The favorable combination of the required properties that white oak possesses accounts for its extensive use for structural parts in boat and ship construction.

A number of hardwoods compare favorably with white oak in most of the required properties, except decay resistance. Red oak, hickory, white ash, rock elm, beech, yellow birch, and hard maple are domestic woods that are equal or superior to white oak in bending strength, shock resistance, and ability to hold fastenings. Because of their low decay resistance, however, they are unsuitable for ship and boat use without thorough preservative treatment.

There is no entirely satisfactory substitute for white oak in members, such as frames, garboards, or chines that must be bent to curvatures of comparatively short radii. Rock elm, and some species of red oak most nearly approach white oak in meeting the combined requirements for such members. Neither red oak nor rock elm, however, is equal to white oak in decay resistance, and rock elm is even more difficult to obtain in required grade than white oak.

4.2.1.2 Teak.--Teak has the well-earned reputation of being the finest of all shipbuilding woods. While equalled or surpassed in each of its individual properties by some other species, teak has a combination of desirable characteristics that can be matched by no other known wood for use in ships and boats. Its relatively high strength, excellent decay resistance, and moderate marine borer resistance combine to give teak this superiority. In practice, however, teak is used much less than its properties would warrant because of its scarcity and high cost. In Navy use, teak is largely confined to the decks of aircraft carriers, battleships, cruisers, and submarines, where no substitute has been found for teak in its ability to withstand submergence, wear, and weathering. To a limited extent teak is also used in Navy small craft.

4.2.1.3 Mahogany (American or True Mahogany) (Swietenia spp), African Mahogany (khaya spp.), "Philippine Mahogany" (Shorea species and Dark Red Lauan).--The species commonly called "mahogany" in ship and boat construction are used where the requirements call for woods that are relatively light in weight, moderately strong, low in shrinkage and swelling, warp resistant, moderately high in decay resistance, and have good fastening strength and wearing properties, are not easily split, and are easy to work. They are used primarily by the Navy for planking and decking in high-speed boats and for joiner work.

The heartwood of true mahogany has high decay resistance. Information on the decay resistance of khaya, tangile, and dark red lauan is not sufficient to justify close comparison with our native woods, but the service they have rendered in boats indicates that their heartwood has moderate decay resistance, with khaya being somewhat superior to both species of Philippine woods.

White lauan, another species of "Philippine mahagony," is suitable for joiner work that will not be exposed subsequently to decay hazard.

4.2.2 Softwoods.

4.2.2.1 Douglas-Fir and Southern Yellow Pine.--Douglas-fir and southern yellow pine are softwoods with the best combination of the properties required for structural parts in ships and boats. The dense heartwood of these two species of wood is generally a little less decay resistant than the heartwood of white oak. As is common to all woods, their sapwood is not decay resistant.

The principal difference in strength between these softwood species and white oak is hardness, shock resistance, shear strength, and ability to hold fastenings. When compared with the two softwoods, white oak is from two-fifths to four-fifths harder and about one-fourth higher in shock resistance, holds fastenings better, has higher shear strength, resists splitting, wears less and bends more easily.

For the above reasons, these softwoods cannot be substituted for white oak in many ship and boat applications, although they have essentially the same static bending strength.

However, because Douglas-fir and southern yellow pine are lower in shrinkage and swelling, resist warping and stay in place better, and are lighter in weight and more available than white oak, they are often specified for such ship and boat applications as planking, decking, deck beams, and other uses where their strength properties are satisfactory.

Of the two softwood species, Douglas-fir is used more widely because of its availability in large sizes, long lengths, and good quality, and its ability to stay in place when installed. Southern yellow pine, however, is preferred where greater abrasion resistance and resistance to splitting are desired.

4.2.2.2 Alaska Yellow-Cedar, Atlantic White-Cedar, Baldcypress, Northern White-Cedar, Western Redcedar, Port-Oxford-Cedar, and Redwood.--The cedars, baldcypress, and redwood are used where the requirements call for wood that is light in weight, low in shrinkage, swelling and warping, high in decay resistance, and easy to work.

The cedars and baldcypress are used principally for planking small boats and high-speed craft. The desirability of these woods for planking is based on their light weight combined with good strength, the exceptionally high natural decay resistance of their heartwood, which is superior to that of white oak, and the ease with which planking strakes can be shaped and sprung to the hull contours.

Baldcypress is heavier and stronger than the cedars, but is more difficult to work. Redwood, although equal in decay resistance to the cedars, is not generally used for boat planking because of its tendency to split and its corrosiveness to metals in salt water, However, because of its somewhat higher strength, abrasion resistance, and slight marine borer resistance, it has been used as worm sheathing on some Navy craft.

4.2.2.3 Sitka Spruce, White Pines, Ponderosa Pine, Western Hemlock, and Noble Fir.--Spruce, pine, hemlock, and fir are very susceptible to decay, and are permitted for use in ships and boats, only for special applications where decay conditions are not generally present. For example, spruce has been used in certain high-speed boats because of its high ratio of strength to weight. White pine is sometimes used for decking of small fast sailboats because of its combination of uniformity of texture and durability, its ability to stay in place, and its light weight.

4.3 STRESS GRADING OF SOLID LUMBER FOR SHIP AND BOAT APPLICATION.--Utilizing the basic stresses for lumber listed in tables 2-1 and 2-2, and the knowledge of loss in strength resulting from occurrence of various strength-reducing defects, lumber can be graded by specialists in such a manner that definite working stresses may be assigned to each grade.

4.3.1 Commercially Stress-Rated Lumber.--Based on various common use applications, a large variety of commercial stress grades have been developed for various species of solid wood. Many of these stress grades are listed and recommended by lumber associations in their grading rule books. Commercially stress-rated lumber is generally considered in the following four use classes:

a) Joists and Planks.--Pieces of rectangular cross section 2 to 4 inches thick and 4 or more inches wide (nominal dimensions), graded primarily for strength in bending edgewise or flatwise but also frequently used where tensile or compressive strength is important. Material 2 inches (nominal) thick is often placed in grades separate from the heavier joists and planks.

b) Beams and Stringers.--Pieces of rectangular cross section, 5 by 8 inches (nominal dimensions) and up, graded for strength in bending when loaded on the narrow face.

c) Posts and Timbers.--Pieces of square or nearly square cross section, 5 by 5 inches (nominal dimensions) and larger, graded primarily for use as posts or columns but adapted to miscellaneous uses in which bending strength is not especially important.

d) Structural Boards.--Lumber 1 inch (nominally) thick used in lightweight trusses or other structural elements, or in glued laminated structural members where the principal stresses are in axial compression or tension. Boards for common purposes such as sheathing, finish, or boxes, are not included.

The principles of stress grading make possible the assignment of bending stresses to posts and timbers, compression or tension stresses to joists and planks and beams and stringers, or stresses for flatwise bending to beams and stringers. Combined applications for any of the use classes are thus made possible. It is likewise possible to formulate combination or all-purpose grades of structural lumber, designed for all the strength requirements of all of these use classes.

Despite differences between normal commercial structural applications and ship and boat applications, these grades of lumber can, with some modification, be used for many ship and boat applications. Modification of commercial stress grades for ship and boat use applications is discussed below.

4.3.1.1 Length Considerations.--Commercial stress grades for most species of wood in general assume that a stress-rated beam will not be significantly altered in length. Consequently a stress-graded piece cannot be cut into smaller lengths without invalidating the assigned stress values. Therefore, in order to use a piece of lumber for a given stress application, it must either be graded in place, or procured to the given final length. For ship and boat application, such a procedure is time consuming and wasteful, particularly if cross sectional dimensions are not altered. A more practical procedure, as far as cutting for length is concerned, is to include a provision in the specifications that requires all stress-graded lumber to be graded in such a manner that the full length of the piece is graded as severely as the center third portion. In this manner, the stress-graded piece can be cut for length, and any of the cut section can generally be considered to have an assigned stress value at least equivalent to the uncut section.

4.3.1.2 Use As a Continuous Span.--Commercial stress grades for most species of wood in general assume that a stress-rated beam will be loaded as a simple span. If the beam is to be used as a continuous span utilizing three or more supports, the assigned stress values are not applicable. Therefore, if beams are to be installed without alteration as a continuous member on three supports, the center two-thirds of the beam must be graded the same as the center one-third for knots and cross grain.

If beams are to be installed as a continuous member on four or more supports, the full length of the beam must be graded the same as the center one-third for knots and cross grain. For all continuous spans, restrictions for visible shakes or checks are applied to the middle half of the height of the piece and for a distance equal to three times the height of the piece (height equals width of wide face) anywhere throughout the length. When beams are graded with equal severity throughout the length, as on continuous beams on four or more supports, cutting of beams for length will not change the assigned stress rating.

4.3.1.3 Cross Sectional Dimensions.--Commercial stress grades for most species in general assume that a stress-rated beam will not be altered in cross section. Altering the cross section of a stress-graded beam will invalidate the assigned stress values. Therefore, if members are to be altered in cross section on installation, stress grading must be accomplished for the piece in its finished state if an assigned stress value is to be met.

4.3.2 Use of Commercial Grades of Nonstress-Rated Lumber.--Various lumber association grading rule books list a large variety of lumber grades intended for structural purposes, which although they are not assigned stress values, could easily be assigned stress values with little additional requirements. Many of these commercial grades have been formulated for a tailored commercial application. Examples of such grades are barge framing, planking and decking, ship decking, and railway car material. For these types of members, stress ratings can normally be assigned by adding a slope of grain limitation and computing the strength ratio. Using these grades of material and "industrial clears" as the basic grade with additional requirements is generally most suitable in fabrication of members of nonuniform cross section and in decking or planking, since these grades have more severe knot limitation both in size and placement than the normal stress-rated grades, which are not intended to be modified in cross-sectional dimensions. Usually a high percentage of these nonstress-rated structural grades will yield suitable stress-graded lumber on regrading.

4.3.3 Special Boat Building Lumber.--Many boat building applications for lumber are such that it is often difficult and sometimes impossible to utilize existing commercial grades of lumber without a great deal of modification. This is particularly true in places where structurally sound material in long length is required, and other important considerations, such as watertightness, fastenings, and weight limitations, are of prime importance. Lumber in such instances is generally purchased for the specific application, assuring that the necessary strength is provided. The lumber is generally selected on a basis of experience and individual judgment as to its adequacy for the requirements of a specific part. Stress values can be easily supplied to such members, since they are small in dimension and practically of clear material. The major basic stress considerations are slope of grain, and frequently, size and location of knots.

4.4 STRESS GRADING OF LAMINATIONS FOR USE IN GLUED WOOD LAMINATES FOR SHIP AND BOAT APPLICATION.--The basic stresses for lumber listed in tables 2-1 and 2-2 are utilized in developing design stresses for wood laminates as well as solid members. If wood laminates are to be utilized at moisture content conditions above 15 percent, the basic stresses listed in table 2-1 apply. If wood laminates are to be utilized and maintained at moisture content conditions at 15 percent or less, the basic stress values listed in table 2-2 apply. Because of the various sizes and shapes, and special purposes for which wood laminates are required in ship and boat application, the grades of laminations (individual layers forming a laminate) in applicable military specifications for wood laminates for marine use are based first on loss in strength due to cross grain. Knot sizes, knot placement, and other strength reducing defects are permitted on the basis of their effect in relation to final laminate width, cross section, and radii of curvature. In special use applications, such as laminated decking, the caulking surfaces and edge grain wearing surfaces must be free from knots.

There are no commercial grades of lumber that can be utilized directly for the grades listed in the applicable military specifications covering wood laminates for marine use. For use in Douglas-fir laminates, however, commercial grades of nominal 1-inch industrial clears with some additional modifications will usually give a high rate of yield of Douglas-fir lamination grades listed in the governing military wood laminate specification. For use in white oak laminates, the commercial grades of Firsts and Seconds, Selects, and No. 1 Common will provide a fair yield of boards of marine laminating grade.

If lumber is properly edge glued with a suitable water-resistant durable adhesive, much higher yields of material of marine laminating grade can be realized from commercial grades of both species of wood.

4.5 PLYWOOD FOR SHIP AND BOAT USE.

4.5.1 Plywood for Navy Use.--The requirements for flat plywood for boat and ship construction are described in Military Specification MIL-P-18066.

Plywood for building boats must be of high quality. Veneers should be all heartwood to provide resistance to decay. Core gaps should be restricted in size to provide added strength and prevent the paint failures that occur when such voids become filled with water. Core voids cause considerable trouble in the long panels used for the hull of a boat. Veneer thickness must be carefully controlled to provide the necessary stiffness and to prevent delamination that occurs during marine exposure if the veneers are too thick. In softwood plywood, veneer thickness should not exceed one-eighth inch. Veneer thickness should not exceed one-tenth inch for mahogany plywood. The veneer quality, and strength and permanence of the glue joints, however, are essentially the same as those established by commercial standards for Exterior-type Douglas-fir plywood or for Type I or Technical-type hardwood plywood, except that the kind of glue for ship and boat use is definitely specified as the phenol or resorcinol type.

Plywood panels longer than 96 inches may be made of two or more smaller panels by scarfing.

Three classes of plywood are provided in the specification. In Class 1, all plies must be of Douglas-fir. Except for requirements that limit splits and gaps more strictly than does commercial practice, the faces must meet the requirements of Grade A, and the inner plies are also limited more strictly than in Commercial Standard CS45. Class 2 provides for mahogany plywood. For the purpose of this specification, the term mahogany includes American mahogany (*Swietenia macrophylla* or *S. mahagoni*), African mahogany (*Khaya ivorensis,* and *K. teysmanniana*). The thickness and the quality of faces and inner plies must conform to the requirements for Grade 1 mahogany, Technical type, contained in Commercial Standard CS35. Class 3 plywood provides for Douglas-fir plywood overlaid with resin-impregnated paper. The paper overlay must conform to the requirements for Type II, medium density overlays, as described in Specification MIL-P-8296.

Overlaid plywood, Class III, must meet certain additional requirements described in Military Specification MIL-P-8296, pertaining to the quality of overlay and that of the bond between the overlay and the face veneers.

4.5.2 Preservative-Treated Plywood.--Requirements for preservative-treated plywood for use in Navy ships and boats are set forth in section 2.14 of Volume II.

4.5.3 Fire-Retardant-Treated Plywood.--Requirements for fire-retardant-treated plywood are set forth in section 2.51 of Volume II.

4.5.4 Commercial Plywood Grades.--Types and grades of commercial plywood that may be suitable for some parts of wood boats and ships are given in four commercial standards that describe specific requirements to which manufacturers generally adhere. Issued by the Commodity Standards Division, Business and Defense Services Administration, U.S. Department of Commerce, Washington 25, D.C., they are:

CS45 -- Douglas-fir Plywood.

CS35 - Hardwood Plywood.

CS122 - Western Softwood Plywood.

CS157 - Ponderosa and Sugar Pine Plywood.

Commercial Standard CS45 provides for four grades of veneers, called A, B, C, and D, and two types of plywood, Exterior and Interior, based on the moisture resistance of the glue bond. The first letter of the grade designation refers to the grade of the face ply and the second to the grade of the back. Inner plies are Grade C or D. The type refers to the resistance or durability of the glue bond under severe exposure conditions. The Exterior type requires essentially that the glue bonds remain intact under prolonged exposure to the elements.

The requirements for Grade A veneer permit no knots, splits, pitch pockets, or other open defects. Patches are permissible if neatly made. Essentially, the grade permits a clear or transparent finish. Grade

B must be free from open defects. Circular plugs, synthetic plugs, sound tight knots up to 1 inch, and tight splits are permissible. The grade must be suitable for painting. Grade C permits knotholes no larger than 1 inch in the least dimension, open pitch pockets no wider than 1 inch, splits no wider than 3/16 inch that taper to a point, knots no larger than 1-1/2 inches in the least dimension, and plugs, patches, and shims. Grade D permits knotholes up to 2-1/2 inches in the maximum dimension, pitch pockets no wider than 2 inches nor longer than 4 inches, and splits no wider than 1/2 inch that taper to a point. Patches, shims, worm holes, and sanding defects are permitted, providing they do not seriously affect the usefulness of the panels.

Tests required for determining the acceptability of the glue bond in Exterior-type Douglas-fir plywood are the cyclic soaking and drying test, the boil test, and a fire test. In the cold soaking and drying test, samples are submerged in water at room temperature for a period of 48 hours and dried for 8 hours at $145^{\circ} \pm 5^{\circ}$ F., followed by two cycles of soaking for 16 hours and drying for 8 hours under the conditions described above. The specimens are soaked again for a period of 16 hours and shear tested while wet. In the boil test, specimens are boiled in water for 4 hours, dried for 20 hours at $145^{\circ} \pm 5^{\circ}$ F., then boiled 4 more hours, cooled in water, and shear tested while wet. After shear testing, the specimens of either test are dried and the percentage of wood failure estimated. The combined average of cold-soak and boil tests must be not less than 60 percent, or 30 percent for an individual specimen.

In the fire test, specimens are subjected to a flame of 800° to 900° C. for 10 minutes in a standard jig. Any delamination due to combustion is considered failure.

The Interior type, as the name indicates, is suitable for use in protected locations where it only rarely gets wet and dries out. Western softwoods other than Douglas-fir may be used in the inner plies. Essentially, the resistance of the glue bond must be such that 2- by 5-inch samples of the plywood will withstand three cycles of wetting and drying consisting of submergence for 4 hours in water at room temperature, followed by drying at a temperature between 100° F. and 105° F. for 19 hours in a drying cabinet to reduce moisture content to a maximum of 8 percent. Delamination of 1/4 inch or more in depth and over 2 inches in length along the edge of the test piece denotes failure.

Commercial Standard CS35, for hardwood plywood, provides for five grades of veneer (Specialty, 1, 2, 3, and 4) and four types of plywood (Technical, I, II, and III). The standards for Specialty grade veneer are based largely on the requirements of the furniture trade to match color and figure, and include special selection and types as agreed upon by buyer and seller. Grade 1 veneer permits minor defects which do not preclude use of a natural finish. Grade 2 veneer permits, in general, only those defects and characteristics that may be satisfactorily covered by painting. Grade 3 veneer permits limited open defects, for example, knotholes not exceeding 3/4 inch in average diameter and splits not exceeding a 3/16-inch width nor extending over half the length of the panel. Grade 4 veneer limits open defects to knotholes no larger than 2 inches in diameter and to splits no wider than 1 inch if they extend no more than one-fourth the length of the panel, and no wider than 1/4 inch if they extend full panel length. It also permits a variety of defects that are not open.

Technical and Type I hardwood plywood require a weatherproof bond essentially equivalent to that required for Exterior-type softwood. In the Technical type, the total thickness of the veneers running in the same direction as the face must be from 40 to 60 percent of the total panel thickness. Type II requires bonds that will withstand 10 out of 15 cycles of wetting and drying in 6- by 6-inch samples, without showing delamination deeper than 1/8 inch or longer than 2 inches. Type III hardwood plywood requires that the samples will withstand two soaking and drying cycles without exceeding the amount of delamination allowed for Type II.

4.5.5 Overlaid Plywood.--Plywood, that is covered, or overlaid, with combinations of paper, synthetic resin, and sometimes fabric is called overlaid plywood.

Overlays, applied primarily to improve painting characteristics and to conceal undesirable characteristics of the wood, are sometimes called masking or medium-density overlays. Such overlays are usually paper, treated or impregnated with 15 to 40 percent of synthetic resin, often phenolic. The surfaces of such overlaid plywood are smooth and free from face checks.

By putting more resin in the paper and changing fabricating techniques, the hardness, wear resistance, and moisture resistance of overlays can be increased. Overlays of this type are often termed high-density and find their greatest use where a smooth, wear-resistant surface is required.

Overlaid plywood is used on ships and boats for deck cabins, pilot houses, and other superstructures.

Standards for both types of the commercial product are described in Commercial Standard CS45.

4.5.6 Metal-Faced Plywood.--The requirements for metal-faced plywood for military use are contained in Military Specification MIL-P-8053. The metal surfacing may be aluminum alloy, zinc-coated steel, phosphate-coated steel, or stainless steel, as specified. For severe service conditions, a two-step process is required for bonding the metal to plywood. An adhesive, conforming to Specification MIL-A-5090, is applied to one surface of the metal facing and cured according to the instructions of the manufacturer. The adhesive-covered metal surface is then bonded to the plywood with a resorcinol adhesive.

4.5.7 Curved Plywood.--For installations that require plywood to be bent or curved in one plane only, the plywood is usually bent to shape after the veneers have been glued as flat plywood. In such cases, the requirements applicable to flat plywood for similar service should be observed.

4.5.8 Molded Plywood.--Although any piece of curved plywood may be properly called molded plywood, the term is usually applied to plywood shapes of compound curvature, formed by molding veneers to a die with the use of fluid pressure. This process has found limited use in naval boats and ships, and marine usage is confined largely to commercial production of small, light-weight boats. The use of molded plywood for boat hulls provides lighter, more rigid construction and the use of fewer framing members, but is limited by the size and cost of production equipment required to fabricate large structures.

The contention is often heard that molded plywood is more brash than flat-pressed plywood of the same species and construction. Recent strength comparisons have shown that there is no significant difference between the two methods of manufacture. The inherent variation in wood properties has a greater effect on strength than does the method of pressing. If the design of a particular craft calls for molded plywood, the general requirements should be the same as those for flat plywood in similar service.

4.6 BOAT AND SHIP HULL SPECIFICATIONS.

4.6.1 Essential Specification Requirements.--In establishing specification requirements for wood parts of vessels, a number of essential elements must be covered to assure obtaining satisfactory parts, to indicate clearly to the contractor what he is expected to furnish, and to to provide a sound basis for inspection. Unless all of the essential elements are covered, personal judgment must be relied on, and this may lead to controversies during inspection. These elements are discussed briefly below.

Species.--In many cases, experience, durability requirements, or other considerations dictate the use of particular species. Where a particular species is required, it should be clearly called for. For example, the combination of bending quality and decay resistance possessed by white oak generally dictates its use for steam-bent frames. If white oak is needed, it is not sufficient to specify oak, since red oak, which is lower in heartwood decay resistance, might be furnished. Species terminology is sometimes variable. Table 1-1 of Volume I relates commercial lumber names to particular domestic species and should be used as a guide for specification terminology.

Grade.--Lumber may be graded commercially in a number of ways, as indicated in section 4.3. For special applications, Navy specifications establish additional grade requirements. Where specific grades are required from considerations of strength, appearance, watertightness, or similar needs, the specification should clearly indicate the grade needed, either by reference to a Navy specification or to specific paragraphs of commercial grading rules. In the interests of economy and availability, the lowest grade adequate for the purpose should be specified. For example, in the case of important structural members, such as deck beams, it is essential that a grade be chosen in which strength-reducing characteristics are limited in such a way as to insure a certain minimum strength loss.

Moisture Content.--As discussed earlier, changes in moisture content affect dimensions and may affect strength, depending on the grade. To reduce dimensional changes, warping, and the like after a wood member has been placed in service, the moisture content of the material should be as near as possible to that expected in service. No distinction should be made between air-seasoned and kiln-dried material, since proper drying practice will result in satisfactory material by either method.

Heartwood, Sapwood.--The heartwood of some species is naturally decay resistant, while the sapwood of all species is low in this property. For use under conditions favorable to decay, therefore, it may be desirable to specify that the material contain some specific proportion of heartwood. For other conditions of use, or where the material is to be preservative treated, sapwood should not be restricted.

Dimensions.--Unless dimensions are fully covered in drawings that are a part of or referred to by the specifications, thickness, depths, and other critical dimensions should be specified. For lumber items, it should be clearly stated whether the widths and thicknesses are nominal or actual.

Plywood.--In plywood, the durability of the adhesive is generally indicated by a type or class designation. This should be included in the specification requirements to assure obtaining plywood with the desired water resistance. If the type of finish to be used is an important factor, or if strength is a factor in the design, the appropriate grade should be specified.

Laminated Members.--The characteristics required in laminated members are set forth in Navy specifications that cover grade of material, type of adhesive, and other features, such as type and location of end joints. All factors that affect properties, particularly grade and adhesive type, should be specified.

4.6.2 Sample Specification Statements.--The following examples are intended only to illustrate the principles outlined in section 4.6.1 and are not to be construed as applying to any particular class of ship or boat, nor as setting up specific requirements for any part.

(A) The main deck beams shall be Douglas-fir of the Select Structural Framing, Joists, Planks, and Small Timbers grade (par. 123a 1900f WCLA Rule Book 15) except that the full length of the beam shall be graded as severely as the center 1/3 and splits, sapwood, and wane shall not be admitted in the finished member. The beams shall be 4 x 6 inches (nominal) in size, crowned in accordance with Bureau of Ships Drawing No.________. The moisture content of any deck beam at the time of installation shall be 13 plus or minus 3 percent.

(B) The plywood for bulkheads shall be Douglas-fir, Exterior type, grade A-C, sanded 2 sides, as described in Military Specification MIL-P-18066 and Commercial Standard CS45. Thickness and number of plies shall, for the various bulkheads, be in accordance with Bureau of Ships Drawing No.____________. The moisture content of the plywood at the time of installation shall not exceed 12 percent.

(C) The frames shall be laminated of white oak in accordance with Military Specification MIL-W-0015154. All laminations shall be of Grade B oak as described in the specification. End joints in individual laminations shall be spaced according to Arrangement 3 of section 3.8 of the specification. Dimensions and forms of individual frames shall be in accordance with Bureau of Ships Drawing No.__________.

CHAPTER 5. BASIC DATA FOR SHIP AND BOAT DESIGN WITH WOOD

5.1 GENERAL.--It is beyond the scope of this publication to deal with general problems of ship design. Rather, this section will discuss methods of design with wood which may be used by the naval architect once he has solved the problems of form, loads, and the like.

During and after World War II, significant advances were made in the design of round-bottom wood hulls. Johnson (*5-2*)[1] has reviewed these developments and shown the advantages to be expected from them. He indicates that design in accordance with traditional classification rules for sawn-frame ships appears to achieve strength by means of mass, rather than by proper utilization of the inherent properties of wood. Thus, in such construction, continuity of strength is achieved by overlapping piece upon piece, with the component structural parts of the ship mutually supporting and clamping each other.

[1] Italic numbers in parentheses refer to references at end of this chapter.

The basic difficulty lies in the fact that the frame futtocks are sawed from planks or timbers and that, to avoid excessive cross grain, the futtocks must be short. To fasten these futtocks together endwise, additional futtocks are added to the first one to act as butt straps and are in turn supported by the first futtock. The continuity of such a frame is somewhat dubious. To assist in holding the frame together, an inner skin or ceiling is introduced so that the frame may be clamped between it and the outside planking (fig. 5-1).

Transverse strength at the keel is achieved by bolting a number of massive longitudinal members (keel, keelson, sister keelson, and rider keelsons) together with transverse floors and frames into a mass of lumber. At the intersection of deck beams and frames, transverse strength is achieved by means of heavy clamps, shelves, and knees that are fastened to the frames, beams, ceiling, deck planking, and each other.

All these details result in a mass of lumber that is used inefficiently. To eliminate such massive construction, however, continuity must be achieved in some other way. For example, if keelsons, floors, and the like are to be eliminated, the frame must be continuous over the keel. If the ceiling is to be dispensed with (for economy and better ventilation), the frame must be continuous--the conventional futtock-frame construction described earlier cannot safely be used.

During World War II, a notable advance in this respect was made in the design of YMS minesweepers. In this class, the frames were made of two sections of steam-bent oak continuous from deck to keel (fig. 5-1). The continuity thus achieved eliminated the necessity for the clamping action afforded by the ceiling. Elimination of the ceiling permitted lightening and simplification of the construction. Also, elimination of the ceiling removed one of the sources of decay--the poorly ventilated bays between frames. In addition to the change in frame construction, a number of other innovations were introduced, including lighter floors, a reduction in the number of timbers at the keel, and substitution of light struts for massive knees. The fine service performance of nearly 500 ships of this class proved the adequacy of the innovations in design.

Extensive developments in laminating and the advent of versatile and durable adhesives during and after World War II opened the way for further improvement in the construction of wood hulls. On the basis of these developments, a postwar minesweeper (the MSO-421 class) was designed. The innovations in this class were even more far-reaching than were those incorporated in the YMS class.

The MSO-421 design was based on the use of glued laminated wood for virtually all of the main structure (fig. 5-2). For example, the ultimate in frame continuity, as compared with frames of the futtock type, was achieved by making the majority of the frames continuous from head to head. By this means, not only was the necessity for ceiling eliminated but, likewise, the necessity for floors and other timbers at the keel. Longitudinal continuity was achieved by the use of a laminated keel more than 100 feet long and a laminated keelson, also more than 100 feet long. In addition, a laminated fashion piece and a one-

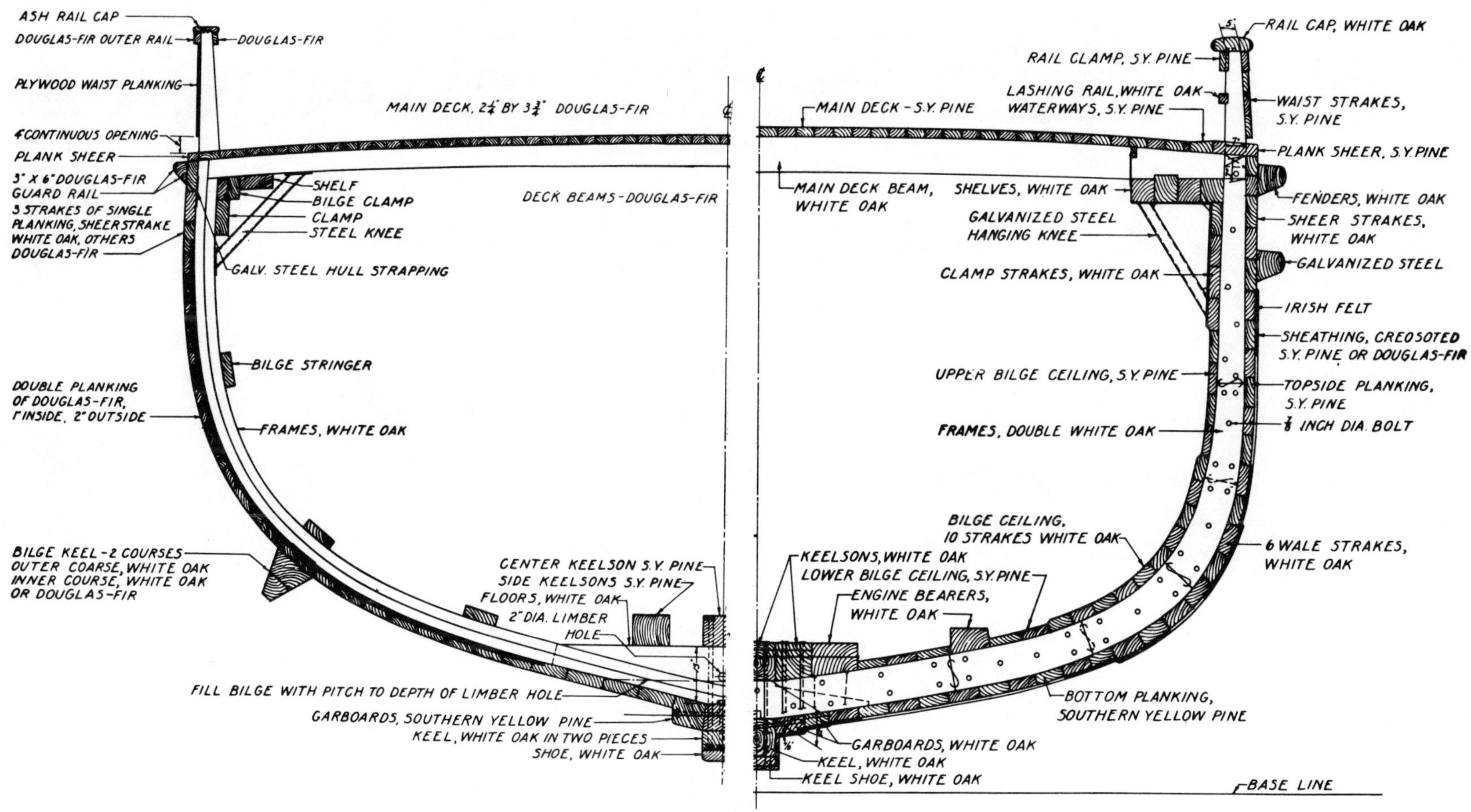

Figure 5-1.--Sections of (left) bent-frame YMS (World War II) minesweeper and (right) sawed-frame YT harbor tug showing framing, planking, ceiling, and decking.

(M 105 854)

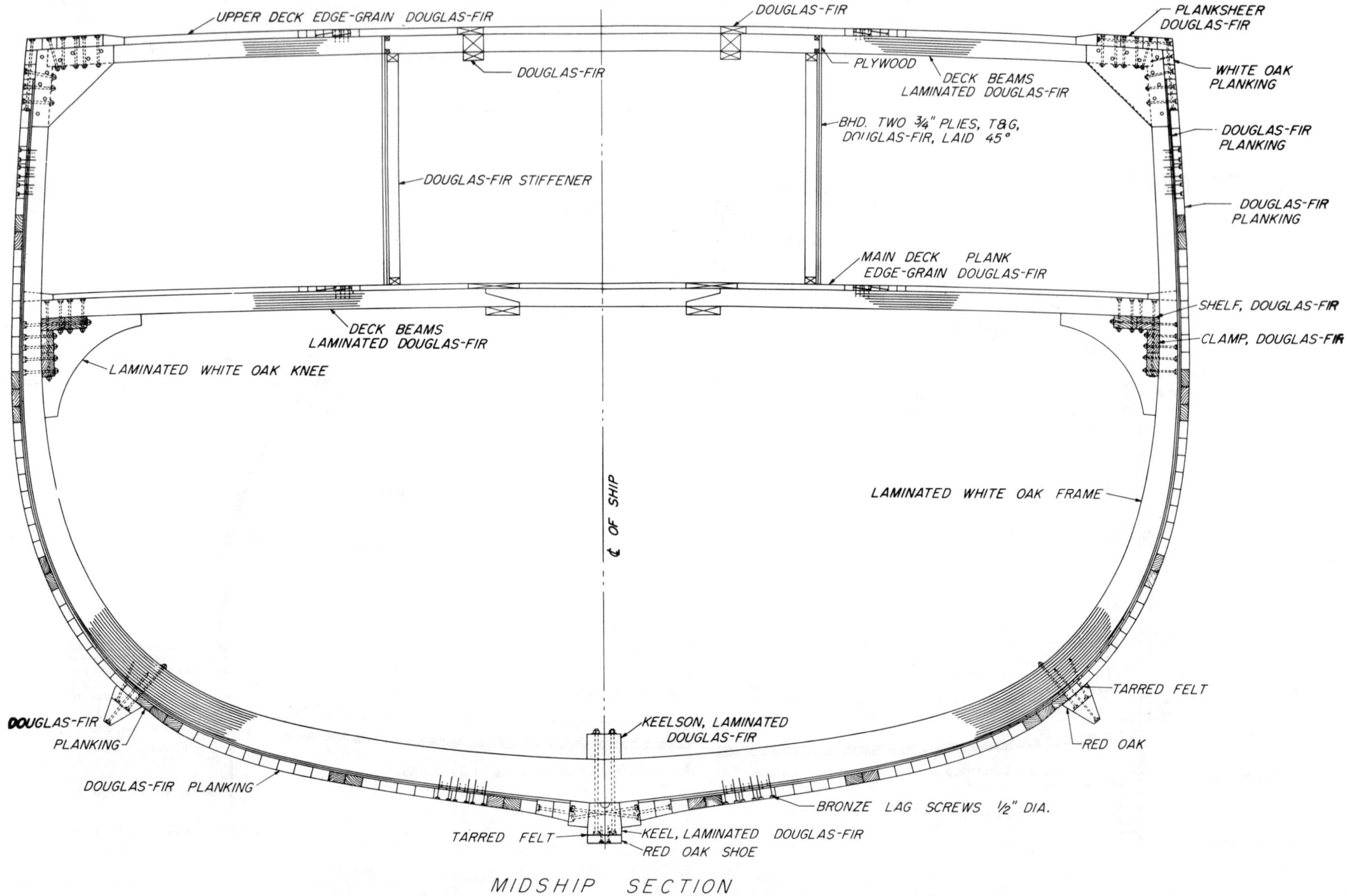

Figure 5-2.--Section of MSO-421-class minesweeper showing framing, planking, and decking.

(M 97617 F)

piece combination horn timber, after-keel, and deadwood were used, as well as laminated deck beams, stringers, girders, knees, bulkhead frames, filler blocks, and other parts. The end result was a vessel that weighs less than 50 percent as much as the same basic ship scantlinged in the traditional manner. This weight saving can be translated not only into an initial saving of material and cost, but also into an increased combat load.

The examples cited above are by no means a complete presentation of the improvements that have been or may be achieved in wood vessels. They are, however, indicative of the improvements that may be made as a result of thought, vision, and the courage to apply the results of research and deviate from time-honored methods of construction. Vision and ingenuity will undoubtedly result in still more improvements.

Small-boat design has likewise greatly benefited from technological advances, notably the increased utility of laminated wood and plywood and the development of waterproof adhesives. Four typical constructions are shown in figure 5-3. The constructions differ in two important respects--round-bottom as opposed to V-bottom framing, and planking of solid wood or of plywood. The result of these fundamental differences in construction is two distinct functional types of boats, one built for heavy-duty loadings and the other for speed and maneuverability.

In general, round-bottom boats are preferred for transport of heavy loads. Typical examples are the 27-foot work boat and the 50-foot utility boat (fig. 5-3). Almost universally, they are built with steam-bent frames, although, especially in the larger boats, laminated frames are practical. Lumber planking, single or double course, is generally used, with at least the outer course running fore and aft, while the inner planking may be diagonally laid for greater hull stiffness. Plywood is not used because of the pronounced double curvature necessitated by the design. Outer planking may be laid edge to edge, outgaged for calking, or it may be lapped clinker fashion (fig. 5-3, B).

V-bottom construction, generally with sawed frames the elements of which are joined with glued and screwed gussets at chine and sometimes also at the keel, facilitates speed and maneuverability because of the planing action made possible by the bottom design. Typical examples are the 24-foot plane personnel boat and the 40-foot rescue boat (fig. 5-3, C and D). Plywood is frequently used for side planking and sometimes also for bottom planking, although fore-and-aft lumber planking is often used on bottoms, especially for such boats as landing craft, to give greater abrasion resistance. For the longer boats of this type, however, double planking laid diagonally, the outer course at right angles to the inner, contributes greater longitudinal rigidity without unduly adding to weight. Lightweight stringers of box- or I-beam design, with plywood webs, are also frequently used for longitudinal stiffness, section 5.4 (5-1).

The sections that follow are intended to furnish the naval architect with basic design data that permit the more efficient design of wood vessels.

5.2 MOISTURE CONTENT OF WOOD IN RELATIONSHIP TO ITS DURABILITY.--It is important that designers and builders of wood hulls understand the role played by moisture content of wood in relation to decay. Decay fungi will not attack wood that is installed dry and kept in the dry condition. Therefore, in ship and boat application, it is important to use dry lumber, and to keep it dry as much as practicable. Keeping lumber dry in the interior of a wood hull can be accomplished through ventilation.

A ventilating system based on a 10-minute rate of change of air, using either a mechanical supply and natural exhaust or a mechanical exhaust and natural supply, is suitable for the purpose in hulls which are decked over. Where mechanical ventilation is impractical, the space should be left open to the extent that all spaces will at least have natural ventilation.

5.3 DESIGN WITH SOLID WOOD.

5.3.1 Design Stresses Based on Commercially Stress-Rated Lumber.--In previous chapters, basic stresses for a number of species are given and it is pointed out that stresses for design were derived by multiplying these basic stresses by a factor (strength ratio) that takes into account the effect of strength-reducing characteristics. Many lumber grades have these characteristics limited in such a way that strength ratios, and thus design stresses, can be assigned.

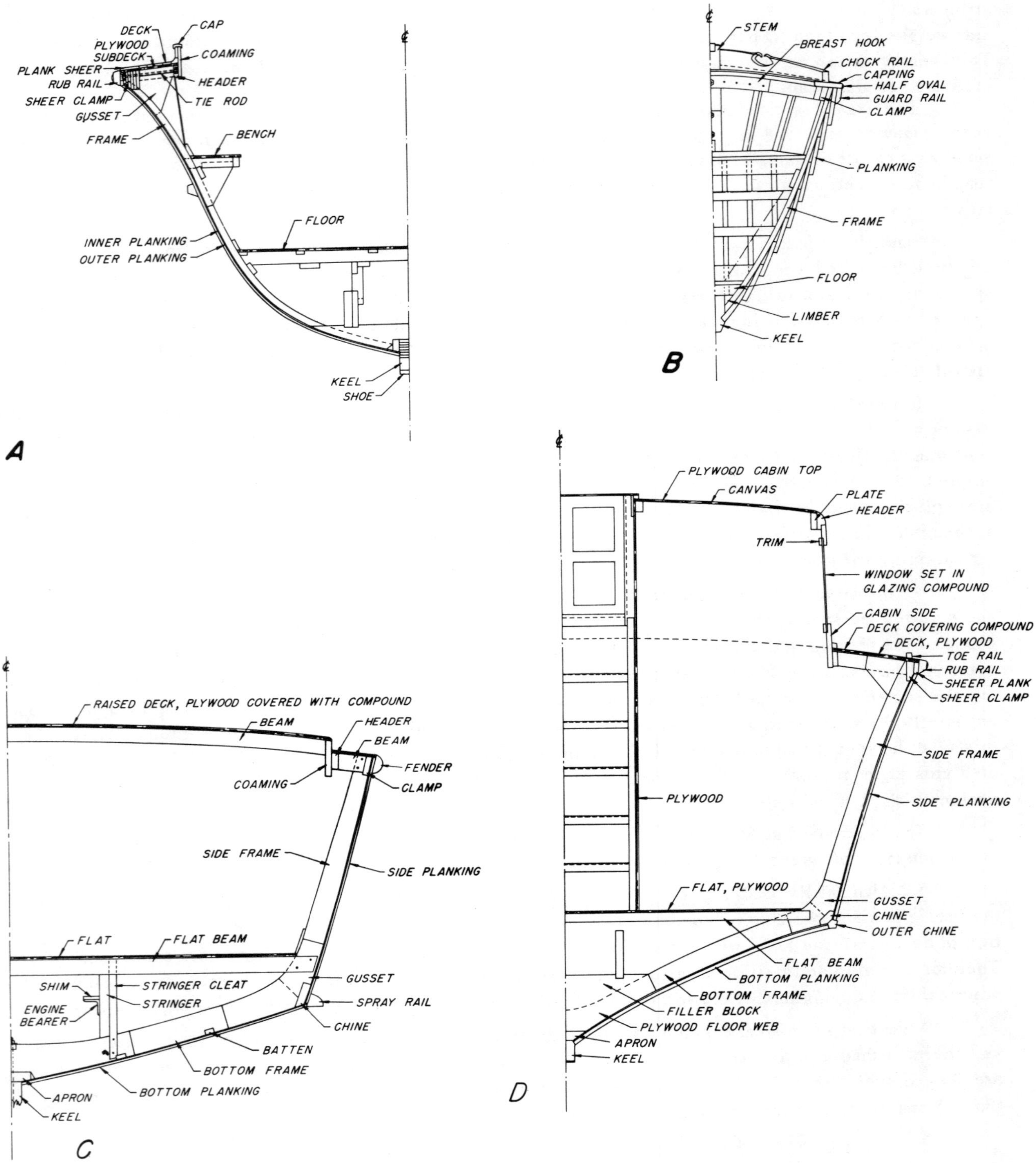

Figure 5-3.--Four types of framing for small boats.. A, 50-foot utility boat, bent-frame, double cedar lumber planking; frame 12, looking forward. B, 27-foot work boat, bent-frame, clinker-planked; view from Station No. 1 looking forward. C, 24-foot plane personnel boat, V-bottom, 1-1/2-inch double plywood planking, typical section. D, 40-foot rescue boat, Mark 2, V-bottom double cedar planking.

Table 5-1 presents design stresses for various grades of a number of species as recommended by the lumber industry for commercial design. These stresses are based on normal loading, which involves a duration of near-maximum load such that an increase of 10 percent is permitted over stresses based on permanent loading. For long-continued loads, therefore, the stresses shown in the table should be multiplied by a factor of 0.90. For other durations, they may be modified as indicated in section 5.3.2.1.

5.3.1.1 Grade Modifications to Make Commercially Stress-Rated Lumber Applicable to Ship and Boat Use.--The stress grades listed in table 5-1 are considered applicable for ship and boat application with the additional grade and special ship and boat construction provisions listed below.

5.3.1.1.1 General Considerations.

5.3.1.1.1.1 Shake and Split.--Because of the difficulties noted in section 3.2.3 connected with shake, those commercial grades that do permit shake, should be modified to eliminate its use for ship and boat application. The same provision should also apply to splits.

5.3.1.1.1.2 Seasoning Checks.--In places exposed to wetting and decay hazard, members are in contact with salt water, and where water-tightness is of prime importance, provisions should be made to reduce occurrence of checks to a minimum.

5.3.1.1.1.3 Decay.--Since there is no satisfactory way of numerically appraising the effect of decay on strength of wood, decay is excluded from most commercial structural grades of lumber. Exceptions to this are usually decay confined to knots and limited decay of the pocket type in the lowest dimension grades of some species of wood. Because of conditions existing on ships and boats which are conducive to spread of decay, however, added specification restrictions are required to eliminate any detectable decay.

5.3.1.1.2 Grading of Simple Span Beams.--The commercial stress ratings presented in table 5-1 for many, but not all [2] species of wood are based on a grading system which assumes that the stress-graded

[2] Douglas-fir posts and timbers, of the grades Dense Select Structural, Select Structural, Dense Construction, and Construction are graded the same for the full length of the piece.

beam will not be cut shorter or reduced in cross section and will be simply supported at the ends and loaded at the center. If a beam is altered in cross section or length, the assigned stress grade for that beam is no longer valid for the smaller resulting sections. Specifications for beams for simple span applications usually should require grading in the final size.

5.3.1.1.3 Grading of Continuous Beams.--In ship and boat construction, beams are often used to span three or more supports. For such uses, the following grade modifications are necessary.

5.3.1.1.3.1 Seasoning Checks.-- In a continuous beam, strength reduction calculations for seasoning checks need to be applied to the middle half of the depth of the piece and for a distance equal to three times the depth of the piece (depth equals width of wide face) anywhere throughout the beam length.

5.3.1.1.3.2 Cross Grain and Knots.--If a beam is to be used as a continuous beam on three supports, and the full length of the graded beam is to be used without modification in cross section or length, the respective commercial grade should be so designated as to require that the center two-thirds of the beam be graded as severely as the center third. If a continuous beam bears on four or more supports, the respective commercial grade requirement needs to be designated in such a manner that the full length of the beam be graded as severely as the center third. In the latter case, cutting of the beam for length on installation will not result in a decreased stress rating for any resulting section.

5.3.2 MODIFICATION OF DESIGN STRESSES FOR SOLID WOOD.

5.3.2.1 Modification for Duration of Load.--Figure 5-4 shows ratios of working stresses for loads of various durations as compared with those appropriate to long-time loading. For shorter durations, working stresses derived directly from the basic stress table may be increased by the ratios shown on the figure. Since the stresses shown in table 5-1 include a 10-percent increase for normal (10 year) duration of load, they must be multiplied by the factor from the chart of figure 5-4 divided by 1.10.

Table 5-1.--Allowable unit stresses for stress-grade lumber under normal loading conditions

(For other loading conditions, the stresses in this table should be modified as described in section 5.3.2.)

Species and commercial grade[1]	Grading rules	Allowable unit stresses				Modulus of elasticity
		Extreme fiber in bending and tension parallel to grain	Horizontal shear	Compression perpendicular to grain	Compression parallel to grain	
1	2	3	4	5	6	7
		P.s.i.	P.s.i.	P.s.i.	P.s.i.	P.s.i.
Bald cypress, Coast Type (Tidewater Red)						
1700 f Grade, J. & P.-B. & S.	Southern Cypress	1,700	145	360	1,425	1,320,000
1300 f Grade, J. & P.-B. & S.	Manufacturers'	1,300	120	360	1,124	1,320,000
1450 c Grade, P. & T.	Association	....	....	360	1,450	1,320,000
1200 c Grade, P. & T.	1958 Structural	....	....	360	1,200	1,320,000
	Stress Grade					
	Standard					
Douglas-fir, Coast Type						
Dense Select Structural, J. & P.[2]		2,050	120	455	1,650	1,760,000
Select Structural, J. & P.[2]		1,900	120	415	1,500	1,760,000
Dense Construction, J. & P.[2]		1,750	120	455	1,400	1,760,000
Construction, J. & P.[2]		1,500	120	390	1,200	1,760,000
Dense Select Structural, B. & S.[2]	West Coast Lumber	2,050	120	455	1,500	1,760,000
Select Structural, B. & S.[2]	Inspection Bureau,	1,900	120	415	1,400	1,760,000
Dense Construction, B. & S.[2]	1960 Standard	1,750	120	455	1,200	1,760,000
Construction, B. & S.[2]	Grading Rules	1,500	120	390	1,000	1,760,000
Dense Select Structural, P. & T.[2]	No. 15	1,900	120	455	1,650	1,760,000
Select Structural, P. & T.		1,750	120	415	1,500	1,760,000
Dense Construction, P. & T.		1,500	120	455	1,400	1,760,000
Construction, P. & T.		1,200	120	390	1,200	1,760,000

(Sheet 1 of 4)

Table 5-1.--Allowable unit stresses for stress-grade lumber under normal loading conditions --Continued

(For other loading conditions, the stresses in this table should be modified as described in section 5.3.2.)

Species and commercial grade[1]	Grading rules	Allowable unit stresses				Modulus of elasticity
		Extreme fiber in bending and tension parallel to grain	Horizontal shear	Compression perpendicular to grain	Compression parallel to grain	
1	2	3	4	5	6	7
		P.s.i.	P.s.i.	P.s.i.	P.s.i.	P.s.i.
Douglas-fir, Inland Type						
Select Structural, J. & P.[2]		2,050	120	455	1,650	1,760,000
Structural, J. & P.[2]		1,900	120	415	1,500	1,760,000
Dense Construction, J. & P.[2]	Western Pine	1,750	120	455	1,400	1,760,000
Construction, P. & T.	Association	1,500	120	390	1,200	1,760,000
Dense Select Structural, P. & T.	1961 Grading	1,900	120	455	1,650	1,760,000
Select Structural, P. & T.	Rules	1,750	120	415	1,500	1,760,000
Dense Construction, P. & T.		1,500	120	455	1,400	1,760,000
Construction, P. & T.		1,200	120	390	1,200	1,760,000
Larch						
Select Structural, J. & P.[2]		2,150	145	455	1,750	1,650,000
Structural, J. & P.[2]		1,900	120	415	1,450	1,650,000
Common Structural, J. & P.[2]		1,450	120	390	1,325	1,650,000
Select Structural, P. & T.[2]		....	....	455	1,750	1,650,000
Structural, P. & T.		....	....	415	1,450	1,650,000
Common Structural, P. & T.	Western Pine	....	....	390	1,325	1,650,000
Dense Select Structural, J. & P.[2]	Association	2,050	120	455	1,650	1,760,000
Select Structural, J. & P.[2]		1,900	120	415	1,500	1,760,000
Dense Construction, J. & P.[2]		1,750	120	455	1,400	1,760,000
Construction, J. & P.[2]		1,500	120	390	1,200	1,760,000
Dense Select Structural, P. & T.		1,900	120	455	1,650	1,760,000
Select Structural, P. & T.		1,750	120	415	1,500	1,760,000
Dense Construction, P. & T.		1,500	120	455	1,400	1,760,000
Construction, P. & T.		1,200	120	390	1,200	1,760,000

(Sheet 2 of 4)

Table 5-1.--Allowable unit stresses for stress-grade lumber under normal loading conditions --Continued

(For other loading conditions, the stresses in this table should be modified as described in section 5.3.2.)

Species and commercial grade[1]	Grading rules	Allowable unit stresses				Modulus of elasticity
		Extreme fiber in bending and tension parallel to grain	Horizontal shear	Compression perpendicular to grain	Compression parallel to grain	
1	2	3	4	5	6	7
		P.s.i.	P.s.i.	P.s.i.	P.s.i.	P.s.i.
Oak, red or white						
Grade I, Class a[3]	Proposed Specification No. MIL-L-2037C, "Lumber, Hardwood, Oak, White, for Ship and Boat Construction," 1961	1,950	165	600	1,400	1,650,000
Grade II, Class a and b[3]		1,400	150	600	1,100	1,650,000
Grade I, Class c[3]		1,950	165	600	1,350	1,650,000
Grade II, Class c[3]		1,350	150	600	1,150	1,650,000
Pine, Southern[4,5,6,7]						
(a) Structural Lumber, wet exposure						
Dense Structural 86(Par. 278) and LL Structural 86(Par. 279), all thicknesses	Southern Pine Inspection Bureau; 1961 Grading Rules	2,400	150	300	1,600	1,600,000
Dense Structural 72(Par. 280) and LL Structural 72(Par. 281), all thicknesses		2,000	135	300	1,400	1,600,000
Dense Structural 65(Par. 282) and LL Structural 65(Par. 283), all thicknesses		1,800	120	300	1,250	1,600,000
Dense Structural (Par. 284) and LL Structural 58(Par. 285), all thicknesses		1,600	105	300	1,150	1,600,000
(b) Structural Lumber, dry exposure, 2 inches thick						
Dense Structural 86 KD and LL Structural 86 KD		3,000	165	455	2,250	1,760,000
Dense Structural 72 KD and LL Structural 72 KD		2,500	150	455	1,950	1,760,000
Dense Structural 65 KD and LL Structural 65 KD		2,250	135	455	1,800	1,760,000
Dense Structural 58 KD and LL Structural 58 KD		2,050	120	455	1,650	1,760,000

(Sheet 3 of 4)

Table 5-1.--Allowable unit stresses for stress-grade lumber under normal loading conditions --Continued

(For other loading conditions, the stresses in this table should be modified as described in section 5.3.2.)

Species and commercial grade[1]	Grading rules	Allowable unit stresses				Modulus of elasticity
		Extreme fiber in bending and tension parallel to grain	Horizontal shear	Compression perpendicular to grain	Compression parallel to grain	
1	2	3	4	5		7
		P.s.i.	P.s.i.	P.s.i.	P.s.i.	P.s.i.
Redwood						
Dense Structural, J. & P.[2]-B. & S.[2]		1,700	110	320	1,450	1,320,000
Heart Structural, J. & P.[2]-B. & S.[2]	California Redwood	1,300	95	320	1,100	1,320,000
Dense Structural, P. & T.	Association	....	....	320	1,450	1,320,000
Heart Structural, P. & T.		....	....	320	1,100	1,320,000

[1]Abbreviations: J. & P., Joists and Planks; B. & S., Beams and Stringers; P. & T., Posts and Timbers; KD, kiln-dried.

[2]The allowable unit stresses for tension parallel to grain "t" and for compression parallel to grain "c" given for these Joist and Plank and Beam and Stringer grades are applicable when the following additional provisions are applied to the grades: The sum of the sizes of all knots in any 6 inches of the length of the piece shall not exceed twice the maximum permissible size of knot. Two knots of maximum permissible size shall not be within the same 6 inches of length of any face.

[3]Class a is planking; b is shaft logs; and c is sawn timbers.

[4]Grade restrictions established under the 1961 Grading Rules apply to the entire length of each piece, and each piece therefore is suitable for use in continuous spans, over double spans, or under concentrated loads without the necessity of regarding for special shear or other special stress requirements. Also the stress-rated grades under these rules are established on a basis that permits cutting graded members to shorter lengths without impairment of stress ratings in the shorter pieces. In addition, the stresses apply to members used either flat or on edge.

[5]A number of other stress-rated grades also appear in the 1961 Grading Rules that could be used for ship and boat application with such modifications as complete restrictions on decay, wane, and more rigid restrictions on warp, pitch, and stain.

[6]Southern pine generally has a large amount of sapwood that is not decay resistant. Therefore, for ship and boat application, sapwood should be greatly restricted or requirements should be added for a thoroughly effective preservative treatment. Southern pine sapwood readily takes an effective preservative treatment.

[7]Longleaf (LL) may be specified by substituting "Longleaf" for "Dense" in the grade name, and when so specified, the same allowable stresses shall apply.

(Sheet 4 of 4)

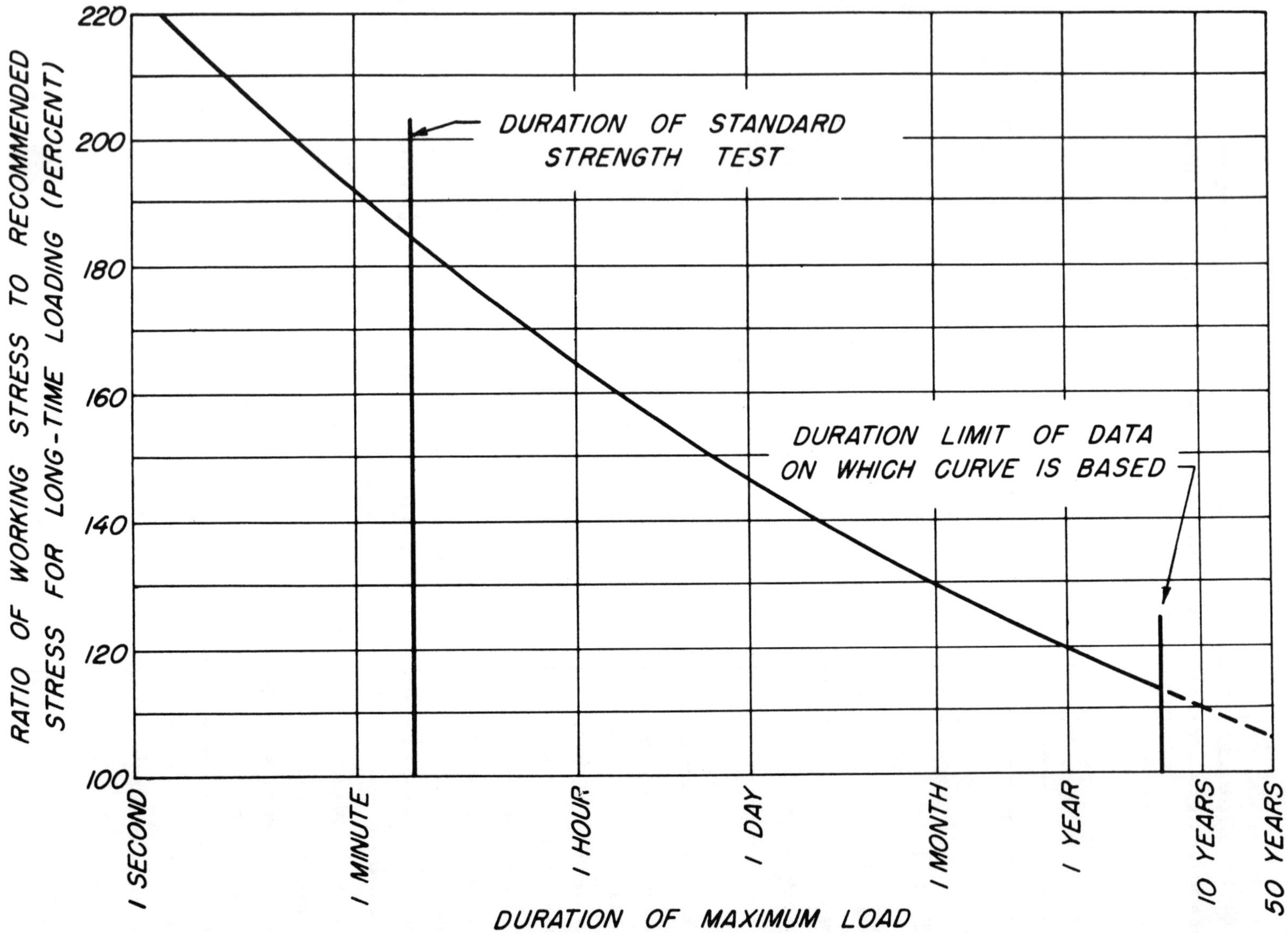

Figure 5-4.--Relation of working stress to duration of load.

(M 89380 F)

In applying a stress increase for conditions of less than full duration of load, the safe stress for the permanent portion of a combined loading must not be exceeded. If the assumed loading conditions involve an infrequent large load and a more frequent smaller load, the working stresses and sizes of structural members should be safe for each of the assumed loads. It is possible for sizes of members to be governed by the permanent load or by some semipermanent load below the maximum load.

Working stress values, except modulus of elasticity, may be used without allowance for occasional impacts up to 100 percent of the static load, provided that the sizes of structural members are safe for the static loadings. Frequently repeated impacts, however, have fatigue effects.

Wood under continuous load takes on a continuing increment of deformation, usually very slowly, but persistently over long periods of time. The set or sag of beams is an illustration. The deformation is greatest when green timbers season under load. It may increase with time to as much as twice the initial deformation without endangering the safety of the timber. This effect is most important if the long-continued load or portion of the load is at or near the maximum design level. If deformation or deflection under long periods of loading must be limited, the necessary extra stiffness can be provided by setting the initial deformation limit at half the long-time deformation limit. Values for modulus of elasticity in table 5-1 will give the initial deflection of a beam, which will increase under long-continued load. The increase may be somewhat greater under varying temperature and moisture conditions than under uniform conditions.

Modulus of elasticity, when used to calculate immediate deflection or deformation caused by imposed loads, is not modified for duration of load effect. When it is used, however, in calculating the strength of a long column, a reduction factor of 3 is applied to the modulus of elasticity to give a working value that is safe for long-time loading. Though the design of long columns is predicted on an initial deflection, most long columns are in a state of unstable equilibrium without deflection. A member in unstable equilibrium is sensitive to long-time loading; that is, it may remain in the unstable equilibrium for a period of time and then suddenly deflect. Since the reduction factor of 3 on modulus of elasticity has been found satisfactory for long-time loading conditions, the working stress for a long column based on that factor may receive the same increases for shorter durations of load that are given to other working stresses. In the case of an intermediate-length column, the increase for short-time loading is most conveniently made after the long-time loading stress for the column is calculated from the appropriate long-time working stresses for compression parallel to grain and modulus of elasticity.

5.3.2.2 Modification for Seasoning Effects.--The strength of clear wood increases as its moisture content decreases below the fiber saturation point, but in structural lumber this increase in strength is largely offset by defects that develop in seasoning. Design values in table 5-1 are suitable for sawn lumber that will be continuously dry in use, as in most covered structures. They may also be used for structures in which the moisture content of the wood is at or above the fiber saturation point; as, for example, when continuously submerged. Under these latter conditions, however, allowable stresses in compression parallel to grain are reduced one-tenth, those in compression perpendicular to grain one-third, and the values of modulus of elasticity one-eleventh.

Because of the offsetting effect of drying defects on timbers of large size, no increase for drying is made for stress in extreme fiber in bending, or in tensile stress for such timbers. For members of small size (2 to 4 inches thick), such as joist and plank grades, however, an increase in the allowable bending and tensile stresses for drying is given. The increase is greater for the higher grades. Since many joists remain continuously dry in use, it is common practice to take advantage of the increase in strength from drying by increasing permissible sizes of knots or other characteristics rather than by increasing working stresses; under this circumstance, the working stress may need to be reduced, if lumber with characteristics of these sizes is to be used under wet conditions. This is the case for many of the joist-and-plank grades of table 5-1, the necessary reductions ranging from 0 for the lower grades to about 12 or 15 percent for the highest grades.

Basic stresses used to obtain working stresses for all grades of 1-inch boards utilized under dry conditions may be increased by one-fourth in bending or tension parallel to grain, one-eighth in horizontal shear, one-half in compression perpendicular to grain, three-eighths in compression parallel to grain, or one-tenth in modulus of elasticity from the values listed in table 2-1. Such increases are included in the values given in table 2-2.

Although drying results in an increase of strength in many structural timbers, the sizes assumed in design may be substantially reduced by the shrinkage that usually accompanies drying. Increases of working stresses for drying are based on strength properties related to the actual dimensions of the dry timbers.

5.3.2.3 Modification for Decay.--Since there is no satisfactory way of numerically appraising the effect of decay on strength of wood, and because of possible conditions existing which are conducive to its spread, wood with any detectable decay is not permitted for use in ship and boat construction. If untreated wood in ships and boats becomes infected, however, some loss of strength may occur before the decay is detected. Stiffness is affected less than breaking strength in a beam, and shock resistance is affected most. A reduction of design stress may therefore be in order, so that the structure can safely carry its design loads for the time until decay is detected and the affected material can be removed. This reduction may range from a few percent up to 25 percent.

5.3.2.4 Modification for Preservative Treatment.--It may be necessary, in establishing allowable working stresses for preservative-treated timber, to take into account possible reductions in strength that may result from the treating process. Tests of treated timber show reductions in stress in extreme fiber in bending and in compression perpendicular to grain that range from a few percent up to 25 percent, depending on the treating conditions. Compression parallel to grain is affected less, and modulus of elasticity very little. The effect on resistance to horizontal shear can be estimated by inspection for shakes and checks after treatment.

5.3.2.5 Modification for Temperature.--Stress values in table 5-1 are applicable to lumber used under ordinary ranges of temperature. Special allowance should be made for lumber subjected to abnormally high temperatures, particularly for long periods of time.

5.3.2.6 Modification for Other Conditions.--In developing the allowable stresses for design given herein, a number of factors are applied to laboratory test data to convert them to a level suitable for use in design. Among these is a factor of safety that is intended primarily to provide a margin for accidental overload. This factor is generally small and is applicable to the usual conditions of design. This factor can, in effect, be increased by using a design stress lower than that recommended in the tables, or decreased by using a higher design stress. Whether or not to adjust the design stresses one way or the other is a matter of judgment for the designer. For example, in ship design, the design loads may be inadequately known, or failure of an individual member may threaten the integrity of a major structure or endanger human life. In such instances, the designer might desire to lower the design stresses to provide a greater margin of safety for unknowns in the service loads or to provide extra safety in a highly critical member. Conversely, in situations where the loads are accurately known or where failure of a member can result only in small damage and repair can easily be made, the use of stresses higher than those of table 5-1 may be justified.

Allowable unit stresses for shear in joint details may be 50 percent greater than the horizontal shear values from table 5-1.

The values in table 5-1 for compression perpendicular to grain apply to bearing surfaces 6 inches or more in length located anywhere in the length of a timber and to bearing surfaces of any length located at the ends of beams or other members. For bearing surfaces shorter than 6 inches located 3 inches or more from the end of a member, the stresses may be increased in accordance with the following factors:

Length of Bearing Inches	Factor of Increase
1/2	1.75
1	1.38
1-1/2	1.25
2	1.19
3	1.13
4	1.10
6 or more	1.00

For stress under a washer or other round bearing area, the same factor may be taken as for a bearing surface whose length equals the diameter of the washer.

5.3.3 Beam Formulas.

5.3.3.1 Solid Beams of Rectangular Cross Section.--Structural beams are usually of rectangular section and of uniform depth throughout their span. They should be so designed that the following stresses do not exceed the allowable unit values: (1) The extreme fiber stress caused by flexural loads; (2) the maximum horizontal shear stress; and (3) the stress in compression across the grain at the end bearing surfaces. Beams should also be so designed that their deflection under load will not exceed the limit fixed by the intended use of the structure. If actual and nominal sizes of lumber differ, the allowable unit values must be used with the actual sizes.

5.3.3.1.1 Extreme Fiber Stress.--Either the ultimate bending strength or the safe load for a wood beam of rectangular cross section may be calculated from the formula

$$M = \frac{Ffbd^2}{6}$$

where M is the bending moment in inch-pounds, F is a form or depth factor, f is the unit flexural stress in the extreme fiber in pounds per square inch, and b is the width and d is the depth of the beam in inches. If the ultimate bending strength is desired, f is the modulus of rupture. If the safe load is desired, f is the safe flexural stress for the kind and grade of material used. In either case, M must be expressed in terms of the load and the span in order to calculate the stress. For example, if a simply supported beam has a total load W uniformly distributed over a span L,

$$M = \frac{WL}{8}$$

If the beam has a concentrated load P at the center of the span,

$$M = \frac{PL}{4}$$

Most ship timbers will be of rectangular cross section, for which the form factor is unity. Beams of I or box section may be designed, without reference to form factor, by the methods outlined in section 5.6.

5.3.3.1.2 Depth Factor.--A depth factor, F, is introduced in the basic design formula for rectangular beams to take into account the somewhat lower unit strength developed in deep beams as compared to shallow beams. In arriving at the basic stress values for bending strength in tables 2-1 and 2-2, a depth factor was assumed that corresponds approximately to a 12-inch depth of beam. In most designs of structural lumber beams of rectangular shape, if working stresses derived from the basic stress values of tables 2-1 and 2-2 are used, no further adjustment for depth is necessary; that is, the basic formula can be used in the form

$$M = \frac{fbd^2}{6}$$

If the depth of a rectangular beam is more than 16 inches, an additional depth factor should be calculated by the formula

$$F = 0.81 \frac{H^2 + 143}{H^2 + 88}$$

where F is the depth factor and H is the height of the beam in inches.

This formula is for use with the basic stress values of table 2-1, and therefore gives a factor of unity when the depth is 12 inches.

5.3.3.1.3 Horizontal Shear in Beams.--The general formula for horizontal shear in beams is

$$q = \frac{VQ}{It}$$

in which all units are in inches or pounds, q is the maximum horizontal shearing stress, Q is the statical moment of the area either above or below the neutral axis and about that axis (as used here, statical moment is a product of an area and the distance from the center of gravity of that area to an axis), V is the external or vertical shear, I is the moment of inertia of the section about the neutral axis, and t is the width of the beam at the neutral axis.

For a rectangular beam b inches wide and d inches deep, the general formula reduces to

$$q = \frac{3V}{2bd}$$

Shearing stress is increased at certain locations in a beam because of stress concentration effects. The most common concentration of stress is at the base of shakes, checks, or splits, where there is an abrupt change of cross section. Since checking occurs to some extent in practically all structural beams, a suitable factor for stress concentration was applied in deriving the basic shear stress values of tables 2-1 and 2-2.

Another shear stress concentration may occur immediately adjacent to points of application of a concentrated load or reaction. Such a concentration is not common, because allowable bearing values in compression perpendicular to grain ordinarily require fairly large bearing areas. Special cases involving applications of load or reaction to areas measuring one-half inch or less in length should be examined for stress concentration effects.

Because the upper and lower portions of a beam checked near the neutral axis act partly as two beams and partly as a unit, a portion of the end reaction is resisted internally by independent action of each half of the beam and consequently is not associated with shearing stress at the neutral plane. In using the basic formula for horizontal shear, in which b is the full width of the beam and q the allowable unit shear stress based on the assumption that checks are present, allowance for two-beam action should be made in calculating the allowable vertical shear V.

For calculating the horizontal shear on the neutral plane in checked beams, the basic shear formula is used and, for the allowable shearing stress, a working stress value is taken from table 5-1,

In calculating the reactions represented by the vertical shear value V in the formula: (1) Take into account any relief to the beam under consideration resulting from the fact that the load is distributed to adjacent parallel beams by decking or other members of the construction; (2) neglect all loads within a distance from both supports equal to the height of the beam; (3) if there are any moving loads, place the largest one at a distance from the support equal to three times the height of the beam; and (4) treat all other loads in the usual manner. If a timber does not qualify for shear resistance under the foregoing procedure, which under certain conditions may be over-conservative, the reactions for the concentrated loads should be more accurately determined by the following equation:

$$r' = \frac{10P'(L - x')\left(\frac{x'}{h}\right)^2}{9L\left[2 + \left(\frac{x'}{h}\right)^2\right]}$$

in which r' is the reaction to be used as due to a load P'; L, the span in inches; x', the distance in inches from the reaction to the load P'; and h, the height of the beam in inches.

5.3.3.1.4 Compression Across the Grain.--The area of end bearing surface should be such that the unit stress in compression perpendicular to grain does not exceed the allowable value given in table 5-1. The stresses in compression perpendicular to grain given in that table apply to bearings of any length at the ends of a beam and to bearings 6 inches or more in length at any other place along the beam. They may be increased when applied to small plates and washers as explained in section 5. 3.2.6. In calculating the bearing area at the ends of beams, no allowance need be made for the fact that, as the beam bends, the pressure upon the inside edge of the bearing is greater than that at the opposite edge. The wood yields enough so that the pressures become equalized before serious damage occurs.

5.3.3.1.5 Deflection of Beams.--When service conditions require that beams be designed for stiffness, the dimensions are determined by applying the usual deflection formulas, the deflection being kept within a prescribed fraction of the span.

Wood beams usually sag in time; that is, the deflection increases beyond what it was immediately after the load was first applied. Green timbers, especially, will sag if allowed to season under load, although partially seasoned material will also sag to some extent. In thoroughly seasoned beams, there are small changes in deflection with changes in moisture content but little permanent increase in deflection.

Calculations from the values for modulus of elasticity in table 5-1 give the initial deflection of a beam. If deflection under long-time load is to be limited, it is customary to design for an initial deflection of about one-half the value permitted for long-time deflection. This can be done by doubling any long-time loads when calculating deflection, by using one-half the values given in table 5-1 for modulus of elasticity, or by any equivalent method.

5.3.3.1.6 Lateral Support of Beams.--Deep, narrow beams are laterally unstable and require restraint against sidewise buckling. The following approximate rules are recommended:

(1) If the ratio of depth to breadth does not exceed 2, no lateral support is needed.

(2) If the ratio is 3, the ends should be held in position.

(3) If the ratio is 4, the piece should be held in line by spiking or bolting as in a vertically laminated beam.

(4) If the ratio is 5, one edge should be held in line, as by decking.

(5) If the ratio is 6, diagonal bridging should be used between beams.

(6) If the ratio is 7, both edges should be held in line.

(7) If a beam is subject to both flexure and compression parallel to grain, the ratio of depth to breadth may be as much as 5 if one edge is held firmly in line, for example, by deck beams. If the dead load is sufficient to induce tension on the under side of rafters, the ratio may be as high as 6.

5.3.3.1.7 Notched Beams.--Beams are often notched at the ends to improve clearance or to bring the top surfaces level with adjacent beams or girders (fig. 5-5). Occasionally they are also notched at intermediate points, either bottom or top, in order to clear other parts of a structure or to receive other members.

When notched on the lower side at the ends (fig. 5-5, A), the strength of a short, relatively deep beam is decreased by an amount depending on the shape of the notch and on the relation of the depth of the notch to the depth of the beam. It is recommended that in designing beams with square-cornered notches at the ends of the beams, the desired bending load be checked against the load obtained by the equation

$$V = \frac{2}{3} \left(\frac{b d_e^2 q}{d} \right)$$

where V is the vertical shear; b, the width of the beam; d_e, the actual end depth above the notch; d, the total depth of the beam; and q, the working stress in horizontal shear.

In setting up this equation, the actual end depth is used as the effective depth to resist shear, and the safe shear stress is reduced by multiplying it by the ratio of the effective depth to the total depth. Experiments made at the Forest Products Laboratory on short deep beams with relatively high shearing stresses substantiate this recommendation. The experiments also showed that, by cutting away the lower corner at the end notches, in order to obtain a gradual change in cross section, the breaking loads were raised very markedly (fig. 5-5, B).

5.3.3.2 Built-Up Beams With Mechanical Fastenings.--In commercial structures beams are sometimes built up from material of relatively small dimensions through the use of mechanical fastenings to utilize small pieces of lumber, reduce seasoning defects, or eliminate knots or other strength-reducing defects. Glued construction is covered in section 5.4.

Provided the individual layers of lumber are full length and width and are well fastened, mechanically fastened built-up members with layers loaded on edge have essentially the same strength as solid beams of the same grade and size. Butt-joints in one or more of the layers, however, reduce strength and stiffness of the piece considerably. It is often difficult to satisfactorily apply fastenings and they may loosen over a period of time. Also, in ship and boat construction, corrosion of fastenings may make the structure ineffective.

Sheathed beams, consisting of two or more horizontal laminations with boards nailed diagonally to the sides, are considerably more limber than a solid beam of the same size. Their strength may be expected to be on the order of 70 to 80 percent of that of a solid beam of the same dimensions.

Horizontally laminated beams, consisting of laminations fastened together with spikes or bolts, have little shear resistance between laminations and cannot be expected even to approach the strength or stiffness of a solid beam of the same size. The use of wood or metal keys or metal connectors to impart shear resistance produces beams with efficiencies on the order of 75 to 85 percent, depending upon the type and number of connections. Stiffness, however, will be considerably less than for a 1-piece beam.

It is very difficult to build up beams through the use of mechanical fastenings that will hold effectively over a period of time, particularly for ship and boat construction, where extroardinary wetting

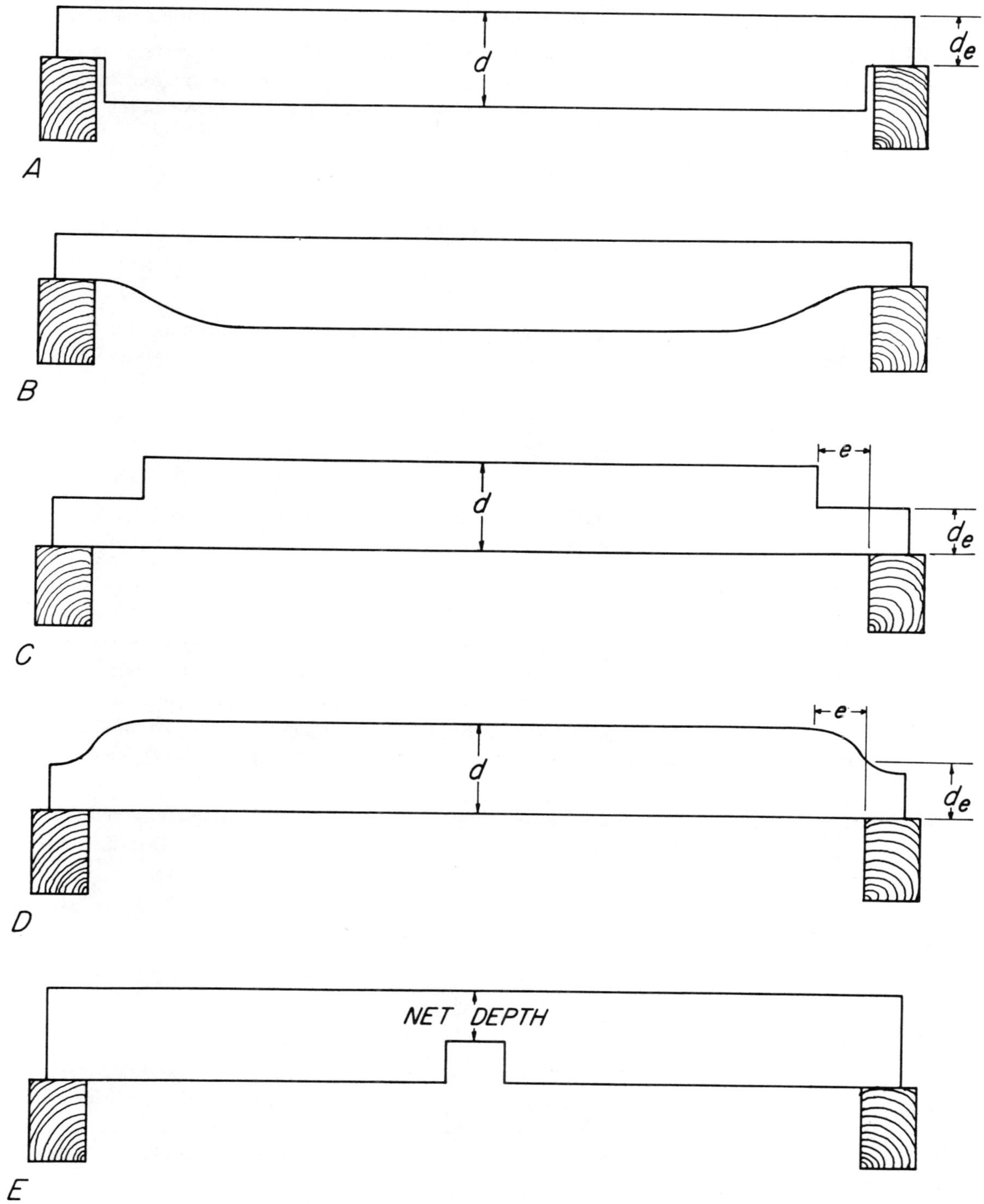

Figure 5-5.--Notched beams: A, Notched at ends to one-half its depth; B, beam with gradual change in cross section at ends to eliminate concentration of stress.

(M 91876 F)

and drying stress, and corrosion to fastenings can occur. Therefore, in present day ship and boat construction, mechanically built-up members are generally not permitted, particularly when glued laminates are available that have fully waterproof bonds and will provide strength equal or superior to that of solid members of the same size, shape, and grade, regardless of the severity of service conditions.

5.3.3.3 Solid Beams of Circular Section.--Such wood members as poles and masts usually have circular cross sections, and sometimes ordinary beams and columns are of circular section. In calculating the bending strength of such members, it is necessary to include a form factor in the beam formula.

The design formula for a beam of circular section is

$$M = \frac{Ff\pi r^3}{4}$$

where M is the bending moment, F is the form factor, f is the unit flexural stress in the outer fiber, and r is the radius of the cross section. The form factor F for a circular wood beam has a value of 1.18, and the formula then becomes

$$M = \frac{1.18\ f\pi r^3}{4} = 0.927\ fr^3$$

The point of maximum stress in a beam of circular cross section depends not only on the location of the load but also on the taper of the beam.

5.3.4 Column Formulas.

5.3.4.1 Rectangular Solid Wood Columns.

5.3.4.1.1 Length Classes.--Solid wood columns fall into three length classes characterized by the type of failure. When the length is less than 11 times the least dimension, failure is by crushing. At lengths between 11 and K times the least dimension, failure is a combination of crushing and lateral buckling. The significance of K is explained in the following paragraphs. Beyond this range, wood columns generally fail by lateral deflection or buckling, behaving essentially as Euler columns.

Since most wood columns are rectangular or square in section, the ratio of unsupported length to least dimension, or the slenderness ratio, is commonly expressed as L/d, in which L is the unsupported length in inches and d the smallest side in inches.

For short columns or posts, that is, the first class of columns with L/d ratios of 11 or less, the same unit stress is recommended for all lengths. For these columns

$$\frac{P}{A} = c$$

in which P/A is the working stress of the column and c is the safe unit compressive stress parallel to the grain for the grade of material and the loading condition under consideration. Basic values of c are given in table 2-1 and for various grades in table 5-1.

For columns of the intermediate class, that is, with L/d ratios between 11 and K, the following formula is recommended:

$$\frac{P}{A} = c\left[1 - \frac{1}{3}\left(\frac{L}{Kd}\right)^4\right]$$

in which P/A and c have the same meaning as before and K is a constant depending upon c and the modulus of elasticity. The value K is the minimum value of L/d at which the column will behave as a Euler column. This value is obtained when

$$\frac{P}{A} = \frac{2c}{3}$$

In the Euler formula for long columns, P/A has the value 2c/3 when

$$\frac{L}{d} = \frac{\pi}{2} \sqrt{\frac{3E}{6c}}$$

where E is the modulus of elasticity. If this value of L/d is designated as K and a reduction factor of 3 is introduced for E, then

$$K = \frac{\pi}{2} \sqrt{\frac{E}{6c}} = 0.64 \sqrt{\frac{E}{c}}$$

Values of E are given in tables 2-1 and 5-1.

For columns within the Euler column class, that is, with L/d ratios equal to K or greater, the following formula based on the Euler formula for maximum load on long columns with a reduction factor of 3 is recommended:

$$\frac{P}{A} = \frac{0.274E}{\left(\frac{L}{d}\right)^2}$$

The values to be substituted for c in the formulas for short and for intermediate columns have equivalent safety factors.

The reduction factor of 3 is applied to the Euler maximum load and in the formula for determining K to convert from laboratory test conditions to the conditions of use (including long-time loading) and to provide a factor of safety. About one-half of the reduction factor is required to take acccunt of such factors as long-time loading and variability; the balance is a factor of safety. Conditions of ordinary hazard to columns as primary structural members are assumed. Conditions of extroardinary hazard may require that a larger safety factor be obtained through a reduced working stress value. Columns to be used continuously dry or subjected to short-time periods of loading may be designed with somewhat less reduction from the Euler breaking load.

The slenderness ratio (L/d) of a solid wood column should in no case exceed 50.

5.3.4.1.2 Side Loads and Eccentricity.--The general formulas for safe eccentric or combined side and end loadings of wood columns are

$$\frac{P/A\left(\frac{6e}{d}\right) + M/S + z\,P/A}{f} + \frac{P/A}{c} = 1$$

for columns with slenderness ratios of 11 or less, and

$$\frac{P/A\left(\frac{15e}{2d}\right) + M/S + z\,P/A}{f - P/A} + \frac{P/A}{c} = 1$$

for columns with slenderness ratios of 20 or more, where P/A is the direct compressive stress induced by axial load; M/S is the flexural stress induced by side loads that are independent of end load; c is the allowable unit working stress in compression parallel to grain for a column of the slenderness ratio under consideration, but with centrally applied axial load and no side load; f is the allowable unit working stress in flexure that is permitted where only flexural stress exists; e is eccentricity; d is the width of the column, measured in the direction of side loads or eccentricity; equivalent to the depth when the column is considered as a beam subject to flexural forces; and z is the ratio of flexural to direct compressive stress when both result from the same loading--it remains constant while the load varies.

Stresses for columns with slenderness ratios between 11 and 20 are determined by straight-line interpolation between the formula for a slenderness ratio of 11 and the formula for a slenderness ratio of 20.

These formulas may be simplified for some conditions of loading by dropping out certain terms; for example, if there are side loads and a concentrically applied end load, e becomes zero and the first term in the numerator of the equations disappears. The formulas can be solved for P/A or for M/S if this will facilitate their use. Solution of the second formula for P/A leads to a quadratic equation with two roots; the root giving the smaller numerical value to P/A should be used.

These formulas are applicable only to columns of square or rectangular cross section.

If the side loads are such that maximum deflection and maximum flexural stress do not occur at midlength of the column, it is generally satisfactory to consider M/S as the maximum flexural stress due to the load or loads, regardless of its position in the length of the column. When the point of this maximum stress is near the end of the column, a slight error on the side of overload will occur.

5.3.4.2 Columns with Side Brackets.--Formulas for eccentric loading assume that the eccentric load is applied at the end of the column. If an eccentric load is applied by a bracket at some point below the upper end, conditions of stress and deflection are somewhat more favorable. An exact solution of the problem is difficult, but the following simple rule is safe and, for brackets in the upper quarter of the length of the column, is quite accurate.

Assume that a bracket load P at a distance a from the center of a column (fig. 5-6), is replaced by the same load P centrally applied at the top of the column plus a side load P_1 applied at midheight.

Calculate P_1 from the formula

$$P_1 = \frac{3aL_1P}{L^2}$$

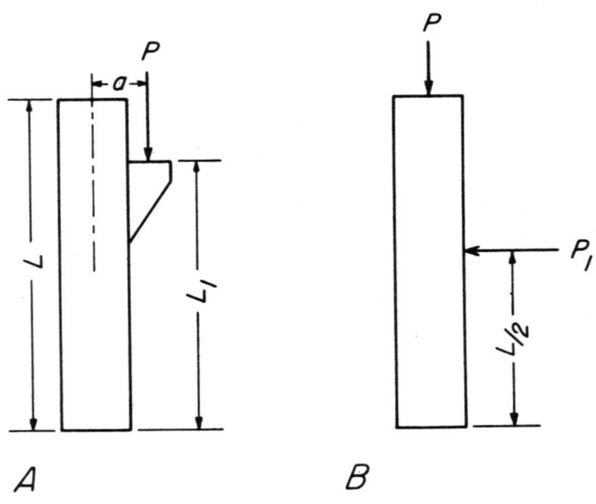

Figure 5-6.--A, Actual bracket loading of a column; B, assumed equivalent loading.

where L is the length of the column and L_1 is the length from the bottom of the column to the top of the bracket. Take care that a, L, and L_1, are expressed in the same units of measurement.

The assumed centrally applied load P can be added to other column loads, and the assumed side load P_1 is used to calculate flexural stress M/S in the formula for side loads.

5.3.4.3 Built-Up Columns with Mechanical Fastenings.--This section refers to built-up columns with pieces joined by nails, bolts, or other mechanical fastenings. Glued laminated columns are discussed in section 5.4.2.2.

5.3.4.3.1 Design of Rectangular Sections.--No arrangement of pieces with any kind of mechanical fastenings will make a built-up column fully equal in strength to a one-piece column of comparable material and dimensions. Arrangements with parallel planks and cover plates (fig. 5-7, A) or with planks boxed around a solid core (fig. 5-7, B) give the following percentages of the strength of a one-piece column of the same dimensions and quality:

L/d ratio	Percentage of strength of one-piece column
6	82
10	77
14	71
18	65
22	74
26	82

For L/d ratios of 10 or greater, these percentages apply to both built-up columns in which the individual pieces are full length and columns in which butt-jointed pieces are used.

Butt-joints in compression members should be perpendicular to the direction of stress and should be accurately cut and closely fitted. In short posts (L/d less than 10), pieces butted end to end fail at 75 to 80 percent of the crushing strength of full-length pieces. If the joint in a short column has full lateral support (such as may occur in the compression members of trusses) and a metal plate not thinner than 20 gage is tightly fitted between the butted pieces, no reduction of compressive strength need be assumed.

A B

Figure 5-7.--Two types of built-up columns: A, Edges tied together with cover plates; B, boxed solid core.

(M 22045 F)

Columns built up through the use of mechanical fastenings should generally not be specified for use in ship and boat construction because of the possibility of fastenings working loose, thus invalidating strength calculations and increasing decay hazard. This is particularly unnecessary in present day ship and boat construction in which glued wood laminates are available that provide strength equal or superior to that of a solid member of equal size and shape, under the most severe service conditions.

5.3.4.4 Spaced Columns.--Spaced columns (fig. 5-8) consist essentially of two parallel compression members separated by spacer blocks and joined through those blocks by means of timber connectors. The compression members and the spacer blocks are often of the same thickness. The spaced column may be used for the direct support of vertical loads, but often appears as a chord or other member in a wood truss.

The compression members in spaced columns are characterized by their slenderness ratio L/d, the ratio of the length, L, to the thickness of one member, d (fig. 5-8). This ratio should not exceed 80. The length L is the full length of a free-standing column, or the length between panel points or other points of lateral support in a truss. A load may be applied directly to the ends of the compression members, in which case the end spacer blocks do not extend beyond the ends of the column. In trusses, however, the end spacer blocks may extend beyond the ends of the column (fig. 5-8), and the load may be applied through them and their connectors to the compression members.

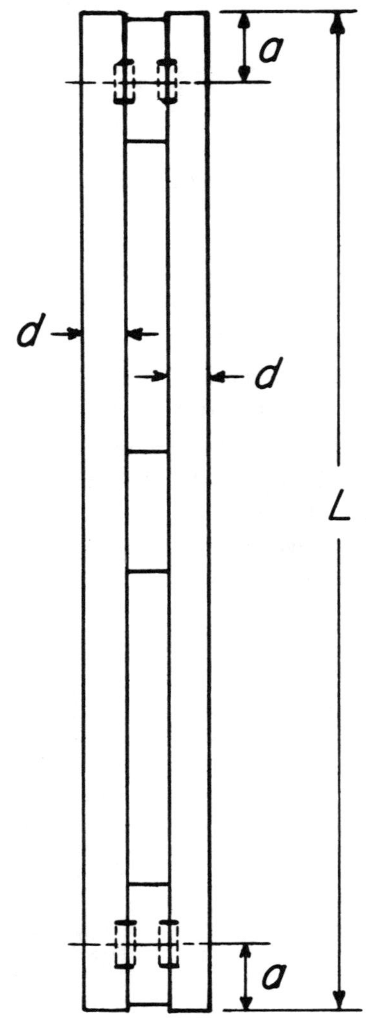

Figure 5-8.--A spaced column made from two compression members joined through spacer blocks by means of timber connectors.

Connectors should be of a size and number to restrain differential movement between the compression members. Since the forces causing differential movements are less in short than in long spaced columns, the connector requirements are correspondingly less. The connector or connectors on each face of each end spacer block should bear safely a load calculated by multiplying the cross-sectional area of one compression member (without deduction for cuts made to receive connectors) by the following unit stress values:

(1) For slenderness ratios of 11 or less, use 0 unit stress.

(2) For slenderness ratios of 60 to 80, use one-fourth of the basic stress value in compression parallel to grain for the species (table 2-1).

(3) For slenderness ratios between 11 and 60, use a unit stress obtained by straight-line interpolation between 0 for a slenderness ratio of 11 and the value for a slenderness ratio of 60.

Spacer blocks joined to the compression members by means of connectors are required at both ends of the spaced column. A spacer block at mid-length or, in long columns, two blocks at the one-third points of length are also required. All spacer blocks must be of equal thickness and at least as thick as the compression members. Their grain should, if possible, run parallel to the length of the compression members. The minimum length of the end spacer blocks is determined by the end distances required by the connectors, and their width should be equal to or greater than the width of the compression members. Middle blocks should be of such size and so fastened as to insure that the compression members will maintain their spacing under load. Spiking or bolting is generally adequate for joining the middle blocks to the compression members. Truss joint members may take the place of end or middle spacer blocks if they meet the requirements for size, location, and number and size of connectors.

Spaced columns are designed from the ordinary column formulas of section 5.3.4.1, considering each compression member as an unsupported simple column and taking the sum of the safe loads for the two compression members. Certain modifications of the formulas are made to convert to spaced-column conditions. Test values show that loads supported by compression members in the short-column class are limited by the net section of the members after deduction of the cuts made to receive the connectors and bolts. Loads on long columns are not so limited. Further, there is a fixity factor that increases the strength of members in the long-column but not in the short-column class. The fixity factor is 2.5 if connectors in the end spacer blocks are at a distance of $\underline{L/20}$ or less from the end of the column, and 3.0 if the connectors are between $\underline{L/20}$ and $\underline{L/10}$ from the end of the column and the end blocks extend to the end of the column. If the compression members of spaced columns are in the intermediate-column class, net-section and fixity effects are intermediate between those for the short-column and long-column classes.

These conditions are most conveniently recognized in design by appropriate modifications of the values for $\underline{c}$ and $\underline{E}$ in the column formulas of section 5.3.4.1. The value of $\underline{c}$ can be reduced by the ratio of net to gross cross-sectional area; this reduction will have the same effect as reducing the area used in the calculation of safe load. The value of $\underline{E}$ can likewise be increased by the fixity factor. If $\underline{R}$ represents the ratio of net to gross section and $\underline{F}$, the fixity factor, the column formulas of section 5.3.4.1 become, for the purpose of spaced-column design,

$$P/A = Rc$$

for short columns,

$$P/A = \frac{0.274\ FE}{(L/d)^2}$$

for long columns, and

$$P/A = Rc\left[1 - \frac{1}{3}\left(\frac{L}{Kd}\right)^4\right]$$

for intermediate columns, where

$$K = \frac{\pi}{2}\sqrt{\frac{FE}{6Rc}}$$

In these formulas, A is the gross cross-sectional area without deduction for connector or bolt cuts.

The compression members in spaced columns are usually of such proportions that critical loads for long columns are governed by the stiffness in a direction perpendicular to the width of those members. If the compression members are not much wider than they are thick, the stiffness in a direction parallel to their width may govern. It can be shown that the strength of a long spaced column is the same for both directions of bending when the ratio of width to thickness of the compression members equals the square root of the fixity factor.

The formulas for spaced columns give safe load values for long-time loading and are subject to increase for shorter durations of load. Increases for drying are reflected in the values chosen for E and c.

5.3.4.5 Columns of Circular Cross Section.--Round and square wood members of the same cross-sectional area will carry the same loads in both bending and compression and will have approximately the same stiffness. To compute the required size of a round column, design first for a square column and then use a round column of a diameter that will give an area equivalent to the area of the square. If the column is tapered, the diameter for use as d and for computing A in the Euler formula should be taken as the diameter at a distance of one-third of the length from the smaller end. This will give a diameter of a round column necessary to prevent failure from buckling. The unit compressive stress at the small end of the column that will result from the calculated load should also be computed, since it must not exceed the allowable stress for a short column.

5.3.5 Combined Loading Formulas.--If flexural stress in the extreme fiber is combined with direct stress resulting from the application of end compression or tension, the combined stress is determined by algebraic addition of the acting stresses. For example, compressive stress is added to bending stress in the upper fiber of a beam and subtracted from bending stress in the lower fiber of the beam. Stresses so combined should not exceed allowable values given by the following formulas.

If a transversely loaded beam is also under compression and is stayed against lateral buckling, the allowable stress is such that

$$\frac{P/A}{c} + \frac{M/S}{f} = 1$$

where P/A is the unit direct stress induced by axial load in compression parallel to grain (P is the axial compressive load and A is the area of the cross section), M/S is the flexural stress induced in bending (M is the bending moment and S is the section modulus, equal to $bd^2/6$ in a rectangular beam), c is the allowable working stress in compression parallel to grain, and f is the allowable working stress in bending. As the end compression approaches zero, the term $\frac{P/A}{c}$ also approaches zero and the preceding equation approaches

$$\frac{M/S}{f} = 1$$

or

$$M = fS = \frac{fbd^2}{6}$$

the basic formula for flexural stress.

If a transversely loaded beam is also in axial tension, the allowable stress is such that

$$\frac{P/A}{t} + \frac{M/S}{f} = 1$$

where P/A is the unit direct stress induced by axial load in tension, t is the allowable working stress in tension parallel to grain, and M/S and f are as previously defined. In many cases t and f have the same value and the equation becomes

$$P/A + M/S = t = f$$

5.4 DESIGN WITH GLUED WOOD LAMINATES.--The design of glued wood laminates in general follows the principles outlined in section 5.3 for solid wood.

5.4.1 Design Stresses.--Basic stresses for wet clear lumber (over 15 percent moisture content) are listed in table 2-1. This table is applicable to wood laminates which are kept in wet conditions or subjected to intermediate wet and dry conditions of use. The basic stresses listed in table 2-2 for dry clear lumber (15 percent or less moisture content) are applicable for wood kept in the dry condition. Design stresses for Douglas-fir and white oak straight laminates, under wet and dry conditions of use are given in table 5-2 for normal loading conditions. The tables list design stresses based on slope of grain, the prime strength ratio consideration. Governing military specifications for these wood laminates specify other permissible defects commensurate with the slope of grain requirement, and which take into account lamination width, lamination thickness, and use applications.

The design stresses listed in table 5-2 can be somewhat misleading as far as ship and boat design is concerned. Douglas-fir cannot be bent to sharp radii of curvatures as white oak of equal thickness. Compared to white oak, Douglas-fir is more prone to splitting, particularly under shock loads, and does not hold fastenings as well. Therefore, although Douglas-fir from a point of view of static bending strength shows higher stress values than white oak, it will show very inferior performance to white oak in use applications, such as frames and stems, where shock resistance and splitting resistance are of prime importance.

5.4.1.1 Scarf Joints.--Since well-bonded scarf joints with a slope not steeper than 1 in 12 are required in laminations (individual layers of wood) used in marine laminates, governed by military specifications, the full cross section of the member may be considered in computing moment of inertia and section modulus for curved and laminated beams or for effective area in axially stressed members.

5.4.1.2 Columns of Different Classes.--The formulas for solid wood columns described in section 5.3.4 apply also to laminated columns.

5.4.2 Modification to Design Stresses.--The modifications described in section 5.3.2 are applicable also, in general, to design stresses for wood laminates. Seasoning effects, however, are already taken into account in governing military specifications for each applicable species. Beams and other laminates subject to bending are generally designed as described in section 5.3.2. Certain special considerations, however, are necessary.

5.4.2.1 Modifications for Moderately Curved Members.--In curved members, the effect of curvature of the laminations on strength should be considered. The design stresses of curved members should be multiplied by the curvature factor

$$F_c = 1 - 2000 \left(\frac{t}{R}\right)^2$$

where t is the thickness of the lamination in inches and R is the radius of curvature of the lamination in inches.

Table 5-2.--Allowable unit stress for structural glued straight laminates under normal loading conditions[1]

(Based on slope of grain measured on surface having maximum slope.) (Three or more laminations.)

Species	Slope of grain measured on surface of lamination having maximum slope	Allowable unit stresses					Modulus of elasticity
		Extreme fiber in bending "f"	Tension parallel to grain "t"	Compression parallel to grain "C"	Horizontal shear "H"	Compression perpendicular to grain "C"	
1	2	3	4	5	6	7	8
		P.s.i.	P.s.i.	P.s.i.	P.s.i.	P.s.i.	1,000 p.s.i.
DRY CONDITIONS							
Douglas-fir (Coast Type)							
Dense	1 in 18	3,000	3,000	2,600	165	455	1,800
Close-grain	1 in 18	2,750	2,750	2,350	165	415	1,800
Close-grain	1 in 15	2,450	2,450	2,100	165	415	1,800
Close-grain	1 in 12	2,250	2,250	1,900	165	415	1,800
Close-grain	1 in 8	1,700	1,700	1,550	165	415	1,800
White oak	1 in 18	2,400	2,400	2,050	230	610	1,600
	1 in 15	2,150	2,150	1,850	230	610	1,600
	1 in 12	1,950	1,950	1,650	230	610	1,600
	1 in 8	1,500	1,500	1,350	230	610	1,600
WET CONDITIONS							
Douglas-fir (Coast Type)							
Dense	1 in 18	2,400	2,400	1,950	145	305	1,600
Close-grain	1 in 18	2,200	2,200	1,700	145	275	1,600
Close-grain	1 in 15	1,950	1,950	1,550	145	275	1,600
Close-grain	1 in 12	1,800	1,800	1,400	145	275	1,600
Close-grain	1 in 8	1,350	1,350	1,150	145	275	1,600
White oak	1 in 18	1,900	1,900	1,500	210	410	1,500
	1 in 15	1,700	1,700	1,350	210	410	1,500
	1 in 12	1,550	1,550	1,200	210	410	1,500
	1 in 8	1,200	1,200	1,000	210	410	1,500

[1]Under long-term loading conditions, values are reduced 10 percent.

5.4.2.2 Modifications for Severely Curved Members.--The stresses in sharply curved members subjected to bending are in error when computed by the ordinary formulas for straight beams. The amount of the error depends upon the relation of the depth of the member to the centerline radius. It is probable that limitations on the sharpness to which laminations can be bent will limit the curvature of such members as frames. The analytical methods applicable to curved beams should be used, however, to determine whether or not the usual engineering formulas may be applied without significant error to deep, sharply curved members of glued laminated wood.

5.4.2.3 Modifications Due to Radial Stresses.--When curved members are subjected to bending moments, stresses are set up in a direction parallel to the radius of curvature (perpendicular to grain). If the moment increases the radius (makes the member straighter), the stress is tension; if it decreases the radius (makes the member more sharply curved), the stress is compression. The stress is a maximum at the neutral axis and is approximately

$$S_R = \frac{3}{2}\frac{M}{Rbh}$$

where M is the bending moment, R is the radius of curvature of the center-line, and b and h are, respectively, the width and height of the cross section.

For compression perpendicular to grain, values of S_R should be limited to those shown in table 5-2. For tension perpendicular to grain, the stress should be limited, for softwoods, to about one-third and, for hardwoods, to about three-eights of the working stress in shear.

5.5 EXAMPLES OF TESTS ON SOLID AND LAMINATED BOAT MEMBERS.--Tests made with members of the size used in Navy 50-foot motor launches furnish a direct comparison of the strength of band-sawed, steam-bent, and laminated frames, floors, and keel stems. The frames were tested both under transverse loads (fig. 5-9) and end loads (fig. 5-10). Three types of frames were tested. They were steam-bent frames connected by bolts to band-sawed floors (fig. 5-11); laminated frames connected by bolts to laminated floors (fig. 5-12); and continuous laminated frames widened in the fore and aft direction, eliminating floors as separate items (fig. 5-13). The laminated floors bolted to laminated frames were superior to the band-sawed floors bolted to steam-bent frames. One-piece laminated floors and frames were superior to either.

Laminated keel-stem assemblies are greatly superior in strength, stiffness, and toughness to conventional keel-stem assemblies with bolted scarf joints. Figure 5-14 shows the wide difference in strength and behavior of the two types under test. (The keel-stem assemblies were tested as shown in figure 5-15.) The joints in bolted keel-stem assemblies started to open at low load, and failure was the result of crushing under the bolts followed by splitting of wood at the point of maximum curvature, as shown in figure 5-16. The laminated stem failed in tension in the outer laminations, followed by splintering, as shown in figure 5-17.

Many laminated structures are of such size and form that they cannot be easily shipped by rail or truck. Fabrication, therefore, is limited either to the building yard or to a laminator who is either near the building yard or can ship by water.

In order to make overland shipment possible, a method of splicing large frames has been devised that permits shipment in sections and assembly at the building yard. The splice consists of a plain glued scarf joint.

The practicability and efficiency of such a splice have been evaluated.

The central section of a spliced frame, including the splice, was tested in bending. The computed stress at failure (modulus of rupture) was on the order of 9,000 pounds per square inch, which is considered satisfactory for the grade of lumber used in the frames tested. More important, the frame section failed because of other factors than the splice. In one test, there was no failure at all in the splice; that is, the splice was capable of sustaining at least as high a load as the unspliced frame.

Thus, there is available a practical and efficient method for splicing large ship timbers. While the procedure was originally considered as a method for splicing new members, it represents also a possible method for repairing damaged frames, deck beams, and other structural members.

5.6 DESIGN OF WOOD-PLYWOOD BEAMS AND GIRDERS.

5.6.1 GENERAL.--The design of a wood-plywood beam or girder is similar to that of a steel plate girder: The flanges are designed to withstand bending moments. The plywood webs are designed to withstand shear. The joints between the webs and the flanges are designed to transfer stresses from the webs to the flanges. Stiffeners and load blocks are provided to distribute concentrated loads and to limit the lengths of the unsupported panels in the webs. The design of such a girder is a "cut and try" procedure, but once a trial section is selected and loads are known, stresses can be calculated by using conventional formulas.

5.6.2 Limiting Considerations.--The overall height of a wood-plywood beam or similar structural member is governed by one of the following considerations: (1) Clearance or other headroom limitations; (2) width of the plywood economically available for use in a beam (this is usually 4 feet or some even fraction, such as 1/2, 1/3, or 1/4, of 4 feet, so that waste will be minimized); or (3) ratio of height to width and span (the member must not be laterally unstable under design loads when it is not restrained).

The choice of the type of cross section that should be used in a given application depends upon several factors. If plywood of sufficient thickness is not available to permit utilization of single-web members (I-beam), members with multiple webs may be required. In certain exposures, however, because of the danger that moisture may condense on or otherwise enter the interior of a box beam, it may be desirable to select an I-beam. The I-beam is easier to inspect for deterioration while in service than a multiple-web beam, because all parts of the I-beam are exposed.

Figure 5-9.--Transverse-load test for a bent boat rib.

(M 43285 F)

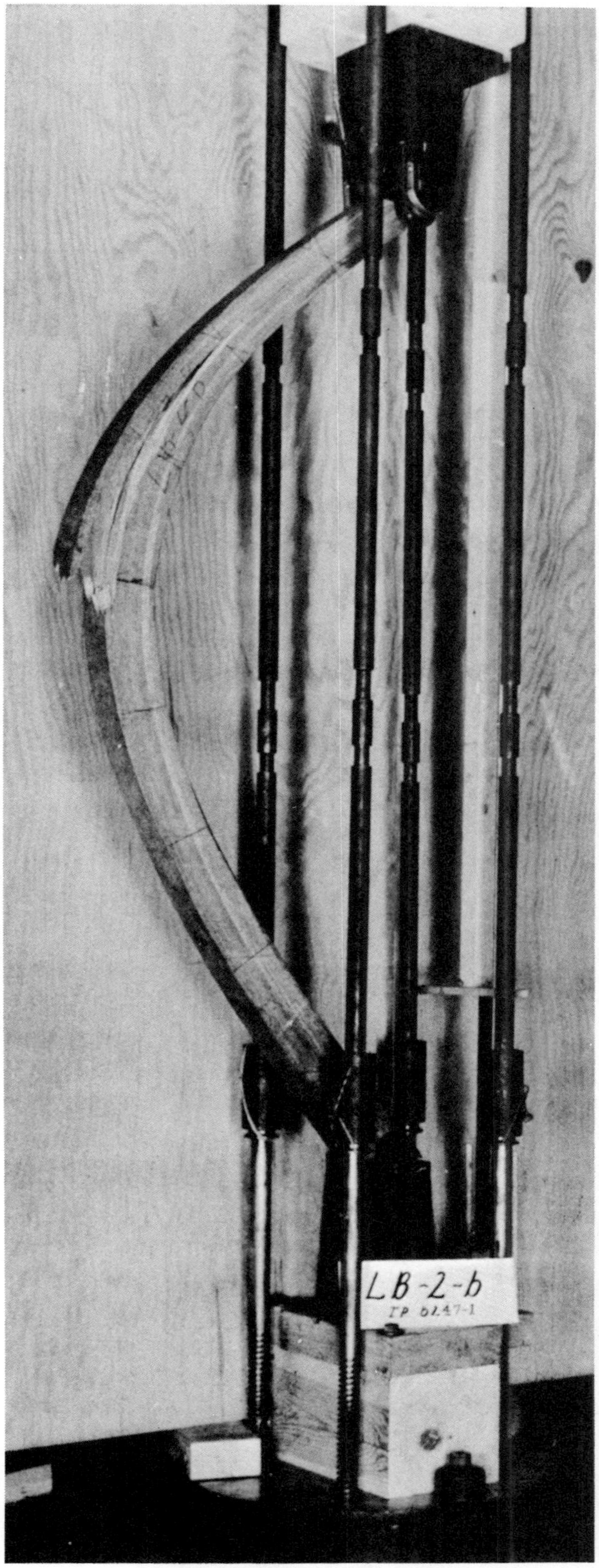

Figure 5-10.--End-load test for a bent boat rib.

(M 43371 F)

Figure 5-11.--Type of steam-bent frame used for strength tests.

(M 43234 F)

Figure 5-12.--Laminated frame bolted to laminated floor as used in strength tests.

(M 44433 F)

Figure 5-13.--Type of continuous laminated frame used for strength tests.

(M 43230 F)

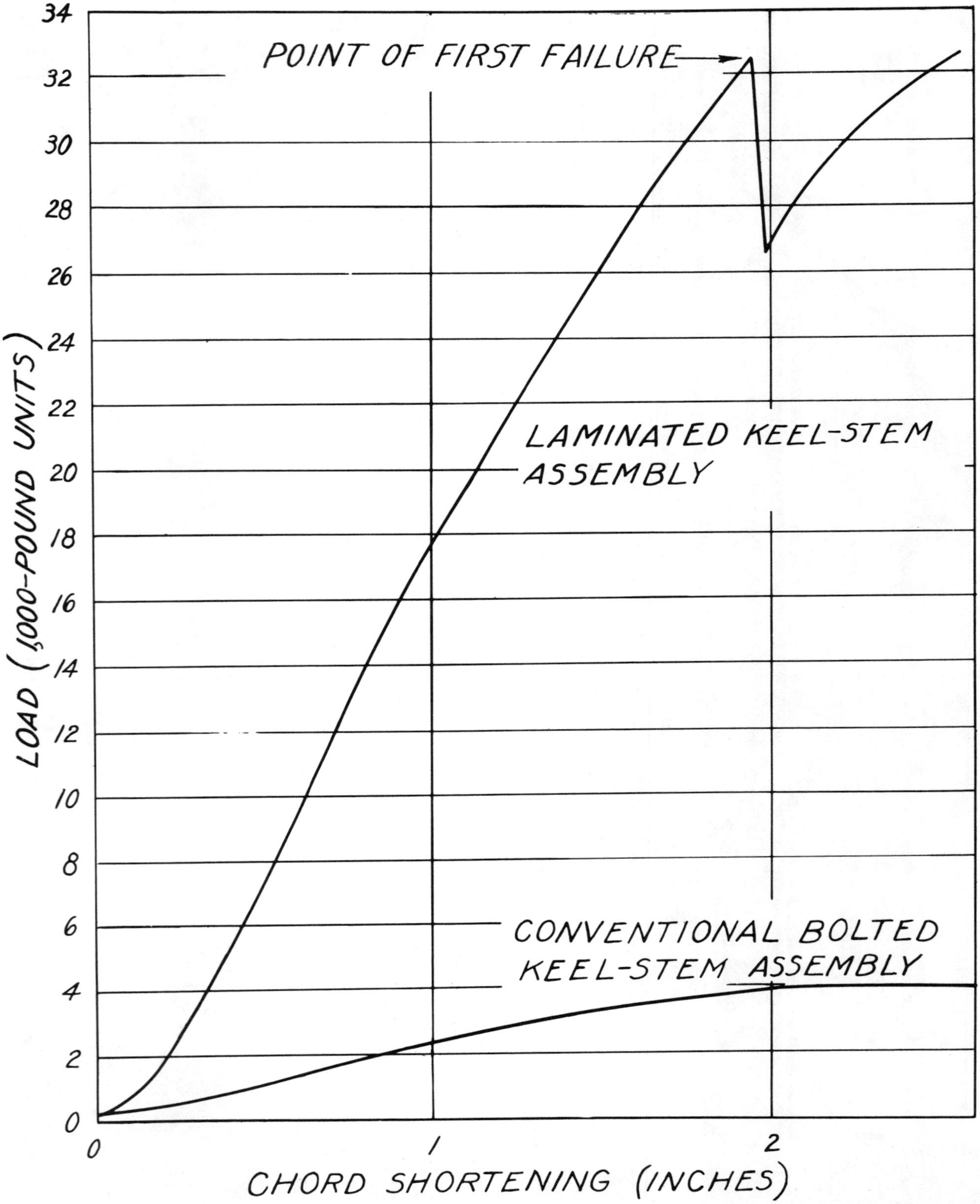

Figure 5-14.--Comparative strength of laminated and conventional keel-stem assemblies as determined in tests.

(M 47642 F)

Figure 5-15.--Method of testing keel-stem assemblies in 1,000,000-pound testing machine.

(M 43856 F)

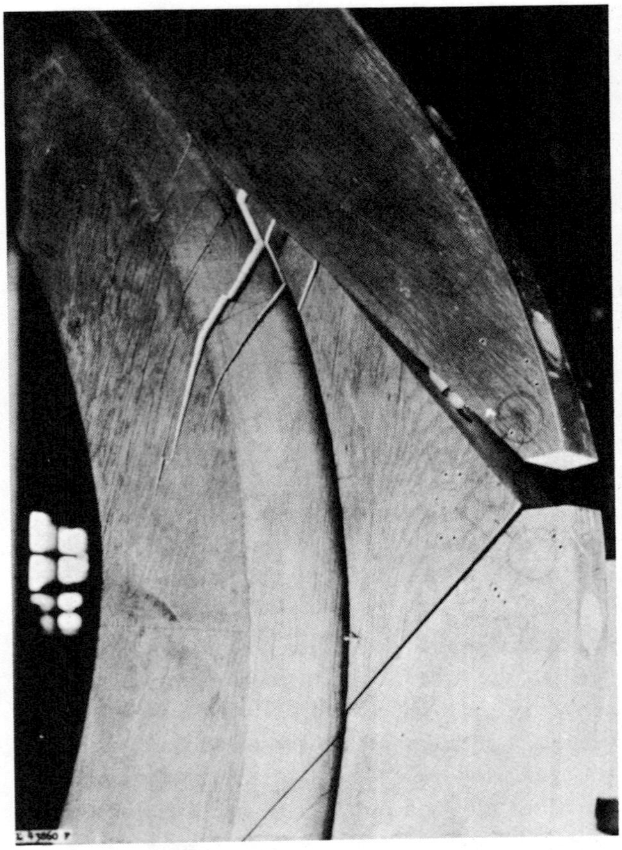

Figure 5-16.--Typical failure of a bolted keel-stem assembly under test.

(M 43860 F)

Figure 5-17.--Typical failure of a laminated stem under test.

(M 43858 F)

5.6.3 Design.--Detailed methods of design for wood-plywood beams and girders are given in references 5-1 and 3-3.

5.7 REFERENCES

(5-1) DOUGLAS FIR PLYWOOD ASSOCIATION
1950. Technical Data on Plywood. Section 7, Form Factors of I-and Box Beams with Plywood Webs. Section 9. The Design of Built Up Beams.

(5-2) JOHNSON, A. MEREDITH
1953. Development in the Hull Design of Wooden Ships. Paper presented at the meeting of the Chesapeake Section, Society of Naval Architects and Marine Engineers. February 26, 1953.

CHAPTER 6. JOINTS AND FASTENINGS FOR SHIP AND BOAT CONSTRUCTION

6.1 GENERAL.--The strength of wood ships and boats depends upon nearly all the parts of the hull proper. Of these parts, none are more important than the joints and fastenings. The strength of the joints is dependent on that of dissimilar materials--metal and wood or glue and wood.

Many types of mechanical and glued joints are used in ship and boat building. The type most distinctive to the construction of watercraft is the nibbed mechanical scarf joint. Figure 6-1 shows four basic types of such scarf joints. They and their variations are used in keels and keelsons, shelves, clamps, and other longitudinals, and sometimes in planking. Their purpose is quite different from that of the glued scarf joint; they are made to permit forming a bolted joint without resort to splice plates commonly used for butt joints. Virtually all are nibbed to avoid the feather edges commonly found in glued joints; the nibs are considered to help in preventing longitudinal slippage, especially if bolt holes develop elongation in use. Hooked and keyed scarfs are installed to help reduce slippage and to produce more watertight joints at the nibs. In glued joints a continuous bond is formed which may be nearly as strong as the wood itself, making nibs and hooks unnecessary; the purpose of scarfing for glued joints is, in fact, to avoid gluing of end-grain surfaces, such as occur at nibs, because such glued joints are invariably weaker than those between side-grain surfaces, and for this reason the longer the scarf in proportion to its slope, the better is the glue joint.

In mechanical scarf joints, the length of the scarf is shown in the plan or specifications and is usually given to the workmen by the foreman who lays out the work. The length of the scarfed area is determined by the ratio of that length to the thickness of the member being scarfed rather than by exact measurement of the scarf in terms of its own length and slope, as is done to produce good glued joints. Wood shipbuilding practice calls for a length of scarfed area equal to six times the depth, or thickness, of the timbers being scarfed together. Since the nibs total about one-half the thickness of the member, the actual slope of the scarf joint thus produced is about 1 in 12, in terms of scarf slope as used in glued scarf joints. It is standard practice to cut each nib to a depth of about 25 percent of the total depth of the member.

The strength of glued scarf joints is discussed in section 3.3.4. Plans or specifications usually prescribe the minimum slope of scarf permitted in a given ship or boat part. Plywood scarfs are used extensively to produce long panels, and the scarf is made with phenol-resin adhesive in a heated press.

Considerable advancement has been made in the design of various types of metal fastenings and connectors for use in connecting wood members. It is not, however, possible with any mechanical fastening to obtain joints that afford strength equal to that of an unjointed piece.

6.2 MECHANICAL FASTENINGS.--Various types of mechanical fastenings are used in wood vessels, including nails, spikes, screws, lag screws, hanger bolts, drift bolts, bolts headed over clinch rings, double-ended bolts threaded and fitted with nuts at both ends, and rivets, which are generally of the copper type headed up over burrs. Modern timber connectors are sometimes used with bolts to increase joint strength (sec. 6.2.3).

The size, number, and spacing of the fastenings to be used have been established by experience and are prescribed for different sizes and kinds of members by plans and specifications. Specific data indicating the strength properties for the various types of fastenings used in large vessels are not available. Even if available, its use in hull design would be limited because the stresses to which the various joints are subjected are usually indeterminate.

The determination of the stress distribution in wood and in the metal fasteners used for fabrication is complicated. The mathematical assumptions made in order to compute it are often so involved that actual tests must be relied upon to provide design data. Such tests have been made on bolts, timber connectors, lag screws, and nails (refs. 6-6, 6-7, 6-10). These data have only limited application in the design of large wood craft but are valuable wherever the loads and stresses which the joint will be required to carry are known.

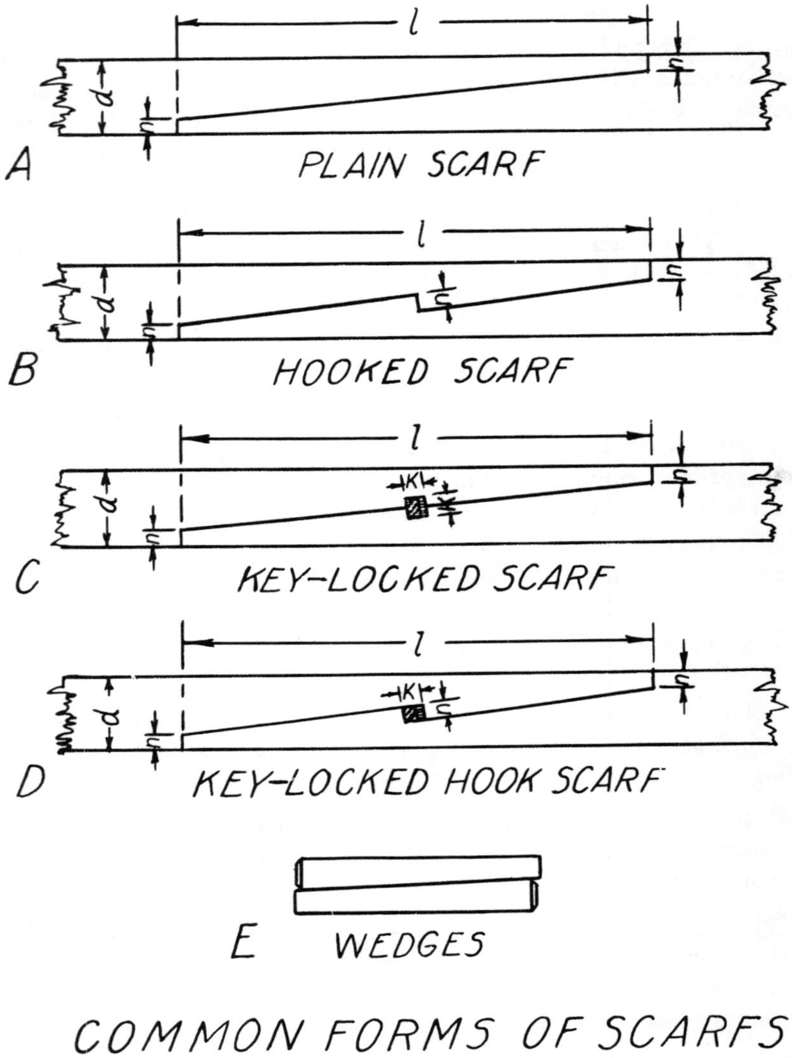

Figure 6-1.--Common types of bolted scarf joints used in boats and ships. The length of the scarf, l, is as a rule 6 times the depth of the timber, d, and the depth of the nibs, n, and cross section of the keys, k, are about 25 percent of d. Keys are double wedges, as shown in E.

(M 58166 F)

6.2.1 Bolts.--Common bolts are used extensively in ship construction and, if properly used, form an effective fastening in many types of joints. High-strength bolts, such as are used in aircraft, carry somewhat higher loads than do common bolts, but their use in shipbuilding is rare. Steel bolts for ship and boat use should be galvanized or otherwise treated to prevent corrosion; usually for underwater fastenings brass or other nonferrous bolts are required.

6.2.1.1 Bearing Strength of Wood Under Bolts. --Basic bolt-bearing stresses are given in table 6-1 for use in calculating the allowable loads for bolted joints acting parallel and perpendicular to the grain (refs. 6-2, 6-10). The average ultimate crushing strength parallel to the grain of small, clear, green specimens was modified for duration of loading, increase in strength with drying, variability of the wood, and a reduction factor for safety. The average proportional limit stress perpendicular to the grain of small, clear, green specimens was modified for increase in strength with seasoning, variability of the wood, and a reduction factor for safety. Special attention was given to the development of the basic stresses for certain species that show abnormal strength, splitting, or drying characteristics.

Basic stresses are listed in table 6-1 for a number of the common species and are applicable for bolted joints in any grade of seasoned lumber destined to be used in a dry, inside location. In locations where the timber is occasionally wet but quickly dried and in locations that are continuously damp or wet, the stresses listed in table 6-1 should be reduced one-fourth and one-third, respectively.

The basic stresses given in table 6-1 are for permanent loading. For loadings of short duration, they may be modified as discussed in section 5.3.2.1.

The allowable loads for correctly spaced and alined bolts bearing parallel or perpendicular to the grain of the wood and with the load applied through metal plates to both ends of the bolt are calculated by multiplying the appropriate basic stress (table 6-1) by a percentage factor (table 6-2) based on the ratio of the length of bolt in the main member to the bolt diameter (L/D). For loads acting perpendicular to the grain, an additional factor, referred to as a diameter factor, is applied to the basic stress values in accordance with the following bolt diameters:

Diameter of bolt	Diameter factor	Diameter of bolt	Diameter factor
Inches		Inches	
1/4	2.50	1-1/4	1.19
3/8	1.95	1-1/2	1.14
1/2	1.68	1-3/4	1.10
5/8	1.52	2	1.07
3/4	1.41	2-1/3	1.03
7/8	1.33	3 or over	1
1	1.27		

6.2.1.2 Allowable Bolt Loads Parallel to Grain.-- The allowable load on a bolted timber joint acting parallel to the grain of the wood through metal plates at both ends of the bolt is calculated as follows:

(1) Select from table 6-1 the basic stress in compression parallel to the grain for the particular species of wood and reduce it in accordance with the service condition if the joint is to be used in other than a dry inside location. Call this stress $\underline{S_1}$.

Table 6-1.--Factors for calculating the allowable strength of mechanical fastenings (permanent loading basis) in seasoned wood

(For loading conditions of shorter durations, allowable lateral load constants and basic stresses for bolts may be modified as described in section 5.3.2.1.)

Species	Specific gravity (G) and powers[1]				Allowable lateral load constant (K)			Basic stresses for bolts[4]		Species group for connector loads[5]
	G	$G^{3/2}$	G^2	$G^{5/2}$	Nails[2]	Screws[2]	Lag screws[3]	Parallel to grain	Perpendicular to grain	
								P.s.i.	P.s.i.	
Ash										
Black	0.53	0.39	0.28	0.20	1,250	2,900	1,900	850	300	....
Commercial white	.61	.48	.37	.29	1,700	4,000	2,200	1,450	500	4
Baldcypress	.48	.33	.23	.16	1,125	2,700	1,700	1,450	300	2
Cedar										
Alaska-	.46	.31	.21	.14	1,125	2,700	1,700	1,050	250	2
Atlantic white-	.35	.21	.12	.07	900	2,100	1,500	750	180	....
Northern white-	.32	.18	.10	.06	900	2,100	1,500	750	180	....
Port-Orford-	.44	.29	.19	.13	1,125	2,700	1,700	1,200	250	2
Western redcedar	.34	.20	.12	.07	900	2,100	1,500	950	200	1
Douglas-fir										
Coast type[6]	.51	.36	.26	.19	1,375	3,300	1,900	1,450	320	3
Rocky Mountain type	.45	.30	.20	.14	1,125	2,700	1,700	1,050	280	2
Hemlock										
Eastern	.43	.28	.18	.12	900	2,100	1,500	950	300	1
Western	.44	.29	.19	.13	1,125	2,700	1,700	1,200	300	2
Larch, Western	.59	.45	.35	.27	1,375	3,300	1,900	1,450	320	3

(Sheet 1 of 2)

Table 6-1.--Factors for calculating the allowable strength of mechanical fastenings--Continued (permanent loading basis) in seasoned wood--Continued

(For loading conditions of shorter durations, allowable lateral load constants and basic stresses for bolts may be modified as described in section 5.3.2.1.)

Species	Specific gravity (G) and powers[1]				Allowable lateral load constant (K)			Basic stresses for bolts[4]		Species group for connector loads[5]
	G	$G^{3/2}$	G^2	$G^{5/2}$	Nails[2]	Screws[2]	Lag screws[3]	Parallel to grain	Perpendicular to grain	
								P.s.i.	P.s.i.	
Oak										
Commercial red	0.66	0.54	0.44	0.35	1,700	4,000	2,200	1,350	500	4
Commercial white	.71	.60	.50	.42	1,700	4,000	2,200	1,350	500	4
Pine										
Eastern white	.37	.22	.14	.08	900	2,100	1,500	1,000	250	1
Red	.51	.36	.26	.19	1,125	2,700	1,700	1,050	220	2
Southern yellow[6]	.59	.45	.35	.27	1,375	3,300	1,900	1,450	320	3
Sugar	.38	.23	.14	.09	900	2,100	1,500	1,000	250	1
Western white	.42	.27	.18	.11	900	2,100	1,500	1,000	250	1
Redwood (old growth)	.42	.27	.18	.11	1,125	2,700	1,700	1,350	250	2
Spruce										
Red	.41	.26	.17	.11	900	2,100	1,500	1,050	250	2
Sitka	.42	.27	.18	.11	900	2,100	1,500	1,050	250	2
White	.45	.30	.20	.14	900	2,100	1,500	1,050	250	2

[1]Specific gravity based on weight and volume when ovendry.

[2]Driven perpendicular to the grain of the wood and loaded either parallel or perpendicular to the grain.

[3]Inserted perpendicular to the grain of the wood and loaded parallel to the grain.

[4]Basic stresses to be used in determining allowable bolt-bearing stresses.

[5]Grouping to be used in determining allowable connector loads. Group 1 woods provide the weakest and group 4 woods the strongest connector groups.

[6]When graded for density, these species qualify for group-4 connector loads.

(Sheet 2 of 2)

Table 6-2.--Percentage of basic stress used in calculating allowable bearing stresses for common bolts when load is applied through metal splice plates

Ratio of bolt length to diameter (L/D)	Bearing stress in percent of basic stress for--						
	Bolts bearing parallel to grain,[1] when basic stress is--			Bolts bearing perpendicular to grain,[2] when basic stress is--			
	750 to 950 p.s.i.	1,000 to 1,200 p.s.i.	1,300 to 2,000 p.s.i.	150 to 180 p.s.i.	200 to 280 p.s.i.	300 to 350 p.s.i.	400 to 600 p.s.i.
1	100.0	100.0	100.0	100.0	100.0	100.0	100.0
2	100.0	100.0	100.0	100.0	100.0	100.0	100.0
3	100.0	100.0	99.0	100.0	100.0	100.0	100.0
4	99.5	97.4	92.5	100.0	100.0	100.0	100.0
5	95.4	88.3	80.0	100.0	100.0	100.0	100.0
6	85.6	75.8	67.2	100.0	100.0	100.0	96.3
7	73.4	65.0	57.6	100.0	100.0	97.3	86.9
8	64.2	56.9	50.4	100.0	96.1	88.1	75.0
9	57.1	50.6	44.8	94.6	86.3	76.7	64.6
10	51.4	45.5	40.3	85.0	76.2	67.2	55.4
11	46.7	41.4	36.6	76.1	67.6	59.3	48.4
12	42.8	37.9	33.6	68.6	61.0	52.0	42.5
13	39.5	35.0	31.0	62.2	55.3	45.9	37.5

[1]For wood splice plates, each of which is one-half the thickness of the main member, the allowable load should be taken as 4/5 those computed for metal splice plates.

[2]No reduction need be made when wood splice plates are used except that the allowable load perpendicular to the grain should never exceed the allowable load parallel to the grain for any given size and quality of bolt and timber.

(2) Calculate the ratio of the length of the bolt in the main member to the bolt diameter (L/D). Select the percentage r of basic stress from table 6-2 for the calculated L/D.

(3) Multiply the basic stress S_1 by the percentage r to obtain the allowable unit stress S_2, which is assumed to be uniformly distributed.

(4) Multiply S_2 by the projected area of the bolt to obtain the allowable load P_1 for one bolt.

When the load is applied through wood splice plates, each of which is one-half the thickness of the center member, the allowable load is obtained by multiplying P_1 by 0.8.

When the bolt holes are properly centered and alined, the allowable load on a number of bolts of the same or different diameters may be taken as the sum of the individual load capacities.

6.2.1.3 Allowable Bolt Loads Perpendicular to Grain.--The allowable load on a bolted timber joint acting perpendicular to the grain of the wood and at both ends of the bolt through metal plates or wood side plates, each of which is one-half the thickness of the main member, is calculated as follows:

(1) Select from table 6-1 the basic stress in compression perpendicular to the grain for the particular species of wood and reduce in accordance with the service condition if the joint is to be used in other than a dry, inside location. Call this stress S_1.

(2) Calculate the ratio of the length of bolt in the main member to the bolt diameter (L/D). Select the percentage r of basic stress from table 6-2 for the appropriate L/D.

(3) Select the diameter factor v for the appropriate size of bolt (sec. 6.2.1.1).

(4) Multiply the basic stress S_1 by the percentage factor r and the diameter factor v to obtain the allowable average unit stress S_2 for the particular L/D.

(5) Multiply the allowable unit stress S_2 by the projected area of the bolt to obtain the allowable load for a single bolt loaded through either wood or metal spliced plates.

When the bolt holes are properly centered and alined, the allowable load on a number of bolts of the same or different diameters may be taken as the sum of the individual load capacities.

The allowable bolt loads, both parallel and perpendicular to the grain, determined by the preceding procedure are of such magnitude as to prevent undue permanent distortion or other evident injury in the joint under long-term loading or repeated application of the load. For loads acting at an angle between the limits of parallel to the grain and perpendicular to the grain, the allowable bolt load may be obtained from the nomograph in figure 6-2.

6.2.1.4 Effect of Quality of Bolt on Joint Strength.--Both the quality of the wood and the quality of the bolt are factors in determining the proportional limit strength of a bolted joint. The percentages of basic stresses given for calculating safe bearing stresses apply to the common commerical bolts used in building construction. For high-strength bolts, such as aircraft bolts, the reduction values listed in table 6-2 would be somewhat larger for the larger L/D ratios.

6.2.1.5 Details of Design with Bolts.--The details of design required in the application of the allowable loads for bolts may be summarized as follows:

(1) A load applied to only one end of a bolt, perpendicular to its axis, may be taken as one-half the symmetrical two-end load.

(2) The center-to-center distance along the grain between bolts acting parallel to the grain should be at least 4 times the bolt diameter. When a joint is in tension, the bolt nearest the end of a timber should be at a distance from the end of at least 7 times the bolt diameter for softwoods and 5 times for hardwoods. When the joint is in compression, the end margin may be 4 times the bolt diameter for both softwoods and hardwoods. Any decrease in these spacings and margins will decrease the load in about the same ratio.

(3) For bolts bearing parallel to the grain, the distance from the edge of a timber to the center of a bolt should be at least 1.5 times the bolt diameter. This margin, however, will usually be controlled by (a) the common practice of having an edge margin equal to one-half the distance between bolt rows and (b) the area requirements at the critical section. (The critical section

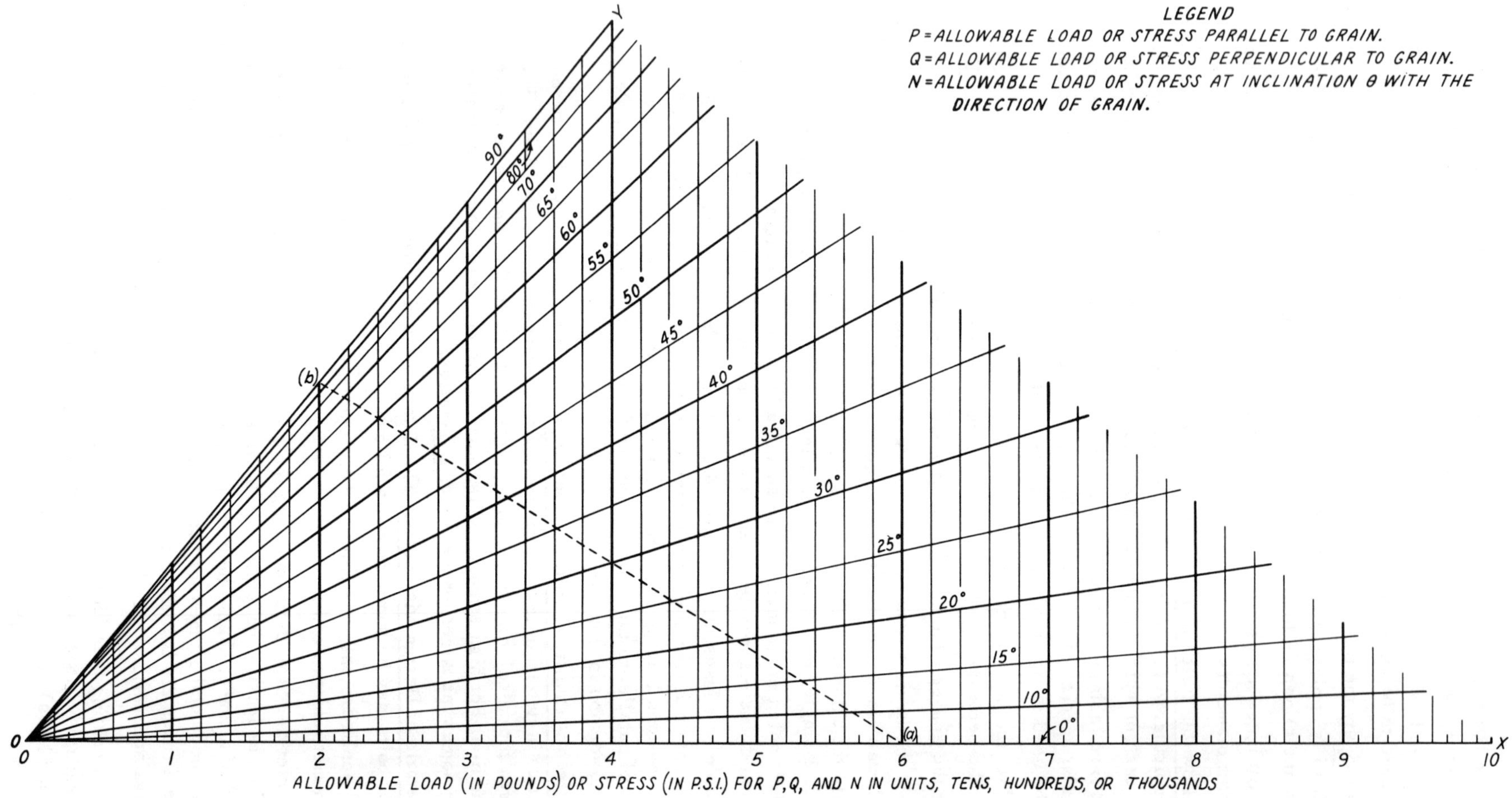

Figure 6-2.--Scholten nomograph for determining the bearing strength of wood at various angles to the grain.

(M 92352 F)

is that section of the member, taken at right angles to the direction of load, which gives the maximum stress in the member, based on the net area remaining after reductions are made for bolt holes at that section.) For parallel-to-grain loading in softwoods, the net area remaining at the critical section should be at least 80 percent of the total area in bearing under all the bolts in the particular joint under consideration; in hardwoods it should be 100 percent.

(4) For bolts bearing perpendicular to the grain, the margin between the edge toward which the bolt pressure is acting and the center of the bolt or bolts nearest this edge should be at least four times the bolt diameter. The margin at the opposite edge is relatively unimportant. The minimum center-to-center spacing of bolts in the across-the-grain direction for loads acting through metal side plates need only be sufficient to permit the tightening of the nuts. For wood side plates, the spacing is controlled by the rules applying to loads acting parallel to grain if the design load approaches the bolt-bearing capacity of the side plates. When the design load is less than the bolt-bearing capacity of the side plates, the spacing may be reduced below that required to develop their maximum capacity.

6.2.2 Drift Bolts.--The ultimate withdrawal load of a round drift bolt or pin from the side grain of seasoned wood is given by the formula

$$P = 6{,}000\ G^2D$$

in which P is the ultimate withdrawal load per lineal inch of penetration; G, the specific gravity based on the weight and volume of the wood when ovendry; and D, the diameter of the drift bolt in inches. This formula is based on the assumption that the bolts are driven into prebored holes with a diameter one-eighth inch less than that of the bolt.

The allowable withdrawal load per lineal inch of penetration should be determined by dividing the ultimate withdrawal load by the factor 5. In lateral resistance, the allowable load for a drift bolt driven into the side grain of wood should not exceed and ordinarily should be taken as less than that for a common bolt of the same diameter (sec. 6.2.1).

6.2.3 Timber-Connector Joints.--When timbers are joined with metal or other types of connectors, the strength of the connector joint depends on the type and size of the connector, the kind of wood, the thickness and width of the member, the distance of the connector from the end of the timber, the spacing of the connectors, the direction of application of the load with respect to the direction of the grain of the wood, and other factors (refs. 6-3, 6-7). Safe working loads for metal connectors in common use--split-ring (fig. 6-3), toothed (fig. 6-4), and shear-plate (fig. 6-5) connectors are discussed in this section.

Toothed connectors carry significantly lower loads than do split rings (compare tables 6-3, 6-4) of the same size. Toothed connectors are generally used where the power tools used to install split rings are not available.

6.2.3.1 Working Loads for Connector Joints.--The safe long-time working loads for the split-ring, shear-plate, and toothed connectors provided in tables 6-3, 6-4, and 6-5 are derived from the results of tests of full-scale joints made from representative species, which were used in conjunction with other basic information on the strength properties and behavior of timber. In establishing these working loads, particular consideration was given to (1) the effect of long-continued loading as against the brief loading period involved in the test of joints (sec. 5.3.2.1) and (2) allowance for variability in timber quality.

The working loads listed in tables 6-3, 6-4, and 6-5 for connectors acting parallel to the grain were derived by applying a reduction factor to the ultimate load found by tests. For split-ring and shear-plate connectors, a reduction factor of 4 gave values that met the provision that working loads for these connectors should not exceed five-eighths of the proportional-limit test load. Because load-slip curves for the toothed connector do not exhibit a well defined-proportional limit, the larger reduction factor of 4-1/2 was applied to the ultimate load.

Figure 6-3.--Split-ring connector assembly--connector, precut groove, bolt, washer, and nut.
(M 32889 F)

Figure 6-4.--Toothed connector assembly.

(M 32890 F)

Figure 6-5.--A, Shear-plate connector assembly with wood side plates; B, shear-plate connector assembly with steel side plates.

(M 92355 F)

Table 6-3.--Safe working loads[1] for split-ring connector and bolt (one connector unit)

Connector unit	Minimum thickness of member: With 1 connector only	Minimum thickness of member: With 2 connectors in opposite faces 1 bolt[2]	Minimum width all members	Group 1 woods 0°	Group 1 woods 90°	Group 2 woods 0°	Group 2 woods 90°	Group 3 woods 0°	Group 3 woods 90°	Group 4 woods 0°	Group 4 woods 90°
1	2	3	4	5	6	7	8	9	10	11	12
	In.	In.	In.	Lb.	Lb.	Lb.	Lb.	Lb.	Lb.	Lb.	Lb.
2-1/2-inch connector	1	1-5/8	3-5/8	1,490	880	1,735	1,025	2,065	1,230	2,395	1,435
and 1/2-inch bolt	1-1/8+	2+	3-5/8	1,785	1,055	2,085	1,230	2,480	1,475	2,875	1,725
4-inch connector and	1-1/8	1-5/8	5-1/2	2,415	1,400	2,795	1,620	3,355	1,945	3,915	2,270
3/4-inch bolt	1-1/4	2	5-1/2	2,775	1,610	3,215	1,865	3,860	2,240	4,500	2,610
	1-1/2	2-5/8	5-1/2	3,380	1,960	3,915	2,270	4,695	2,725	5,480	3,175
	1-5/8+	3+	5-1/2	3,445	1,995	3,985	2,310	4,780	2,775	5,580	3,235

[1]The safe working loads apply to seasoned timbers in dry, inside locations for a long-continued load. It is assumed also that the joints are properly designed with respect to such features as centering of connectors, adequate end-distance and suitable spacing. For loading conditions of shorter duration, loads may be modified as described in section 6.2.3.2.

[2]A three-member assembly, with two connector units would therefore take double the safe working loads indicated in columns 5 through 12.

Table 6-4.--Safe working loads[1] for one-toothed connector and bolt (one connector unit)

Connector unit	Minimum thickness of member		Minimum width all members	Load when angle of load application to grain is--							
	With 1 connector only	With 2 connectors in opposite faces: 1 bolt[2]		Group 1 woods		Group 2 woods		Group 3 woods		Group 4 woods	
				0°	45° to 90°	0°	45° to 90°	0°	45° to 90°	0°	45° to 90°
1	2	3	4	5	6	7	8	9	10	11	12
	In.	In.	In.	Lb.	Lb.	Lb.	Lb.	Lb.	Lb.	Lb.	Lb.
2-inch connector with	1	1-5/8	2-5/8	780	520	900	600	1,000	665	1,100	735
1/2-inch bolt	1-1/8+	2+	2-5/8	860	570	990	660	1,100	735	1,210	805
2-5/8-inch connector	1	1-5/8	3-5/8	1,170	780	1,350	900	1,500	1,000	1,650	1,100
with 5/8-inch bolt	1-1/8	2	3-5/8	1,295	865	1,495	995	1,660	1,105	1,825	1,220
	1-3/8+	2-1/2+	3-5/8	1,460	975	1,690	1,125	1,875	1,250	2,060	1,375
3-3/8-inch connector	1	1-5/8	4-5/8	1,520	1,015	1,755	1,170	1,950	1,300	2,145	1,430
with 3/4-inch bolt	1-1/8	2	4-5/8	1,665	1,110	1,920	1,280	2,135	1,425	2,350	1,565
	1-1/2	2-5/8	4-5/8	1,910	1,270	2,200	1,465	2,445	1,630	2,690	1,795
	1-5/8+	3+	4-5/8	2,055	1,370	2,370	1,580	2,630	1,755	2,895	1,930
4-inch connector	1	1-5/8	5-1/2	1,835	1,220	2,115	1,410	2,350	1,565	2,585	1,725
with 3/4-inch bolt	1-1/8	2	5-1/2	1,985	1,320	2,290	1,525	2,540	1,695	2,795	1,865
	1-1/2	2-5/8	5-1/2	2,235	1,490	2,575	1,720	2,865	1,910	3,150	2,100
	1-5/8	3+	5-1/2	2,385	1,590	2,750	1,835	3,055	2,035	3,360	2,240

[1]The safe working loads apply to seasoned timbers in dry, inside locations for a long-continued load. It is assumed also that the joints are properly designed with respect to such features as centering of connectors, adequate end-distance and suitable spacing. For loading conditions of shorter duration, loads may be modified as described in section 6.2.3.2.

[2]A three-member assembly, with two connector units would therefore take double the safe working loads indicated in columns 5 through 12.

From tests of connectors under loads bearing parallel and perpendicular to grain, a relationship between the two directions has been established from which allowable loads were derived for perpendicular bearing. For perpendicular bearing, greater dependence was placed on factors such as the load at proportional limit and at given slips of the joint than on ultimate load.

The figures quoted as the ratios between working loads and the loads found in test are in no instance true factors of safety. For example, the reduction factor of 4 for split-ring and shear-plate connectors includes allowances for duration of stress and for variability as well as a margin for safety. Thus, after the values from test are multiplied by a factor of 9/16 as an allowance for a long-continued load, and by 3/4 to cover variability of the wood, the actual factor of safety for a connector joint is on the order of 1-3/4 (4 x 9/16 x 3/4 = 1-11/16) if the working load acts over a long period. The tests from which working loads were derived were on specimens carefully made from seasoned material, under favorable conditions, and by experienced workmen. A lower standard of workmanship, or seasoning subsequent to the fabrication of a joint made in green or unseasoned timber, would further reduce the indicated factor of safety. Detailed design data for connectors may also be found in references 6-3, 6-9, and 6-11.

Table 6-5.--Safe working loads[1] for one shear-plate connector and bolt (one connector unit)

Connector unit	Minimum thickness of member: With 1 connector only	Minimum thickness of member: With 2 connectors in opposite faces: 1 bolt[2]	Minimum width all members	Group 1 woods 0°	Group 1 woods 90°	Group 2 woods 0°	Group 2 woods 90°	Group 3 woods 0°	Group 3 woods 90°	Group 4 woods 0°	Group 4 woods 90°
1	2	3	4	5	6	7	8	9	10	11	12
	In.	In.	In.	Lb.	Lb.	Lb.	Lb.	Lb.	Lb.	Lb.	Lb.
2-5/8-inch connector	....	1-5/8	3-1/2	1,470	852	1,704	988	2,045	1,186	2,386	1,384
with 3/4-inch bolt	....	2	3-1/2	1,785	1,035	2,069	1,200	2,483	1,440	2,667	1,680
	1-5/8+	2-5/8+	3-1/2	1,889	1,096	2,191	1,271	2,629	1,525	2,667	1,779
4-inch connector	....	1-3/4	5-1/2	1,901	1,102	2,204	1,278	2,645	1,533	3,086	1,789
with 3/4-inch bolt	....	2	5-1/2	2,120	1,229	2,458	1,426	2,950	1,711	3,442	1,996
	....	2-5/8	5-1/2	2,486	1,442	2,882	1,672	3,458	2,006	4,035	2,340
	....	3	5-1/2	2,705	1,568	3,137	1,819	3,763	2,182	4,391	2,547
	1-3/4+	3-5/8+	5-1/2	2,851	1,653	3,306	1,917	3,967	2,301	4,562	2,684
4-inch connector	....	1-3/4	5-1/2	1,901	1,105	2,204	1,278	2,645	1,533	3,086	1,787
with 7/8-inch bolt	....	2	5-1/2	2,120	1,229	2,458	1,426	2,950	1,711	3,442	1,996
	....	2-5/8	5-1/2	2,486	1,442	2,882	1,672	3,458	2,006	4,035	2,340
	....	3	5-1/2	2,705	1,568	3,137	1,819	3,763	2,182	4,391	2,547
	1-3/4+	3-5/8	5-1/2	2,851	1,653	3,306	1,917	3,967	2,301	4,628	2,684

[1] The safe working loads apply to seasoned timbers in dry, inside locations for a long-continued load. It is assumed also that the joints are properly designed with respect to such features as centering of connectors, adequate end-distance and suitable spacing. For loading conditions of shorter duration, loads may be modified as described in section 6.2.3.2. The tabulated loads apply for metal or wood side plates, except that, for 4-inch shear plates with metal side plates, the parallel-to-grain load may be increased 5, 11, and 18 percent for groups 2, 3, and 4 woods respectively. The allowable loads for all loading except wind should not exceed 2,900 pounds for 2-5/8-inch shear plates, or, for 4-inch shear plates, 4,970 and 6,760 pounds with 3/4- and 7/8-inch bolts respectively. For wind loading, the corresponding allowable loads should not exceed 3,870, 6,630, and 9,020 pounds.

[2] A three-member assembly, with two connector units would therefore take double the safe working loads indicated in columns 5 through 12.

6.2.3.1.1 Tables of Safe Working Loads.--For convenience and simplicity in design, the more important commercial species were classified into four groups in accordance with their strength in connector joints; all species in any one group take the same working values. The group numbers of these species are given in table 6-1; group 1 woods provide the weakest and group 4 woods the strongest connector joints.

In tabulating the safe working load for a connector joint of any number of members, the unit is one connector with a bolt in shear. For any joint assembly in which more than one connector unit is used in the contact faces with the same bolt axis, the total safe working load is the sum of the safe working loads of each connector unit. For example, in tables 6-2, 6-3, and 6-4, minimum actual thickness of the members is given for a joint assembly of three members employing two connectors in opposite faces with a common bolt; this assembly is equivalent to two connector units, and, therefore, the safe working load will be twice the corresponding value shown for a one-connector assembly. The loads as given apply only when the joints are properly designed with respect to such features as centering of connectors, adequate end distances, and suitable spacing of connectors.

6.2.3.2 Modification of Working Loads.--Some of the factors that affect the safe working loads of connectors were included in deriving the tabular values. Others require modification of the values listed in accordance with the provisions outlined in the following paragraphs. The modifications given in sections 6.2.3.2.1 and 6.2.3.2.2 are essentially the same as those for lumber (sec. 5.3.2).

6.2.3.2.1 Wind Loads.--In designing for wind forces acting alone, or acting in conjunction with dead and live loads, the safe working loads for the various connectors may be increased by the following percentages, provided the number and size of connectors is not less than that required for the combination of dead and live loads alone (sec. 6.2.3.2.7).

Type of connector	Increase
	Percent
Split-ring connector, any size, bearing in any direction	50
Shear-plate connector,[3] any size, bearing parallel to grain	33-1/2
Shear-plate connector,[3] any size, bearing perpendicular to grain	50
Toothed-ring connector, 2-inch, bearing in any direction	50
Toothed-ring connector, 4-inch, bearing in any direction	25

[3] See footnote 1, table 6-5.

6.2.3.2.2 Impact Forces.--Impact may be disregarded up to the following percentage of the static effect of the live load producing the impact (sec. 6.2.3.2.7).

Type of connector	Impact Allowance
	Percent
Split-ring connector, any size, bearing in any direction	100
Shear-plate connector,[4] any size, bearing parallel to grain	66-2/3
Shear-plate connector,[4] any size, bearing perpendicular to grain	100
Toothed-ring connector, 2-inch, bearing in any direction	100
Toothed-ring connector, 4-inch, bearing in any direction	50

[4] See footnote 1, table 6-5.

One-half of any impact load that remains after disregarding the percentages indicated should be included with the other dead and live loads in obtaining the total force to be considered in designing the joint.

6.2.3.2.3 Factor of Safety.--The procedures described for increasing the allowable loads on connectors for forces suddenly applied and forces of short duration do not reduce the actual factor of safety of the joint but are recommended because of the favorable behavior of wood under such forces. The differentiation among types and sizes of connector and directions of bearing is due to variations in the extent to which distortion of the metal, as well as the strength of the wood, affects the ultimate strength of the joint.

6.2.3.2.4 Special Design Considerations.--Conditions of design may be encountered, with respect to the kind of load on a structure and the period of its continuation, that are neither "long continued" nor "suddenly applied." These conditions require or justify special consideration and possible modifications, other than those that have been indicated, of the working loads listed in tables 6-3, 6-4, and 6-5. For such conditions, it may be assumed that 90 and 80 percent of the stress that causes failure in 5 minutes (time usually assumed for wind load) will cause failure in 50 minutes and 10 hours, respectively (sec. 5.3.2).

6.2.3.2.5 Exposure and Moisture Condition of Wood.--The loads listed in tables 6-3, 6-4, and 6-5, apply to seasoned timbers used where they will remain dry. If the timbers will be more or less continuously damp or wet in use, two-thirds of the tabulated values should be used. The amount by which the loads should be reduced to adapt them to other conditions of use depends upon the extent to which the exposure favors decay, the required life of the structure or part, the frequency and thoroughness of inspection, the original cost and the cost of replacements, the proportion of sapwood and the durability of the heartwood of the species if untreated, and the character and efficiency of the treatment if treated. These factors should be evaluated for each individual design. Industry recommendations for the use of connectors when the lumber is neither continuously wet nor continuously dry are given in reference 6-3.

6.2.3.2.6 Quality of Lumber.--With the recommended safe loads for connectors, it is assumed that the wood at the joints is clear and relatively free from checks, shakes, and splits. The wood should be either free from knots, or, if knots are assumed to be present in the longitudinal projection of the net section within a length from the critical section of one-half the diameter of the connector, the area of the knots should be subtracted from the area of the critical section. It is also assumed that cross grain at the joint does not exceed a slope of 1 in 10.

6.2.3.2.7 Loads at Angle with Grain.--The safe working loads for the split-ring and shear-plate connectors for intervening angles of 0° to 90° between direction of load and grain may be obtained by the nomograph in figure 6-2. With the toothed connectors, the safe working load at an inclination to the grain of 0° to 45° may be obtained with the previously mentioned formula; but from 45° to 90° it is equal to the working load perpendicular to the grain.

6.2.3.2.8 Thickness of Member.--The relationship between the loads for the different thicknesses of lumber is based on test results for connector joints. The least thickness of member given in tables 6-3, 6-4, and 6-5 for the various sizes of connectors is, in general, the minimum that should be used. The loads listed for the greatest thickness of member in each type and size of connector unit are the maximum loads to be used for all thicker material. The loads for wood members of thicknesses intermediate to those listed can be obtained by direct interpolation.

6.2.3.2.9 Width of Member.--The width of member listed for each type and size of connector is the minimum that should be used. When the connectors are bearing parallel to the grain, no increase in load occurs with an increase in width over the minimum. When they are bearing perpendicular to the grain, the load increases about 10 percent for each 1-inch increase in width of member over the minimum widths required for each type and size of connector, up to twice the diameter of the connectors. When the connector is placed off center and the load is applied continuously in one direction only, the proper working load can be determined by considering the width of member as equal to twice the edge distance (the distance between the center of the connector and the edge of the member toward which the load is acting), but the distance between the center of the connector and the opposite edge should not be less than one-half the permissible minimum width of the member.

6.2.3.2.10 End Distance and Spacing.--The working load values in tables 6-3, 6-4, and 6-5, apply when the distance of the connector from the end of the member (end distance e) and the spacing (s) between connectors in multiple joints are not factors in the strength of the joint (fig. 6-6, A). When the end distance or spacing for connectors bearing parallel to the grain is less than that required to develop the full load, the proper reduced working load for design may be obtained by multiplying the working loads in tables 6-3, 6-4, and 6-5, by the appropriate strength ratio given in table 6-6. For example, the load for a 4-inch split-ring connector bearing parallel to the grain, when placed 7 or more inches from the end of a Douglas-fir tension member that is 1-5/8 inches thick, is 4,780 pounds. When the end distance is only 5-1/4 inches, the strength ratio obtained by direct interpolation from the values given in table 6-6 is 0.81 and the load equals 0.81 times 4,780 or 3,870 pounds.

6.2.3.2.11 Placement of Multiple Connectors.--Preliminary investigations of the placement of connectors in a multiple joint, together with the observed behavior of single connector joints tested with variables that simulate those in a multiple joint, furnish a basis for some suggested design practices.

When two or more connectors in the same face of a member are in a line at right angles to the grain of the member and are bearing parallel to the grain (fig. 6-6, C), the clear distance (c) between the connectors should not be less than one-half inch.

When two or more connectors are acting perpendicular to the grain and are spaced on a line at right angles to the length of the member (fig. 6-6, B), the rules for the width of member and edge distances used with one connector are applicable to the edge distances for multiple connectors. The clear distance between the connectors (c) should be equal to the clear distance from the edge of the timber toward which the load is acting to the connector nearest this edge (c).

In a joint with two or more connectors spaced on a line parallel to the grain and with the load acting perpendicular to the grain (fig. 6-6, D), the available data indicate that the clear spacing between adjacent connectors (c) should not be less than 1 inch and that the total load used should be equal to the full load of one connector, plus one-third this amount for each additional connector. In a joint of this type, somewhat more favorable results are obtained in tests if the connectors are staggered so that they do not act along the same line with respect to the grain of the transverse member.

The placement of connectors in joints with members at right angles to each other is subject to the minimum limitation of either member. It is virtually impossible to set up general rules regarding the alinement, spacing, and margin of connectors to cover all possible directions of the applied load. The designer must rely upon a sense of proportion and fitness in applying the rules set forth to a condition of loading that is within the limits discussed.

6.2.3.2.12 Cross Bolts.--Cross bolts placed at or near the end of timbers joined with connectors or at intermediate panel points will provide additional safety. They may also be used to reinforce members that have, through change in moisture content in service, developed checks to an undesirable degree.

6.2.3.2.13 Net Section.--The stress in the net area (whether in tension or compression), which is the area remaining at the critical section after subtracting the projected area of the connectors and bolt from the full cross-sectional area of the member, should not exceed the safe stress of clear wood in compression parallel to the grain. In using this stress, it is assumed that knots do not occur in the longitudinal projection of the net section within a length of one-half the diameter of the connector from it.

6.2.4 Wood Screws.--The common types of wood screws have flat, oval, or round heads. The flathead screw is most commonly used if a flush surface is desired. Ovalhead and roundhead screws are used for appearance or when countersinking is objectionable. Besides the head, the principal parts of a screw are the shank, thread, and core (fig. 6-7).

6.2.4.1 Withdrawal Resistance of Wood Screws.--The resistance of steel wood screws to withdrawal from the side grain of seasoned wood varies directly with the square of the specific gravity of the wood. Within certain limits, it also varies directly with the depth of penetration and the diameter of the screw. The type of metal used does not as a rule materially affect withdrawal resistance.

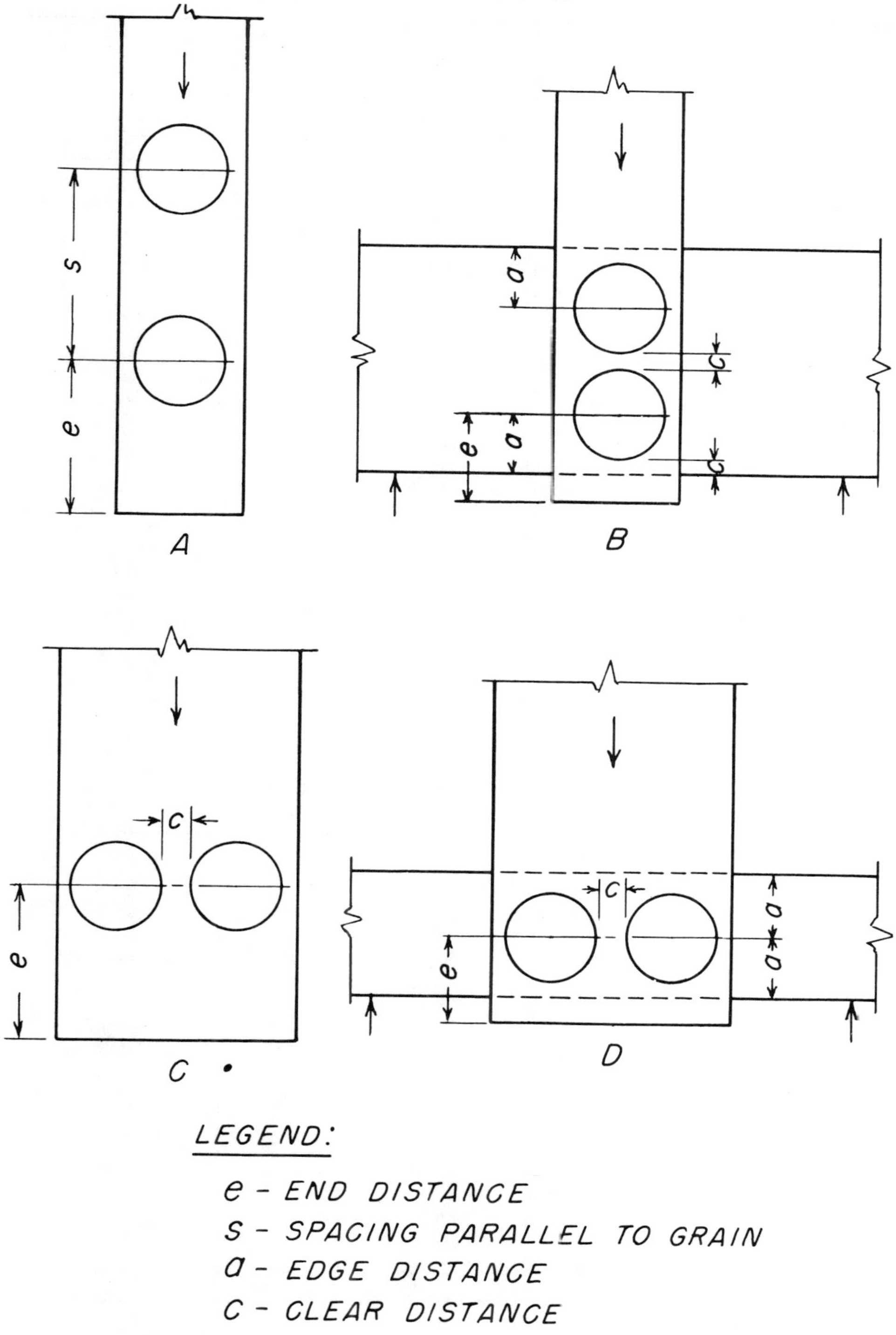

Figure 6-6.--Types of multiple-connector joints: A, Joint strength dependent upon end distance e and connector spacing s; B, joint strength dependent upon end e, clear c, and edge a distances; C, joint strength dependent upon end e and clear c distances; D, joint strength dependent upon end e, clear c, and edge a distances.

94

(M 39253 F)

Table 6-6.--Strength ratios for connectors at various longitudinal spacings and end distances[1]

Type of connector	Diameter of connector	Spacing[2]	Strength ratio	End distance[3] Tension member	End distance[3] Compression member	Strength ratio
1	2	3	4	5	6	7
	In.	In.	Percent	In.	In.	Percent
Split-ring	2-1/2	6-3/4+	100	5-1/2+	4+	100
	2-1/2	3-3/8	50	2-3/4	2-1/2	62
Split-ring	4	9+	100	7+	5-1/2+	100
	4	4-7/8	50	3-1/2	3-1/4	62
Shear-plate	2-5/8	6-3/4+	100	5-1/2+	4+	100
	2-5/8	3-3/8	50	2-3/4	2-1/2	62
Shear-plate	4	9+	100	7+	5-1/2+	100
	4	4-1/2	50	3-1/2	3-1/4	62
Toothed-ring	2	4+	100	3-1/2+	2+	100
	2	2	50	2		67
Toothed-ring	2-5/8	5-1/4+	100	4-5/8+	2-5/8+	100
	2-5/8	2-5/8	50	2-5/8		67
Toothed-ring	3-3/8	6-3/4+	100	5-7/8+	3-3/8+	100
	3-3/8	3-3/8	50	3-3/8		67
Toothed-ring	4	8+	100	7+	4+	100
	4	4	50	4		67

[1] Strength ratio for spacings and end distances intermediate to those listed may be obtained by interpolation, and multiplied by the loads in tables 6-3, 6-4, and 6-5, to obtain design load. The strength ratio applies only to those connector units affected by the respective spacings or end distances. The spacings and end distances should not be less than the minimum shown.

[2] Spacing is distance from center to center of connectors.

[3] End distance is distance from center of connector to end of member.

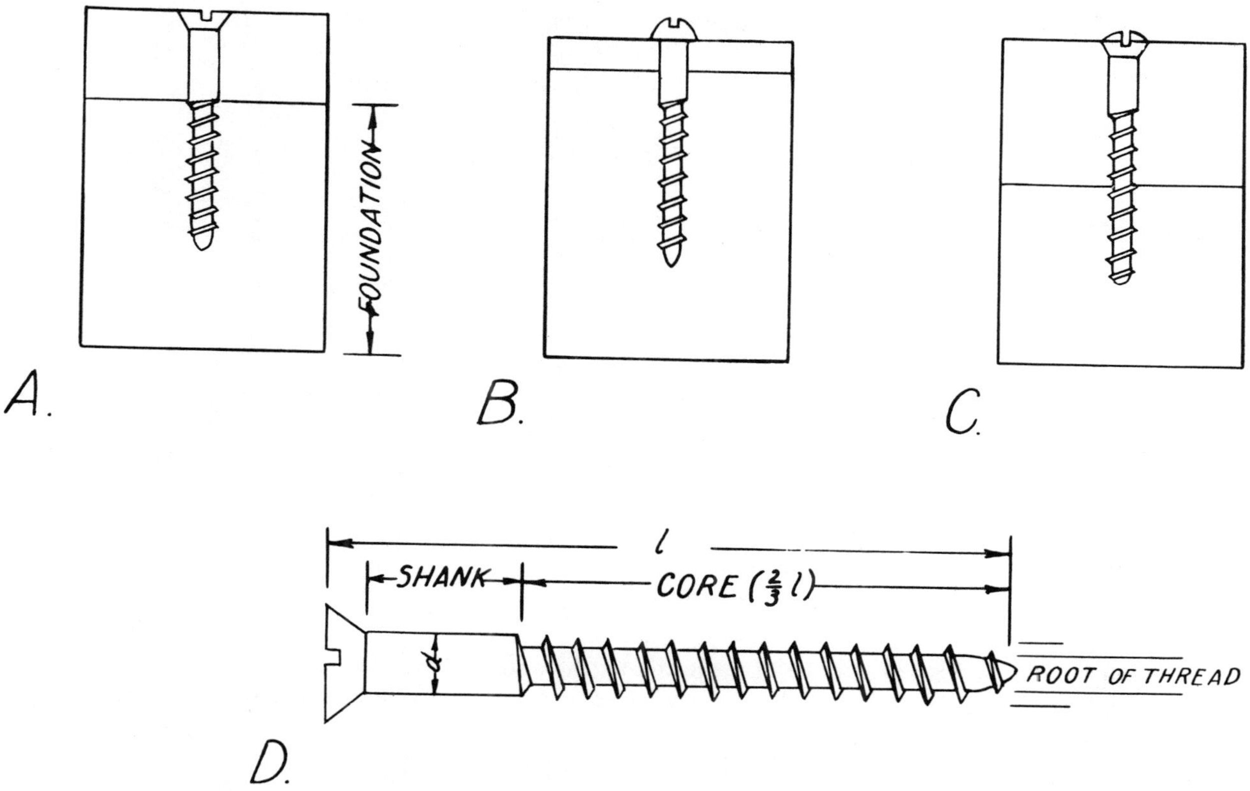

TYPES OF WOOD SCREWS
A, FLAT-HEAD; B, ROUND-HEAD; C, OVAL-HEAD.
A SHOWS A SCREW PROPERLY FASTENING AN ATTACHED MEMBER TO ITS FOUNDATION.
B AND C SHOW INCORRECTLY INSERTED SCREWS.
D SHOWS PRINCIPAL PARTS AND DIMENSIONS OF A WOOD SCREW.

Figure 6-7.--Types of wood screws: A, Flathead; B, round-head; C, ovalhead; D, parts and dimensions, l representing length and d shank diameter. Screws inserted as shown in A have maximum resistance to direct withdrawal; B, maximum lateral resistance; C, relatively low in both.

(M 58167 F)

The allowable withdrawal load of common wood screws inserted into the side grain of seasoned wood may be expressed as follows:

$$p = 2,370\ G^2D$$

in which p is the allowable withdrawal load (one-sixth of the ultimate load) per lineal inch of penetration of the threaded portion of the screw; G, the specific gravity based on weight and volume when ovendry; and D, the shank diameter of the screw in inches. Specific gravity values raised to the second power are given for a number of widely used wood species in table 6-1.

This formula applies when the screws are inserted into lead holes with a diameter of about 70 percent of the root diameter of the threads for softwoods and about 90 percent for hardwoods. The root diameter for most sizes of screws averages about two-thirds of the shank diameter.

The equation values are applicable to screws of the following sizes:

Screw length	Gage limits
Inches	
1/2	1-6
3/4	2-11
1	3-12
1-1/2	5-14
2	7-16
2-1/2	9-18
3	12-20

For lengths and gages outside of these limits, the actual values are likely to be less than the equation values. The withdrawal loads of screws from end grain average 75 percent of side-grain loads.

6.2.4.2 Lateral Resistance of Wood Screws.--The allowable load in lateral resistance for steel wood screws in the side grain of seasoned wood is given by the formula:

$$p = KD^2$$

in which p is the allowable lateral load in pounds, D the diameter of the screw shank in inches, and K a constant depending on species. Values for converting screw gage into D^2 are given in the following tabulation:

Screw number or gage	D	D^2
0	0.060	0.0036
1	.073	.0053
2	.086	.0074
3	.099	.0098
4	.112	.0125
5	.125	.0156
6	.138	.0190
7	.151	.0228
8	.164	.0269
9	.177	.0313
10	.190	.0361
11	.203	.0412
12	.216	.0467
14	.242	.0586
16	.268	.0718
18	.294	.0864
20	.320	.1024
24	.372	.1384

Values of K for a number of species are given in table 6-1. They apply to wood at about 15 percent moisture content. Loads computed by substituting these constants in the equation are expected to allow a slip of from 0.007 to 0.01 inch, depending somewhat on the species and quality of the wood.

The lateral resistance of screws is affected by the properties of the metal used to make them.

The formula applies when the depth of penetration of the screw into the block receiving the point is not less than 7 times the shank diameter and when the cleat and the block holding the point are approximately of the same density. This depth of penetration gives an ultimate load of about 6 times the recommended allowable load. For lesser depths of penetration, the ultimate load is reduced about in proportion to the reduction in penetration. When penetration depth in the holding block is 4 times the shank diameter, the maximum load will be less than 4 times the load expressed by the formula, and the proportional-limit load will be approximately equal to that given by the formula. When screws hold metal to wood, allowable loads can be increased by about 25 percent.

The formula applies when the part of the lead hole receiving the shank, for hardwoods such as oak, has the same diameter as the shank, and that receiving the threaded portion has the same diameter as the root of the thread. For softwoods such as southern yellow pine and Douglas-fir, the lead hole should have about seven-eighths the diameter of the shank and root, respectively.

6.2.5 Lag Screws.--Lag screws are commonly used because of their convenience, particularly where it would be difficult to fasten a bolt or where a nut on the surface would be objectionable. Lag screws range from about 0.2 to 1 inch in diameter and from 1 to 16 inches in length. The threaded portion ranges from 3/4 inch with the 1- and 1-1/4-inch screws to half the length for all lengths greater than 10 inches. The equations given herein for withdrawal and lateral loads are based on lag screws having an average yield point of about 45,000 pounds per square inch and an average tensile strength of 77,000 pounds per square inch. For lag screws made of metal having greater or lower yield points and tensile strength, the withdrawal loads should be adjusted in proportion to the tensile strength and the lateral loads in proportion to the square root of the yield-point stresses.

6.2.5.1 Withdrawal Resistance of Lag Screws.--The allowable load in direct withdrawal of lag screws from seasoned wood may be computed from the equation:

$$p = 1,500\, G^{3/2}\, D^{3/4}$$

in which p is the allowable withdrawal load (one-fifth of the ultimate load) in pounds per inch of penetration of the threaded portion; D, the shank diameter in inches; and G, the specific gravity of the wood based on weight and volume when ovendry (ref. 6-6).

Lag screws, like wood screws, require prebored holes of the proper size (fig. 6-8). The lead hole for the shank should be of the same diameter as that of the shank. The diameter of the lead hole for the threaded portion varies with the density of the wood: For lightweight softwoods, such as the cedars and white pines, 40 to 70 percent of the shank diameter; for Douglas-fir and southern yellow pine, 60 to 75 percent; and for dense hardwoods, such as the oaks, 65 to 85 percent. The smaller percentage in each range applies to lag screws of the smaller diameters, and the larger percentage to lag screws of larger diameters. Lead holes slightly larger than those recommended for maximum efficiency should be used with lag screws of excessive length.

In determining the withdrawal resistance, the allowable tensile strength of the lag screw at the net (root) section should not be exceeded. Penetration of the threaded portion to a distance about 7 times the shank diameter in the denser species and 10 to 12 times the shank diameter in the less dense species will develop approximately the ultimate tensile strength of the lag screw.

The resistance of a lag screw to withdrawal from end grain is about three-fourths that from side grain.

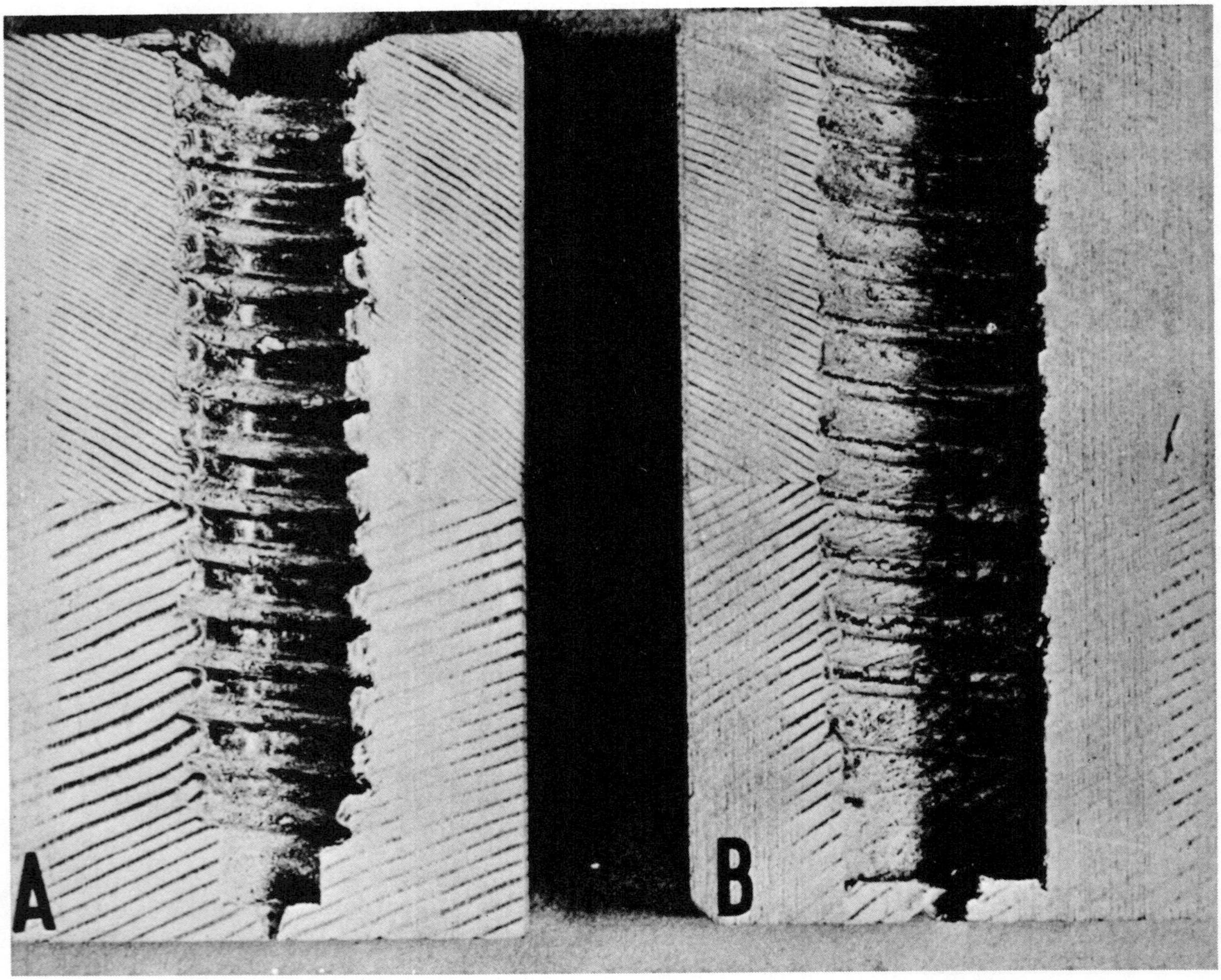

Figure 6-8.--A, Clean-cut, deep penetration of thread made by lag screw turned into a lead hole of proper size; B, rough shallow penetration of thread made by lag screw turned into over-sized lead hole.

(M 24964 F)

6.2.5.2 Lateral Resistance of Lag Screws.--The allowable lateral loads for lag screws inserted in the side grain and loaded parallel to the grain of a piece of seasoned wood can be computed from the equation:

$$p = KD^2$$

in which p is the allowable lateral load parallel to the grain in pounds, K a constant depending on the species, and D the shank diameter of the lag screw in inches. Values for K for a number of species may be found in table 6-1. The values given by this formula apply when the thickness of the attached member is 3.5 times the shank diameter of the lag screw. For other thicknesses, the computed allowable loads should be multiplied by the following factors:

Ratio of thickness of member to shank diameter of lag screw	Factor
2	0.62
2-1/2	.77
3	.93
3-1/2	1.00
4	1.07
4-1-2	1.13
5	1.18
5-1/2	1.21
6	1.22
6-1/2	1.22

When the lag screw is inserted into the side grain of wood and the load is applied perpendicular to the grain, the allowable load given by the lateral-resistance formula should be multiplied by the following appropriate factors:

Shank diameter of lag screw	Factor
Inches	
3/16	1.00
1/4	.97
5/16	.85
3/8	.76
7/16	.70
1/2	.65
5/8	.60
3/4	.55
7/8	.52
1	.50

For other angles of loading, the allowable loads may be computed from the parallel and perpendicular values by the use of the Scholten nomograph for determining the bearing strength of wood at various angles to the grain (fig. 6-2).

Example: P, the allowable load parallel to grain is 6,000 pounds, and Q, the allowable load perpendicular to the grain, is 2,000 pounds. N, the allowable load at an angle of 40° to grain is found as follows: Connect with a straight line 6,000 pounds (a) on line OX of the nomograph with the intersection (b) on line OY of a vertical line through 2,000 pounds. The point where this line (ab) intersects the line representing the given angle 40° is directly above the allowable load, 3,285 pounds.

Working values for lateral resistance as computed by the preceding methods are based on complete penetration of the shank into the attached member but not into the foundation member. When the shank penetrates the foundation member, the following increases in allowable loads are permitted:

Ratio of penetration of shank into foundation member to shank diameter	Increase in load
	Percent
1	8
2	17
3	26
4	33
5	36
6	38
7	39

When lag screws are used with metal plates, the allowable lateral loads parallel to the grain may be increased 25 percent, but no increase is allowed for load applied perpendicular to the grain.

Lag screws should preferably not be driven into end grain, because splitting may develop under lateral load. If lag screws are so used, however, the allowable loads should be taken as two-thirds the lateral resistance of lag screws in side grain to loads acting perpendicular to the grain.

The spacings, end and edge distances, and net section for lag-screw joints should be the same as those for joints with bolts of a diameter equal to the shank diameter of the lag screw (sec. 6.2.1.5).

6.2.6 Nails and Spikes.--For several reasons, common wire nails are not extensively used in ship and boat building. The most important reason is that they tend to lose withdrawal resistance seriously in wood that undergoes wide changes in moisture content. Alternate wetting and drying of wood can reduce the withdrawal resistance of such nails by as much as one-half unless they are clinched or riveted. Besides the loss in strength, this loosening is serious in watercraft because it allows joints to loosen and leak.

Various other types of nails, however, have been developed for ordinary or special uses, including annular grooved, spirally grooved, and coated or etched nails. Under conditions involving changes in the moisture content of the wood, annular grooved and spirally grooved nails provide considerably greater withdrawal resistance than the common wire nail.

Corrosion is another serious shortcoming of common wire nails. To overcome it, they are generally coated with zinc or cadmium, or made of copper or brass. The less corrosive metals are not so strong as steel, but corrosion seriously weakens steel nails in far less time than the life of the craft, hence their immediate strength advantage is lost. Even non-corrosive coatings are not complete guarantees against corrosion.

The fact that lead holes must be bored for nails in the denser woods widely used in boat building, such as white oak for frames and keels, also largely offsets the labor-saving advantage of nails over screws and bolts.

To overcome the loss in withdrawal resistance that comes with frequent wetting and drying of the wood, boat nails are often used rivet fashion, the driven ends being peened over burrs. Riveting prevents gradual withdrawal of the nail as moisture content fluctuates in the wood parts held together, thus reducing joint distortion and thereby maintaining a tighter joint. Some types of boat nails are also shank-grooved (fig. 6-9).

The following discussion of nails is based on tests of common steel wire nails as ordinarily used in general construction. Adaptation of test results to nails as used in ships and boats is indicated where possible.

6.2.6.1 Withdrawal Resistance of Nails.--The resistance of a nail to direct withdrawal from a piece of wood is intimately related to the specific gravity of the wood, the diameter of the nail, and the depth of penetration. The surface condition of the nail at the time of driving also influences the immediate withdrawal resistance.

For bright, common wire nails driven into the side grain of seasoned wood that will remain dry, or into unseasoned wood that will remain wet, the allowable withdrawal load is given by the formula:

$$p = 1,150\, G^{5/2} D$$

in which p represents the allowable load (one-sixth of the ultimate load) per lineal inch of penetration in the member holding the nail point; G, the specific gravity of the wood based on over-dry weight and volume; and D, the diameter of the nail in inches. The average specific gravity of a number of the common species raised to the 5/2 power is listed in table 6-1. The diameters of various penny or gage sizes of nails are given in table 6-7.

Table 6-7.--Sizes of bright common wire nails

Penny	Gage	Length	Diameter D	$D^{3/2}$
		Inch	Inch	Inch
2d	15	1	0.072	0.0193
4d	12-1/2	1-1/2	.098	.0307
6d	11-1/2	2	.113	.0380
8d	10-1/4	2-1/2	.131	.0474
10d	9	3	.148	.0570
12d	9	3-1/4	.148	.0570
16d	8	3-1/2	.162	.0652
20d	6	4	.192	.0841
30d	5	4-1/2	.207	.0942
40d	4	5	.225	.1068
50d	3	5-1/2	.244	.1205
60d	2	6	.262	.1342

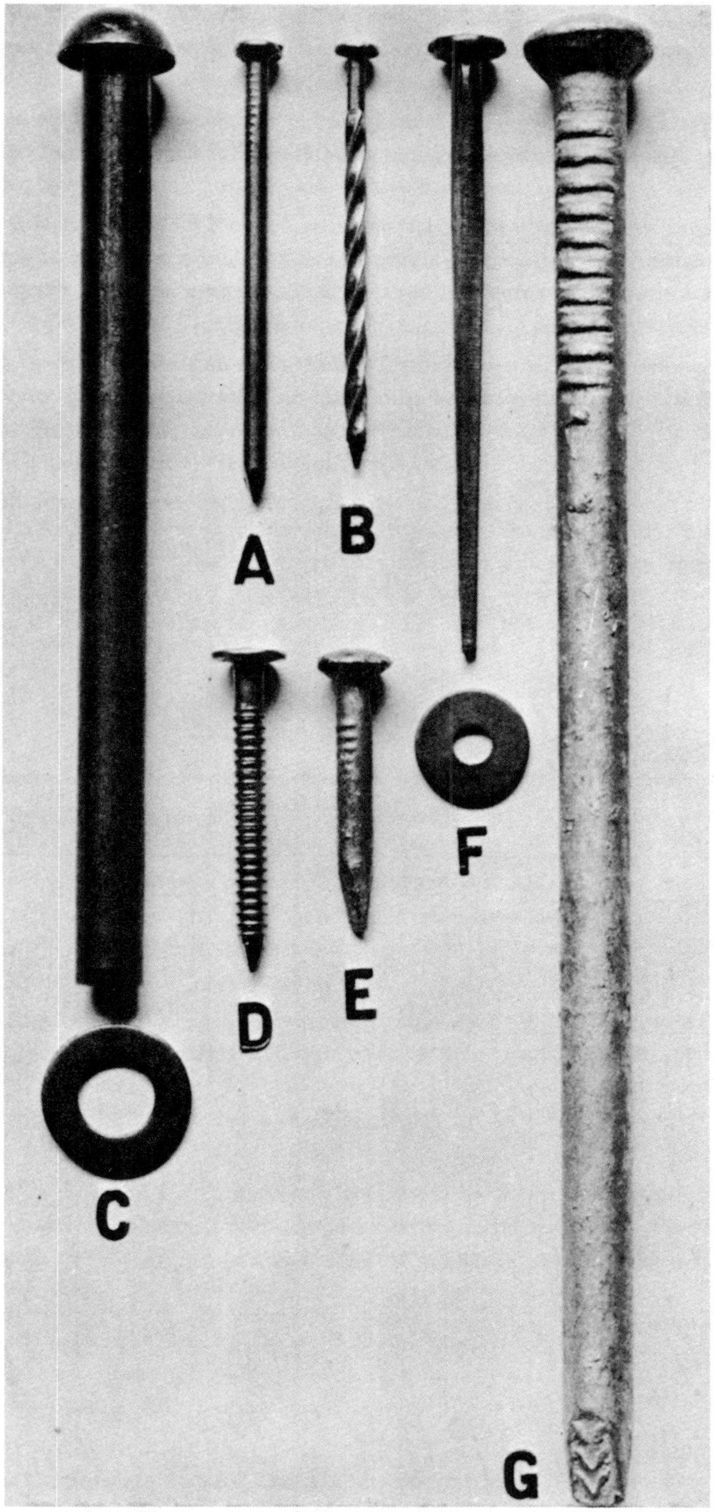

Figure 6-9.--Types of nails. A, Common steel wire nail; B, spiral-grooved steel wire nail; C, bolt-type brass rivet with burr; D, cadmium-plated boat nail with helically grooved shank and diamond point; E, chisel-pointed, zinc-coated boat nail; F, square-cut copper boat nail with rivet burr; G, zinc-coated boat spike with chisel point.

(M 105 826)

6.2.6.2 Factors That Affect Withdrawal Resistance of Nails.

6.2.6.2.1 Surface Condition.--The surface condition of nails can be modified during manufacture to improve their withdrawal resistance. Methods used include surface coating, surface roughening, or deformation of the shanks.

6.2.6.2.2 Surface-Coated Nails.-- Withdrawal resistance of nails with noncorroding coatings, such as zinc, may be increased, but extreme irregularities of the coating may actually reduce it. The advantage of zinc-coated nails is usually eliminated by repeated wetting and drying of the wood.

6.2.6.2.3 Surface-Roughened Nails.--Chemically etched nails give somewhat higher withdrawal resistance than other coated nails and remain superior under varying moisture conditions. Sand-blasted nails perform in much the same manner as chemically etched nails.

6.2.6.2.4 Form of Shank.--Nails with spirally grooved, annular grooved, and other irregular shanks (fig. 6-9) have generally greater withdrawal resistance than common wire nails, especially if driven into green wood that subsequently seasons. In general, annular grooved nails sustain larger withdrawal loads, and spirally grooved nails sustain greater impact withdrawal work values than the other nail forms.

Square-cut boat nails are usually riveted, which greatly increases their withdrawal resistance.

6.2.6.2.5 Nail Composition.--In general, the withdrawal resistance of copper nails is somewhat comparable to that of the common wire nails when pulled soon after driving. Their comparative freedom from corrosion, however, warrants the assumption that they will have a somewhat longer service life. Nails of other corrosion-resisting alloys, such as aluminum, brass, and stainless steel, are rarely used in boats. The inherent tensile strength of the alloy is usually great enough not to be a factor in withdrawal resistance.

6.2.6.2.6 Clinched Nails.--The withdrawal resistance of clinched nails is from 45 to 170 percent higher than that of unclinched nails. It varies with the species of wood, the size of nail, and the direction of clinch with respect to the grain of the wood. Nails clinched across the grain have approximately 20 percent more resistance to withdrawal than nails clinched along the grain.

6.2.6.2.7 Nails in Plywood.--The nailing characteristics of plywood are not greatly different from those of solid wood except for plywood's greater resistance to splitting when nails are driven near an edge. The nail-withdrawal resistance of plywood is from 15 to 30 percent less than that of solid wood of the same thickness, because fiber distortion is less uniform in plywood.

6.2.6.3 Lateral Resistance of Nails.--The allowable load for a bright, common wire nail in lateral resistance when driven into the side grain of seasoned wood is given by the formula:

$$p = KD^{3/2}$$

where p represents the allowable lateral load in pounds per nail; K, a constant; and D, the diameter of the nail in inches. Values of K for a number of the common hardwoods and softwoods are listed in table 6-1. Values for converting the size of common wire nails from penny to $D^{3/2}$ are listed in table 6-7.

These loads apply only for conditions where the side member and the member holding the nail point are of approximately the same density and where the depth of penetration of the nail in the member holding the point is not less than 10 times the diameter of the nail for dense woods and 14 times the diameter for lightweight woods. When metal is held to wood, an increase of about 25 percent can be applied to the allowable lateral nail load.

The safe lateral loads for end-grain nailing should be about 60 percent of those computed for side-grain nailing in lightweight species and slightly higher in denser species. The allowable lateral loads should be reduced 25 percent for unseasoned wood that will remain wet or will be loaded before seasoning takes place.

The erratic behavior of a nailed joint that has undergone one or more moisture content changes makes it difficult to recommend an allowable lateral load for a nailed joint under these conditions. If inspection shows that a loss in joint strength occurred during drying, the joint should be reinforced with additional nails. Simply setting the original nails is not sufficient.

Riveting greatly increases maximum loads nails can carry, but usually maximum load is reached only after the joint has been excessively distorted. For loads within allowable distortions, the advantage of riveting is much smaller. The chief advantage of riveting is that it maintains a tighter joint under normal loadings. Similarly, deformed-shank nails carry somewhat higher maximum lateral loads than common wire nails, but both types of nails perform similarly at small distortions in the joint.

Nails used to fasten two members to a third, as, for example, the inner and outer shear clamps to the frame heads of a boat, work in double shear and thus can carry about twice the lateral loads they can in single shear (two members fastened together).

6.2.6.4 Spikes.--Common wire spikes are manufactured in the same manner as common wire nails. They have either a chisel point or a diamond point and are made in lengths of 3 to 12 inches. For corresponding lengths (3 to 6 inches), they have larger diameters than the common wire nails, and beyond the sixtypenny size they are usually designated by inches of length.

The allowable withdrawal and lateral resistance formulas and limitations given for common wire nails are also applicable to spikes, except that, in calculating the withdrawal load for spikes, the depth of penetration should be reduced by two-thirds the length of the point.

6.3 GLUED JOINTS.

6.3.1 GENERAL.--Glued joints in ships and boats must develop the full strength of the wood under all conditions of service to be satisfactory. The production of such joints requires special waterproof adhesives and exacting bonding techniques.

Glued joints most commonly will be between pieces whose grain directions are parallel (as between the laminations of a beam) or between solid or laminated boards and plywood, with the face grain of the plywood either parallel or at right angles to the grain direction of the wood. In special cases, material in lumber thicknesses may be glued with the grain direction of adjacent pieces at an angle. If the angle is large, dimensional changes resulting from moisture changes will set up large stresses in the glued joint and the strength of the joint may be considerably reduced over a period of time. Data are not available from which to predict the magnitude of the strength reduction in such joints. Their use, therefore, is not suggested in ship construction, where considerable variations in moisture content are to be expected.

In joints connected with plywood gusset plates, this shrinkage differential is minimized because plywood shrinks and swells much less than solid wood. Typical gusseted joints are those used in V-bottom frames.

It is practically impossible to make glued joints between end-grain surfaces sufficiently strong or durable to meet the requirements of even ordinary service. With the most careful gluing possible, not more than about 25 percent of the tensile strength of the wood can be obtained in butt joints. For end-to-end joining, therefore, scarf joints must be used to obtain a high proportion of the strength of the wood.

Joints of end grain to side grain are also difficult to glue properly. Furthermore, they are subjected in service to severe stresses as a result of unequal dimensional changes in the two members of the joint with changes in moisture content. It is therefore necessary to use irregular shapes of joints, dowels, tenons, or other devices to reinforce such a joint in order to bring side grain into contact with side grain to secure larger gluing surfaces. All such joints should be carefully protected from changes in moisture content.

For these reasons, joints between end-grain surfaces and between end-grain and side-grain surfaces should not be used, especially where they are expected to carry load. Data by which to design end-grain to side-grain joints are not available.

6.3.2 Joints Between Pieces with Grain Parallel.--The allowable shear stress on the glue area for joints between pieces having parallel-grain gluing is equal to the allowable shear stress parallel to grain for the weaker species in the joint. This value will be found in tables 5-1 and 5-2 and should be used only when uniform stress distribution in the joint is assured.

6.3.3 Joints Between Wood and Plywood.--The allowable shear stress on the glue area between wood (solid or laminated) should be taken from table 3-9.

6.3.4 Joints Between Pieces with Grain at Angle.--The allowable shear stress on the glue area between pieces whose grain directions make an angle of other than 0° and 90° joints may be found by means of the diagram of figure 6-2, using allowable values for 0° and 90° joints as in the two preceding sections. When the angle between the grain direction of the adjacent pieces does not exceed 15°, the stress allowed for parallel-grain gluing, section 6.3.1, may be used without correction.

6.4 REFERENCES

(6-1) FREAS, A. D., and SELBO, M. L.
1954. Fabrication and Design of Glued Laminated Wood Structural Members. U.S. Dept. Agr. Tech. Bull. No. 1069.

(6-2) GOODELL, H. R., and PHILLIPS, R. S.
1944. Bolt-Bearing Strength of Wood and Modified Wood: Effect of Different Methods of Drilling Bolt Holes in Wood and Plywood. Forest Products Laboratory Report No. 1523.

(6-3) NATIONAL LUMBER MANUFACTURERS ASSOCIATION
1960. National Design Specification for Stress Grade Lumber and Its Fastenings. 68 pp.

(6-4) NEWLIN, J. A., and GAHAGAN, J. M.
1930. Tests of Large Timber Columns and Presentation of the Forest Products Laboratory Column Formula. U.S. Dept. Agr. Tech. Bull. No. 167, 44 pp., illus.

(6-5) NEWLIN, J. A., HECK, G. E., and MARCH, H. W.
1934. New Method for Calculating Longitudinal Shear in Checked Wooden Beams. Transactions, American Society of Mechanical Engineers, Vol. 56, No. 10.

(6-6) NEWLIN, J. A., and GAHAGAN, J. M.
1938. Lag Screw Joints: Their Behavior and Design. U.S. Dept. Agr. Tech. Bull. No. 597.

(6-7) SCHOLTEN, J. A.
1944. Timber Connector Joints, Their Strength and Design. U.S. Dept. Agr. Tech. Bull. 865.

(6-8) SMITH, C. B., and VOSS, A. W.
1948. Stress Distribution in a Beam of Orthotropic Material Subjected to a Concentrated Load. National Advisory Committee for Aeronautics, Tech. Note No. 1486, 37 pp., illus.

(6-9) TIMBER ENGINEERING COMPANY
1952. TECO Design Manual for TECO Timber Connector Construction. Washington, D. C. 27 pp., illus.

(6-10) TRAYER, G. W.
1932. Bearing Strength of Wood Under Bolts. U.S. Dept. Agr. Tech. Bull. 332.

(6-11) U.S. FOREST PRODUCTS LABORATORY
1955. Wood Handbook. U. S. Dept. Agr. Agriculture Handbook No. 72, 523 pp., illus.

(6-12) WOOD, L. W.
1956. Formulas for Columns with Side Loads and Eccentricity. Forest Products Laboratory Report No. 1782, 21 pp., illus.

WOOD: A MANUAL FOR ITS USE AS A SHIPBUILDING MATERIAL

VOLUME IV

BOAT AND SHIP CONSTRUCTION TECHNIQUES

Prepared by

Bureau of Ships

Department of the Navy

With the cooperation of

Forest Products Laboratory

Forest Service

United States Department of Agriculture

DEPARTMENT OF THE NAVY
BUREAU OF SHIPS
WASHINGTON 25, D. C.

IN REPLY REFER TO

During World War II and the Korean action, more than 40,000 wooden ships, boats and landing craft were added to the Fleet. In the ensuing years, many additional wooden ships and boats have been put into service, incorporating the most recent knowledge of wood as an engineering material. A continuing program of improvement in wood laminating, extreme service adhesives and development of wood preservatives for protection against marine borers has been largely responsible for the satisfactory performance of these craft.

Because of the many new developments in marine uses of wood, a need has arisen for a comprehensive, up-to-date, collection of information based on recent investigations and tests. This new publication, Volume IV of "Wood: A Manual For Its Use As A Shipbuilding Material," the fourth and final volume of the series, should prove most useful to shipyards, naval architects, Supervisors of Shipbuilding, Inspector of Naval Material, and officers and those concerned with all phases of wooden ship and boat design, construction, or maintenance.

R. E. Harris

R. E. HARRIS

COASTAL MINESWEEPER,
MSC – 284 AT CRUISING SPEED,
PRELIMINARY TRIALS

ACKNOWLEDGMENTS

In an industry so old as shipbuilding, many men, most of them unknown, have contributed to our present day knowledge and advancement of the art and science. A notable early attempt to organize the improvement of shipbuilding was made by King James I in 1605 when he chartered a Shipwrights Guild with supervision over all shipbuilding in England, and which all English shipwrights were required to join. The Guild made significant progress under the leadership of Master Shipwright Phineas Pett who introduced the use of scale models and mathematically developed plans, making possible the building of two ships alike.

Later, the French took up the banner of progress and carried it well into the 1800's, surpassing the British. This era of shipbuilding witnessed the highest development of the wooden sailing ship in construction, speed, and beauty. During this time, the American forest contributed enormous quantities of shiptimbers, planking, decking, masts, and spars. American shipwrights achieved a reputation for skilled workmanship held to this day.

Shipbuilding began to emerge from an art to a science during the 19th century. Several schools of naval architecture were founded in England, the United States, and Germany, followed by such technical societies as the Institution of Engineers and Shipbuilders in Scotland, the British Institution of Naval Architects, and Institute of Marine Engineers, and in the United States, the Society of Naval Architects and Marine Engineers.

Thus, the modern science of wood shipbuilding was developed over the centuries. Most significant of the achievements during the 20th century has been the adoption of adhesive-bonded wood as an engineering material in the construction of wooden hull ships. The bonding together of small pieces of wood into large structural members or panels in either straight or curved form results in a far more efficient use of wood than was formerly possible. This development is the first basic change in wooden shipbuilding in 2,000 years.

Previous editions of this manual have mentioned many individual shipbuilders, researchers, and other contributors who furnished information regarding modern wood ship and boat building practices. The present edition represents the combined efforts of virtually all American technical societies, research agencies, trade associations, and universities which foster progress in wood technology, naval architecture, and timber engineering. It is the Bureau's desire to give full recognition to these groups and to the part played by the lumber, laminating, plywood, adhesives, and wood preservation industries in modern wood shipbuilding. Without their cumulative contributions to the advancement of wood as an engineering material, this manual would not be possible.

The basic manuscript for "Wood: A Manual for Its Use as a Shipbuilding Material" was prepared by the Forest Products Laboratory, Forest Service, United States Department of Agriculture, at the request of the Bureau of Ships, Department of the Navy. A very large share of the credit for the material presented in the manual is due the Laboratory, which has been foremost in the field of forest products for over 50 years. The Laboratory's individual staff members who have contributed to the manual are recognized authorities in their individual fields. Thanks are also due the staff of the Division of Forest Disease Research of the United States Forest Service, and the Timber Engineering Company Laboratory for their advice and guidance.

The manuscript was reviewed and edited for the Bureau of Ships by members of its staff. The final arrangement of the material for publication has been made under the direction of Mr. John G. Kuenzel who, until his retirement in early 1962, supervised wood products development for the Bureau of Ships.

TABLE OF CONTENTS

CHAPTER 2. GLUING AND LAMINATING (Continued)

CHAPTER 3. FABRICATION OF WOOD BOATS AND SHIPS

WOOD: A MANUAL FOR ITS USE AS A SHIPBUILDING MATERIAL

INTRODUCTION TO VOLUME IV

Purpose and Scope. The three chapters comprising Volume IV cover techniques applicable to the construction of wood hull boats and ships. Included are discussions of building ways, bending and machining wood and plywood, gluing, laminating, joints and fastenings, fabrication, preservative treatment, painting, and finishing.

The complete Manual is composed of four volumes, as follows:

Volume I Basic Wood Technology Applicable to Boat and Shipbuilding.

Volume II Techniques and Practices Applicable to Preservation and Storage.

Volume III Technical Data Applicable to Boat and Ship Design.

Volume IV Techniques Applicable to Boat and Ship Construction.

CHAPTER 1. BENDING AND MACHINING OF WOOD AND PLYWOOD

1.1 GENERAL.—Many of the members that make up the structural parts of a ship are curved. Since wood—available as naturally curved portions of trees, such as that found in crotches or stumps—is obviously limited it is necessary and desirable to bend, shape, or saw straight lumber to meet the various contour requirements of the vessel. Curved members also can be made by glue laminating (sec. 2.2). Shaping and sawing are wasteful of material and, because of the resulting cross grain, do not produce a piece of uniform strength along its length or of as great strength as is otherwise obtainable.

1.2 BENDING OF WOOD.—Within limits fixed by the thickness and species of wood used and the radius of curvature required, properly selected and prepared stock can be bent to produce parts of acceptable strength although weaker than those made of naturally curved wood or curved glue-laminated members. Wood bending has long been practiced, but the large percentages of breakage that occur in many bending operations give evidence that current practices often fall short.

Many members of wood ships can be satisfactorily bent, especially those of smaller types of ships which are relatively light. Some parts can be bent into place cold; for example, planking is usually forced or sprung into place in this manner unless the bends or twists are severe, in which case it is necessary to steam the planks or soak them in hot water before forcing them into position. Frames that do not require severe bending can be steamed and bent in place by fastening one end to the keel and forcing the part against stringers or ribbands outlining the ship's curvature. One disadvantage of this method is the tendency of the frame to change shape while cooling. For this reason it is preferable to hold the frame in position by temporary fastenings, such as clamps, until it has cooled and dried. Then the permanent fastenings can be installed. Severely bent parts must be curved over a form on a bending table, or in a bending machine, and fixed at the shape desired by cooling and drying them before fastening them in place. It is necessary to use metal tension straps when producing severe bends in order to counteract the tensile stresses set up on the convex side of the bent piece (sec. 1.2.5).

The production of bent wood parts may be divided into the following steps: (1) selection of stock; (2) seasoning of stock; (3) preparation of bending blanks; (4) softening of the wood; (5) the bending operation; and (6) fixing the bent shape.

1.2.1 Selection of Stock.—Although the wood of all species can be bent or sprung to some extent, where severe bends are necessary in solid stock only hardwoods can be used. Oak, elm, ash, hickory, and beech long have been used as bending stock. The white oaks are preferred for bent ship parts because of their greater resistance to decay.

Not all wood of even the preferred bending species is suitable. The grain of the wood should not exceed a slope of 1 in 15, except when bending straps are used and the bend is not too severe, a slope of 1 in 12 may suffice. Steeper slope of grain causes excessive breakage in bending; irregularities of grain around knots and burls may be as important as the grain of the piece as a whole, depending on their location. The wood must be free from decay. High-density wood is desirable because of its superior strength properties, but, with proper bending technique, wood of high density or of a particular rate of growth—as determined by the number of annual rings per inch of radius—is not essential to success in bending. Defects such as knots, wormholes, and surface checks tend to cause failures, but small, sound knots near the ends, moderate surface checking, and scattered wormholes are allowed. Wood containing shakes or pith is not suitable for bending. Generally, it is preferable to make the bend with the bark side of the piece acting as the tension face.

1.2.2 Seasoning of Stock.—Most shipbuilders prefer green stock for bending, but wood can be bent at moisture content values ranging from the green condition to 12 percent. Green stock possesses some advantages; it is free from surface checks, heats more quickly, and requires heating only to prepare it for bending. Seasoned stock requires some added moisture in the outer zones, as well as heat. Wood with an exceptionally high moisture content is not suitable for severe bending because hydrostatic pressure may develop on the side in contact with the form and cause objectionable compression failures. The optimum moisture content for bending purposes is probably in the vicinity of the fiber saturation point, 30 percent moisture content; but, although wood can be dried to this moisture content, it cannot be maintained in this condition for any period of time.

Wood for bending should be piled and protected in accordance with the recommendations in Volume II, Chapter 1, so that as little surface checking as possible will develop.

1.2.3 Preparation of Bending Stock.—The preparation of material for bending depends to some extent on the bending method used. If a tension strap with end blocks is to be used, the blank must be trimmed to uniform length, with the ends cut square. In all cases, the depth or thickness of the blank should be kept to a minimum, making due allowance for subsequent shrinkage during drying. To prevent side buckling, the width should be greater than the depth. Blanks wide enough for several units of the part being made are sometimes bent before being sawed to the required unit width. By surfacing the inner side, that next to the form, and the two adjacent faces, the danger of excessive compression failure will be reduced. Compression failure is likely only when a strap is used and, up to a certain point, is not objectionable.

1.2.4 Softening the Wood.—When wood is both wet and hot its plasticity is increased. This phenomenon is more marked in hardwoods than in conifers, thus hardwoods can be bent more successfully. The usual practice is to steam the wood at atmospheric pressure in a steam box, using wet steam, or to submerge it in hot—preferably boiling—water. From a technical standpoint, wood exposed to saturated steam at atmospheric pressure can be bent as well as wood softened with boiling water. The discharge of high-pressure steam into a steam box, however, can produce, under certain conditions, a drying atmosphere, thereby removing rather than adding moisture to the wood. Such drying is less likely to occur where low-pressure steam is used and where water is allowed to stand in the bottom of the steam box. The use of boiling water eliminates the possibility that the surfaces will lose moisture during the heating process.

The steaming or soaking time used by shipbuilders varies widely, but about 1 hour per inch of thickness is ordinarily sufficient. Green wood can be heated more rapidly than dry wood and does not require the addition of moisture to prepare it for bending. Oversteaming or overboiling is undesirable because the mechanical properties of the wood may be impaired by excessive heating. Some chemicals plasticize wood and have been used to a limited extent to soften it for bending purposes, but no chemical treatment has been found that is as effective as treatment with steam or boiling water.

1.2.5 The Bending Operation.—After it has been softened, the stock should be bent as quickly as possible to minimize the loss of heat and moisture from the inner and outer surfaces of the piece, which must withstand the maximum stress and which lose their plasticity first as it cools and dries. If the bend is made without a tension strap, the outer as well as the inner surface of the piece is stressed; if a strap is used, most of the stress is concentrated in the inner surface of the piece.

Mild bends can be made over frames or forms without using a strap. One end of the heated blank is fastened in place and the blank is then pulled against the form and clamped to it at intervals; sometimes it is necessary to clamp it only at the ends. A similar method is employed when members are bent to shape in the boat. Such bends are possible because wood will stretch 1 or 2 percent without failure. Excessive breakage usually can be corrected either by (1) more restrictive grading of the bending stock, or by (2) greater control over the bending process.

Severe bends require the use of a strap with end fittings or blocks to restrain the tension side, which is the outside of the bend (figure 1-1). The strap and end blocks hold the wood rigidly while it is being bent, compressing the inner, or concave, section of the piece and preventing the outer, or convex, section from stretching. By thus restricting the natural reaction of the wood to the bending stresses, a condition is produced within the plasticized wood that is called "upset." Upset is measured by the difference between the length of the outer, or convex, side and that of the inner, or concave, side, accomplished principally by compressing the inner side. This condition makes possible much more severe bends than could otherwise be produced. It is not practical to apply pressure to the ends of long, slender pieces with blocks because such end pressure may cause the pieces to buckle. Tension straps must be fastened at both ends of the bent section with clamps or some other device.

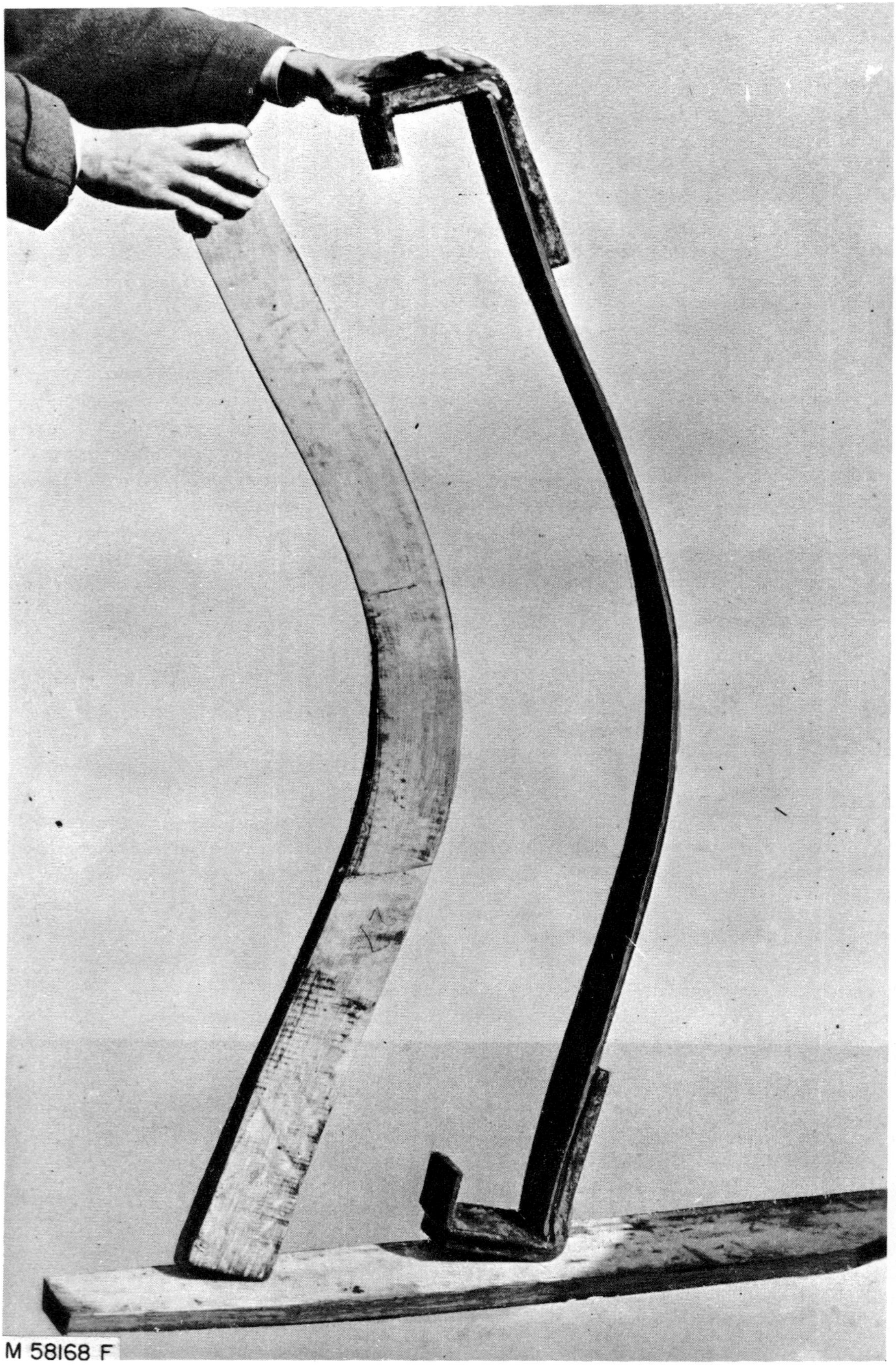

Figure 1-1.—Bent member and tension strap used at a shipyard. This member is resawed to form two floor timbers. The strap is straightened prior to reuse.

If the end pressure is applied at the start of the bend, as it should be, some means must often be provided to relieve the increasing end pressure that builds up as bending progresses. The end pressure can be kept more or less constant as bending progresses by means of a screw adjustment, or by placing (between the ends of the stick and the end blocks of the strap) blocks of wood that absorb the pressure and are crushed as bending progresses. Such "cushion" blocks should be placed so that pressure is exerted on side grain. The apparatus must be designed so that the end blocks will not overturn (figure 1-2), thus releasing the end pressure or permitting the piece to acquire a reverse bend (figure 1-3). As bending progresses, the center part of the piece tends to back away from the form. To counteract this tendency, pressure to hold the portion already bent against the form must be provided by means of clamps or dogs set into the table in combination with screws, wedges, or jacks.

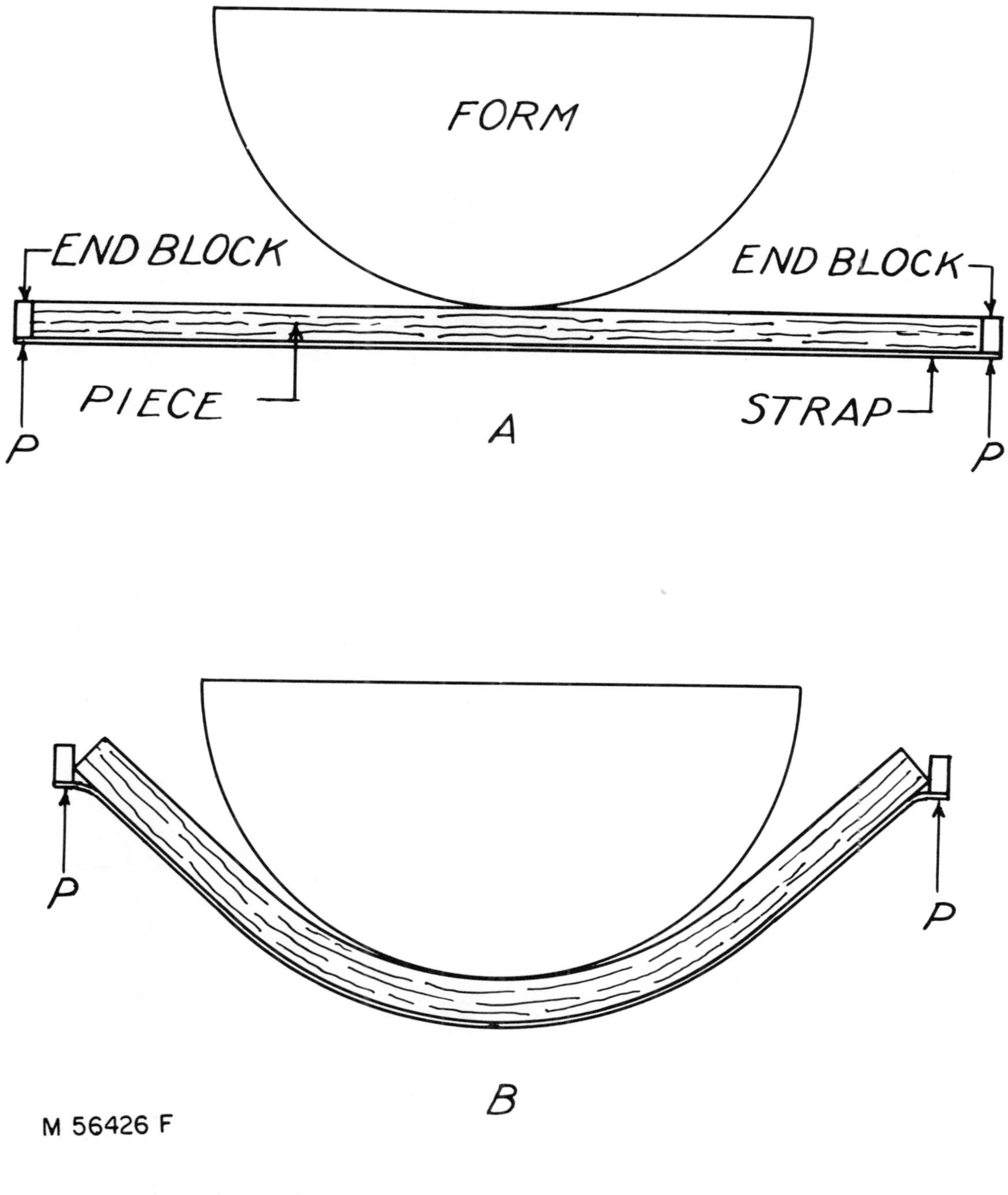

Figure 1-2.—Diagrammatic sketch of a bending strap having ineffective short end blocks. A, before pressure (P) is applied; B, after pressure is applied. Note how end blocks tip over and slide off.

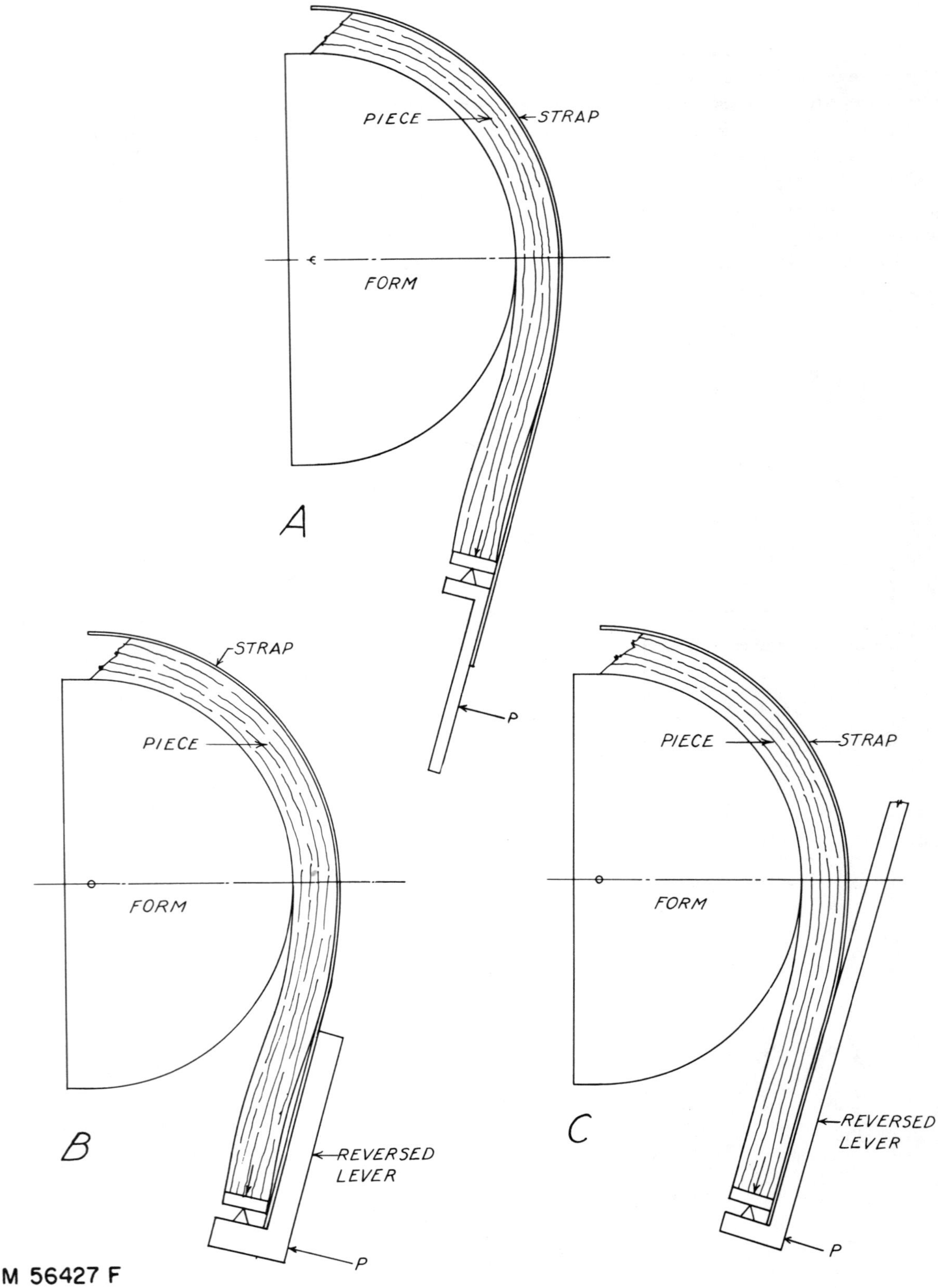

M 56427 F

Figure 1-3.—A, Diagrammatic sketch showing undesirable reverse bend near end of piece with one type of bending gear; B, diagrammatic sketch showing undesirable reverse bend near end of piece caused by too short a reversed lever arm on the end block; C, sketch showing proper application of the reversed lever principle, thus avoiding objectionable reverse bend near end of piece.

In making severe bends, the essential feature of the whole operation is the maintenance of strap tension and end pressure during the formation of the bend. The reversed lever principle (figure 1-3, C) is one of the most effective ways of controlling pressure and strap tension. The reversed lever prevents the end blocks from overturning (figure 1-2) and, if the bending force is applied close to the end of the piece of wood, limits the end pressure to the amount generated by the piece of wood itself. In figure 1-3, A, the force P needs to be increased as the bend progresses, applying increasing end pressure to the stick. This principle is embodied in the bending apparatus shown in figure 1-4. The dimensions and other construction details of this device can be adapted to the requirements of bent parts of various types and sizes.

Figure 1-4.—Apparatus for bending 4- by 6-inch stock for motor-launch frames. The end blocks, equipped with reversed levers, are prevented from overturning, and the end pressure is automatically regulated. The nuts engaging bolts welded to the ends of the main strap allow the end blocks to be fitted snugly to the stock at the start of bending, and the release of end pressure as necessary during the progress of the bend. At the completion of the bend, the ends of the inner strap are secured by means of metal straps and turnbuckles, and after wood stays are nailed on, the bent member can be removed from the main strap and the form.

A typical shipyard bending apparatus is shown in figure 1-5. This apparatus employs a cast-iron slab, wood form, block and tackle, power winch, steel strap, end blocks, dogs, and wedges. End pressure developed during the bending operation is absorbed by blocks of balsa set between the ends of the piece and the steel end blocks.

1.2.6 Radius of Curvature.—Even though stock has been carefully selected, properly softened, and bent by the best-known technique, there is a limit to the extent to which it can be bent. Hardwoods, such as oak and ash, can, in extreme instances—with careful selection of stock and careful manipulation—be bent to a radius as small as 3 to 4 times the thickness of the piece without excessive failure. For softwoods, which do not bend so readily, the limit is about 18 times the thickness.

1.2.7 Fixing the Bent Shape.—To fix the bend permanently, the wood must be cooled and dried under restraint. Cooling and drying cause the wood to lose its plasticity and to become stiff. The bent pieces can be left on the form to cool and dry; or, if they are bent on a bending table, one or more stay-laths can be nailed across the piece, connecting its ends, before it is removed. The bent piece is then allowed to season, with the stay-laths in place, until it will hold its shape.

Figure 1-5.—Bending apparatus employing a single bending strap, wood form, grid-like cast-iron slab, block and tackle, and power winch. The piece being bent is fixed firmly at one end with dogs and wedges and bent against the form by power exerted at its other end.

Sometimes tension failures take place on the outer side of the piece during this drying process. To prevent such failures, a minor strap can be applied to the piece under the tension strap before it is bent. After the bending is completed, a chain is attached to the ends of this strap to hold the piece in shape and to maintain sufficient end pressure during seasoning to prevent tension failure. When a single strap is used (figure 1-5), it is left on the bent piece and a restraining chain is attached to the ends of the strap until the piece has cooled and dried sufficiently to set the bend (figure 1-6).

Sometimes bent parts check while cooling. Such checking is likely to be severe in stock that is bent while green or only partially seasoned. Heat stored in the bending blank during the softening process causes rapid evaporation from the surfaces before the interior has lost any appreciable moisture. Under these conditions, the shrinkage of the surface layers is restrained by the wet interior, thereby fostering the development of surface checks. Any scheme that would permit the part to cool without much drying might reduce the amount of surface checking. The bent members can be sprinkled with or submerged in cold water, or covered to retard drying.

1.3 BENDING OF PLYWOOD.—Plywood used by the shipbuilder is ordinarily bought flat and bent to whatever form is required for the particular use to which it is to be put. The amount of curvature that can be introduced into a flat piece of plywood depends on numerous variables, a few of which are moisture content, direction of grain, thickness and number of plies, species and quality of the veneer used in making the flat plywood, and the technique applied in producing the bend.

The use to which the bent plywood part will be put should govern its thickness, number of piles, and direction of grain. Once these factors are established, the method of producing the bend usually can be determined.

Figure 1-6.—Bent frames for motor launches setting in their bending straps. Restraint during the setting period is provided by a restraining chain attached to both ends of the bending strap.

Bent plywood parts are more likely to be changed in shape by subsequent moisture changes than are molded plywood parts which have been glued up in a curved shape. Bent plywood pieces are limited, therefore, to those parts which are supported at frequent intervals by a rigid framework.

<u>1.3.1 Radius of Curvature.</u>—The radius of curvature to which plywood or veneer can be bent without fracture decreases when:

1. Thickness is decreased.
2. Proportion of veneer laid with grain at right angles to plane of the curvature is increased.
3. Moisture content and/or temperature of the plywood is raised.
4. Quality of veneer is increased (particularly by minimizing cross grain).
5. Mechanical arrangement of bending is improved.

In most applications of plywood to ships and boats it is not practical to approach closely the breaking radius of the material, as considerable allowance must be made for variations in veneer quality and the supporting medium. The types of plywood commonly used in boat-building vary somewhat in quality of veneer, especially as regards slope of grain, knots, and other defects which, while acceptable for flat plywood, increase the amount of breakage when bent to more or less severe curvatures. Technical information on the curvatures to which plywood of various types can be successfully bent is extremely limited. Because of the great diversity in species, quality, and arrangement of veneers and the thickness of plywood, the minimum practical bending radii for specific uses have generally been determined by trial and error. There is, however, a growing interest on the part of industry and the Navy to establish reliable information on bending limits that can be applied in boat design.

Bending to extreme curvatures is inadvisable because of the effect on subsequent performance of the part. Even though breakage does not occur in bending, severe curvature of the piece sets up stresses in the plies which reduce the ultimate strength of the part.

Fortunately, only mild curvature is required for most purposes in boatbuilding for which plywood can be bent safely while dry and cold. Little difficulty is experienced in bending plywood decking, for example, to the sheer of the vessel. The slight amount of double curvature introduced by deck camber can be accommodated by cold bending.

Where more severe bends must be made, the material for bending should be selected with care. The material selected should have faces free from knots, burls, excessively sloping grain, and core and crossband plies that are tightly butt jointed (without core gaps). The use of two or more species of wood in the same plywood panel does not preclude bending. Plywood that is to be bent to a small radius may be laid with the face grain parallel or up to 45° to the axis of curvature. For plywood five-eighths inch thick or thicker, and of seven or more plies, this requirement is not so important because it bends nearly as well in the perpendicular as the parallel-to-grain direction.

Wood is least flexible when cold and dry, somewhat flexible when hot and dry or water soaked at room temperature, and most flexible when boiled or steamed. Soaking in water at room temperature can be expected to produce a degree of flexibility about midway between that of dry and that of hot, water-soaked plywood. Only exterior, marine, and boat hull types of plywood can be softened by water treatment without affecting adversely the strength of the glue bond. Bending dry plywood over a hot mandrel is an intermediate treatment more effective with thin than with thick material.

When only a portion of a flat sheet of plywood is to be bent to a severe radius of curvature, it is common practice to soak only that portion by sponging it until thoroughly wetted with hot or cold water before bending. A strip of wet cloth laid on the area to be bent will accomplish the same result as repeated sponging. If the entire piece is bent to a small radius, or if it is desired to reduce the force necessary to hold a piece of varying radius of curvature while attaching it to the supporting frame, the whole sheet should be steamed or soaked. In either instance, the sheet can be bent over a form—preferably heated to facilitate the bending and reduce the drying time—similar to that shown in figure 1-7.

An illustration of how thicker plywood can be bent to a more moderate single curvature on an unheated male form is shown in figure 1-8. The plywood was steamed for about 3 hours and then clamped to the form and allowed to dry.

1.3.2 Compound Curvature.—No simple criterion is available for predetermining whether a surface of compound curvature can be covered with flat plywood. Soaking thin plywood and the use of heat during application are aids to manipulation. Some compound curvature can be imparted to flat plywood by die molding. Figure 1-9 illustrates one method of bending a flat sheet of plywood to double curvature; the plywood is first soaked in hot water and then dried between heated male and female forms that have been attached to a hydraulic press.

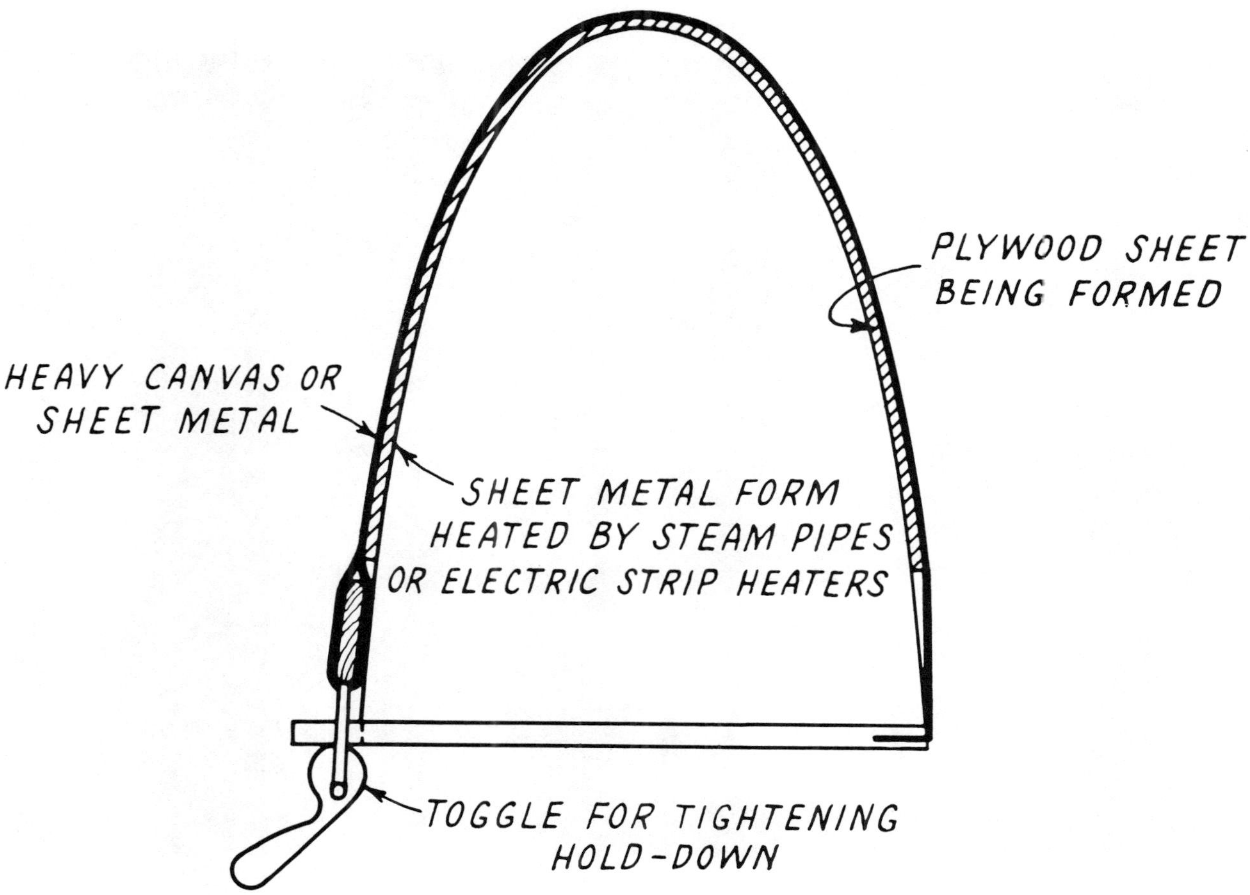

Figure 1-7.—Steam-bending form for plywood.

Under some bending conditions, plywood of five plies is less susceptible to fracture than three-ply material of similar thickness. The thickness of veneer is important, particularly that of the face veneer.

In preforming plywood by any wetting and drying technique, there is a certain amount of springback, for which allowance must be made if accuracy is desired. Springback depends on thickness, moisture content, temperature, bending time, ratio of radius to thickness, and other factors. As an allowance for springback, some operators bend soaked plywood on heated forms having a radius of curvature 20 percent less than the desired radius of the bend.

Figure 1-8.—Form for bending steamed plywood to moderate curvature.

1.3.3 Molding Plywood, Bag Method.—Molded plywood is produced by several techniques often referred to as "bag molding," "autoclave molding," or "tank molding." Perhaps the most inclusive is the term "fluid-pressure molding."

The fundamental procedure is the same for all processes in common use. In principle the technique consists of attaching temporarily—by staples, tape, clips, or some other means—superimposed layers of strips or sheets of glue-coated veneers to a mold of the desired shape, and molding these into a unit structure by the application of heat and fluid pressure through a flexible, impermeable bag or blanket. All the processes are relatively simple and provide a means by which plywood of simple or compound curvature, and of constant or varying thickness, can be produced in any arrangement of plies. Flat plywood can also be made by bag molding.

The same principles of balanced construction that apply to flat plywood are applicable to molded plywood. For maximum resistance to warping, all plywood should be symmetrical about the center plane of thickness. In this connection, symmetry involves species, number of plies, thickness of plies, and direction of grain. In theory, a symmetrically constructed panel with alternate plies laid at 90°, with respect to direction of grain, would have maximum dimensional stability. In practice, however, a construction with alternate plies at 90° to each other is often impossible in pieces of pronounced compound curvature.

It also is impossible to make flat strips of veneer conform to appreciable compound curvature before the application of fluid pressure. Fluid pressure forces the veneer to conform to the surface of the mold; therefore, the greater the compound curvature at right angles to the grain of the strips, the narrower the strips must be to avoid wrinkles.

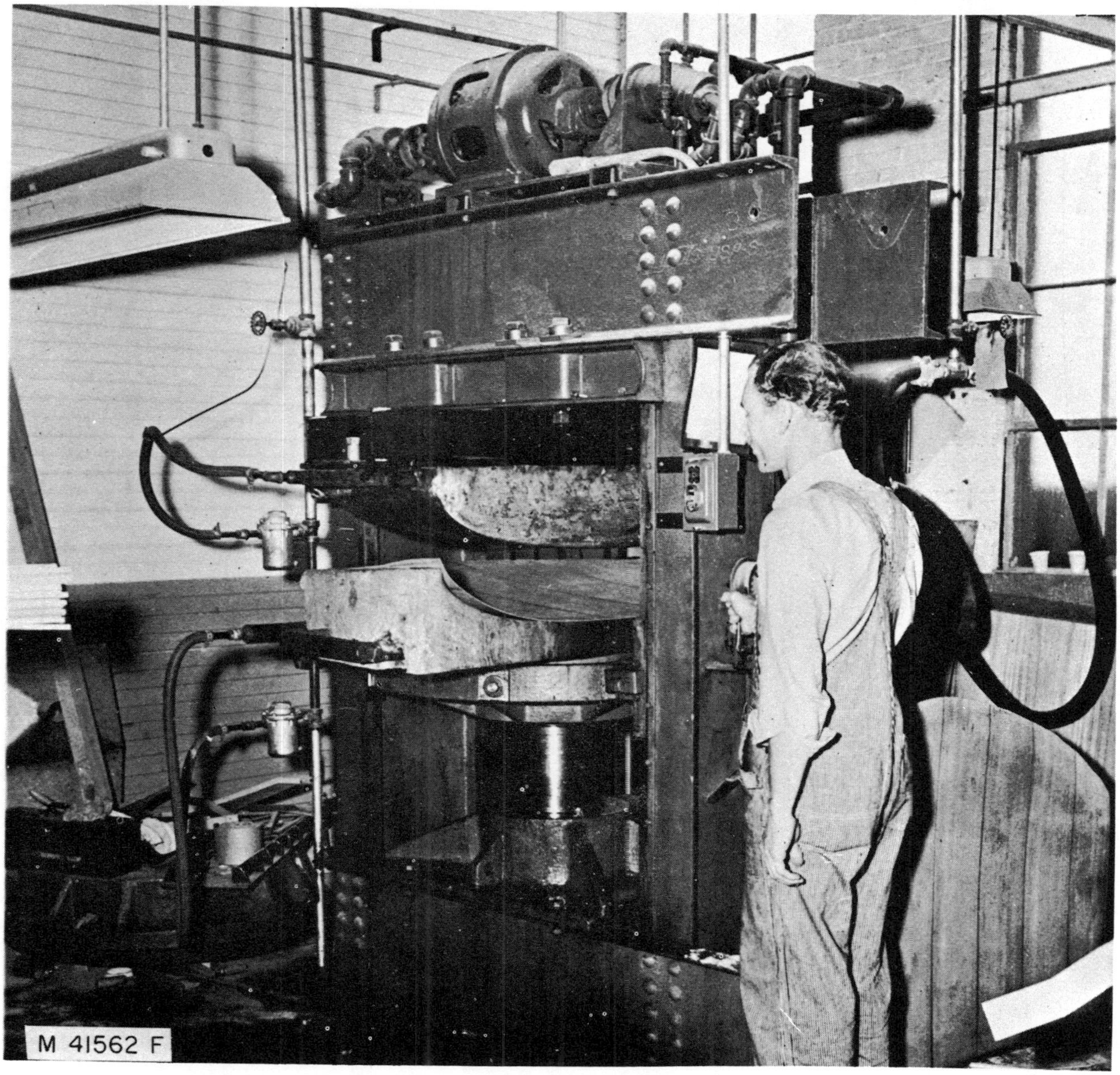

Figure 1-9.—Bending flat plywood to desired shape between heated plates.

1.3.3.1 Equipment.—The manufacture of bag-molded plywood of boat quality requires the use of considerable equipment. Means of supplying fluid pressure and heat must be provided. Normally, an autoclave or cylindrical pressure tank of 3- to 12-foot diameter and 10 to 60 feet in length— which will withstand an internal working pressure of 30 to 120 pounds per square inch—is used. In figure 1-10, C, and E, the use of an autoclave in bag molding is illustrated diagrammatically. Occasionally, the means of supplying heat and pressure is combined with the mold as illustrated in figure 1-10, B, and D. A mold capable of withstanding the desired internal fluid pressure is expanded by the pressure fluid.

All bag molding of boat-quality plywood requires heat. This often can be supplied most economically by steam, either directly or indirectly, by heating water or air, hence a boiler, or an adequate supply of steam from an existing steam line is required. When air is used for a steam-air mixture, or to provide

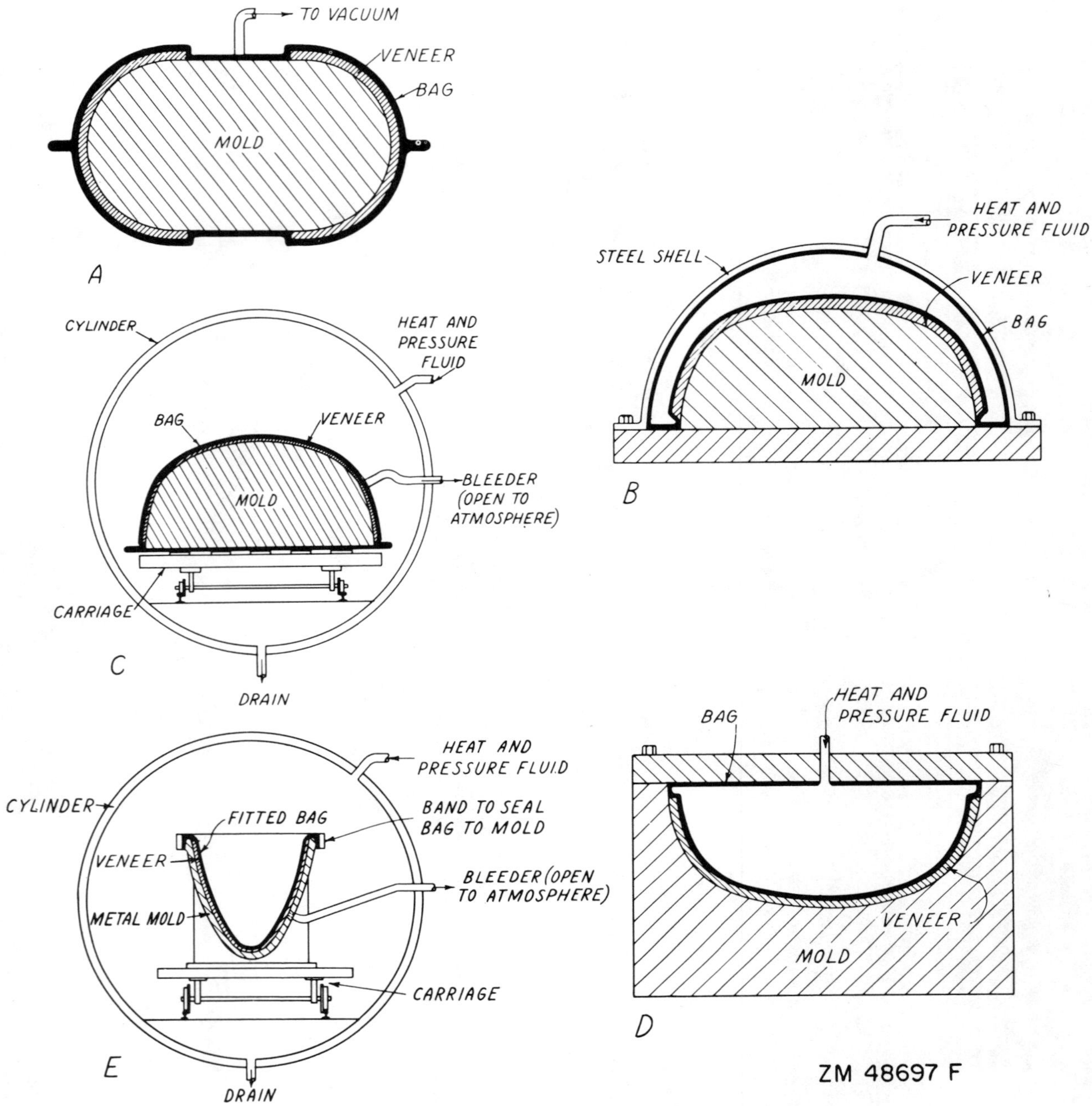

Figure 1-10.—Five methods of forming bag-molded plywood. A, Vacuum bag process using atmospheric pressure at room temperature to set glue; B, male mold with completely enclosed bag for application of heat and fluid pressure; C, male mold and bag used in a compression chamber; D, female mold with completely enclosed bag; E, female mold and half bag in chamber.

pressure on hot water, or as a combined heating-pressure medium, a compressor and receiver are required. The size and capacity of the compressor and the steam generator depend entirely upon the size and number of autoclaves or units being operated.

A vacuum pump may be used either to induce air pressure or to check the bag over an assembly for leaks. Vacuum alone produces insufficient pressure for most bag-molding operations.

Any operation involving the use of veneer presupposes the use of some of the ordinary veneer trimming tools, such as a saw, clipper, shaper, or router, as well as a glue spreader and veneer conditioning racks.

In some bag-molding operations careful control of the moisture content of the veneer throughout the process is desirable. Adequate air-conditioning equipment is then an additional requirement.

1.3.3.2 Molds.—The forming of any piece of bag-molded plywood requires a mold of some type. Molds, sometimes called forms, dies, or mandrels, are broadly classified as male or female. Male molds (figure 1-10, A, B, and C) are the desired shape for convex surfaces, while female molds (figure 1-10, D and E) have the proper shape for concave surfaces.

Common mold materials are wood (solid or plywood), metal (steel, cast iron, or low-temperature alloys), plastic materials, and cements. The choice of mold materials will depend largely on the shape of the item to be molded, the quantity desired, availability, and advantages and disadvantages of considered materials.

Wood molds are normally of the male type and have the obvious advantage of presenting a tacking surface often necessary in temporarily attaching the strips of veneer. Male wood molds may be grooved to allow the insertion of stiffeners to be glued to the molded plywood in a single operation. Wood molds are prone to distort somewhat after repeated heating and cooling, particularly if allowed to become wet—a condition which may result from leaks in the bag. The temperature penetration in wood molds is relatively slow as the mold usually is several inches thick, and the molded plywood must be heated essentially from the bag side.

Metal molds, usually more costly than those of wood, may be less expensive to use because of their long life and stability. In common practice, metal molds are female in type. The smoother surface of the piece being molded is on the convex side of the piece.

Some attempts have been made to use concrete, cements, and casting resins in mold construction, but to date these materials have proved practical in relatively few uses.

1.3.3.3. Bags or Blankets.—The purpose of the bag is to provide a flexible impervious barrier between the fluid under pressure and the mold. The piece being molded is pressed between this flexible bag and the rigid surface of the mold. The full fluid pressure is applied at right angles to the surface of the bag, regardless of the shape.

Bags are classified as full bags or half bags (blankets). A full bag is a complete envelope of impervious flexible material (figure 1-10, A and C) clamped shut at one end or side, and having a connection, usually called a bleeder, to allow entrapped air to escape to the atmosphere. A full bag may be completely closed, similar in principle to a basketball bladder (figure 1-10, B and D), having only a tube connection for inflation. A half bag, or blanket, is a sheet which normally fits the mold without wrinkling and is sealed by some temporary means to the edges of the mold (figure 1-10, E). The bleeder may be attached to the mold or to the bag. Full bags are normally used over male molds and half bags are used on female metal molds.

Most bag-molding operations at present require bags made of specially compounded natural or synthetic rubber. The useful life of a bag depends largely on the heating medium used, the temperature of the cycle, the size of the bag, and the care used in handling. Cellophane has been used to some extent for bags.

1.3.3.4 Glues.—It is possible to use a variety of glues for bag molding; however, typical bag-molded parts require long assembly periods, and have definite use requirements that normally limit the choice of glues to the hot-setting resin types. Glues most generally preferred are those that are dry at the time the veneer is assembled on the mold and may be pressed at any time within 7 days after application.

The molding of curved parts often requires that the flat strips of veneer slip slightly during molding. Some glues aid this slipping by acting as a lubricant while passing through the plastic stage when heated. This peculiar characteristic and the somewhat critical relation between temperature, time, and pressure in the curing of the glue make close cooperation between operator and glue supplier advisable. Specially compounded glues are available which have been developed for bag molding. These glues have been formulated to produce good bonds at the normal moisture content values and temperatures used in hot-pressing plywood at fluid pressures of 30 to 100 pounds per square inch commonly used in bag molding.

1.4 MACHINING OF WOOD

1.4.1 General.—Among the important properties that affect the utility of any wood are its machining properties, which affect all woodworking operations. In these, as in other classes of properties, different woods vary widely and a given wood may give good results in some operations, fair results in others, and poor results in still others. The "workability" of any wood, therefore, cannot be judged by one operation, but depends upon the sum total of all of them. In an operation there are several factors, both in the wood itself and in the machine, that affect the results, and these results may be good or bad depending upon the conditions under which the work is done. Some woods machine well under a relatively wide range of conditions, while others require exacting techniques if good results are to be obtained. Table 1-1 sums up the results of extensive machining tests on southern hardwoods.

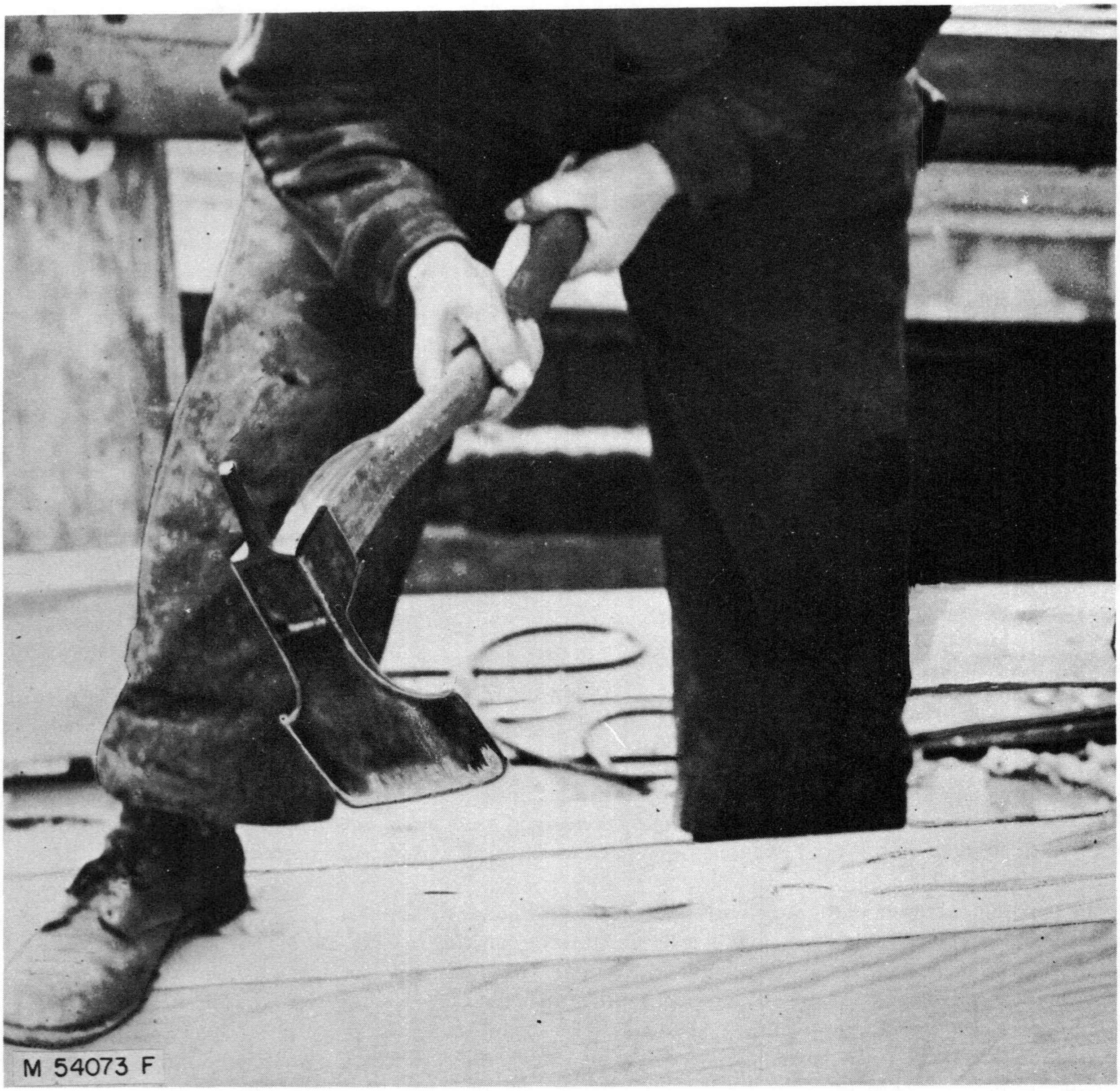

Figure 1-11.—Adzing a scarf joint.

Planing is the most important of the machining properties shown in table 1-1. The quality of planing that is required will depend largely upon the end use of the part. Where exacting uses require first-class finish, adjusting the machine to give at least 16 knife cuts per inch will help greatly. Proper cutting angles also are important. For hardwoods the most commonly used cutting angle is 20° and for softwoods 30°.

Figure 1-12.—Mortising caprail to fit tenons of gunwale stanchions.

Table 1-1.--Some machining and related properties of hardwoods

Kind of wood	Planing	Shaping	Turning	Boring	Mortising	Sanding	Steam bending	Nail splitting	Screw splitting
	Perfect pieces	Good to excellent pieces	Good to excellent pieces	Good to excellent pieces	Fair to excellent pieces	Good to excellent pieces	Unbroken pieces	Pieces free from complete splits	Pieces free from complete splits
	Percent	Percent	Percent	Percent	Percent	Percent	Percent	Percent	Percent
Ash	75	51	79	94	62	75	67	65	71
Basswood, American	64	9	68	75	51	17	2	79	68
Beech	...	21	90	99	93	49	75	42	58
Birch	63	53	80	98	97	34	72	32	48
Buckeye	...	6	58	75	18	...	9	...	...
Cottonwood	21	3	70	70	52	19	44	82	78
Chestnut, American	74	24	87	91	72	64	56	66	60
Elm	33	11	65	94	75	66	74	80	74
Gum, black	48	23	75	82	24	21	42	65	63
Hackberry	74	10	77	99	70	...	94	63	63
Hickory	...	19	84	100	98	80	76	35	63
Magnolia	65	25	79	69	32	37	85	73	76
Mahogany	80	68	89	100	100	...	41	68	78
Maple, hard	54	62	82	99	95	38	57	27	52
Maple, soft	41	22	76	80	36	37	59	58	61
Oak, chestnut	...	23	90	100	100	75	85	49	70
Oak, red	91	21	84	99	100	81	86	66	78
Oak, white	87	28	85	95	100	83	91	69	74
Pecan	88	31	89	100	100	...	78	47	69
Sweetgum	51	21	86	92	58	23	67	69	69
Sycamore	22	8	85	98	96	21	29	79	74
Tupelo	...	43	79	62	35	34	46	64	63
Walnut, black	62	34	91	100	98	...	78	50	59
Willow	52	5	58	71	24	24	73	89	62
Yellow-poplar	70	12	81	87	63	19	58	77	67
Average	61	25	79	89	70	45	62	62	67

Such manual tools as saws, adzes, chisels, and planes are still standard equipment in the shipbuilder's kit. A great deal of shaping work is done with them, especially for final fitting of large, heavy parts rough-shaped with power equipment. It is often necessary, for example, to fair the bevel of a frame after it is in position on the ship with an adz because the original bevel cut made with a bevel band saw was only an approximation of the correct, graduated bevel between angles of cut established at given points along the frame. Scarf joints are often finished smooth with an adz (figure 1-11). A chisel is used for many fitting operations (figure 1-12 and 1-13) and planes are used to trim shutter-strake planks to final fit.

Power equipment used by ship and boat builders may be either portable or stationary. Portable power tools include hammers and augers that are either electrically or air driven. The chief drawback of such tools is the necessarily lengthy cable or hose that connects them to the source of power. On the other hand, stationary machinery requires additional handling of lumber enroute from the lumber pile to the vessel Both types have their advantages for particular uses, and the well-managed shipyard has its production system arranged to take best advantage of both. In the following paragraphs brief descriptions of the more common power-driven woodworking tools and machines are presented.

While most shipbuilding woods grown in the United States are easy to work, there are some exceptions; certain species of tropical woods also are difficult to work or machine.

Often some other factor, such as availability, durability, or exceptional strength, makes it necessary to use a wood that is comparatively hard to machine. Certain shaping operations necessary in shipbuilding, such as the cutting of variable bevels and the scarfing of solid wood and plywood, require a degree of accuracy in machining that emphasizes the necessity for skilled workmanship.

Figure 1-13.—Finishing a stem rabbet to final bevel with a chisel.

1.4.2 Woodworking Tools and Machinery

1.4.2.1 Bandsaws.—The bandsaw is constructed in a wide variety of sizes and types depending on its use. The type most adaptable for the shipyard is the bevel bandsaw, which is designed to facilitate the sawing of variable bevels in frames. The bevel bandsaw also is used, of course, for other types of cuts ordinarily made with a bandsaw. Bevel saw cuts are produced either by changing the angle of the table from the horizontal or by controlling the angle of the blade. Bevel saws with tilting tables are widely used in shipyards, but the types permitting variation of the blade angle (figure 1-14) have a certain advantage in that the constantly horizontal table facilitates handling of heavy pieces of wood while cutting a variable bevel. Changes in bevel may be controlled manually by means of a wheel, or on larger machines the saw head-rig may tilt slowly under electric power thus changing bevel as the piece is fed through the saw. The range in bevel is ordinarily 45° left and 15° right. The speed of saw is usually from 5,000 to 7,000 feet per minute.

Figure 1-14.—Shipyard bandsaw with variable-angle blade.

1.4.2.2 Circular Saws.—The circular saw has many uses in the shipyard. While it is employed principally for ripping, beveling, crosscutting, resawing, and mitering, various other operations—such as grooving, dadoing, rabbeting, molding—also can be performed on this machine by using special attachments and set-ups. Circular saws are made in a number of types and sizes, most common of which are the universal saw and the variety saw. The universal type is equipped with two arbors which permit mounting both rip and crosscut blades, either of which is brought into use by simply turning a hand wheel that controls the positions of the arbors. The variety saw employs a single arbor and is generally fitted with mortising and boring attachments. Circular saws are used to rough-cut flitches into planking, frame stock, beams, and similar thick material.

A portable version of the circular saw (figure 1-15) is finding extensive use in shipyards for rapid on-the-job cutting to length, rough cutting of scarf joints, and other shaping operations.

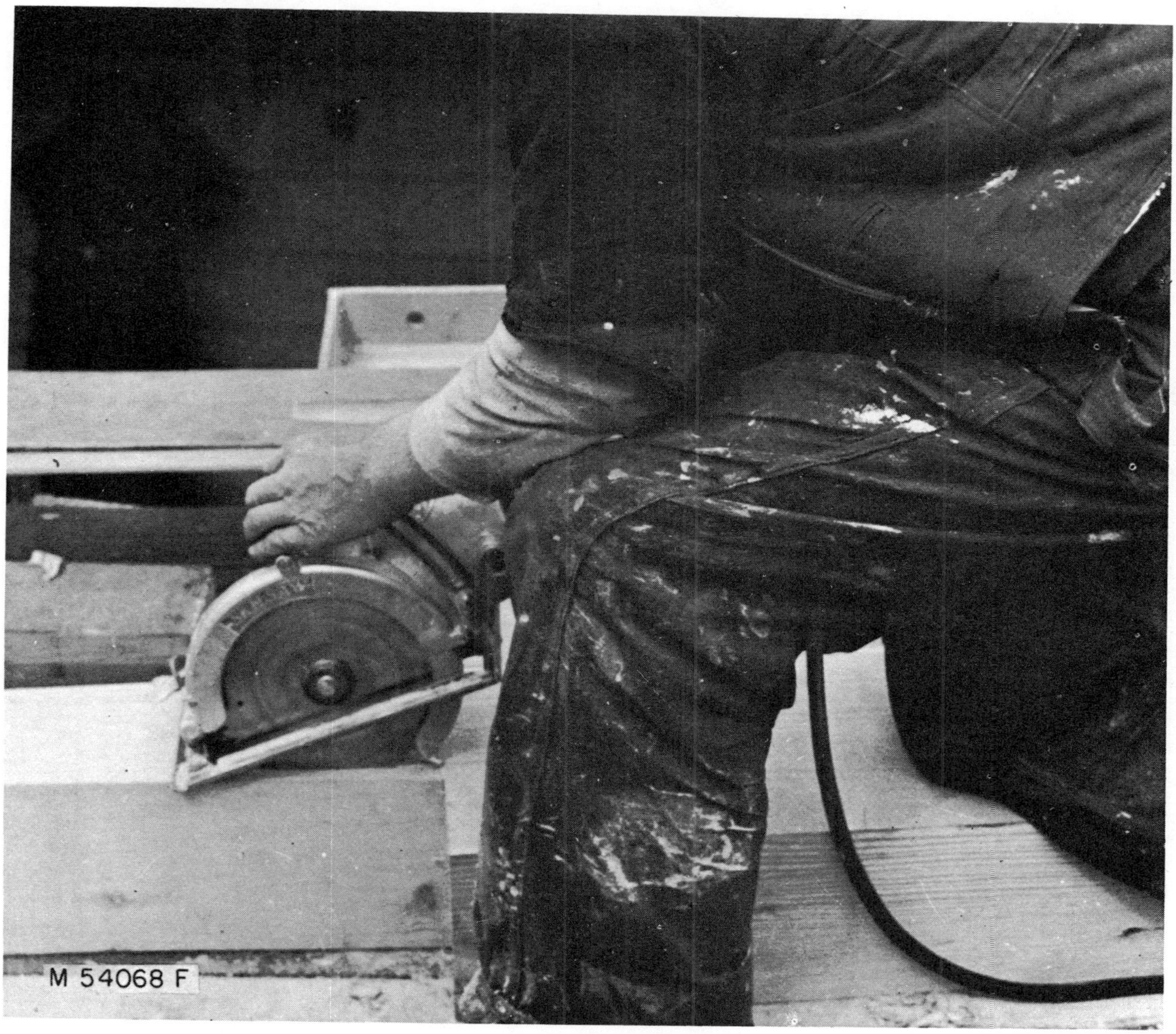

Figure 1-15.—Rough-cutting a scarf joint with a portable electrically driven circular saw.

1.4.2.3 Jig Saws.—The jig saw differs radically in construction from the bandsaw. It is more adaptable for cutting small, sharp curves because much smaller and finer blades may be used. Inside cutting is accomplished on the jig saw, since the blade is easily removed and inserted through an entrance hole bore in the stock. The principal use of the jigsaw in a shipyard is for cabinet work, plywood deck fittings, and similar parts made of thin stock requiring cuts of sharp curvature.

1.4.2.4 Jointers.—The jointer, also called a hand planer because it is fed by manual power, is used to cut square edges or true surfaces on keel, frame, decking, and other stock. It can also be used for tapering and beveling straight sections, such as top timbers, floors, keelsons, and clamps.

1.4.2.5 Surfacers.—The surfacer is a power planer used for planing surfaces of flat stock, reducing stock to thickness, and, sometimes, for scarfing thin stock. It may have only one cutterhead, in which case only one surface of the piece is planed at a time, or it may have a double cutterhead for planing two surfaces simultaneously. The surfacer is widely used to finish surfaces of ship parts ranging from large keel sections and deck beams to light planking and joinery (figure 1-16). Portable surfacers are available for planing beams and other members (figure 1-17).

Figure 1-16.—Surfacing laminated PT boat frames.

1.4.2.6 Shapers.—The shaper is used mainly in trimming, shaping, and molding stock irregular in outline, and also for scarfing solid stock (figure 1-18) and wide sheets of plywood. Various types are available. Spindle speed is usually about 7,000 revolutions per minute. In some shipyards it is employed to cut variable bevels on frames or light, two-part keels subsequently joined together so that the bevels form planking rabbets; for such uses special wood jigs are employed to guide the cut along a predetermined bevel. Extension tables are often used (figure 1-19).

Figure 1-17.—Power hand-surfacer.

1.4.2.7 Sanders.—Various types of sanders are employed for woodworking, among them the drum sander, the belt sander, and the combination disk and spindle sander. The combination type is adaptable to the majority of sanding operations needed for ship joinery.

Portable sanders of the disk type are used extensively in boat yards to sand planking and deck surfaces before they are painted.

1.4.2.8 Routers.—Routers of the portable type are used widely in shipyards for cutting plywood and thin solid stock to shape and for cutting scarfs in plywood and trimming plywood decking and bulkhead sheathing. The common types consist of an electrically- or air-driven spindle, to which a cutting blade and collar are fixed. The collar serves to guide the router along a template edge. Portable routers have head speeds of 15,000 to 50,000 revolutions per minute, the air-driven machines being slightly faster and more compact.

Stationary routers have a vertical head that operates above the stock and are used for jobs similar to those performed on the shaper. The cuts, however, are usually lighter.

Figure 1-18.—Cutting a scarf joint in a keel.

1.4.2.9 Boring Machines.—The conventional boring machine is designed to hold and operate various wood bits. The single-spindle, manually operated borer is adaptable to most operations. The table may be tilted up to 45° and rotated through a complete revolution, making it possible to bore at any angle or combination of angles.

Portable boring machines find their most common use in ship assembly operations, rather than in preboring of parts. Power borers, either pneumatically or electrically driven, are used to bore bolt, drift-pin, hatch-nail, and treenail holes through keels, keelsons, planking, decking, ceiling, beams, shelving, and other structural parts.

A special boring machine is used to cut shaft holes in shaft logs and sternposts. In some yards manually operated shaft-log borers are still used, but power borers have come into general use. Small borers are sometimes powered by an electric motor. Shaft-log boring is an exacting operation; the hole must be true so that bushings and bearings will fit precisely.

Figure 1-19.—Keel section partly scarfed with a shaper. Note extension table equipped with rollers.

CHAPTER 2. GLUING AND LAMINATING

2.1 GLUING

2.1.1 General.—Before World War II the use of glues in ship and boat building was limited almost entirely to plywood, ship's furniture, and other joinery. Available glues either required high curing tempertures that limited their use largely to the hot-press gluing of thin plywood, or they lacked adequate water resistance for boat construction. Durability under great variations in climatic conditions is important for glues used in ship and boat building. Information has been accumulated during recent years regarding performance of adhesives, and new adhesives have been formulated that offer decided advantages over the older types.

The development of resorcinol and phenol-resorcinol adhesives in the early forties stimulated great expansion in the use of glued wood products for marine service. These products included laminated members, such as stems, keels, frames, and deck beams, made from seasoned lumber selected for the particular requirements, as well as glued gusset constructions for more efficient transmission of stresses from one member to another.

2.1.2 Types of Adhesives for Bonding Wood.—Numerous types of glues and adhesives are used to bond wood parts together. Very few, however, are suitable for such extreme service conditions as marine use. Animal or hide glues are widely used in the furniture industry, although in recent years urea-resin adhesives have found extensive use. Some use has been made of polyvinyl resin adhesives in furniture manufacture. Vegetable or starch glues have been used primarily for furniture-type plywood, and soybean glue has found extensive use for interior-type softwood plywood. In laminated construction for protected service, casein glue is used almost exclusively, although in buildings where high humidity prevails for long periods of the year more water-resistant adhesives have been found necessary.

All these glues have different degrees of water resistance. Vegetable-starch glue has about the lowest resistance and polyvinyl resin, animal, soybean, casein, and urea-resin follow more or less in that order. None of these glues and adhesives has sufficient durability under severe marine conditions to make it suitable for marine use.

Adhesives used for marine laminating operations must be listed on the latest issue of Qualified Products List, QPL-397, as conforming to the Type II, Class 2 requirements of Military Specification MIL-A-22397. The curing times and temperatures should not be less than those listed on QPL-397 for the particular adhesive used. Further restrictions on curing-chamber and glue-line temperatures may be required when bonding wood treated with preservative.

Phenol-resin adhesives have a long background of successful use in exterior-type plywood. They are formed by the reaction of phenol or cresol with formaldehyde. Most of these adhesives are alkaline in reaction and require temperatures from 240° to 320° F. for adequate curing. Because of the high curing temperature required, these adhesives are not well adapted for assembly gluing or for laminating lumber. Phenol-resin adhesives have been used, however, for nearly two decades to produce waterproof, exterior-type, hot-pressed plywood, and the excellent performance of such joints under exposures involving moisture and heat indicates what may be expected of phenolic-type adhesives.

Some phenol-resin adhesives have been formulated to set at intermediate to low temperatures by use of acid catalysts. They have not been used nearly so extensively as the hot-press phenols. Weathering and soaking-drying tests on white oak indicate that highly acid adhesives damage the wood adjacent to the glue lines and to date are not considered for this purpose.

Melamine resin adhesives are produced by the reaction of melamine and formaldehyde and are available either as hot-setting or intermediate-temperature-setting types. Test data on the latter type accumulated over the past 10 years indicate that the durability of their joints, when set at 190° F., is similar to that of joints made with resorcinol resins. Melamine resine adhesives have not gained much use in this country, however, for laminating of structural members and their suitability for marine service has not been definitely established, probably because of their high cost.

In laminating and assembly gluing for marine service, the resorcinol and phenol-resorcinol adhesives have been used almost exclusively. Consequently, these will be discussed in more detail than the other adhesives mentioned.

2.1.3 Resorcinol-Formaldehyde Resin Glues.—Resorcinol-formaldehyde resin adhesives have the high quality and durability of the phenol resins, but cure at much lower temperatures. They are produced by the reaction of resorcinol with formaldehyde and are marketed as liquids consisting of partly polymerized resin in a water-alcohol solution. The solids content of the solution is usually from 60 to 70 percent by weight. The adhesives are dark red and make dark joints when set. A hardener, usually paraformaldehyde but sometimes formalin, and a filler (commonly walnut shell flour) are mixed with the resin before use. A mixture of powdered hardener and filler is usually furnished by the manufacturer.

Test data and service records indicate that, when properly used, resorcinol adhesives compare favorably with phenol-resin glues in resistance to moisture, high temperatures, chemicals, and micro-organisms. They are very nearly neutral in reaction.

2.1.4 Phenol-Resorcinol-Formaldehyde Resin Adhesives.—Phenol-resorcinol-resin combinations set at substantially lower temperatures than do hot-setting phenols. Some of the more recent formulations have curing requirements very similar to those of straight resorcinol-resin adhesives. The phenol-resorcinol glues are generally made by combining resorcinol resin with phenol resin in various proportions. Like the resorcinols, the phenol-resorcinol glues are marketed in liquid form, with a separate powdered hardener that is mixed with the resin prior to use. A filler, commonly walnut shell flour, often is added with the hardener. In common with the straight resorcinol adhesives, these modified adhesives are nearly neutral or slightly alkaline in reaction. The durability of well-made phenol-resorcinol-resin glue joints, based on experience over about a decade, appears to be essentially equal to that of resorcinol and hot-press phenol-resin glue joints.

2.1.5 Prerequisites for Good Glue Joints.—A good glue joint must exceed the strength of the wood under all conditions of service. To produce glue joints of this quality, the conditions involved in the gluing operation must be carefully controlled so as to obtain a continuous, thin, uniform film of glue in the joint, with adequate adhesion to both surfaces of the wood. These conditions involve:

1. Proper moisture content of the wood (considerable range in moisture content permissible).
2. Properly prepared wood surfaces (machined smoothly within a reasonably short period before gluing).
3. Glue of good quality, properly prepared (meets applicable specifications).
4. Good gluing technique (mixing, spreading, clamping, and curing done properly).

Satisfactory and unsatisfactory joints resulting from improperly controlled gluing conditions are shown in figure 2-1.

2.1.6 Gluing Characteristics of Wood.—Since the standard for a good glue joint is that it be stronger than the wood, it naturally follows that the stronger, denser species of wood will require stronger glue joints than less dense, weaker species. Dense species of wood generally shrink and swell more than the less dense ones and thus exert greater stresses on the glue joints with given moisture changes. Consequently, the denser species of wood, such as the oaks, will require higher-quality joints and more care in gluing.

Certain species of wood, such as teak, may be more difficult to glue than would be expected from their density because of oiliness due to extractives, or other foreign matter. The oiliness of such species may vary a great deal, however, from log to log, so that it is difficult to generalize about such effects. The heartwood of most species is more difficult to glue than the sapwood of the same species. These effects of heartwood and extractives have not been fully investigated, but such problems generally can be overcome satisfactorily by taking special care in the various steps of the gluing operation.

2.1.6.1 Gluing of Preservative-Treated Wood.—As a general rule, wood treated with preservatives Volume II, Section 2.1 is somewhat more difficult to glue than untreated wood of the same species. Strong

Figure 2-1.—Poor and good joints made with resorcinol glue under different gluing conditions. A, excessively thick glue line caused by too long assembly period; B, too short assembly period (joint pressed before glue had reached desired viscosity); C, glue not adequately cured; D, well-made joint that developed high percentage of wood failure when broken.

and durable bonds have been consistently produced, however, on pressure-treated wood of species widely used in ships and boats, notably oak. The production of satisfactory glue joints in preservative-treated lumber usually requires that the treated lumber be surfaced immediately before it is glued, that higher curing temperatures or longer curing periods be used, and that gluing be more closely controlled than is necessary for untreated lumber.

It is good practice to surface the lumber to uniform thickness before treatment, hence less of the treated material needs to be removed in the final surfacing for gluing. For waterborne preservatives, which introduce a great deal of moisture into the wood, the stock must be kiln dried after treatment to remove the excess moisture. The kiln drying must be carefully done to avoid development of excessive checking and warping in the stock.

With oil-borne preservatives, the moisture content of the wood is not changed appreciably if the stock is well dried before treating. A conditioning period of two weeks or more is nevertheless often desirable to permit excess solvent to volatilize. A great deal of oil on the surface of the wood after planing may prevent adequate gluing and such oil should be wiped off before spreading the glue. Excess oil on the surface of treated wood is usually not a problem where moderate retentions are used, although this will vary considerably between species. Some species exude or "bleed" oily preservatives more than others. A light oil, as a rule, causes fewer gluing difficulties than heavier oils since it is more readily volatilized. Treating specifications usually specify the type of oil to be used with the different preservatives.

Glues of moderate curing requirements should be used for bonding treated wood, since prolonged heating at elevated temperatures may adversely affect its strength. Current military specifications limit the curing of glued treated wood to a maximum temperature of 170° F.

2.1.7 Moisture Content of Wood for Gluing.—The seasoning of wood for ship and boat parts is described in Volume II, Chapter 1. In general, it is desirable to produce a glued wood member at a slightly lower moisture content than that expected in service, thereby avoiding not only surface drying in use, but a serious change in the entire member. Gluing of wood at a moisture content close to that expected in certain parts of boats or ships in service is not generally feasible. It is advisable, however, to glue at a moisture content that will ensure a good glue bond and also approximate service conditions.

Wood used for laminating of ship and boat members should have a moisture content within the range of 10 to 15 percent. With the types of glues suitable for ship and boat use (resorcinol and phenol-resorcinol), excellent bonds can be obtained on lumber in this moisture-content range. In any laminated assembly, therefore, the pieces glued together will not vary in moisture content more than 5 percent. This control of moisture content will tend to reduce or eliminate damage from unequal swelling or shrinking of the pieces glued together. Unequal swelling or shrinking of pieces glued together causes stresses on the glue joints that may result in separation of laminations and distortion of the finished member.

Glues used in ship and boat building generally are applied as water-alcohol suspensions, and the moisture content of the wood is slightly increased near the glue line. Usually, however, parts are thick enough so that the over-all increase in moisture content resulting from the water introduced with the glue is insignificant.

2.1.8 Machining Wood for Gluing.—Wood should be conditioned uniformly to the desired moisture content before it is machined for gluing. Changes in the moisture content of stock after it has been machined produce distortion and surface irregularities which are objectionable from a gluing standpoint; consequently, lumber should be surfaced just before it is glued.

For glue joints of maximum strength, all gluing surfaces should be machined smooth and true. Torn or damaged wood fibers should be avoided; they are usually due to improper maintenance or operation of cutting tools. Torn fibers reduce joint strength and cause shallow wood failures. Sawed surfaces are not recommended for gluing where maximum joint strength is required. Matchers and molders are less adaptable to the surfacing of laminations for gluing than single head cabinet surfacers. The final surfacing should be light cut (section 2.2.8.5.3.5). Knives should be kept sharp; dull ones compress or do other damage to wood fibers. Glazed or burned areas indicate damaged fiber.. Clipping or sniping at the ends of boards usually indicates improper adjustment of the chip breaker or pressure bar.

Roughening of surfaces by tooth planing, scratching, or sanding with coarse sandpaper is of no benefit to the glue bond. Sanding generally removes more wood at edges and soft areas than at denser areas, and thus causes the thickness of the sanded board to be less uniform than that of a well-planed board.

2.1.9 Storage of Resorcinol Glues.—Resorcinol and phenol-resorcinol resins tend to thicken with age and the thickening is accelerated at higher temperatures. Consequently, the resin should be stored in a cool place. Freezing, however, is sometimes undesirable and storage temperatures of from 32° to 70° F. are usually recommended. To avoid evaporation of solvent, the resin should be stored in tightly sealed containers. Adequate facilities should be available to store adhesives in accordance with specifications.

Liquid hardeners usually are shipped in stoppered bottles that prevent evaporation. Powdered hardeners should be stored in air-tight containers to avoid appreciable change in their composition.

All shipments of resin and hardener should be dated as received and care taken always to use the oldest lot of each glue first. It is normally recommended that the hardener for a given shipment of resin glue be used only with this particular shipment of the resin.

2.1.10 Mixing and Applying Glue.—Woodworking glues generally are not furnished ready for use but must first be mixed with catalysts, hardeners, fillers, or solvents. Extreme care should be taken in preparing any glue. Manufacturers' directions for mixing should be carefully followed. Scales of the proper sensitivity are essential. Containers used for weighing the ingredients should be clean; a separate container is usually used for each ingredient. To avoid waste, mix only as much glue as can be conveniently used during the expected working life of the glue. Never add over-age glue to a fresh batch.

Small amounts of glue can be mixed by hand, but for larger batches a mechanical mixer is needed. Various types of mixers have been used successfully. Only the mechanical-type mixer should be used for marine laminates. The dough-type mixer (figure 2-2), equipped with a mechanism for turning the paddle in a double rotary motion at two or three different speeds, has been used with excellent results for many types of glues. Too rapid stirring may foam the glue, while slow stirring will require excessive mixing time and thus shorten the pot life of the glue. For the type of mixer shown in figure 2-2 a paddle speed of 60 revolutions per minute is usually satisfactory. Portable electric drills equipped with suitable stirring devices also have been used successfully for mixing resorcinol and phenol-resorcinol glues. Since the speed of such drills is very high, the paddles must be small to avoid too violent agitation. Mixing should be continued until a smooth, homogeneous, easily spread mixture is obtained.

Figure 2-2.—Three-speed, dough-type, electric mixer with 3- and 8- quart bowls and two sizes of paddles for mixing glue.

Mixing bowls made of steel, zinc, copper, brass, or aluminum are suitable for use with the approximately neutral glues (such as resorcinols and phenol-resorcinols) recommended for laminating or assembly gluing of parts in wood ships and boats. In warm weather a water-jacketed pot may be used to cool the glue mixture and maintain a long working life. Placing the glue bucket in a tank having cold water running through it also is helpful in prolonging the working life of the glue in hot weather.

With resorcinol and phenol-resorcinol glues, it is usually most convenient to add the powdered hardener while slowly stirring the liquid resin in a mixing bowl. Rapid stirring at the start is likely to cause loss of hardener. A total mixing time of 3 to 10 minutes is usually sufficient.

Since some resins tend to develop heat when mixed with a hardener, the manufacturer's instructions for keeping the materials cool during mixing should be followed. Precooling the resin overnight in a refrigerator before it is mixed is helpful in avoiding undesirable heating. For glues that do not heat up excessively during mixing it usually is desirable to permit the resin to warm up to room temperature before it is mixed. Many manufacturers recommend that the mixed glue be allowed to mature for about 20 to 30 minutes before it is spread.

Many synthetic-resin glues contain ingredients that may injure the skin and other body tissues. Workers, therefore, may need rubber gloves, aprons, and goggles for adequate protection. If toxic fumes are involved, good ventilation should be provided; exhaust fans may be needed. Information on precautionary measures needed in using any particular glue should be obtained from the glue manufacturer.

Mixers, spreaders, and other equipment used with all glues should be thoroughly cleaned several times a day. Glue must be cleaned off before it reaches the insoluble hardened stage. Special care should be taken to clean all equipment when changing from one type of glue to another, as catalysts or other ingredients used with one type may cause objectionable reactions in another type.

2.1.11 Working Life of Mixed Glue.—Setting or curing of resorcinol and phenol-resorcinol glues is a result of chemical reaction and loss of solvent due to evaporation or diffusion. There is a definite relation between the temperature and the rate at which a chemical reaction proceeds, as well as between evaporation and temperature; both are speeded by increasing temperature. The mixed glue will have a longer working life if it is kept cool until spread on the wood.

Different formulations of the same type of glue vary considerably in working life at the same temperature. At 70° to 75° F., the working life of the straight resorcinol glues is generally in the range of 2 to 5 hours and that of phenol-resorcinol glues from about 2 to 8 hours. The glue will set faster under constant agitation on the spreader than when in the pot; consequently, the work should be arranged so that the spreader is not run idle for an appreciable length of time. Glues that have become too thick for proper spreading should be discarded.

2.1.12 Spreading of Glue.—To make satisfactory glue joints, it is necessary to apply the correct amount of glue and to spread it evenly over the entire joint surface. In laminating, these requirements can usually be most easily met by using a double-roll machine spreader with rubber-covered rolls and fitted with doctor rolls to control the glue spread (figure 2-3). For assembly gluing, however, where various shaped pieces are often involved, it is frequently necessary to spread the glue by hand. Mohair paint rollers work very well for hand spreading. The rollers have to be cleaned extremely well after use; otherwise, they will become hard and unusable. Brushes are used to spread glue on small, irregular areas. Tampico fiber brushes are more suitable for spreading wood glues than soft bristle brushes.

Application of glue to only one of the two mating surfaces of a joint (single spreading) is satisfactory for many types of work. When spreading glue on long laminations, especially if relatively long assembly periods are necessary, double spreading (coating both mating surfaces of a joint with glue) is recommended. In double spreading, one-half of the glue required should be applied on each face.

The amount of glue required varies somewhat with gluing conditions and the species of wood being glued. The higher spread should be used with long assembly periods, excessively sloping or end-grain surfaces, scarf-joint surfaces, or imperfectly planed surfaces. Lower spreads may be satisfactory when the assembly period is short and the wood used is not too absorbent.

Figure 2-3.—Typical glue spreader with rubber-covered spreader rolls and doctor rolls.

The glue spread can be checked by weighing a board of a convenient size before and after spreading and dividing the weight of the glue by the area of the board. Glue spreads should be checked frequently. Test pieces should be the same thickness as the material being glued or laminated. If a test piece 6 inches wide and 26½ inches long (total area of both faces 2.2 square feet) is used and glue is applied on both faces of the test piece, the weight of glue in grams is numerically equivalent to the number of pounds per thousand square feet of surface area spread. If, for example, a test piece of this size increases 30 grams in weight when passed through the spreader, the rate of spread is 30 pounds per 1,000 square feet. If double spreading is done, the number of grams of glue on the test piece must be multiplied by 2 to obtain the rate in pounds per 1,000 square feet of joint area.

The glue spreader should be adjusted and the glue spread checked before the spreading operation is started. When adequate gluing pressure is applied at the proper time, a thin bead of squeezed-out glue along the edges of the joint indicates that sufficient glue has been applied. Cleaning schedules should be closely observed. It is particularly important to clean the spreader before the glue sets.

2.1.13 Assembly Time in Gluing.—The interval between spreading of the glue and the application of full gluing pressure is called "assembly time." If pieces of wood are coated with glue and exposed freely to the air, a much more rapid change in consistency of the glue occurs than if the pieces are laid together as soon as the glue has been spread. The condition of free exposure is called "open assembly," the other "closed assembly."

Experience has shown that for each glue there is an optimum assembly time during which gluing pressure must be applied in order to obtain the best glue bonds. With resorcinol and phenol-resorcinol glues the range between minimum and maximum permissible assembly time is rather wide and may be two hours or more at 70° F. For some of these glues a minimum assembly time of 30 minutes and a maximum of 2½ hours at 70° F. is recommended when gluing white oak. With species that absorb the glue or solvent from the glue faster, shorter assembly periods should be used. At low temperatures a considerably longer assembly time may be allowed than under hot, dry conditions. For any specific glue formulation, the manufacturer should furnish instructions for permissible assembly periods at different temperatures and for different species.

2.1.14 Gluing Pressure.—The application of adequate and uniformly distributed gluing pressure to the joint at the proper time is essential to the production of consistently high-quality joints. Pressure brings both wood surfaces into intimate contact with the glue, smoothes the glue to a continuous, uniformly thin film between the wood layers, and holds the layers in this position until the glue has set. The best results are obtained when the pressure is distributed uniformly over the entire joint area. Fluid pressure, such as is used in bag-molding processes with thin veneers, most nearly accomplishes these results. In laminating, retaining clamps spaced at regular intervals are generally used, with cauls laid between the clamps and the glued assembly to distribute the clamp pressure to other parts not directly under the clamps. Cauls must be thick enough to distribute the pressure uniformly, as well as being true and even. Clamps must be sufficiently close together to produce adequate pressure between as well as under the points of contact.

Firehose inflated with compressed air has been used between the cauls and the assembly, although it is not common practice. The hose must be of sufficient diameter that the portion in contact with the assembly is fully as wide as the assembly. Caul boards are also sometimes placed between the firehose and the glued assembly; in this case the hose need not necessarily contact the entire width of the caul board if the latter is sufficiently thick so it does not deflect across its width. Several firehoses side by side can also be used to obtain the desired pressure (figure 2-4). The pressure in the hose may be read by means of a hydraulic pressure gage, and the total load computed from the following relationship: Load = pressure x contact area between hose and caul.

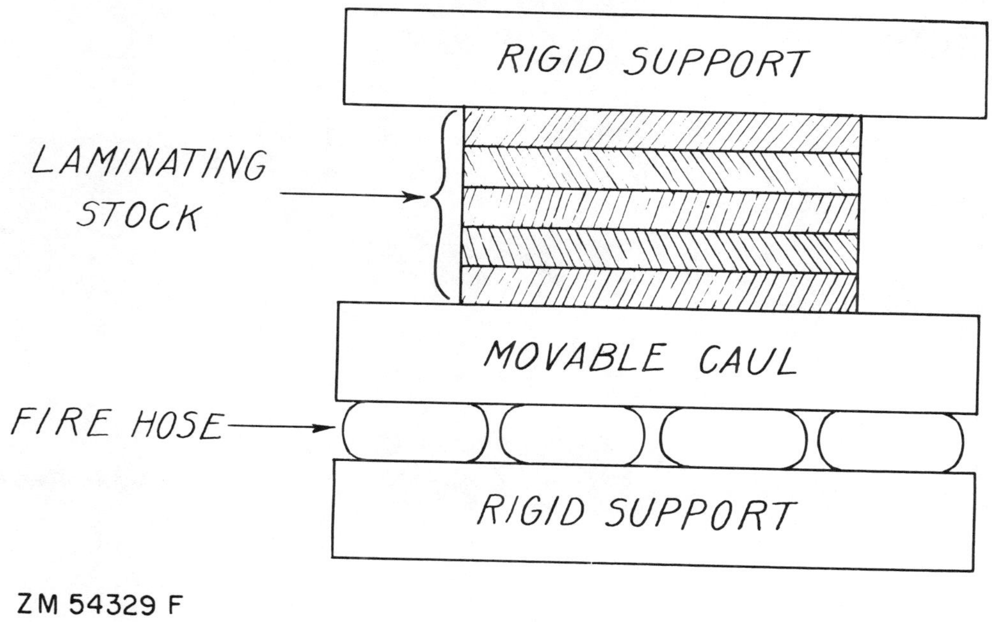

Figure 2-4.—Cross-sectional plan of a firehose press.

If, for example, for each hose shown in figure 2-4, it is assumed that 2 inches of circumference of the inflated hose bear on the caul, and the gage shows 100 pounds of pressure in the hose, then the load delivered per 12 inches of length of a single hose will be 100 x 2 x 12, or 2,400 pounds. The four hoses shown in figure 2-4 would, accordingly, deliver 4 x 2,400, or 9,600 pounds per foot of length in contact with the caul. If the width of block to be laminated is 5 inches, this load per foot of length must be distributed over 60 square inches (5 x 12) of glue joint.

Gluing pressure = 9,600/60 = 160 pounds per square inch. A heavy, rigid caul would be required to prevent undesirable deflections in this type of pressure application.

The amount of pressure required to produce a strong joint varies over a wide range, depending upon the species, the construction, and the consistency of the glue. Species of high crushing strength require and withstand higher pressures than woods of low strength. With dense woods like oak, satisfactory results may be obtained with pressures in the range of 150 to 200 pounds per square inch. For softwood species, pressures of 125 to 175 pounds per square inch are usually satisfactory. The successful use of light pressures presupposes that the wood surfaces are true and accurate as to fit or that they make intimate contact under small loads. The minimum pressure permissible for any assembly is one that will insure close contact between the wood surfaces and hold the members in close contact until the glue has set.

C-clamps and bar clamps are used to apply pressure to smaller glued parts. For laminating of larger parts, the clamp illustrated in figure 2-5 is sometimes used.

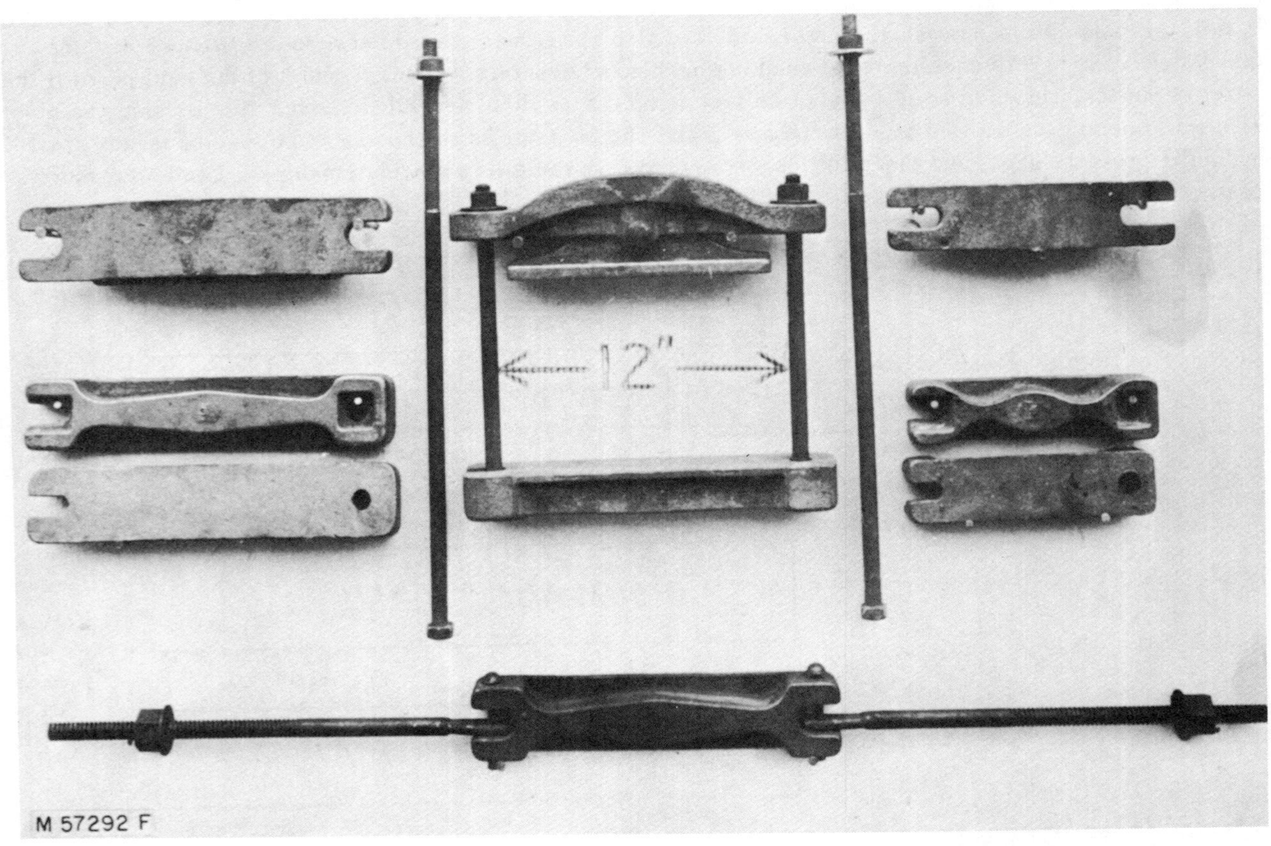

Figure 2-5.—One type of rocker-head clamp used to apply pressure to laminated assemblies. (see fig. 2-19).

With jack screws or clamps, the use of a torque-indicating wrench or lever arm which shows the force applied is recommended (figure 2-6, B). The amount of pressure exerted by a screw when a known torque is applied may be measured by a hydraulic device called a compressometer. The compressometer is illustrated in use in figure 2-6, A and in the drawing in figure 2-7.

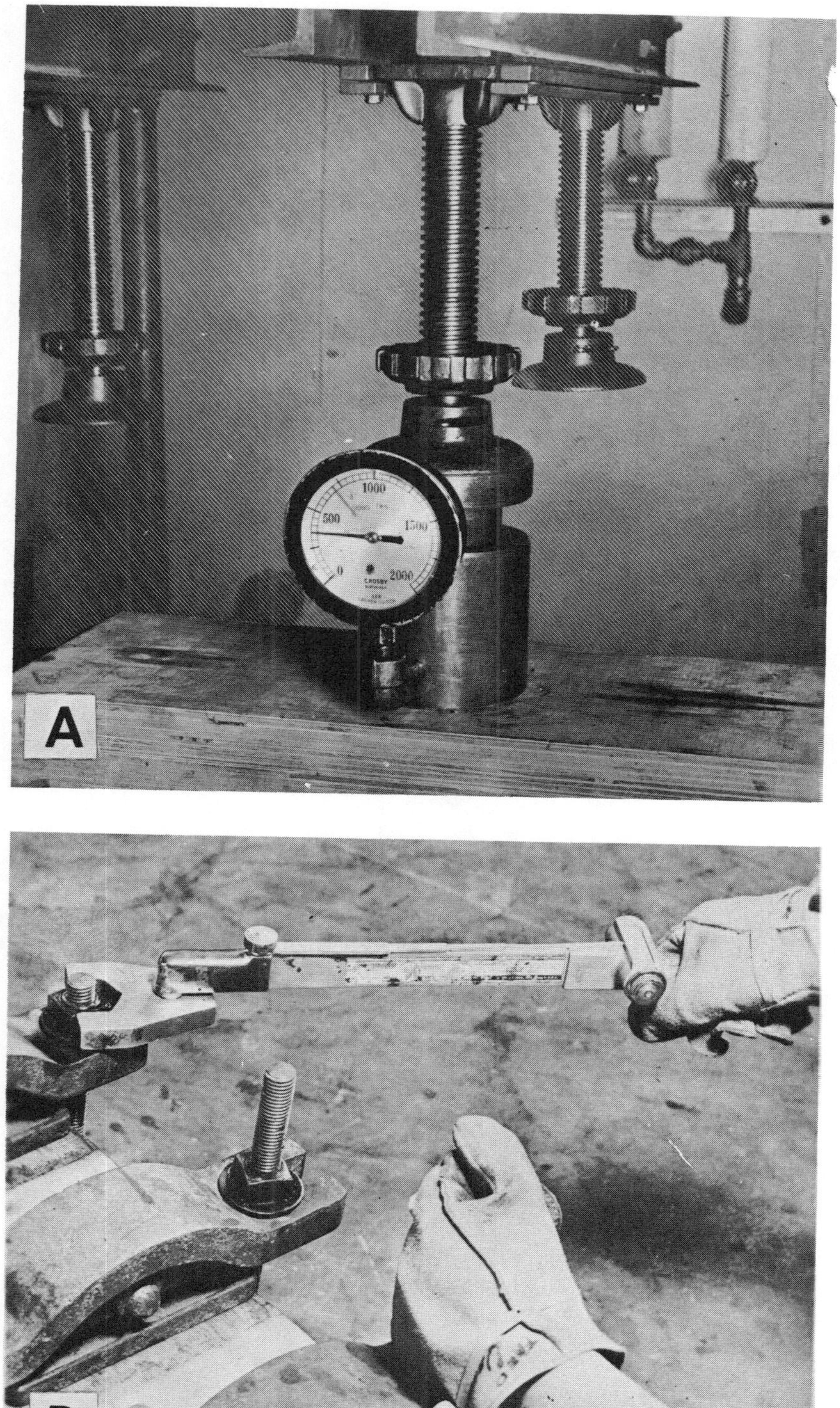

Figure 2-6.—A, compressometer for measuring pressure applied to clamps, and B, torque wrench.

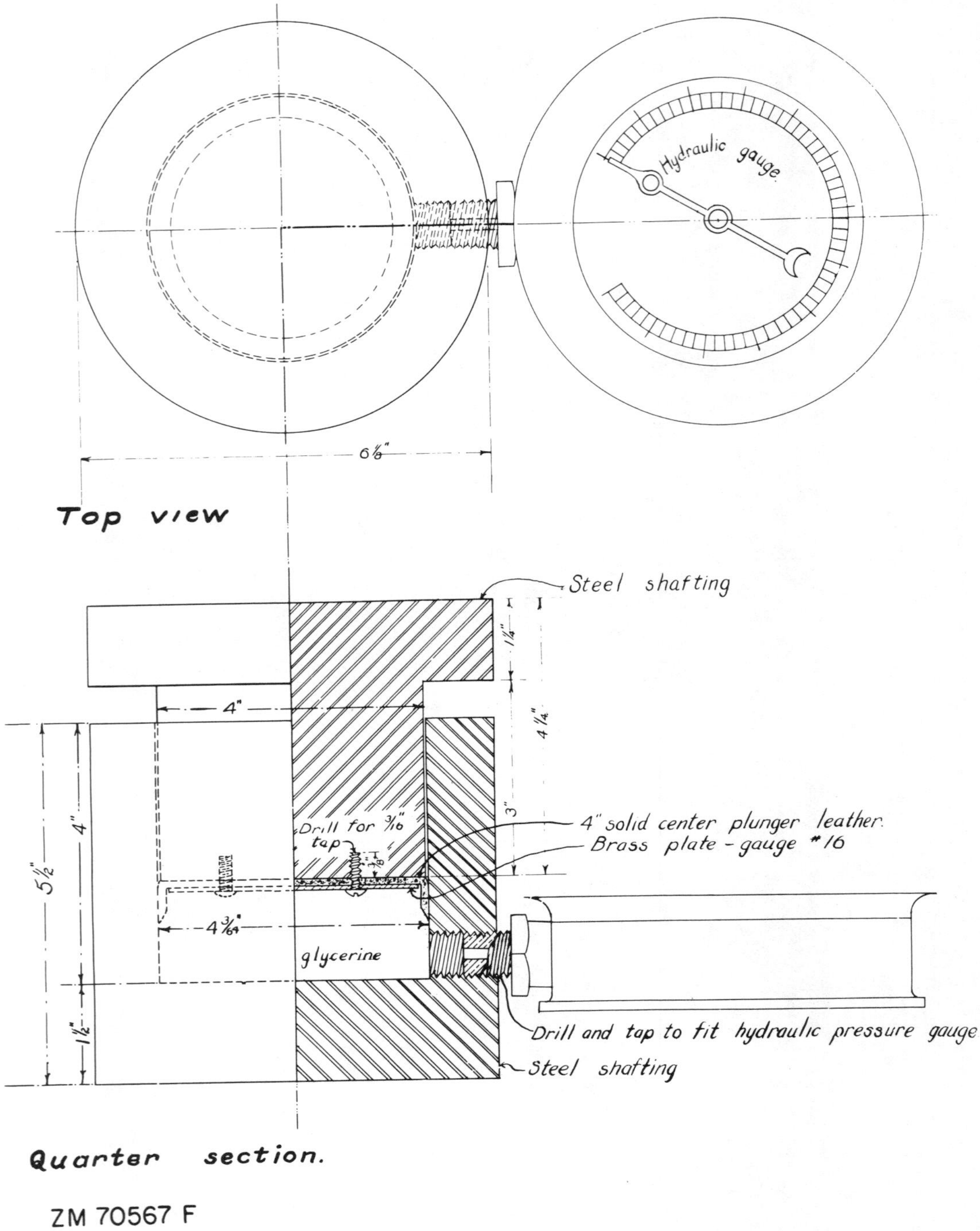

Figure 2-7.—Detail drawing of compressometer.

2.1.15 Curing of Glues.—Curing of resorcinol and phenol-resorcinol glues is a function of both time and temperature. At slightly elevated temperatures, impractically long curing periods are required, particularly with dense species. Higher temperatures increase the rate of cure. Elevated temperatures are utilized for lamination of heavy timbers and for various assembly-gluing operations to cure the glue in the shortest practical time. The glues are spread and the assemblies laid up and clamped at ordinary room temperatures, but are later placed in heated chambers, or in special assembly jigs heated by steam, hot air, electricity, high-frequency currents, and other means. All such operations except high-frequency heating involve the problem of heat transfer through the wood to obtain the required temperature at the glue joint.

In all assembly-gluing and laminating operations carried on at elevated temperatures, excessive drying of the wood should be prevented during the heating period. For this reason heated chambers must be properly humidified.

2.1.16 Final Conditioning.—After gluing pressure has been released, a conditioning period is sometimes required to allow time for completion of the curing reaction and for the solvent from the glue to distribute itself throughout the assembly. Glued members that have been fully cured at elevated temperatures under controlled relative humidity conditions do not normally require further conditioning treatment. If room temperature cures are used, as is done in assembly gluing operations, several days or weeks may be required before the glue is sufficiently cured to permit appreciable stressing of the members. During this period it is very important that the assembly glued members be protected from severe mechanical stresses or drying conditions, which could cause glue bonds to rupture. Good paint protection will reduce the rate of moisture changes and thus alleviate drying stresses on the wood and the glue joints.

2.1.17 Glue-Joint Quality Test.—Before using a new glue on production, it is desirable to make some tests to determine its quality. Tests for the quality of glue joints in laminated members are usually made by the block-shear and delamination test methods. Figure 2-8 shows the shear test specimen, shear tool, and testing machine used to conduct block-shear tests.

The more important points of procedure in the test are summarized as follows:

1. For test specimens, use wood that is straight grained, free from defects, and of high density. Test the species to be used on production. Condition the wood to a moisture content similar to that of the production material, usually averaging about 12 to 14 percent for material to be used under outdoor or exterior conditions.

2. Cut this material into pieces about 1 by 2½ by 12 inches, in size, or of such other width and length as the testing specification calls for so as to provide at least four specimens of the dimensions shown in figure 2-9, A. The specimen shown in figure 2-9, B often is more practical to use. Surface the pieces smoothly to uniform thickness and glue promptly after surfacing.

3. Glue at least two joints for each test if the specimen shown in figure 2-9, A is used. If specimen type B is used, three pieces of the dimensions given under point 2 would provide about the same number of joint tests.

4. Spread the glue evenly on the mating surfaces at the rate recommended by the glue manufacturer and apply pressure uniformly to the joint within the assembly period limitations specified for the glue. It is good practice to glue some samples at the shortest and some at the longest assembly period recommended. The quantity of the glue spread can be determined by weighing the pieces immediately before and after spreading.

5. Apply a pressure of 150 to 200 pounds per square inch and leave the test blocks under pressure for a period and at a temperature recommended by the manufacturer for production gluing. Condition the test blocks for about six days at room temperature before testing.

6. Cut the glued blocks into specimens of the form and dimensions shown in figure 2-9, and test on a machine equipped with a shearing tool (figure 2-8). Apply the load to the specimens at a rate of 0.025 inch per minute, plus or minus 25 percent.

7. Record, for each joint or shear block tested, the breaking load and the approximate percentage of wood failure occurring over the glue-joint area. Compute the breaking load in terms of pounds per square inch of glue-joint area. Both the shear strength and wood failure must meet the values required by the specification under which the glued product is being produced.

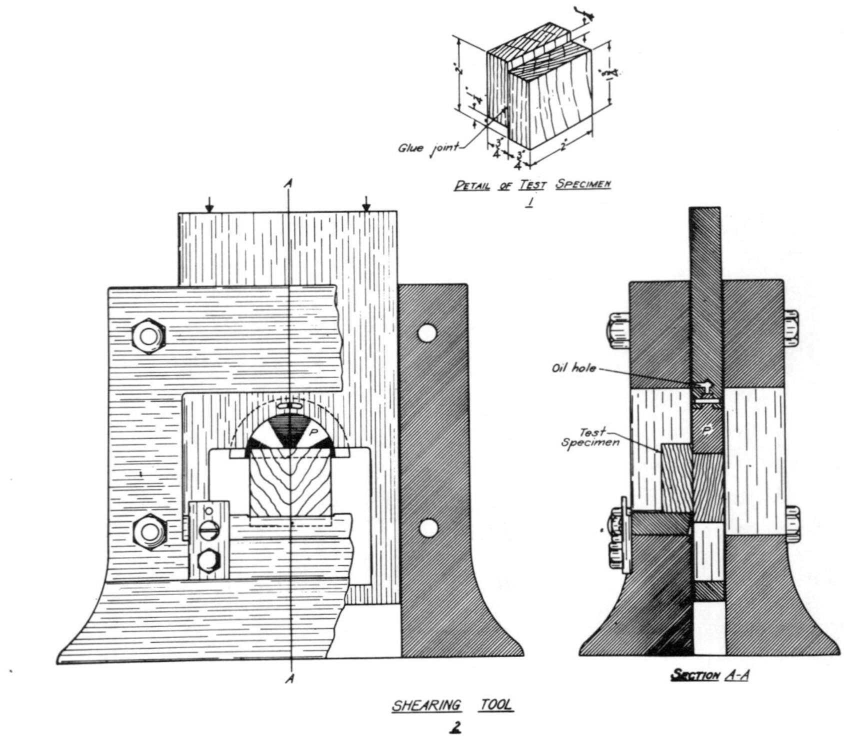

M 41775 F

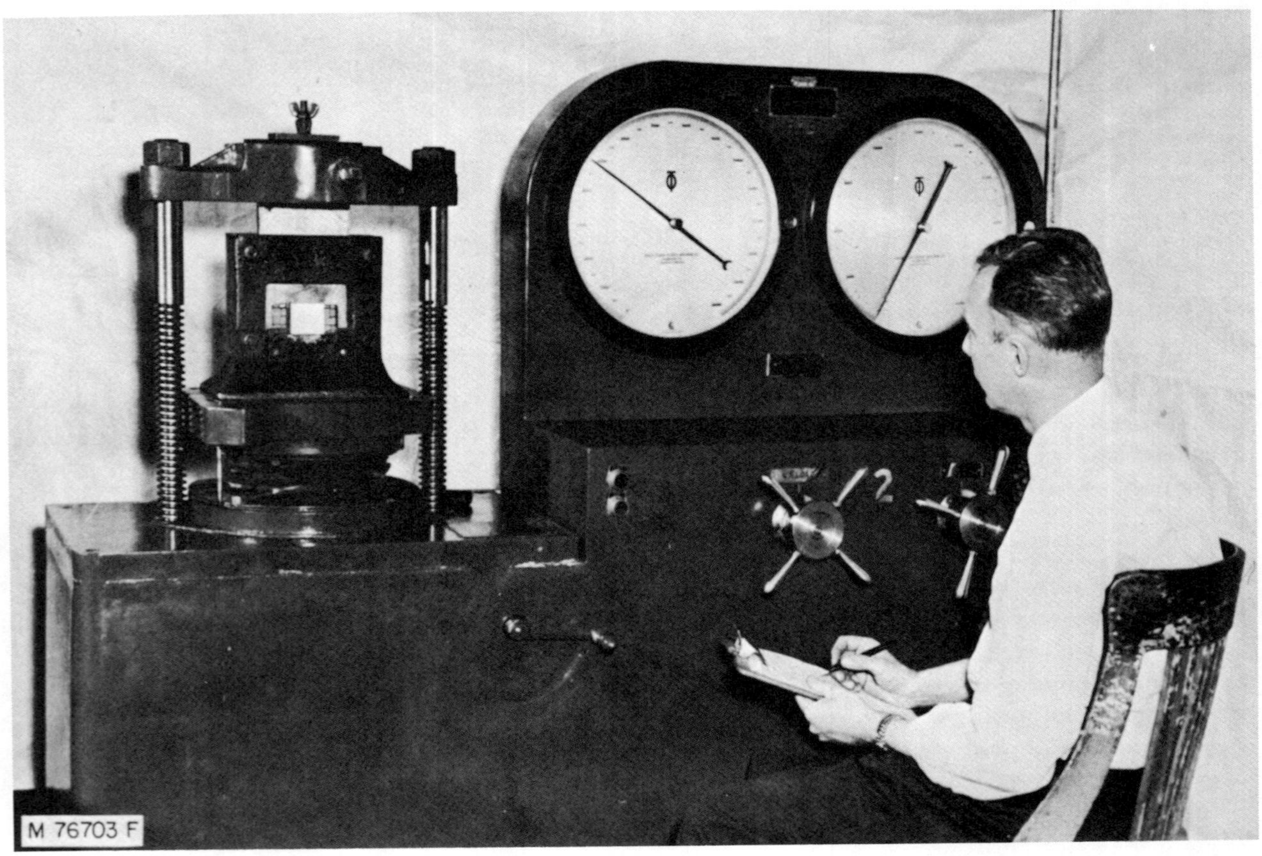

Figure 2-8.—Shear-test specimen and tool (upper) and machine used (lower) for block-shear tests.

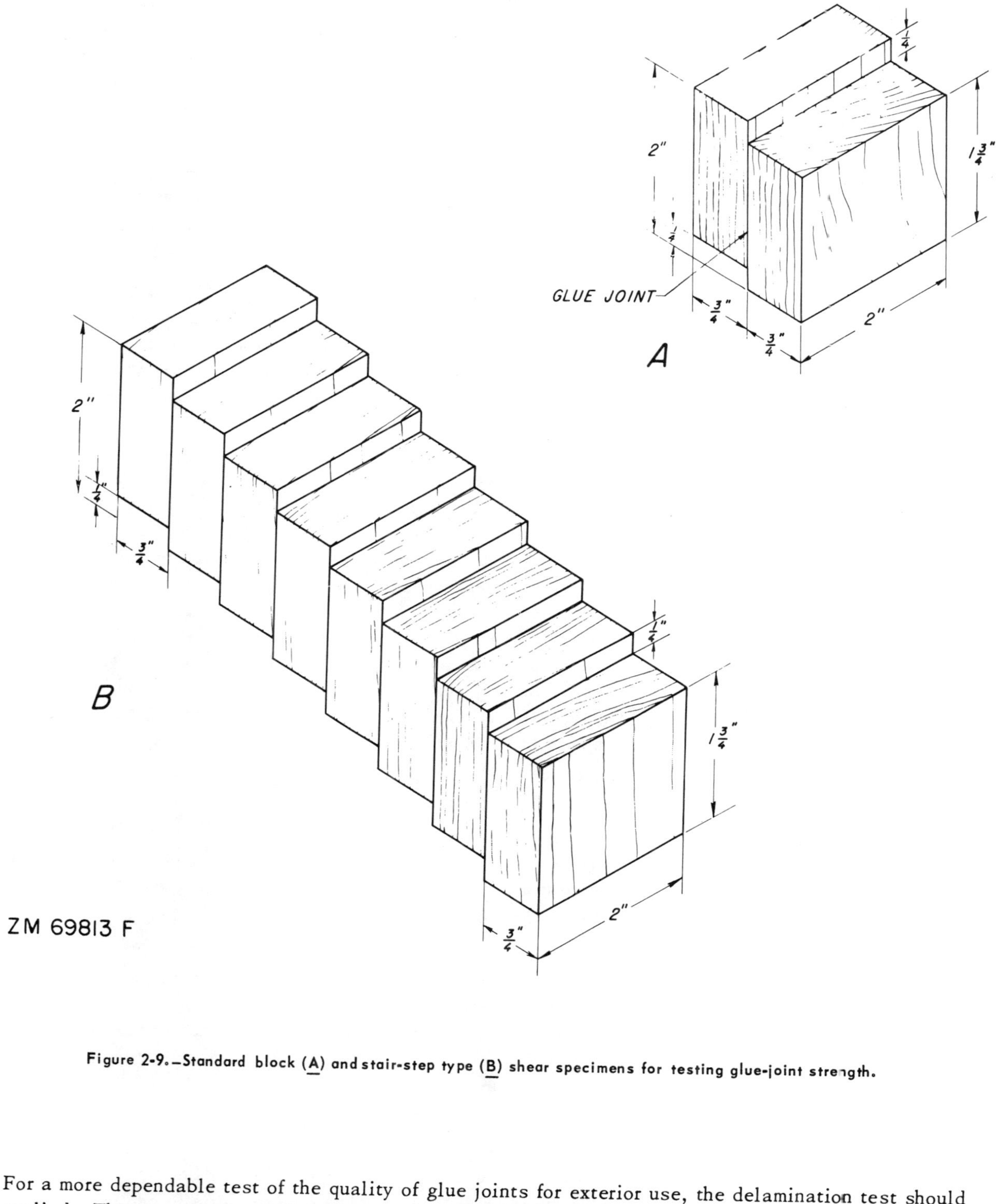

Figure 2-9.—Standard block (A) and stair-step type (B) shear specimens for testing glue-joint strength.

For a more dependable test of the quality of glue joints for exterior use, the delamination test should be applied. The test specimens for this method should be about 6 by 6 inches in cross section and made up of laminations about 3/4-inch thick and long enough to provide at least three 3-inch-long sections. It is usually most economical to make the beam of sufficient length to provide material for both the block-shear test and the delamination test. The stair-step type of shear specimen is used for this shear test. The 3-inch-long sections (along the grain) for the delamination test should be cut with a high-speed saw or similar tool, so that the end-grain surfaces are smooth.

The procedure for the test is as follows: Place the test specimens in an autoclave or other type of pressure vessel and immerse them in water at room temperature, 65° to 80° F. The specimens should be weighted down to keep them submerged, and should be separated by stickers or other means in such a manner that all end-grain surfaces have free access to the water. Draw a vacuum of at least 25 inches of mercury (at sea level) and maintain it for two hours, then release the vacuum and apply a pressure of 80, plus or minus 5, pounds per square inch for two hours.

Repeat this vacuum-pressure cycle with the specimens still immersed, making a two-cycle impregnating period requiring a total of eight hours. Dry the specimens for 88 hours in air at 80° to 85° F. and 25 to 30 percent relative humidity, circulated at a rate of 500, plus or minus 50, feet per minute. While drying, the specimens should be at least two inches apart, with their end-grain surfaces parallel to the stream of air.

Run the entire soaking-drying cycle three times to comprise a total test period of 12 days. At the end of the final drying period, measure the total length of open glue joints (delamination) on the end-grain surfaces of the specimens to the nearest 1/20 inch (0.05 inch). Failure in the wood due to checking or other causes should not be regarded as delamination. The amount of delamination should not exceed that prescribed by the specification under which the glued material is produced.

2.2 LAMINATING

2.2.1 General.—Laminated wood parts were first used by the Navy on a limited scale during World War II. Their performance was excellent, and, as a result, the laminating of Navy ship and boat parts was greatly expanded during the Korean emergency. Laminated skegs of landing craft, for instance, which are fully exposed to the water and to frequent drying when the boats are out of water, are reported to have far outperformed their solid counterparts. Such service was made possible by the development of very durable glues that cure at temperatures attainable in heavy assemblies. Ship and boat parts that have been produced by laminating include keels up to 12 by 12 inches by 120 feet, stems varying in cross section from 12 by 12 by 12 by 60 inches, frames varying in cross section from 6 by 6 inches at the deck to 6 by 15 inches at the keel and about 70 feet in length, rudder foundation blocks 4 by 4 by 2 feet, and breasthooks, bulwark stringers, shelves, garboards, deck beams, deck girders, shaftlog girders, etc. As many as 2,200 laminated members are used in a ship such as a large wood minesweeper.

The term "glued laminated construction," as applied to structural members, refers to material glued up from smaller pieces of wood, in either straight or curved form, with the grain of all laminations approximately parallel to the length of the member. It is thus basically different from plywood, in which the grain direction of adjacent plies is usually at right angles. The laminations may be of any practical thickness or length, of narrow pieces glued edge to edge to make wide ones, of different wood species, or of pieces bent to curved form during gluing—all of which afford infinite choice in design, subject only to economic factors involved in production and use.

The requirements for laminated members in ships and boats vary considerably for the particular parts or items involved. Navy specifications covering the specific items describe the quality of products required and permissible defects. It is very important that the laminated items be produced under adequate control of the laminating process to ensure satisfactory service for marine use. The following paragraphs discuss good laminating techniques based on the discussion of glues and gluing in section 2.1.

2.2.2 Materials for Laminating.—Laminated members of practically any size and shape can be built up from boards. This permits utilization of lower-grade material, inasmuch as defects can be cut out. Use of board-thickness stock also accelerates and simplifies seasoning. The opportunity afforded by laminating to produce structural elements that vary in cross section in accordance with strength requirements, and of practically any curvature, is of particular importance in ship and boat building because parts of so many different shapes and sizes are needed.

In selecting and preparing lumber for laminating, it is essential to consider the requirements of individual laminations as well as the end use of the finished member. Sharply bent members require laminations that are thinner and of better-grade stock than do flat or slightly curved assemblies. Defects such as cross grain, pitch, large knots, holes, and decay contribute to breakage of bent laminations and also reduce the effective glue-joint area. Such material must be limited in accordance with the strength requirements of the product. The amount of sapwood must be kept to a minimum or the wood must be preservative treated if timbers of good durability are required.

2.2.3 Segregation for Grain.—Because plainsawed and quartersawed boards shrink and swell different amounts with equal changes in moisture content, it is good practice to use only one type of material in the same assembly. It is therefore advisable to separate the plainsawed from the quartersawed lumber. Boards should be classified as quartersawed according to predominant ring direction. It usually is most convenient to do the sorting as the stock leaves the rough planer. Segregation for ring direction is more important with dense than with light species.

2.2.4 Moisture Content of Lumber for Marine Laminating.—Lumber used for laminating ship and boat members should be between 10 and 15 percent moisture content. Lumber within this moisture-content range has been found satisfactory for practical use in service, and has also been found to bond satisfactorily with marine service resorcinol and phenol-resorcinol adhesives.

Members laminated at this moisture content must be protected by surface coatings or paint, particularly when the boats are being built in areas, or during seasons, of low humidity. To avoid checking, therefore, during ship construction it is usually necessary to apply a good surface coating on laminated members as soon as they are made. Several coats of high-grade aluminum paint will provide good protection.

It is important that the moisture content of the different laminations of any one assembly, and throughout the cross section of each lamination, be fairly uniform (section 2.1.7). A resistance-type electric moisture meter is a convenient means for spot checking the moisture content of surfaced stock. The meter should be checked for accuracy at frequent intervals by the standard oven test.

2.2.5 Preliminary Surfacing.—Rough planing usually is the most practical first step in machining stock for laminating. This operation, in addition to equalizing the thickness of the boards, will disclose many natural and seasoning defects otherwise difficult to detect.

Reducing lumber to fairly uniform thickness by rough planing facilitates subsequent operations, such as ripping, edge gluing, scarfing, and final surfacing. To maintain a balance of seasoning stresses that may exist in the lumber, equal light cuts should be made on each face of the boards. A double surfacer will usually be satisfactory for this operation. Before proceeding with the machining operations of material for laminating, careful planning of the sequence of operations always is advisable. The sequence in ripping and cross cutting to eliminate defects and reduce stock to desired size will be determined by layout and facilities of the individual plant. In some boards, less waste is incurred if ripping precedes the cross cutting; for other boards, it is advantageous to reverse the process, depending on the type of defect.

2.2.6 Steps in Preparing Laminates.—

2.2.6.1 Edge Joints.—Lumber is edge glued to obtain laminations of the required width. Joints machined with a jointer or a molder are preferable to sawed joints for good edge gluing, even though joints of the former type involve more loss of material and are more difficult to prepare on boards of appreciable length. Straight-line ripping, however, is required before edge surfacing to provide straight edges on the strips of lumber to be edge glued. The glue may be applied by brush, mohair paint roller, or, preferably, by a single-roll mechanical spreader. Bar clamps or piling clamps are often used to apply pressure in edge gluing. Such clamps will permit assembly of the stock on trucks for ready removal to a heated chamber for curing. Clamp carriers also are used for edge gluing, although the width of the carrier limits the length of the boards. Clamp carriers can be enclosed and heat applied to speed the cure of the glue.

In assembling edge-glued joints, it is important to provide sufficient gluing pressure along the entire glue line. Best distribution of pressure is obtained if the wider boards are placed on the outside, next to the clamps, and the narrower strips are placed inside the pack. For outside boards not less than 4 inches wide, a maximum clamp spacing of 14 inches is recommended; if the outside boards are less than 4 inches wide, closer clamp spacing should be used. Tightening of clamps should begin at the center of the package and proceed progressively toward each end.

Well-made edge joints are as strong as the wood itself. Edge joints in the same plane in adjacent laminations, however, increase the possibility of vertical cleavage if any imperfect joints are present. When laminations are laid up, therefore, it is good practice to offset the joints by at least the thickness of one lamination. Edge-glued and solid boards should be alternated in successive laminations whenever possible.

Several methods of gluing stock to be band-sawed for deck beams in small craft are shown in figure 2-10. A special double tongue and groove is employed in A. The material is glued into wide boards, from which individual beams are sawed to the camber of the deck. Sample edge gluing is shown in B. The preferred, and by far the best method of laminating the stock to the proper curvature in shown in C.

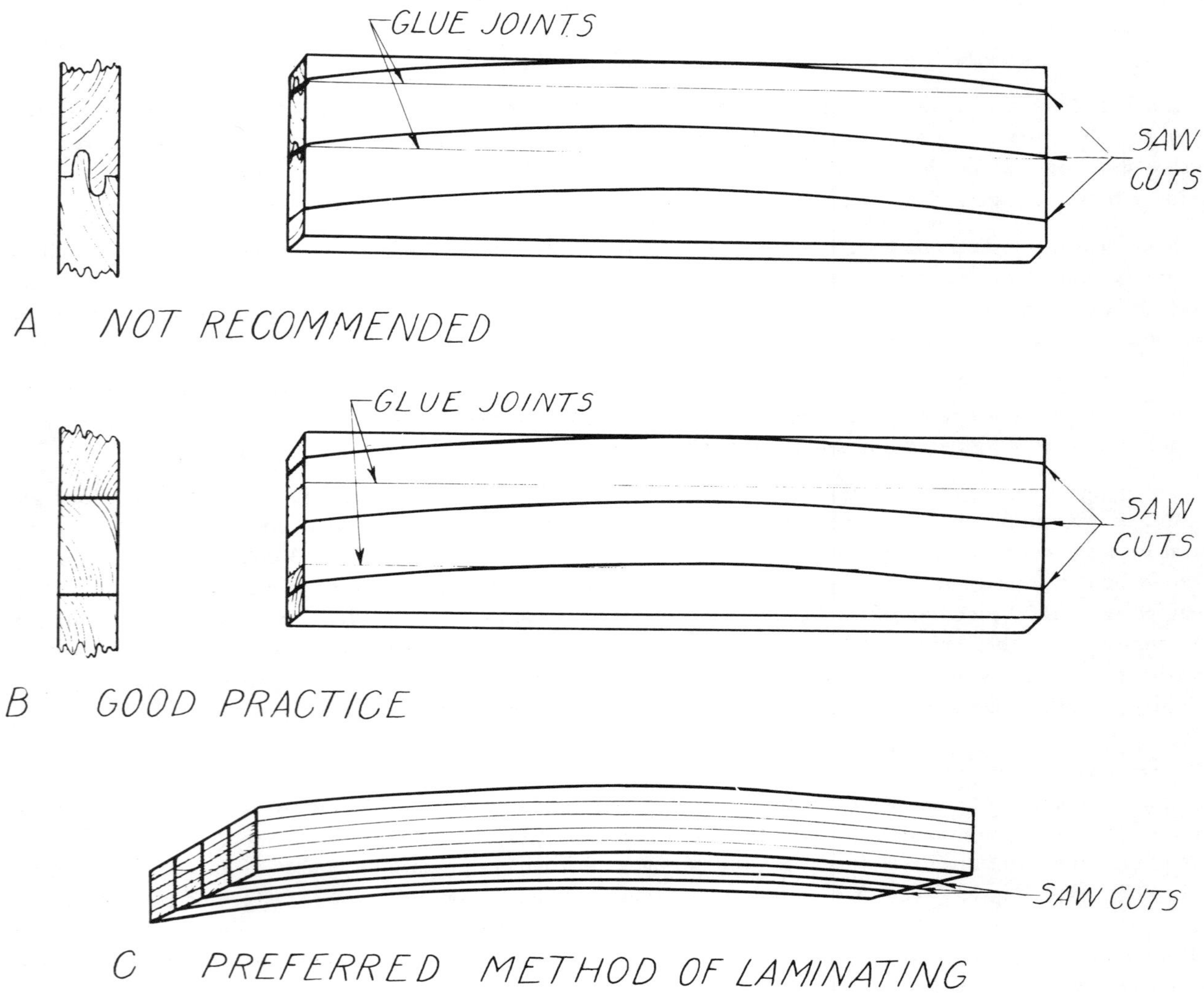

M 56350 F

Figure 2-10.—Three methods of fabricating deck beams. A, beams are bandsawed from wide pieces formed by gluing double tongued-and-grooved joints—not recommended because of the difficulty of getting the accurate fit essential to good gluing; B, simple edge gluing, preferred to A because of ease of fitting joints; C, laminating, a superior method of forming the beam to the required camber.

2.2.6.2 End Joints.—End joints are employed for better utilization of the lumber and to obtain laminations of the required length. There are various types of end joints, such as the plain scarf, hooked or stepped scarf, and finger joints of numerous designs (figure 2-11). Glued scarf joints with relatively long slope are probably the only end joints that can develop a strength approximating the tensile strength of the wood. The plain glued scarf joint with a slope not steeper than 1 in 12 is almost exclusively specified for marine laminating, due to the fact it can be most readily and consistently made strong and durable, with joint strength 85 to 95 percent that of adjacent solid wood. Scarf joints of short slope and other end joints with low strength are occasionally useful, however, in facilitating the handling – as full-length pieces – of laminations to be used in short columns or low-stressed members. Such joints must be spaced sufficiently to produce laminated members of adequate strength. The stepped scarf offers some advantage in case of positioning the parts to be end jointed, but the strength is reduced in proportion to the depth of the step.

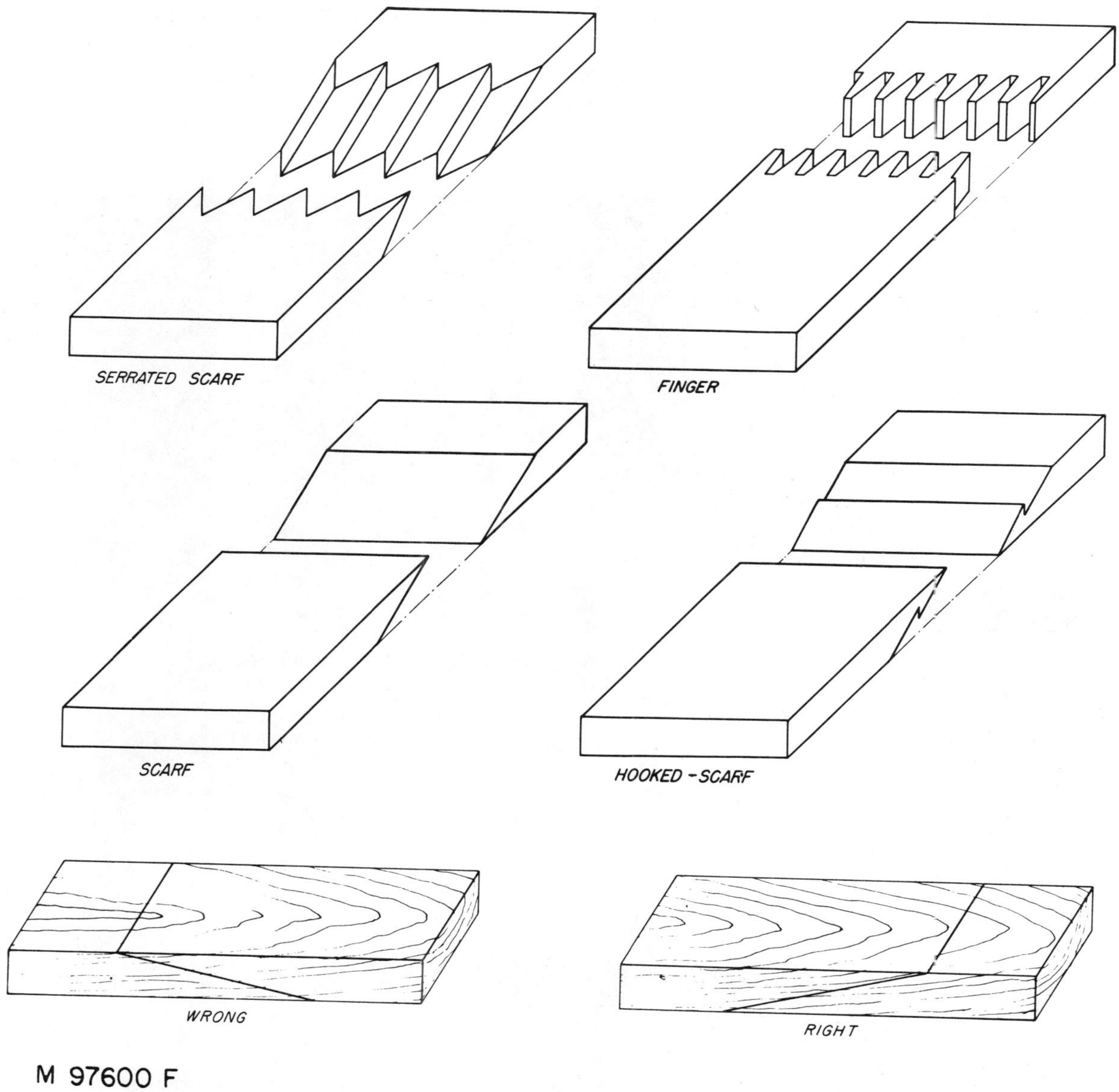

Figure 2-11.—Typical glued end joints used in laminating, and wrong (cut against grain) and right (cut with grain) ways to cut scarf joints through wood.

Glued end joints that must have high tensile strength should be cut so that the joint surfaces are smooth and fit true, and should be prepared and glued before the final surfacing is done. This will ensure uniform thickness throughout the full length of each lamination containing end joints and will facilitate speedier assembly during the final gluing operation. Surfaces for scarf joints can be prepared on special scarfing machines, planers, shapers, single-end tenoners, or saws. Surfaces prepared on machines equipped with cutterheads fitted with sharp knives generally are superior to sawed surfaces, especially with lumber of dense hardwood species, and are preferred for producing satisfactory scarf joints. Other types of end joints, such as the serrated scarf and finger joints, require special machines and cutterheads for their preparation.

Machines for cutting scarf joints are usually made to the user's specifications, but at least one type is commercially available. In this machine several boards to be scarfed are laid edge to edge and clamped to a flat bed, both at the extreme end of the boards and near the heel edge of the scarf (figure 2-12). The traveling cutterheads do not cut completely through the boards, but leave about 1/20 inch at the feather edge which is subsequently broken off (figure 2-13). The advantage of this method is that any stresses in the boards are prevented from causing the feather edge of the scarf to be distorted during cutting, which might result in a poorly fitted joint. Closeup of cutterheads and a general view of the machine are shown in figure 2-12.

Figure 2-12.—Scarf cutting machine. Inset (upper left) shows cutterheads and inset (lower right) shows method of clamping boards to be scarfed at feather edge and at heel.

Figure 2-13.—Boards scarfed to within 1/20 inch of their thickness.

It is desirable to have the slope of a scarf with, rather than against, the grain of the wood (figure 2-11). Steep slopes have an advantage in that they result in less waste of wood, but they also are weaker and may contribute little or nothing to the final strength of the laminated member. There is also the risk that they will break when long laminations are handled in assembling and when bent to curved form. Glued scarf joints introduce added stiffness to a board at the scarf area that somewhat increases the difficulty of drawing such areas into form when curved assemblies are fabricated.

In the gluing of a scarf joint or any other type of end joint with sloping surfaces, it is important to provide some method to prevent end slippage, such as blocking or clamping the ends of the boards. Blocking or clamping should also be arranged to minimize side slippage and thus prevent unnecessary waste of material. Efforts should be made to keep the parts in proper alinement, as illustrated by figure 2-14. A small amount of overlap, as illustrated, is desirable and ensures that the joint will receive adequate pressure. If the members slip endwise excessively during the pressing operation, a condition will exist as illustrated in figure 2-14, C. When the members are in this position, the joint will not receive sufficiently uniform pressure and erratic joint strength may be expected. The condition illustrated by figure 2-14, B, is to be preferred to that in C. In case the members are in a position illustrated by B, it is possible that the scarf joint will receive higher pressure than is intended and some crushing may result, but probably the quality of the glue joint will not be adversely affected.

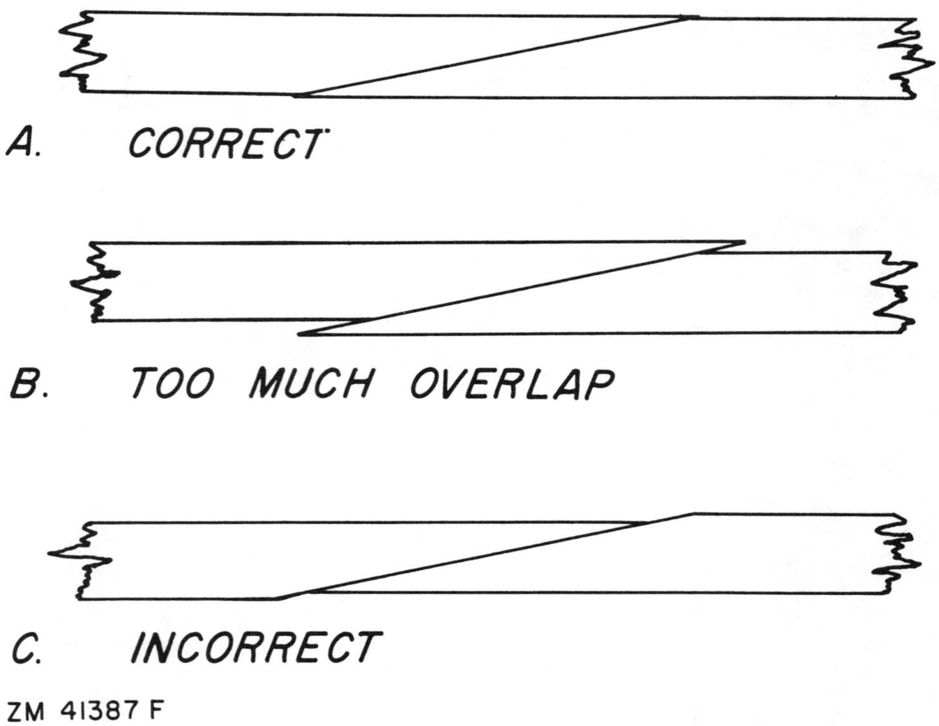

Figure 2-14.—Alining scarf joints for gluing. A, correct, with slight overlap; B, incorrect, too much overlap; C, incorrect.

A small dowel often is used for alining the scarf joint while it is glued. A jig for use when drilling the dowel holes, to ensure that the scarfed surfaces are uniformly overlapped, is illustrated in figure 2-15. Part A of this figure shows how the jig is set to accommodate scarfed boards of a certain thickness. The thickness of the feeler gage shown in the sketch determines the amount of overlap. After the top member has been adjusted to the proper position, as shown in A, the feeler gage is removed. Figure 2-15, B, shows the jig in use. The details of the construction of the jig and for adjusting the distance between the top and bottom member may be varied to suit the user's needs.

To obtain uniform spreading of all end-joint areas, it is recommended that glue be applied to both contacting surfaces. Because glue penetrates readily into end grain, a fairly heavy spread should be used. In gluing scarf joints, it is important that the adhesive be allowed at least the minimum assembly time recommended by the manufacturer before gluing pressure is applied. It usually is more convenient to give the joints an open than a closed assembly period. The minimum open assembly required also is much shorter, usually about one-half of the minimum closed assembly that would be needed. Ordinarily it is advantageous to use somewhat greater gluing pressure for scarf joints than for side-grain to side-grain joints.

Various means of heating are employed for curing glue in scarf joints. Curing chambers with insulated, highly vaportight enclosure equipped with heating coils, steam spray, and forced air circulation with proper controls give uniform and reliable results where size of members and quantity of scarf joints merit use of this method. Infrared heat lamps are growing in use as a reliable means of producing satisfactory scarf joints. When the lamps are used in banks enclosed in a metal frame and placed on each side of a group of scarfed boards mounted one above another and properly clamped, with a thin metal sheet separating each of the scarfed boards, the heat is conducted to the center of the stock and the efficiency of the heating process is considerably increased.

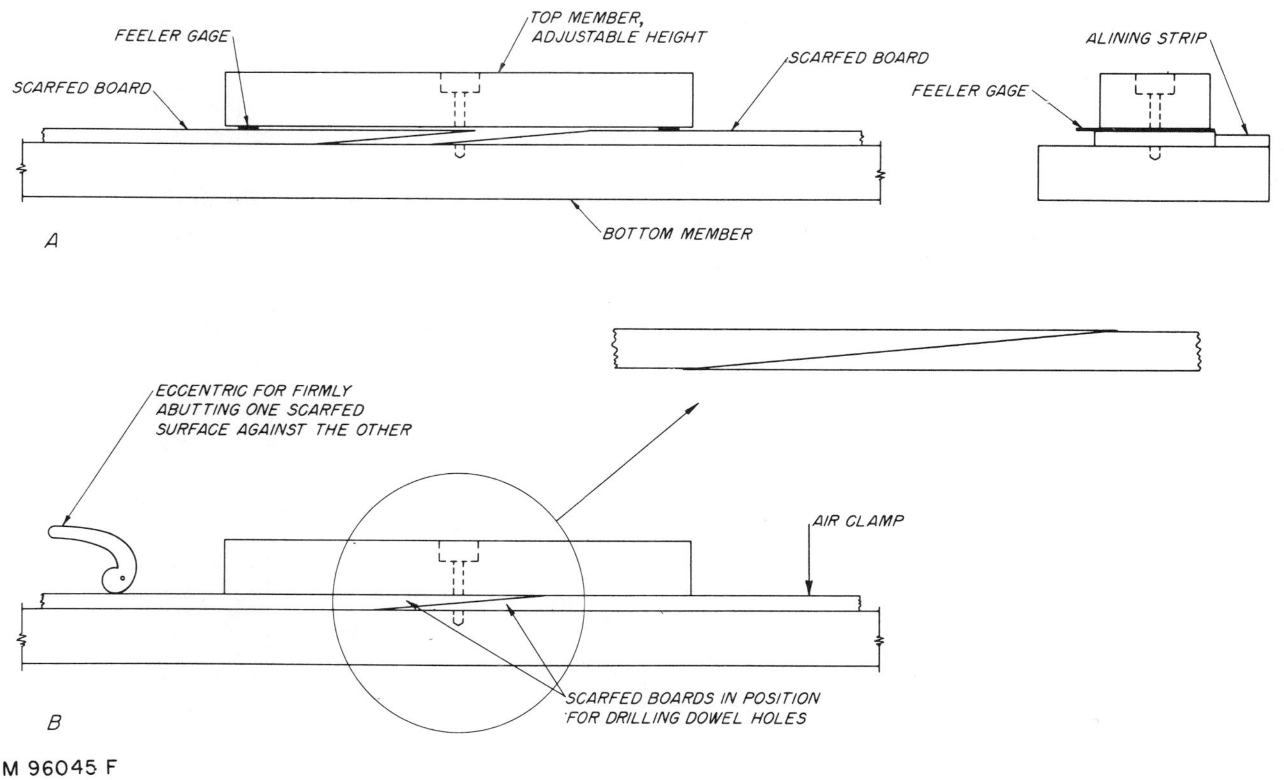

Figure 2-15.—Sketch of jig for alining scarf joints while drilling hole for positioning dowel.

High-frequency curing may be adapted for quick setting of the glue in scarf joints, particularly when narrow boards are end jointed and parallel heating is practical. Where wider boards (about 4 inches or more) are end jointed, stray-field or perpendicular heating is probably more applicable. Unless great care is exercised, scarf joints produced by the high-frequency method may not be of uniformly high quality. The press and generator must be adequately shielded to prevent radiation, which may interfere with radio and television. Cost of equipment is usually greater where high frequency curing is employed than with other methods. The requirement for proper alinement of the scarf surfaces is equally important regardless of the curing method used in setting the glue.

Hot presses have been used in some plants for curing the glue in end joints. Severe surface drying of the lumber is usually a problem with this method. When several scarfs are glued side by side in a hot press, it is important that all boards are of equal thickness and that the overlap is uniform. Silicone rubber sheets between the platens and the boards help to compensate for slight inaccuracies in machining, but increase the required heating period.

Flexible heating pads made of nicrome wire imbedded in silicone rubber have also been used to cure glue in scarf joints. This method is more efficient than when heat is applied from the outside by use of infrared lamps, but the initial cost of equipment usually is higher. When several joints are pressed in one package, the pads serve as separators between the different joints. The silicone rubber is more flexible under prolonged or repeated heating than ordinary rubber and thus compensates for slight inaccuracies in machining. When several joints are pressed in one package, fairly heavy cauls must be used on top and bottom of the package.

2.2.6.3 Final Surfacing.—The final surfacing before the stock is glued is one of the most important operations in the production of laminated members. Surfaced lumber should be flat and smooth to permit even spreading of the glue and to obtain intimate contact between the joint surfaces when uniform gluing pressure is applied. Before scarfed lumber is surfaced, it is necessary to remove surplus glue at the scarf joints; otherwise, skips in dressing are likely.

Final surfacing can be done most satisfactorily on a double cabinet-type surfacer equipped with well-fitted cutterheads mounted in ball bearings. Single-head cabinet surfacers are preferable to matchers and molders for the surfacing of laminations for gluing. The finish cut should be comparatively light, taking not more than 1/16 inch from each face. Rates of planer feeding that result in 20 to 30 knife marks per inch on the stock have been satisfactory for the final surfacing of laminations for gluing. The stock should be tested frequently with slip-on gages throughout the length of the piece. The thickness of a planed board should be the same when measured on either edge. Micrometers are best suited for tolerance measurements, but usually require too much time to be practical for routine checking. Differences in thickness of a lamination for gluing should not exceed 0.010 inch. The surfacing should be of such quality that knife marks are hardly perceptible.

Cup in boards after the final surfacing should not exceed 1/96 of the ratio of the board's width to its thickness (1/96 x W/T = allowable cup). For boards 6 inches wide this would allow 1/4 inch of cup in a board 1/4 inch thick, 1/8 inch in a board 1/2 inch thick, and 1/16 inch in a board 1 inch thick.

The layover period between surfacing and laminating should be made as short as practical, preferably not more than a day. Lumber stored too long after final surfacing may change in moisture content and may develop raised grain. Such lumber will not be in optimum condition for gluing.

2.2.6.4 Thickness of Laminations.—In straight assemblies, the thickness of laminations is governed mainly by the dimensions of available stock; in curved members bending radius is the deciding factor. The practice of steaming thick lumber to permit bending to sharp curvature has no application in a gluing operation. The minimum radius to which dry, straight-grained, clear lumber can be bent without breaking is approximately 40 to 60 times its thickness, varying with species of wood and the quality of individual boards. In general, hardwoods bend to somewhat shorter radius than do softwoods of the same thickness. Lumber containing knots and other defects cannot be bent to so sharp a curvature as clear lumber. Laminating specifications for particular species of wood give in detail the permissible radius of curvature for given lamination thicknesses, and should be closely adhered to.

2.2.7 Gluing Laminated Members.—Successful fabrication of laminated members depends in great part on following the correct procedure when gluing the assemblies. Good practice requires that the laminations be arranged in the proper order in a place that is convenient for feeding them to the glue spreader. It is necessary to spread the glue properly and place the laminations on the jig or gluing bed and apply adequate pressure uniformly and quickly to avoid excessive drying of the glue before application of pressure is completed. With resorcinol and phenol-resorcinol glues, a spread of 60 to 70 pounds per 1,000 square feet of joint area usually is satisfactory.

Before gluing is started, the laminations should be stacked in the order desired in the glued member, but with the bottom lamination of the glued assembly on top of the stack. Outer surfaces of top and bottom laminations should be covered with wax paper. The laminations then should be passed through a double-roll spreader that applies one-half the glue required on each face, and assembled in the same order. This procedure will minimize the chance of interchanging laminations, which might upset the predetermined pattern for end-joint spacing. If workmen are assigned to definite stations, confusion and delay in laying up of long assemblies can be avoided. The use of roller conveyors at the glue spreader may save manpower, but rollers used to carry glue-coated laminations should be double-cone-shaped to avoid appreciable disturbance of the glue film. The spread laminations should be assembled promptly, so that the glue surfaces are not exposed to the atmosphere, thus reducing evaporation of solvent and permitting longer assembly periods.

Normally, the lumber should be at room temperature before the glue is spread, especially when glues not requiring elevated curing temperatures are used. Since sizable assemblies heat rather slowly, minimum pressing periods are obtained if the lumber is at the desired glue-curing temperature when the assembly is laid upand clamped. Cold lumber will permit the use of longer assembly periods, such as may be required for very large members, but longer pressing or heating of the clamped assembly then becomes necessary.

Wood surfaces to be glued should be clean and free from crayon marks, grease, oil, or other materials that might interfere with good bonding.

In the usual gluing of curved, laminated members, the template is first placed on the jig and the jig stops are set to the form of the template (figure 2-16).

It is impractical to place retaining clamps on the assembly sufficiently close together to apply uniform gluing pressure over the glue joints. Cauls are placed between the laminated assembly and the retaining clamps, therefore, to distribute the pressure over the area between the clamps. These cauls may be of hard, dense wood, such as white oak, or of metal, and should bear against both faces of the assembly. Wide assemblies usually need short clamp spacings to obtain the required average pressure.

It is advantageous to use thick cauls except when curved members of short radius are made. Two or more thin cauls on each face of an appreciably curved assembly will distribute the pressure more evenly than one. On straight members, or members with a very mild radius of curvature, the distance between clamps should not be greater than seven inches center to center. On members with a radius of curvature so short that 3/4-inch or thicker white oak cauls cannot be applied, clamps should not be more than 4 inches apart on the concave surface of the bend. Fire hose inflated with compressed air is sometimes used between the caul and the assembly to provide more uniform pressure. The retaining clamps are tightened snugly before the air pressure is turned on.

Figure 2-16.—Template for U-shaped minesweeper frame on laminating jig or table as guide for setting jig stops.

Various types of jigs are used for gluing curved members. For small members, those illustrated in figure 2-17 are usually the most convenient. For large assemblies, such as the U-frames in the MSO-421 minesweeper, horizontal jigs are more convenient. The type illustrated in figure 2-18 is probably the one most widely used. It is made of heavy steel I-beams equipped with stops that can be set to the pattern of different templates. Steam pipes and jets are sometimes placed under the jig to furnish heat for curing; heat can also be circulated from a radiator by a strong fan.

Figure 2-17.—Universal jig used to glue bent laminated assemblies.

Various means are used to apply gluing pressure to laminated assemblies. If only a moderate curing temperature is needed, hydraulic presses or screw presses may be used to make straight members. In gluing curved members, retaining clamps usually serve to best advantage, since pressure must be applied progressively along the assembly to permit slippage of laminations longitudinally. Unless heat is supplied at the assembly jig or table, it is important that the assemblies under pressure be readily movable to a curing chamber.

Metal covers for jigs are sometimes used while the assembly is being heated. These covers are so constructed that they can be lifted with an overhead crane and stacked out of the way while the jig is being filled or unloaded. One type of clamp illustrated in figure 2-19 permits the gluing of two laminated frames one on top of the other and thus makes more efficient use of the jig. An important feature of this clamp is that the pressure is self-equalizing across the assembly. A single-bolt type of clamp made of two sections of I-beam with a bolt through their middle, has two so-called rockerheads attached to the upper rails that equalize pressure across the assembly. The advantage of the single-bolt clamp is that much less time is required to apply gluing pressure. An air-operated nut runner is generally used to tighten the clamp. A block and tackle is often used to pull a curved member into the jig.

Figure 2-18.—Jig for gluing U-shaped minesweeper frame. The smaller members shown in the jig were clamped elsewhere and brought to the jig by overhead crane for curing of glue.

Even more complicated jigs are required for laminating members such as garboards for the larger wood ships. These members are both curved and twisted and require clamping in two directions to bring them to the proper shape. For the jig to produce these garboards one yard built a mock-up bottom frame and keel assembly of a wood ship upside down. This jig then served for both port and starboard garboard sections.

Tops and bases of retaining clamps are made of cast iron, steel, or aluminum alloy. The initial cost is considerably higher when clamps of steel or aluminum alloy are used, but they deteriorate much less under conditions of high temperature and humidity than do cast iron ones, and there is less breakage. Retaining-clamp bolts may be procured with V-threads or square threads. The initial cost of V-threaded bolts is usually less than for the square-threaded ones, but V-threads are easily damaged and do not give such satisfactory service over long periods of use. Air-operated or electric nut runners facilitate rapid tightening of nuts on retaining clamps and speed production.

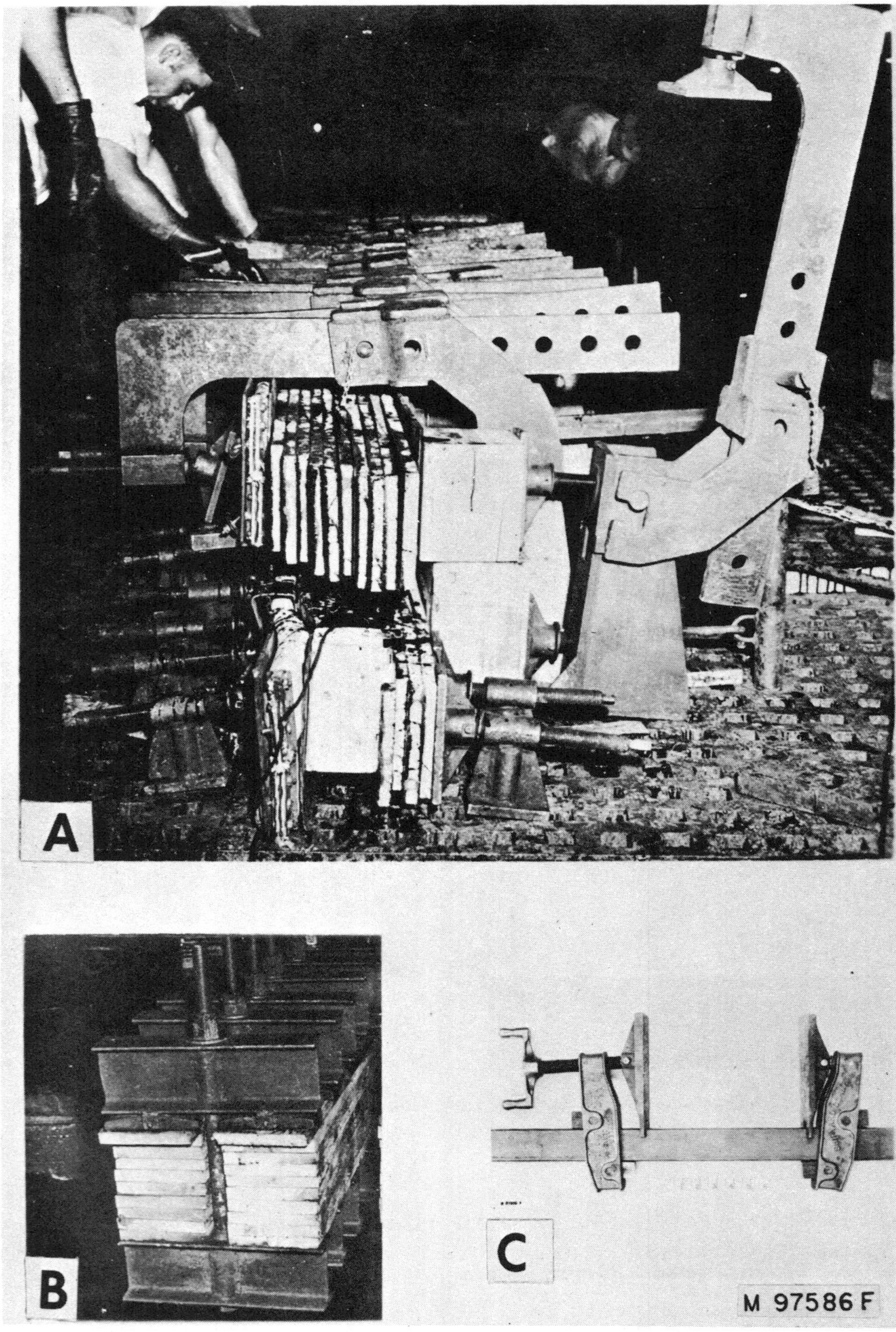

Figure 2-19.–Types of clamps used to apply pressure to laminated assemblies. Upper (A) shows type used on two assemblies simultaneously. B, single-bolt I-beam type with rockerheads on upper I-beam jaw. C, screw clamp with rockerhead jaws.

2.2.8 Curing of Laminated Members.–Curing of resorcinol and phenol-resorcinol glues is discussed in section 2.1.16. Different resorcinol or phenol-resorcinol glues require different glue-line curing temperatures and curing periods. These are specifically defined in applicable laminating specifications for different wood species.

Various means have been devised for applying heat to laminated assemblies to facilitate curing the glue. Each has advantages and disadvantages, and some have distinct limitations. Moderate heating to temperatures of about 140° F. to 170° F. with proper humidification can be done in a warm room. A canvas or synthetic-rubber canopy, dropped over the assembly to enclose automatically controlled heating and humidification equipment, involves a relatively low expenditure and may serve satisfactorily (figure 2-20). Portable plywood or steel enclosures are also being used for this purpose.

Figure 2-20.—Placing canvas cover over clamped assemblies in preparation for curing the glue.

Steam heat generally is employed for curing the glue in laminated members. Large electrical heaters also have been employed. In either case, some method of humidification is needed to maintain a humidity that will prevent loss of moisture from the laminated member. Humidification usually is most conveniently provided with steam jets automatically controlled. Figure 2-21 shows an enclosed heating and blower system. Temperature and humidity are automatically controlled. During curing the jig and laminated assembly are enclosed by a vaportight cover to prevent excessive escape of heat and humidity.

Heated chambers are used to cure laminated assemblies at relatively high temperatures. They should be vaportight and, in order to conserve heat, should be made of materials that have good insulating characteristics. Steam coils usually are the most convenient means for supplying heat, and should have sufficient capacity to raise the temperature of the chamber to the desired level in 1 to 2 hours. The curing chamber should be equipped with steam or water sprays for maintaining the proper humidity during the heating cycle and with cold-water sprays for humidification during the cooling period.

In a curing operation where a glue-line temperature of 190° F. is required, it is most practical to operate the chamber at 210° to 215° F. At temperatures much above 215° F., it is difficult to maintain the proper relative humidity and the use of such temperatures is not recommended. Whenever assemblies of considerable size are to be heated in a curing chamber, it is usually most practical to maintain the chamber temperature about 20° above the required glue-line temperature. If a smaller differential is used, longer curing cycles that may be impractical will be required.

Where treated wood is laminated, current specifications limit the curing temperature of the glue to 170° F.

Radiant heat supplied by infrared lamps or electric heaters has been used for curing glue in laminated assemblies of smaller size, but these means usually are less satisfactory than heated chambers because of the difficulty of properly controlling humidity at the surface of the wood. Infrared lamps and strip heaters might be useful for curing glues in small members of species in which relatively little checking will occur when appreciable changes in moisture content are encountered. Radiant heat does not penetrate beyond the surface of the wood and heating of the interior is substantially by conduction.

2.2.8.1 Curing of Glue by High-Frequency Heating.—By using high-frequency dielectric heating equipment, all parts of the glue lines in a large assembly can be heated in a relatively short period. This is a distinct advantage over the slow process of transferring heat by conduction from an exterior source. Due to various technical difficulties in production, high-frequency curing has not as yet been found satisfactory for curing laminated members for marine use. In a boat or ship, very few members are exactly alike and building of jigs for high-frequency curing would be very costly.

Figure 2-21.—Enclosed heating unit and blower for furnishing heat to laminated assembly. During curing the jig is covered with a metal cover such as those shown in background. Edge of cover fits into a water-filled gutter around edge of jig, making a water-sealed joint.

2.2.8.2 Rate of Heating in Large Laminated Assemblies.—When heat is applied to the exterior of a laminated assembly, the rate at which it is transferred from the surface to the center varies somewhat with the species and the moisture content of the wood and depends a great deal on the dimensions of the assembly. Figures 2-22 and 2-23 show the rate at which members of white oak and Douglas-fir assembled in wood cauls and clamps heat in a curing chamber maintained at the temperatures indicated. The temperatures listed in each instance are curing chamber temperatures. The graphs indicate the approximate glue-line temperature, assuming that the curing chamber temperature, is uniform around the laminated assembly. Heating periods required for other sizes can be estimated from the curves or can be calculated by the use of formulas. When several assemblies are clamped in one package the rate of heat transfer to the interior of the package may be considerably increased by inserting sheets of aluminum or similar conductive material between the assemblies. This procedure is especially advantageous when several wide assemblies are clamped in one package.

Curing of resorcinol or phenol-resorcinol glue is usually specified as a certain number of hours at a certain temperature; for instance, 6 hours at 150° F. or 10 hours at 140° F. The curves in figures 2-22 and 2-23 show that a considerable period is required to heat a member of appreciable size to a certain temperature. Actually, curing starts at room temperature; but the rate of setting increases as the temperature rises. Some curing is therefore imparted to the glue during the heating up as well as during the cooling process. A member of small cross section can be heated rather rapidly to a temperature such as 150° F. and will also cool down rather rapidly when the source of heat is shut off. A large member, however, heats up slowly and also cools down slowly. Consequently, when a curing of 6 hours at 150° F. is specified, there will be more excess heating when a large member is cured than for a small member.

Although the outer area of a laminated assembly heats up rapidly as the temperature of the curing-chamber atmosphere is raised, hours are often required before the glue line at the center of the assembly reaches the temperature specified to cure the glue. During the cooling period, however, the outer area cools first and hours may again elapse while the center or inner glue lines remain at elevated temperatures. Consequently, cooling can often be started before the required curing time at the innermost glue line has elapsed.

The use of thermocouples and a potentiometer is recommended for checking the interior temperature during the curing, particularly when laminating the first few members of a particular size and shape.

In addition to placing thermocouples at various points within a member, it also is desirable to place some on the outside surface. During the heating-up period, the wet-bulb temperature should be kept slightly lower than the surface temperature of the member. If this is done, condensation on the members will be avoided.

After heating has been carried on for the required period, it is necessary to cool the member to room temperature without appreciable drying in order to avoid surface checking. Ordinarily, cooling is accomplished by shutting off the steam supply to the heating coils and to the steam spray. Water vapor usually will escape from the chamber so rapidly during the cooling period, however, that the relative humidity will become very low and the surface of the wood will dry and check. It is important, therefore, to control the relative humidity during the cooling period.

A satisfactory method of maintaining a high humidity during the cooling is to inject a fine water spray or mist into the circulating air of the curing chamber. This is done conveniently by using spray nozzles attached to the cold water supply. Use of compressed air in the water-spray line helps to avoid clogging of spray nozzles. The introduction of water spray or mist, however, should be controlled to avoid wetting of the glued members. These precautions are most important when laminating members of dense species such as oak than members of softwood species such as Douglas-fir.

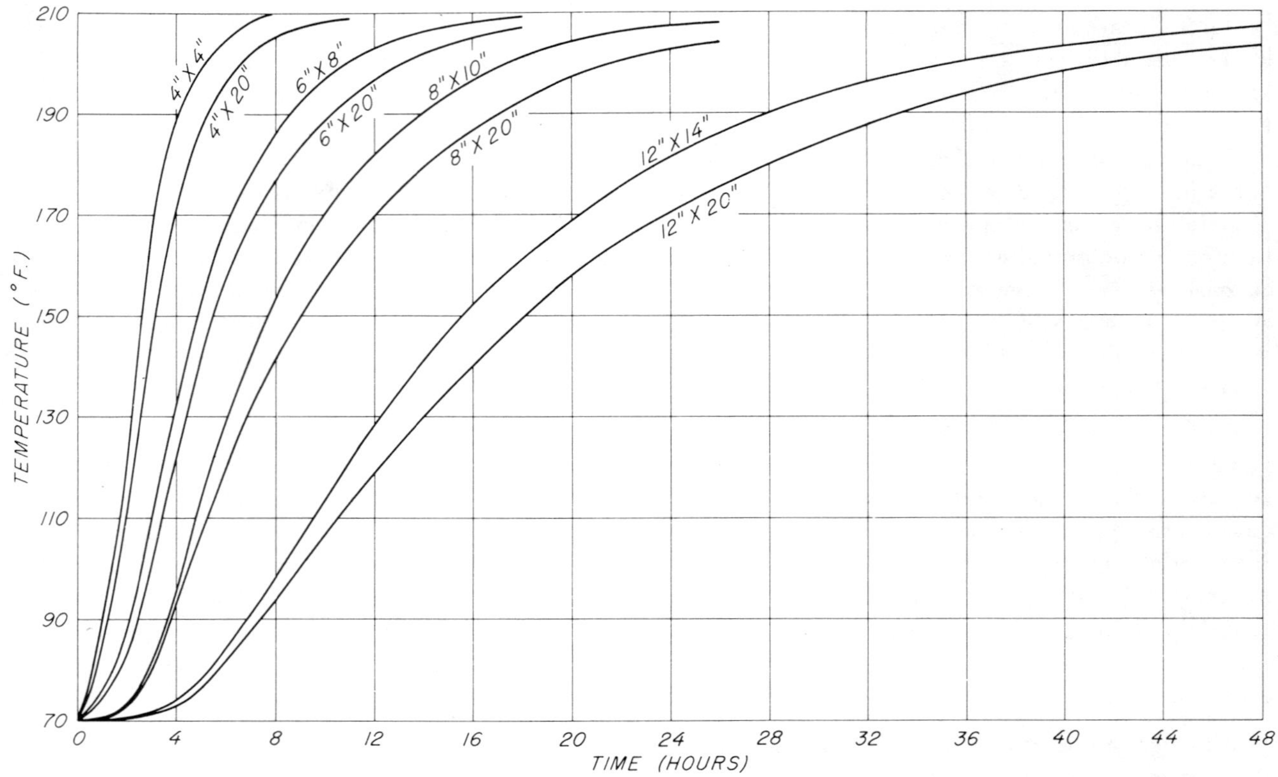

M 89994 F

Figure 2-22.—Glue-line temperatures at center of laminated members. Curing temperature, 210° F. Relative humidity, 80 percent.

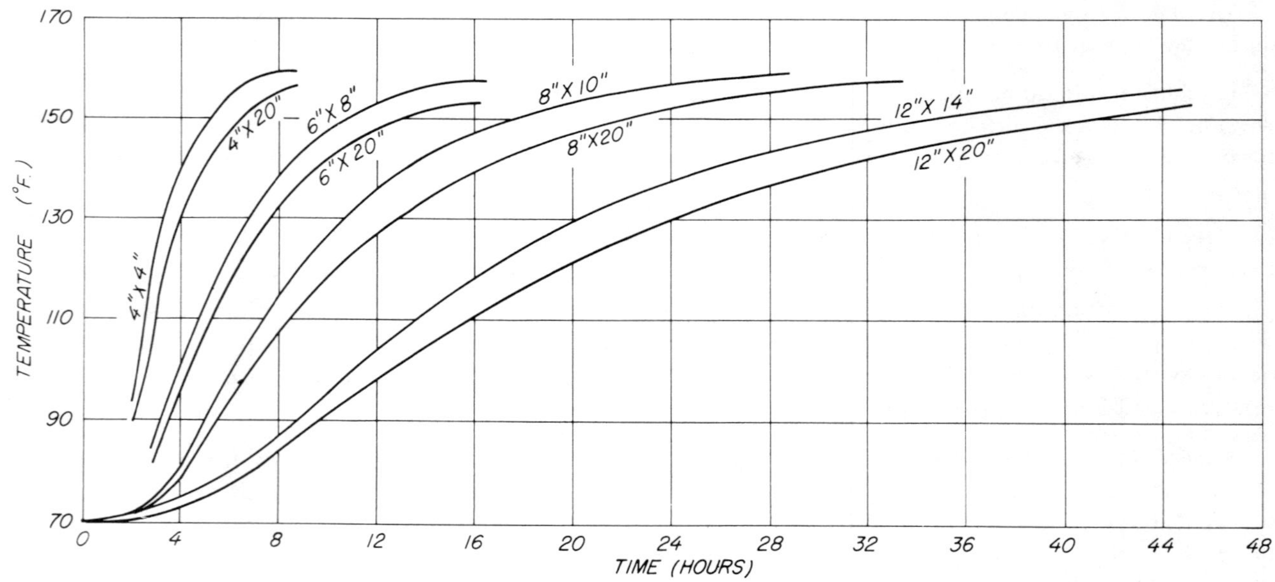

M 89995 F

Figure 2-23.—Glue-line temperatures at center of laminated members. Curing temperature, 160° F. Relative humidity, 80 percent.

2.2.8.3 Humidification of Curing Chambers.—When wood is heated for a number of hours, in an atmosphere of low relative humidity, it dries and tends to shrink. This shrinkage may cause checking of the wood or opening of the glue joints along the edges before the glue has cured sufficiently to withstand the drying stresses. If a high relative humidity is maintained in the curing chamber the wood will not dry, hence will not shrink and check, and the glue joints will remain tight. It is desirable to provide a relative humidity that will maintain the moisture content of the wood in a laminated assembly without appreciable change. Figure 3-2 in Volume I indicates, for example, that, a relative humidity of 80 to 85 percent is required to maintain a moisture content of 10 to 12 percent in wood heated to 210° F. At a curing temperature of 210°, such a relative humidity requires a wet-bulb temperature of 199° to 202°. At a curing temperature of 170° F. a wet-bulb temperature of 161° is required to maintain a relative humidity of 80 percent and about 12 percent moisture content in the wood.

2.2.8.4 Handling and Surfacing Laminated Members.—Provision should be made for the mechanical handling of heavy laminated members by overhead hoists, cranes, or other lifting equipment. The weight of a laminated member with the clamps and cauls may sometimes be several tons and special consideration must be given to the handling problem. Overhead hoists may be most effective, but floor cranes can also be used to advantage if they are designed and sized to span the gluing beds. If trucks or dollies are used, a flat and preferably level floor is necessary. Figure 2-24 illustrates use of overhead cranes for handling U-shaped minesweeper frames. Dollies of the hydraulic type that can be raised and lowered are convenient for supporting U-shaped frames that are being bandsawed or surfaced. For outdoor handling or movement from laminating shed to the ways, a crane mounted on a truck is convenient (figure 2-25).

Figure 2-24.—Cured laminated 7-shaped frame being lifted from jig with overhead crane.

Figure 2-25.—Laminated frame being moved by crane mounted on truck.

Most cured glues tend to dull the knives of surfacing equipment, so that it is undesirable to use finishing planers or jointers for the surfacing of glued products. If dulling is experienced, it may be advisable to use carbide-tipped knives in the surfacing equipment. Figure 2-26 shows a type of power-operated portable device that contains several sets of sharp-toothed rollers which break up the squeezed-out hardened glue into small particles that are easily removed. This device reduces the wear on the knives of the planer (figure 2-27) used to produce a smooth surface satisfactory for the finished member. Timber sizers or shapers also may be used to advantage if the dulling effect of the glue squeezeout is not too objectionable.

2.2.8.5 Plant Requirements for Laminating.—Laminated items produced in a plant vary considerably in size and complexity according to the type and size of boats or vessels produced. Most of the equipment listed in this section is considered essential for high-quality laminating of ship and boat members. The actual volume and size of production will determine the quantity of equipment needed.

Figure 2-26.—Power-operated portable device for removing surplus or squeezed-out glue from laminated assembly.

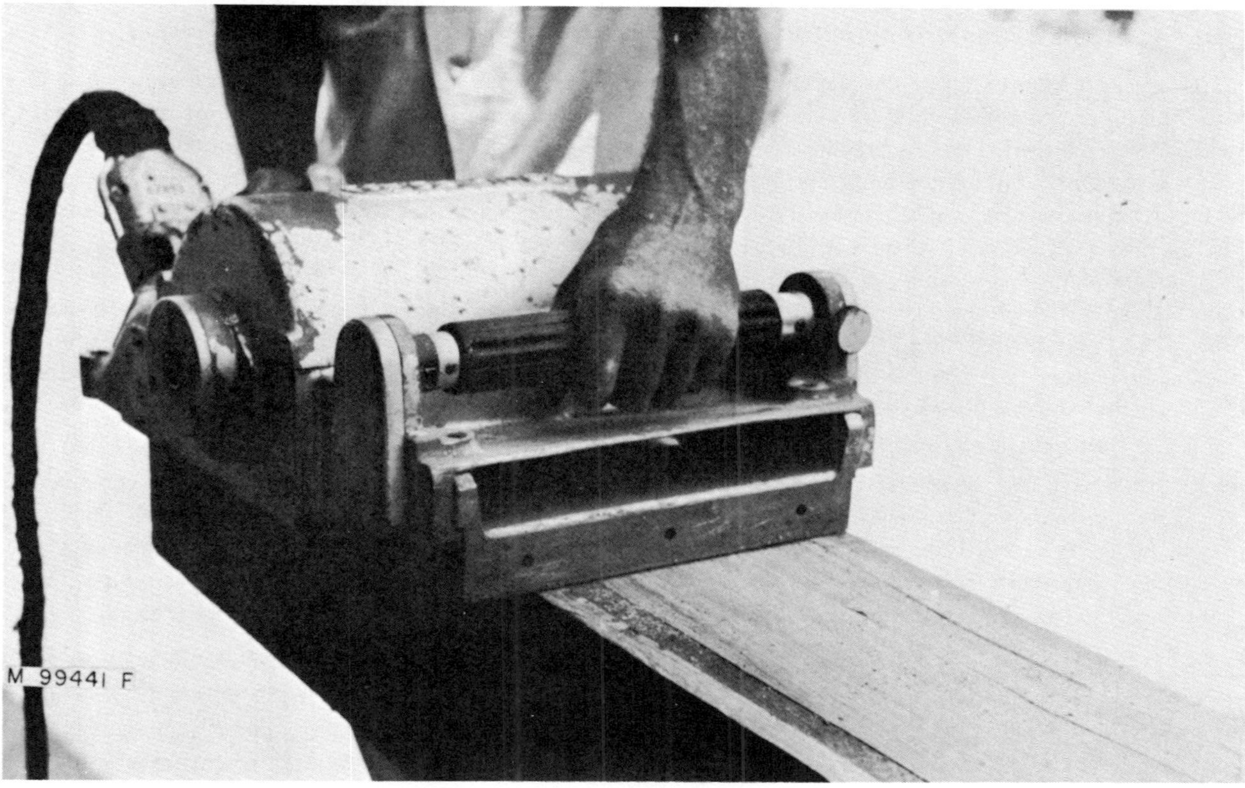

Figure 2-27.—Power-operated portable planer for finish surfacing of laminated member.

2.2.8.5.1 Plant Space.—Space should be provided for the inspection of lumber and for dry assembly of laminations to the final package size before they are glued. The space provided should permit placing the laminations in the correct position for the required spacing of scarfs and for assembling the proper number of pieces for each package. Space required for spreading glue is, as a rule, more than twice the length of the longest member to be fabricated.

2.2.8.5.2 Storage Buildings.—Facilities must be provided for the dry storage of kiln-dried lumber in enclosed and sometimes in heated buildings. This includes adequate space for storage of dressed laminations.

2.2.8.5.3 Laminating Equipment.—The minimum equipment necessary to produce wood laminates for marine use, and the procedures for determining the adequacy of each item are listed in the following paragraphs. It is important that all of the items listed be thoroughly checked for adequacy. A full knowledge of the capabilities of the equipment and its functions will avoid production difficulties encountered in the fabrication of this material. In a process so critical as the production of wood laminates for marine use, consistently reliable results cannot be obtained on equipment that will not give uniformly reliable performance. Frequent periodic checks of the equipment must be made to maintain the specified quality standard.

2.2.8.5.3.1 Straight-Line Ripsaw.—A straight-line ripsaw is necessary to cut lumber to required width and to remove defects.

2.2.8.5.3.2 Rough Planer.—A rough planer is needed for initial surfacing of both sides of kiln-dried lumber before it is cross cut and edge ripped to remove defects, and to reduce the stock to approximately uniform thickness for edge-joining and end-joining operations. Either a single- or a double-cutterhead planer is satisfactory, but it must provide true flat surfaces to give a proper reference bed for surfacing edges for gluing.

2.2.8.5.3.3 Trimsaws.—Trimsaws are necessary for trimming boards to the desired length and for removal of defects.

2.2.8.5.3.4 Scarfing Equipment.—Scarfing equipment is required to form laminations of the desired length and to produce consistently the quality of scarf required in the laminating specifications.

2.2.8.5.3.4.1 Scarf Cutter.—An acceptable scarf cutter produces a good cut with no digging, tearing of grain, chipping, or peeling between annual rings. The scarf cutter also must produce a uniform cut across the board without splitting. At present the most satisfactory methods for consistently producing satisfactory scarf joints are on a cabinet surfacer used with a form to hold the lumber, on a modified surfacer, and on an Onsrude-type veneer scarfer. A modified radial-arm saw with a bar-type cutterhead has also been used, but, in general, it has not produced consistently satisfactory cuts and is difficult to keep in proper maintenance. If a surfacer is used in preparing scarfs, a relative cut to the approximate angle of the surface should be made by sawing or surfacing so that when the final scarf surface is prepared only a minimum cut is required. (If care is not taken, an initial cut by a saw will result in splitting of the feathered edge.) Forms used to produce scarfs on a surfacer should be of sufficient length to assure uniform contact with the bed of the machine without danger of canting or clipping.

2.2.8.5.3.4.2 Clamps.—The clamping assembly must be capable of applying and holding 150 to 200 pounds per square inch of pressure on all points of the glued assembly. The most reliable and economical method for applying pressure to scarf joints during the curing process is in a package assembly. In this method, a number of scarf assemblies are laid one above the other, using a thin metal plate between each layer to prevent layers from sticking together, to conduct heat, and to distribute pressure. Three calibrated rocker-head clamps apply pressure to heavy oak cauls placed at the top and bottom of the package.

2.2.8.5.3.4.3 Equipment for Curing Scarf Joints.—The method used to cure scarf joints must consistently produce acceptable quality scarfs of maximum strength. The curing temperature at all points of the glue line should be not less than that required to set the glue. At no point in the curing chamber or glue line should the temperature exceed 215° F., except for short periods, such as during high-frequency heating. The curing chamber and glue line temperatures may also require further temperature restrictions when bonding wood treated with preservative. Relative humidity should be maintained to the extent that no appreciable change in moisture content of wood occurs. Also, no degrade of the wood should result. Curing has been done most satisfactorily in a chamber or under curing blankets. Hot platens and dielectric heating methods, however, can produce acceptable quality scarf joints when proper procedures are used.

2.2.8.5.3.5 Cabinet Surfacer.—The cabinet surfacer used for final surfacing of laminations must be capable of finishing stock to a maximum tolerance of ±.005 inches. Certain species of wood, such as teak, require even closer tolerances to avoid difficulty in production. Lumber that is rough surfaced to approximate thickness before final surfacing and laminating is desirable. The maximum material to be removed by surfacing should not exceed one-sixteenth inch for each face. Knives should be carefully jointed to eliminate revolution marks. Planer speed should be regulated to ensure a minimum of 20 and a maximum of 30 knife cuts per inch. The cutting angle for hardwoods should be less than for softwoods. The recommended cutting angle for oak should not exceed 15° and for Douglas-fir 20°. A double-head cabinet surfacer with ball-bearing cutterheads has proved most satisfactory for dressing laminations. For limited production, a single-head surfacer with ball bearings is satisfactory if tolerances for thickness can be maintained. If four-side matchers are used, particular attention must be given to the condition of the bed plates and pressure bars. It is very difficult to hold the necessary tolerances on molders or matchers.

2.2.8.5.3.6 Shaping Equipment.—Either a marine-type bandsaw or a conventional 36-inch bandsaw capable of carrying a 1½-inch blade is suitable. Filing room equipment adequate for servicing heavy duty bandsaws should be provided.

2.2.8.5.3.7 Additional Equipment.—A heavy-duty grinder is needed for maintenance of cutterheads and saws. In addition, a band resaw is desirable when thin laminations are needed for laminating sharply bent items. Such a saw can also be used for resawing items that can be glued in multiple widths.

2.2.8.5.3.8 Laminating Jigs.—Laminating jigs must be capable of withstanding load under clamp pressures and during conditions as outlined in applicable specifications. All jigs must be of rigid construction and of a type and size consistent with the bulk of the material to be processed.

2.2.8.5.3.9 Clamps for Laminate Assembly.—Application of correct, uniform pressure on a glue line is the prime requisite to the formation of a good bond. All clamps must be spaced and tightened uniformly to the load that will deliver the required glue line pressure for the specific clamp spacing and glue line area. For Douglas-fir, the proper glue line pressure is 150 ±25 pounds per square inch, and for white oak, it is 175 ±25 pounds per square inch. To achieve this pressure, clamps must be capable of applying and holding the specified pressures and withstanding repeated heating in the curing chamber without warping.

The clamps must be equipped with rocker-type equalizing pressure heads to distribute pressure evenly across the whole member. To apply and maintain pressures uniformly along the length of a glue line, each individual clamp must have provisions for regulating the load and easy means for determining this load. At present, the most reliable clamps are those that use a threaded shaft system to apply pressure. This type of clamp can be easily checked for adequacy through the use of a compressometer or torque wrench. Acme threaded shafts, although generally a little more expensive, will provide longer service and have less tendency to bind and hang under heavy load.

2.2.8.5.3.10 Curing Chamber.—The chamber used for curing laminates must provide proper uniform heat and humidity to the material. Therefore, the walls of the chamber must be insulated for heat and vapor, and the construction of the chamber must allow air to circulate freely about the curing material, distributing uniform heat and humidity. The temperature needed to cure the adhesive at all points of the glue line and to keep the average moisture content of the wood essentially the same as it was before the period was begun must be maintained. No degrade of the wood should result from the curing process. Several types of chambers have been successfully used, such as lumber dry kilns, specially built stationary curing chambers, or portable curing chambers.

2.2.8.5.3.11 Torque Wrenches.—At least two torque wrenches, calibrated to the particular type of clamps used, should be available.

2.2.8.5.3.12 Laminating Room.—A clean, well-kept laminating room is essential for good gluing. All woodworking equipment must be equipped with dust collection systems. Temperature controls and adequate ventilating devices are needed.

2.2.8.5.3.13 Handling Apparatus.—Adequate handling equipment is required for moving laminated members around the laminating shop and yard without damage.

2.2.8.5.4 Maintenance Equipment.—Facilities should be available for maintenance of equipment, including a grinding room adequately equipped for maintenance of cutterheads and saws.

2.2.8.5.5 Testing Equipment.

2.2.8.5.5.1 Drying Oven.—A drying oven, capable of maintaining a temperature of 105° C. (215° F.), must be available for use in determining moisture content and moisture distribution in lumber by the standard oven test.

2.2.8.5.5.2 Weighing Apparatus.—A suitable weighing apparatus is needed to weigh test samples of lumber in calculating the moisture content of wood. It should be accurate to within 1 percent and indicate the weight of samples to the nearest 0.01 gram.

2.2.8.5.5.3 Moisture Meter.—An accurate moisture meter is required to determine quickly the amount of moisture in lumber. Some moisture meters tend to become inaccurate when batteries run down or when room humidity conditions are adverse. Moisture meters should therefore be checked for accuracy against oven tests at periodic intervals.

2.2.8.5.5.4 Potentiometer and Thermocouples.——A potentiometer and thermocouples are needed to determine accurately the glue line temperature in scarf joints, laminate assemblies, and edge joints. The accuracy of these instruments must be checked with a laboratory thermometer and a hot water bath. Their calibration should include a full temperature range from 60° F. to 212° F. Although a potentiometer is generally used to measure glue line temperature, it can be used also to determine ambient dry bulb and wet bulb chamber temperature conditions. This is done by using two units next to each other with a wet sleeve over one, as is done with a normal wet- and dry-bulb system.

2.2.8.5.5.5 Chemical Kit.—A chemical kit, applicable only for use on various species of oak, is required to segregate white oak from red oak and heartwood from sapwood.

2.2.8.5.5.6 Compressometer.—An accurate compressometer or equally effective portable apparatus to measure compression loads should be available to calibrate torque wrenches and clamps. The apparatus to measure compression loads must be calibrated and certified for accuracy by a test laboratory.

2.2.8.5.5.7 Block Shear Testing Apparatus.—A suitable testing apparatus is necessary to conduct block shear tests on material prior to submittal for Government inspection. The shear test tool must be similar to that described in military specifications for wood laminates for marine use. The load measuring device should be accurately calibrated and should include a damping apparatus that permits readings of ultimate loads under conditions of sudden failure. It should be capable of being read accurately to at least 100 pounds.

2.2.8.5.5.8 Small Bandsaw.—A small bandsaw is needed for preparation of casehardening samples, moisture determination samples, and for preparation of block shear test specimens.

2.2.8.5.5.9 Miscellaneous Laboratory Equipment.—Miscellaneous equipment includes such items as thermometers, gauges, and micrometers.

2.2.8.5.6 Edge-Joining Equipment.—Producing high-quality edge joints for use in marine laminates requires adherence to close tolerances in equipment and methods. The extent of marine laminating, the type of marine laminates being produced, and the availability of equipment and experienced personnel will determine the practicability of using edge-joined material. It will also determine whether the laminator should do his own edge joining or purchase edge-joined material from a manufacturer.

2.2.8.5.6.1 Equipment for Preparing Surfaces.—Equipment for producing surfaces for edge joining should be of reasonably late design and must be in first class condition. The equipment must be capable of preparing edge-joint surfaces on the longest length of edge-glued material expected to be used in production. If a straight chain-feed edge-jointing rip saw is used for this purpose, the ball bearings on the saw arbor and the grooves and track on the chain must be in first class condition. The saw must be well balanced and fitted. For limited production, surfacing of edge joints may be accomplished satisfactorily on a molder that is kept in excellent condition. All edge joints should be carefully checked for smoothness and evenness of cut. Assurance also should be made that the planes of the edge-joint surfaces are perpendicular to the planes of the flat surfaces of the lamination, and that the prepared surfaces will match when assembled.

2.2.8.5.6.2 Clamps for Edge Joining.—Application of the correct, uniform pressure on a glue line is a prime requisite to the formation of a good bond. This requires that all clamps be uniformly spaced and uniformly tightened to the proper load which will deliver the required glue line pressure for the particular clamp spacing and glue line area. In edge gluing lumber for width, the maximum clamp spacing measured on centers should not exceed 14 inches. Closer clamp spacing is often necessary to distribute more uniformly the load supplied by the clamps to the glue line, and to assure that the clamp loading heads do not crush the wood. The most satisfactory methods for applying pressure to edge-joint glue lines are stock-type clamp carriers and stacking clamps. Stacking clamps generally have been found much more adaptable to producing edge joints for laminations for marine use because of their low initial cost, versatility in clamping various lengths of material (particularly lumber over 8 feet in length), ease of handling and storage, low maintenance costs, and ability to hold material flat. They also generally require less extensive curing facilities. Both stacking clamps and clamp carriers have threaded shafts for applying pressure to each individual clamp head, which can be readily calibrated to determine the load delivered for a specific torque.

2.2.8.5.6.3 Curing Chamber.—The curing chamber used to cure edge joints should be similar in design and operation to that used for laminated assemblies, described in paragraph 2.2.8.5.3.10.

2.2.8.5.7 Calibration of Various Equipment.—Certain equipment used in marine laminating operations require calibration to assure correct performance. Calibrating such equipment should be in accordance with the following recommended procedures.

2.2.8.5.7.1 Clamps.—In calibrating clamps with threaded shafts, an accurate compressometer and torque wrench are required. The compressometer is inserted between the pressure heads of the clamp. It may be necessary to place white-oak cauls between the pressure head of the clamp and the bearing surfaces of the compressometer as filler. To compress the compressometer, the clamps are tightened with the torque wrench. The applied load can be read on the dial gage of the compressometer in pounds, and the corresponding torque on the torque wrench can be read in foot-pounds. A full range of load readings should be taken. All clamps should be well greased to aid in preventing hanging or binding becuase of an abnormally high coefficient of friction on the screw. Care should also be taken to assure that the compressometer dial actually reads total load in pounds. If the dial reads fluid pressure in pounds per square inch, this reading must be converted to total load readings in pounds. The observed torque readings are plotted on graph paper against the corresponding load readings to determine the safe capacity for the clamp. A straight line can usually be drawn through the first group of points. Then at a specific load, the points deviate from a straight line. The point on the graph where this deviation occurs is known as the yield point or elastic limit, and is the point at which weakening at the metal takes place. Clamps should not be loaded beyond the yield point. A suitable number of clamps of each type used should be calibrated to assure adequate representation.

2.2.8.5.7.2 Torque Wrenches.—Torque wrenches are calibrated in conjunction with clamps through use of an accurately calibrated compressometer as outlined in the calibration procedure for clamps.

2.2.8.5.7.3 Compressometer.—The compressometer, or some equally effective portable device for measuring compression loads, should be calibrated by comparing various loads indicated by it with those of a standard test machine, which in turn has been recently verified for accuracy by one of the methods listed in the American Society for Testing and Materials (ASTM) standard E4-50T, "Verification of Testing Machines."

Loads should be calibrated at frequent intervals throughout the full range of the load-indicating dial of the compressometer. For visual comparative purposes, the corresponding loads shown by the compressometer should be plotted on a graph against those indicated by the standard test machine.

2.2.8.5.7.4 Temperature and Humidity Instruments.—All temperature and humidity recording and indicating instruments used in curing laminate assemblies, scarf joints, and edge joints should be calibrated by means of a hot-water bath and an accurate laboratory thermometer. This calibration should also include potentiometer and thermocouple units. The calibration range should run from approximately 60° F. to approximately 212° F. Calibration of indicating and recording temperature units is accomplished by cleaning all temperature elements thoroughly and immersing them in a large glass container of distilled water at a temperature of approximately 60° F. Place an accurate laboratory thermometer, graduated in increments of not larger than 1° F., in the water bath and stir the water until the indicating or recording instruments show an equilibrium with the temperature of the water. Then record the temperature indicated on the laboratory thermometer and corresponding readings on the other instruments. Raise the temperature of the water in the container an increment of approximately 15° F. When the instruments show readings that are in equilibrium with the temperature of the water, again record the temperature indicated on the laboratory thermometer and the instruments being calibrated. In like manner, the calibration should be extended to a useful range of approximately 212° F.

2.2.8.5.7.5 Weighing Mechanisms.—All weighing mechanisms should be calibrated for accuracy with the use of standard weights.

2.2.8.5.7.6 Moisture Meters.—To check the accuracy of a moisture meter, take some readings with the instrument on a few typical pieces of lumber. Within a short time thereafter, cut sections out of the boards where the moisture readings were taken; weigh the small sections with a weighing mechanism (section 2.2.8.5.5.2) and dry in an oven kept at 105° C. (215° F.) until there is no appreciable loss in weight (a 1½-inch-square block of 1-inch nominal material dries in approximately 6 to 8 hours).

$$\text{Percent of Moisture Content (Based on weight of ovendry wood)} = \frac{\text{wt. before drying - wt. after drying x 100}}{\text{wt. after drying}}$$

2.2.8.5.7.7 Block Shear Testing Apparatus.—Any of the methods listed in American Society for Testing and Materials Designation E4-50T, "Verification of Testing Machines," is considered acceptable for calibration or verification of the shear test machine. Another permissible alternate would be to compare various loads registered by the laminator's apparatus with those of a standard test machine that has been recently verified for accuracy. A full range of loads indicated by the dial of the test apparatus being calibrated should be verified. A graph should be drawn in which corresponding values of the block shear test apparatus and the verification apparatus are plotted against each other.

CHAPTER 3. FABRICATION OF WOOD BOATS AND SHIPS

3.1 GENERAL.—The many technical advances that have affected the use of wood in recent years have produced marked changes in ship and boat building. The most recent has been the newly developed laminating process, which within a decade has evolved from a shipyard novelty to a widely accepted method of making virtually any structural part of a hull. The process has had a deep and permanent impact upon hull design, as notably typified by the MSO minesweepers. Other developments that have had significant effects upon ship and boat building include improved plywoods, wood preservatives, and various other new sheet materials such as hardboards and metal-faced plywood.

With all these developments, however, the basic problems inherent in the building of wood boats and ships remain relatively unchanged. A laminated keel or frame, while perhaps smaller in cross section than its solid wood counterpart, nevertheless, is only as strong as the wood from which it is made. Plywood decking, even though it shrinks and swells across its surface less than does solid wood decking, does absorb moisture and swell – notably along its edges – and therefore must be properly protected. Preservative-treated sheathing, while well protected from decay and other wood-destroying organisms, is affected dimensionally by moisture when continuously in water for long periods. In like manner, many other properties of wood remain relatively unchanged regardless of the form to which modifying processes convert it.

New materials have markedly affected the architectural, as well as the engineering, design of round-bottom hulls (Volume III, Chapter 5), notably by permitting use of smaller structural members through lamination and by facilitating production of curved surfaces for hulls and superstructures. The conventional round-bottom sawed frame, with its many carefully shaped futtocks that result in numerous water-trapping joints and excessive use of wood, has been largely outmoded by laminated frames that extend continuously from sheer to sheer. Round-bottom hulls utilizing steam-bent frames, however, retain an important place for the smaller ships as well as for numerous classes of boats, although laminated parts can be successfully bent.

The greatest effects upon architectural design brought about by new materials have been in V-bottom hulls. These lightweight, high-speed craft are built for service where speed and maneuverability are prime assets, as in assault landings and short-run attack operations. The light weight and uniform stiffness of plywood are used to advantage in such hulls for planking, decking, bulkheads and brackets, gussets, stiffeners for frames, boxbeam webs of girders and other longitudinals, and similar structural parts. Waterproof glues are essential to assure permanent strength and stiffness. Preservative treatment lengthens greatly the service life of these lightweight constructions, and is in turn practical only because of the availability of nonleaching preservatives that are odorless and paintable. Lamination is employed for such parts as keels, skegs, frames, stems, shaftlogs, and other members.

The purpose of this chapter is to guide ship and boat builders in the proper use of wood, in its various forms, as a shipbuilding material. No attempt is made to present in detail any particular method of ship or boat construction. Such specialized operations as the laying down of the hull lines on the mold loft floor, fairing, determination of proper declivity, spar making, calking, and many other operations are referred to only incidentally as they concern the use of wood in shipbuilding; no effort is made to describe or discuss them in detail, or to set forth requirements as to workmanship and the like. Such topics are adequately handled in other publications or in plans and specifications for individual types of ships and boats. Emphasis throughout this chapter is on the properties and characteristics of wood, so that it can be fully and properly utilized.

3.2 LOFTING OPERATIONS

3.2.1 Mold Lofts.—In one form or another, mold lofts are universally employed by shipbuilders to lay down full-size, accurately measured lines for every part of the hull form. From these lines the builder obtains the templates, patterns, and molds that guide him in the multitude of wood shaping operations necessary to produce the parts of a boat. The mold loft is traditionally a special floor in a building large enough to accommodate full-scale layouts (figure 3-1). The floor is usually painted a light color before use. Sometimes it is necessary to lay down a boat with a length greater than that of the available mold-loft floor. In such cases, a skillful loftsman superimposes the forebody on the afterbody to obtain the desired results.

Eastern white pine has long been the preferred species for mold-loft floors. It is not, however, so available as formerly. Sugar pine, western white pine, or ponderosa pine, selected for uniform texture, should prove suitable for mold-loft floors. In addition to the foregoing softwoods, yellow poplar, a hardwood, has the desired low shrinkage, good texture, and light color that suit it for such use. Mold-loft floors of Douglas-fir plywood are also in use, because of their good resistance to shrinking and swelling and the large panel sizes in which they are obtainable, making it possible to eliminate many of the cracks in a board floor.

For small boats, the mold loft may actually be a large plywood panel, which can be placed on supports for convenience in using it, or it may be made of specially treated heavy paper, which can be rolled and stowed away when not in use. The latter type is especially convenient when a number of different small boats are being made in the same shop and mold-loft floor space is not available for all.

Figure 3-1.—Painted wood floor in a modern mold loft.

3.2.2 Patterns, Molds, Templates.—Plywood has largely replaced solid wood as the material from which patterns, molds, and templates are made. Eastern white pine was long a favorite species for molds because of its ease of working and dimensional stability. Solid wood is seldom used nowadays for anything but long, simple shapes, such as beam molds or keel patterns. Light weight, availability in large sheets, relative freedom from shrinking and swelling, and resistance to breakage are the properties that make plywood well suited for this use. Three-ply Douglas-fir plywood one-fourth inch thick or less is usually employed.

When large templates are needed, it is customary to join the margin pieces of the template required by means of gussets and strips extending from side to side, generally as X-braces (figure 3-2). By so doing, plywood is conserved and the resulting template is light in weight.

In using templates to mark off ship or boat parts, an efficient workman will avoid waste by selecting lumber that is as nearly the size of the mold as is available. Sometimes, waste can be avoided by careful placement of the mold; for example, two or more parts of the same dimensions can frequently be scribed on a piece of lumber. Pieces containing natural crook can be effectively utilized for bent members. Skillful laying down can also utilize lumber that contains objectionable knots, cross grain, compression failures, or other defects, by so placing the mold that the defects are either eliminated from the finished part, or so located in it that they do not impair its strength objectionably.

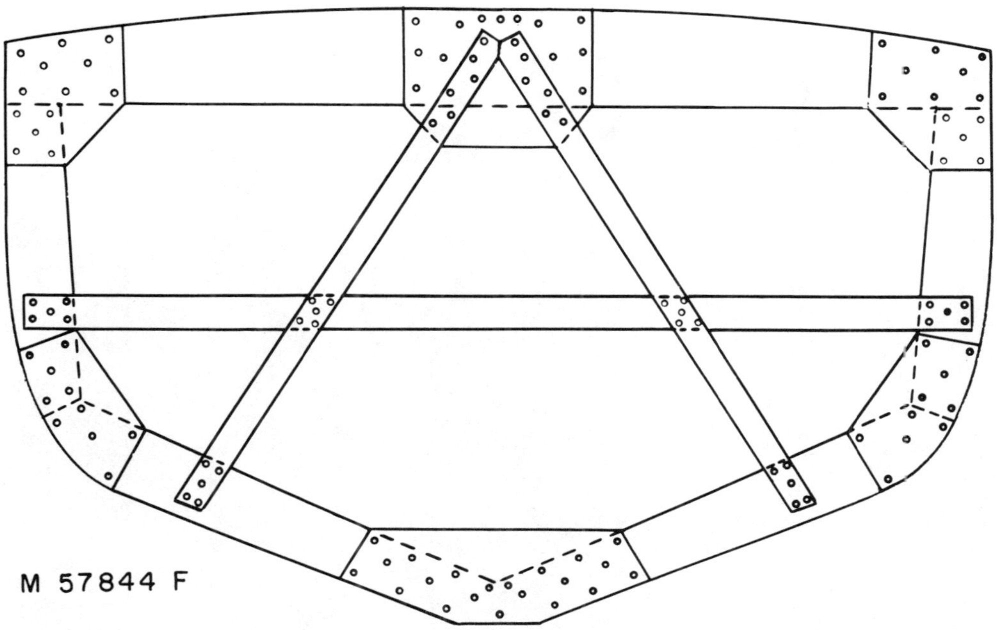

Figure 3-2.—Skeleton template of bulkhead. Margin pieces are cut to the required shape from lines on the mold-loft floor, joined by gussets, and braced by solid wood battens, clinched nails or glue being used for fastening.

3.3 BUILDING WAYS.—To facilitate assembly operations and launching, ships and boats are built on properly designed ways. These range from heavy piling under keel blocks for large vessels to special jigs for use in volume production of the smaller types of boats.

3.3.1 Ways for Large Ships.—Among the various pieces of equipment used for building and launching ways are: piling, cap logs, ground ways, cradles or slippers, keel blocking, shoring, and wedges. Long life is desirable for piling, and it is therefore usually made of pressure-creosoted Douglas-fir or southern yellow pine because of their availability in straight, long sizes and ability to take preservative treatment.

Requirements for cap logs are decay resistance and high strength; Douglas-fir and southern yellow pine structural timbers are used because they are available in the sizes and qualities required and can be given preservative treatment; all-heart white oak, when available, can be used, because of its natural decay resistance.

The preferred woods for ground ways are Douglas-fir, southern yellow pine, and western larch in structural stress grades because their heartwood is moderately decay resistance and can be pressure treated with preservatives. For cradles or sliding ways, woods used are dense southern yellow pine, Douglas-fir, and western larch. Requirements for keel blocks (figure 3-3) are not exacting except for the important blocks directly in contact with the keel, which should be white or red oak because of the requirement for high compressive strength across the grain and the ability to withstand rough usage. Steel plates are sometimes used between the keel and these blocks. Other blocks may be of any of the denser hardwoods or softwoods, such as Douglas-fir, southern yellow pine, western larch, beech, birch, maple, or elm. They are frequently obtained from clear, sound parts of rejected timbers or cutoffs. Requirements for shoring lumber are not exacting, any commercial species that can be obtained in the necessary sizes being suitable. Wedges should be made of dense hardwoods of good quality, such as red or white oak, hickory, beech, birch, maple, elm, or pecan, which have the desired hardness, impact resistance, and freedom from splitting under heavy blows.

Figure 3-3.–Keel blocking used on ways of MSO minesweeper. Note use of steel plate under keel to avoid crushing of edges against blocking of inadequate compressive strength.

3.3.2 Keel Posts and Jigs for Small Boats.–Smaller craft are built under cover and are comparatively easy to handle during all stages of construction. Various methods of fabricating them have accordingly been developed. Figure 3-4 shows a simple keel-post arrangement on which the keel of a V-bottom boat is laid. Some method of bracing is necessary to hold the boat's frames, bulkheads, and other parts in position during construction (figure 3-5). Such keel posts are used again and again, and must hence be placed with care on a solid foundation and properly braced. Since the keel frequently has some curvature, keel posts must be built to varying height to accommodate this curvature. They can be made of any structural wood available, as in most boat shops decay is not a serious hazard.

Some builders find it convenient to erect the frames and plank the hulls of small craft in an upside-down position (figure 3-6). For such construction a jig can be used; the heads of the frames are securely fastened to it, and lateral jig members are fastened to the frames in the way of the chine notches of V-bottom frames to hold the frames rigidly vertical. With such an assembly system the keel is, of course, laid after the frames are set in position. For larger boats, an overhead crane mechanism is used to turn the hull upright after planking is in place (figure 3-7). Figure 3-8 shows a rack with special frame supports used for PT boats. Ponderosa, eastern white, and sugar pine, or similarly easily worked species that do not swell and shrink much, are suitable for such jigs.

Figure 3-4.—Keel posts with a V-bottom keel and stem in position.

3.3.3 Staging, Scaffolding, Shoring.—Much lumber is used in shipyards for the purely auxiliary purpose of facilitating the work of putting the hull together. Staging of many kinds is needed to hold parts temporarily in place while other parts are fastened and to provide ladders, ramps, and walkways that enable workmen to perform the many operations involved in assembly. Shoring is necessary to hold the hull skeleton in place on the ways while building proceeds.

While common grades of lumber are suitable for many of these jobs, it must always be borne in mind that the safety of the workman as well as the progress of the work depends upon adequate supports. This is especially true of scaffold plank on which the workmen walk. These must be capable of carrying any reasonable load safely, in short, a high stress grade of lumber is required. They should be carefully examined before use. Supporting members of scaffolds need not be of such a high grade if they are adequately braced so that loads they carry impose chiefly compressive stresses.

Staging used to hold frames, stems, and similar parts in place while sheer clamps and other longitudinals are installed is usually loaded in tension or compression, and the better common grades of softwood lumber are generally suitable. Shoring also carries principally compressive loads, hence No. 2 common softwood lumber is suitable if used with judgment.

Figure 3-5.—Supporting framework for V-bottom boat erected upright on keel posts.

3.4 JOINTS AND FASTENINGS.—Joints and their fastenings are vital to good boat and ship construction. Inefficiently designed or poorly made joints can critically affect seaworthiness and service life. Design fundamentals of joints and fastenings are given in Volume III, Chapter 6. Good workmanship and materials are of equal importance.

Mechanical joints — those made with nails, bolts, screws, dowels, and similar fastenings — of ships and boats are made in certain ways developed chiefly from long experience, that dictates the kind, size, location, and number of fastenings needed. For this reason, detail drawings that specify the kind, size, number, and positioning of mechanical fastenings in a particular type of hull must be rigidly followed. But this in itself is not enough. Good workmanship is a topmost requisite of mechanical joints. Faying surfaces must be accurately made, holes of proper sizes drilled smoothly and cleanly, and fastenings installed with precision. Wherever called for, bedding compound, calking, and other materials needed to improve watertightness and forestall decay hazards must be applied.

Glued joints have only recently become practical for use in wood ship and boat building as a result of the development of waterproof glues (Chapter 2). The design of glue joints, therefore, has been developed largely from experience with types of structures other than ships and through research and test work. Good workmanship and expert use of materials are just as essential to the forming of good glue joints, however, as they are to the production of efficient mechanical joints. The principles of good gluing described in Chapter 2 should always be followed.

3.4.1 Nailed Joints.—Common wire nails are seldom used to fasten structural parts of ships and boats together. Among the reasons for this are the fact that moisture-content changes in wood, accompanied by shrinking and swelling, tend to lower the holding power of common nails, and that the complex stresses exerted upon a hull in the water cause joints to "work" in ways that require fastenings that cannot loosen.

Special types of boatnails are often used in small boats to fasten planking to frames, as well as in other parts. Spikes are used frequently for such purposes as fastening planking of ships to the frames. Volume III, Section 6.2.6 gives design data for various types of nails and spikes by which joint strength can be calculated.

The density of the wood used often affects the type of nail or spike that should be used and the method of installing it. The denser, harder woods tend to split more readily when a nail is driven into them, hence blunt-pointed nails should be used and a pilot hole drilled first. In dense woods it is especially necessary to drill pilot holes; these are slightly smaller than the diameter of the nails used, in order to avoid splitting or to permit use of copper or aluminum nails. Such drilling, however, reduces the labor-saving advantage of nails over screws and bolts.

Figure 3-6.—Inverted construction, fastening keel to frames.

Figure 3-7.—Turning a PT boat right side up during construction.

3.4.2 Screwed Joints.—Screws generally assure much stronger and more permanent joints than do common nails or spikes. Volume III, Section 6.2.4 gives design data for calculating the strength of screwed joints in wood and describes requirements for joints, such as size of lead holes and depth of penetration of the screw into the members being joined, that will develop maximum strength.

Straight-shank, tapered-point drills are available that provide the proper size of hole for shanks and tapered threaded portions of wood screws and can be used with countersinks to make the complete hole in a single drilling operation. Lead holes are especially necessary when driving screws into dense woods, such as oak and teak, in order to avoid breaking the screw.

In ship and boat building, lead holes are drilled and screws are driven chiefly with portable power drills. Hand drills are used only for jobs requiring special care of finish, very small screws, or under other unusual circumstances. Driving is facilitated by lubricating the screws with soap or beeswax.

Screws should always be turned in. They should never be started or driven with a hammer, because this tears the wood fibers and injures screw threads, thus seriously reducing the load-carrying capacity of the screw.

Lag screws (Volume III, Section 6.2.5) also require lead holes, and lubrication is recommended, especially for unusually long lag screws or dense woods. Information on recommended lead-hole sizes is given in Volume III, Section 6.2.5.1. Lag screws should be turned in with a wrench, not driven with a hammer.

Figure 3-8.—Jig used for inverted assembly of PT boats.

3.4.3 Bolted Joints.—The most common type of joint used in ship and boat building for joining major structural parts together is the bolted joint. One widely used joint is the nibbed scarf, the design of which is discussed in Volume III, Section 6.1. It is found in many longitudinal members, from keels to deck clamps and margin pieces. Bolts are also used to fasten planking to frames, decking to deck beams, and foundation members of superstructure and ship fittings to hull framework.

Critical to joint strength are bolt holes. Too large a hole causes nonuniform bearing of the bolt; too small a hole may cause the wood to split when the bolt is driven in. Bolt holes should fit neatly, so that the bolt can be driven with light taps. The smoother the hole, the stronger the joint will be and the less the hole will be deformed in use.

General shipbuilding practice is to drill bolt holes in large members 1/16 inch smaller than the diameter of the bolt. This ensures a tight fit in green timbers 6 by 6 inches or larger, and tends to compensate for rough holes. In dry members, or in smaller ones, such as laminated keels or frames, the holes are too small and splitting is likely when bolts are driven into them; this class of material, the bolt holes should be of the same diameter as the bolts.

Rough holes result from the use of dull bits and from improper drill speeds and feed rates. A twist drill, preferably sharpened by machine at a 118° angle and run at a speed that corresponds to about 1,350 revolutions per minute divided by the diameter of the drill, produces smooth, uniform holes. Carbon steel drills should be operated at about one-half the speed of high-speed drills. The rate of feed should be such that the drill cuts rather than tears the wood and produces shavings not chips.

The same requirements for lead holes apply to drift bolts. Normally, holes for drift bolts are one-eighth inch smaller than the diameter of the bolt, except for long holes in dense wood, in which the bolt may buckle unless they are full size.

Requirements for positioning of bolts with relation to edges and ends of members to be joined are given in Volume III, Section 6.1.

Nuts are usually turned down as tight as possible without crushing the wood fibers under the washer. Crushed fibers are, of course, a point of weakness, since they are severed from the surrounding wood. Moreover, since some slight swelling of the wood can be expected when the ship is put into service – especially if glued laminated members are used – excessive tightening of nuts can result in further crushing of wood. On the other hand, nuts should not be left loose in the belief that subsequent swelling will tighten them adequately. Where wood not fully seasoned is used, bolts often have to be tightened after the vessel is in use.

Bolted scarf joints in keels, stems, and similar parts exposed to the water are often fitted with stopwaters (Section 3.5.1.3). Nibbed scarf joints also often have keys that serve to lock the two members firmly against each other at the nibs. In making nibbed scarf joints, care should be taken in cutting the nibs not to run the saw too deep, as this will produce a weak point in the timber. Scarfs are usually sawn with a band or circular saw and finished with an adz and plane to produce smooth, well-matched faying surfaces (Section 1.4.2).

Keys for keyed scarfs are made of heartwood of a decay-resistant species such as white oak or black locust. They are wedge-shaped so that the two parts can be driven snugly together from opposite sides of the joint. They are then sawed off flush with the surface and wedged at the small ends so that they cannot back out.

3.4.4 Connector Joints.–Metal connectors (Volume III, Section 6.2.3) serve to take shear load between joined members off the bolt and distribute it over a larger area of the members, thus reducing unit loads and permitting greater total loads. In general, connectors are of two types – those which are driven into the wood, and for the purpose are equipped with teeth or claws, and those which fit into precut grooves.

Connectors find relatively little use in ships and boats, but have definite potentialities. One use made of them in the MSO minesweepers is in the centerline wood blocking between deck beams, where they are used to transfer shear loads on the blocking to the longitudinal member that runs beneath them, thus stiffening the deck beams at centerpoint against lateral buckling when the ship encounters heavy seas. There are, of course, many other points in a ship's construction where connectors might prove feasible, such as various parts of engine and machinery foundations.

Split-ring connectors are the most widely used type. Grooves are usually made with a special cutterhead that simultaneously cuts the bolt hole. Connectors should be installed as soon as possible after grooves are cut to avoid possible distortion of the groove through moisture content changes in the wood. Members consisting of several pieces can be grooved and drilled for connectors and bolt holes before they are assembled and installed.

In bolted or connector joints, the nuts should be retightened after a period of use.

3.4.5 Glued Assembly Joints.–Glued assembly joints are used principally in V-bottom hulls, notably in frames to connect the parts at chine and keel. To form these joints, plywood gusset plates are glued to adjoining frame members and pressure is applied with screws or carriage bolts, which are left in place after the glue cures. Screws and bolts are commonly spaced 4 to 6 inches apart; thin plywood requires closer spacing than thicker plywood. Gluing practices regarding preparation and use of glues, pressures, curing times, etc., as given in Chapter 2, are recommended for such joints. Plywood gussets should be cut so that the face grain of the plywood will be at about the same angle to both frame parts.

Design considerations for glued assembly joints are given in Volume III, Section 6.3. While strength test data are not available for spliced joints made with plywood gussets glued to solid wood members, such as frames, experience indicates that such joints are adequately strong. Gussets are usually placed on both sides of the joint to assure balanced construction, but in small craft they are often placed between the pieces to be joined.

3.5 SHAPING AND ASSEMBLY

3.5.1 General.—In the following paragraphs, various shaping and assembly operations employed to produce ships and boats are discussed. The discussion is necessarily limited, and should be interpreted rather as a guide than as the only acceptable method of performing the various operations. The comments are based primarily upon fabricating operations observed in a number of representative ship and boat plants in the United States, viewed in the light of general information on wood.

3.5.1.1 Keels.—The keel is usually the first part of the ship to be fabricated. It is desirable to have the keel made up of a single long timber, and lamination provides a means of doing this for even large vessels. Otherwise, for large craft it is necessary to select a number of timbers as long as possible and fasten them together, keeping the number of joints at a minimum.

The procurement of sound, all-heart keel stock is often difficult, particularly for the larger keels. Since a large part of the keel of a ship is continuously submerged in water, it is questionable whether sapwood should be excluded from at least the portion that is in water. Experience indicates that keels rarely prove troublesome because of decay; deterioration from marine-borer attack is much more likely, and even the most decay-resistant species native to the United States are not immune to marine borers unless protected by copper sheathing or wood sheathing thoroughly impregnated with a suitable preservative. Provided a good preservative is applied to the exposed parts of a keel, relatively nondurable species should prove satisfactory. Well-treated sapwood should also serve. Admission of less durable species and sapwood for use in keels greatly broadens the sources of such lumber and reserves the supplies of white oak heartwood and other species of good decay resistance for ship parts where the decay hazard is more critical (Section 3.6).

Owing to the growing scarcity of high-quality, long-length stock, it is desirable to utilize standard lengths wherever possible, bearing in mind that there may be some exceptional cases where the position of the scarfs is definitely fixed for some structural reason.

The keel used on the conventional round-bottom vessel is, from the woodworking standpoint, relatively simple to produce. Necessarily, it must be sided and molded to correct dimensions; other than this, the principal shaping operations involved are scarfing (Section 3.4.3) and rabbeting. It is particularly important to finish the top and bottom faces of each keel section true, because the frames, deadwood, stem, keel shoe, and sometimes other parts are fastened to these faces and, snug fit is necessary. These surfaces can generally be finished on a single-head planer or, for larger keels, with a portable planer.

Rabbets are usually cut into the keel by routing, chiseling, or adzing, working carefully to the lines drawn on the surface in order to assure the proper bevel for the garboards and planking at the various stations. Unless they are cut to the correct angle, planking will not fit tightly and a troublesome joint will develop in spite of subsequent calking. The joint between keel and garboard is obviously one of the most critical in the entire ship. To speed this operation, rabbets can often be rough cut with a portable circular saw and finished to the proper bevel with an adz, wood chisel, or router. Stopwaters are a necessary in keel joints that extend upward into the bilges.

Glued laminated keels are made by the techniques described in Chapter 2. In ship assembly, the only operations necessary are siding, rabbeting, drilling of holes for frame bolts, cutting of seats for half frames such as the bow frames used on MSO minesweepers, and the joining of the keel to the skeg and stem by means of bolted scarf joints. For small craft, laminated keel-stem assemblies are sometimes made complete with shaftlog, deadwood, horn timber, and related parts. It is usually practical, however, to laminate the keel separately. Bedding compound is liberally applied to faying surfaces of the joints between keel, stem, and skeg when they are assembled.

Keels of small vessels, especially V-bottom types such as landing craft, are often built of two parts, the keel proper and a keel apron. This simplifies rabbeting, since variable bevels can be cut along the edges of the two parts to form a rabbet when they are assembled. Bevels can be adzed, or they can be cut with a vertical spindle shaper, using a jig to guide the part as it is moved past the cutterhead to cut the variable bevel. One type of jig used to control the bevel on such keel parts is shown in figure 3-9. When the two parts are joined, bedding compound is liberally applied to the faying surfaces; or a canvas liner is embedded in the compound, as in keels for 36-foot LCVP's, to insure watertightness (figure 3-10).

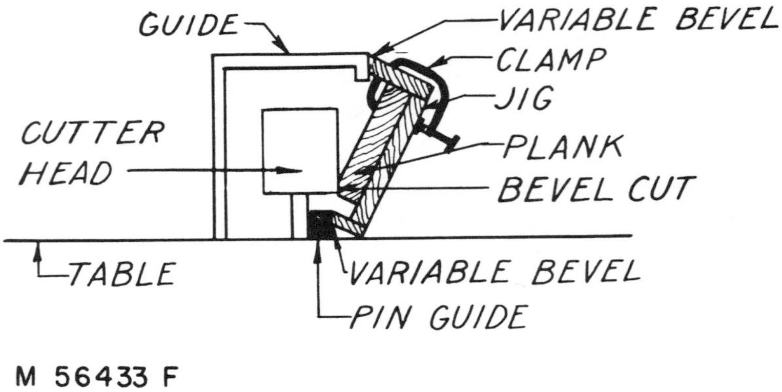

Figure 3-9.—Jig used to cut variable bevels on two-part keels for planking rabbets.

Figure 3-10.—Two-part keel for LCVP, showing canvas between keel and apron.

For comparatively short keels of this type, sections of the parts may be scarf-jointed together before being beveled, final assembly of the two parts then resulting in a finished keel. For longer boats, two-part keels may be made up in sections and scarf-jointed together as is a solid wood keel. Keel joints should, of course, always be stop-watered to prevent water from entering the hull.

Ordinary glues should not be used in conventional bolted scarf joints of solid wood keels. White lead or marine glue serves much better in preventing water leakage than do glues that set to a hard film and subsequently are liable to crack and craze. Marine glues are generally made of rosin, pine tar, denatured alcohol, and a drying oil, such as tung oil or linseed oil. While a good marine glue has some adhesive strength, its principal function is to make seams and joints between wood members waterproof. To do this, it should remain tacky in all climates. So-called bedding compounds are also widely used.

Figure 3-11.—Limber hole in top of laminated keel of MSO minesweeper.

For keels of large vessels, white oak is generally preferred; dense southern yellow pine is second in preference, and dense Douglas-fir is third. In small boats white oak is again the first choice. Mahogany and Sitka spruce are used also, mahogany having advantage because of its high decay resistance, a property in which Sitka spruce is low. Keels of boats that are intermittently in and out of water are exposed to a much greater decay hazard than those that are continuously in water, hence require more decay-resistant wood.

While the keel itself is seldom seriously damaged by decay, provision is generally made in ships for the drainage of bilge water that inevitably collects around it. In many ships this is done by providing limber holes adjacent to the keel. In some ships, a drainage limber in the form of a groove is provided on the top of the keel itself (figure 3-11), which helps prevent collection of water at the joint between keel and frame and clears the frame bays.

3.5.1.1.1 Skegs.—The skeg is as a rule the aftermost part of the keel of a vessel when the keel projects outside the skin of the ship. Its primary purpose is to protect the vessel's one or more propellers (figure 3-12), but it often serves also to stiffen the after portion of the hull (figure 3-13). In certain types of vessels, the skeg serves as the gudgeon to receive the lower pintle on the rudder blade. In LCVPs the skeg is a separate member attached to the under side of the keel amidships and extending aft below the keel to act as a tail skid in landing operations in order to protect the propeller rudder, and the after planking.

Glued laminated skegs have proved more satisfactory than solid ones for certain types of craft, notably the LCVP's due to the fact that they are laminated in a dry condition and thereby change less in shape when in use.

Figure 3-12.—Laminated skeg of LCVP, designed to protect propellers in landing operations.

3.5.1.2 Frames.—The frames determine the shape of a vessel's hull, contribute to transverse strength, and provide members for fastening planking, ceiling, keelson, shelves, and clamps. The frame top timbers serve in some types of ships as bulwark stanchions.

The shaping and assembly of frames is one of the more intricate processes in ship and boat building. Depending upon the type of frame employed, the pieces must be carefully sawed or bent to the proper curvature so that, when planking is attached, the hull will be fair. Even laminated frames need to be shaped (figure 3-14). A minimum of adzing or dubbing after erection of the frame is desirable to reduce labor and to avoid any possibility of weakening the frame.

Types of solid wood frames used in ships and boats can be broadly classified in four categories: sawn frames for round-bottom vessels, bent frames for round-bottom vessels, sawn frames for V-bottom craft, and bent, or web, frames for V-bottom craft. There are variations of these frame types, but in general such modifications do not depart markedly from the basic designs. Each of these four types has its advantages and disadvantages, depending upon the type of ship for which it is used.

The sawn frame as used in round-bottom ships is by far the most wasteful of wood. Compared with the round-bottom bent frame, it is inefficient in that it makes use of so small a part of the inherent strength of the futtocks. The method of fabrication is responsible for this shortcoming. The large number of joints, the necessity of assembling it double in order to stagger the butt joints, and the need of producing curved members from straight flitches, thus producing weakening cross grain, all contribute to its relative inefficiency. The use of such double frames adds greatly to the weight of the vessel.

On the other hand, the round-bottom bent frame is employed only in relatively small dimensions because of the difficulty of bending larger sections and the need of clear bending stock, hence its maximum utility is governed by these limitations. Even doubling of the bent members in depth fails to develop the strength that an equal solid member would provide.

Lamination makes possible the production of ship frames that are free of many of the disadvantages of both sawn and bent round-bottom frames for ships. There is no practical limit to the size of laminated frames insofar as their application in ship building today is discernible. Being continuous, they have uniform strength throughout their length. Even bending is simplified, since each lamination is bent to the proper curvature, thus practically eliminating the danger of breakage that is present when solid wood frames are bent. Ship designers can take fuller advantage of the strength of wood by using smaller sizes. These advantages go far toward offsetting the added cost of laminating (Section 2.2). For these reasons, sawn frames have practically passed out of the picture as far as large ships are concerned.

Figure 3-13.—Skeg of MSO minesweeper, used chiefly for stiffening.

The sawn V-bottom frame has obvious production advantages over the V-bottom bent, or web, frame, being simpler to produce. The bent frame requires careful fitting of filler blocks which are glued to the bent members and longitudinals.

Figure 3-14.—Bandsawing laminated minesweeper frame.

3.5.1.2.1 Sawn Frames for Round-Bottom Vessels.—Sawn frames of round-bottom vessels are made of heavy timbers in sections, called futtocks, butt-jointed together. In order to stagger the butt joints, the futtocks are doubled up with adjacent futtocks through-bolted together (figure 3-15).

In the bow and stern sections of a round-bottom ship, sawn frames are beveled to the requirements of the planking and are hence called bevel frames. Midships frames that do not require such beveling are called square frames. Bilge futtocks of bevel frames, which must be sawed to the bilge curvature and molded to variable bevels as well are the most difficult parts of frames to shape, and a bevel bandsaw is essential for this operation. It is customary to shape all futtocks with a bandsaw.

At sections of the bow and stern where the construction does not permit full frames, half frames are used which are fastened to the deadwood, keelson, or horn timber. At the extreme bow of the vessel, half frames are usually canted instead of being at right angles to the keel; this is done to avoid the extreme bevel and its accompanying weakness that would otherwise be necessary to provide a fair faying surface for planking.

Futtocks and floor timbers are generally assembled into the complete frame on a temporary platform built at the head of the building ways so that, when assembled, they can be conveniently moved into position on the keel. The most important phase of assembly is to get the butts of futtocks to fit tightly together; usually it is necessary to refit them with a saw before through-bolting is done. Timber connectors (Volume III, Section 6.2.3) are inserted in circular grooves specially cut into adjoining faying surfaces of pairs of futtocks to increase joint strength. When the futtocks have been bolted together, one or more temporary cross pieces connecting the top timbers, called spalls, are attached and the frame is ready for positioning on the keel.

The wood generally preferred for round-bottom sawn frames is white oak. Southern yellow pine, Douglas-fir, and western larch are, however, suitable for use. All four of these woods have moderate decay resistance and good holding power for fastenings. All are readily available commercially except western larch.

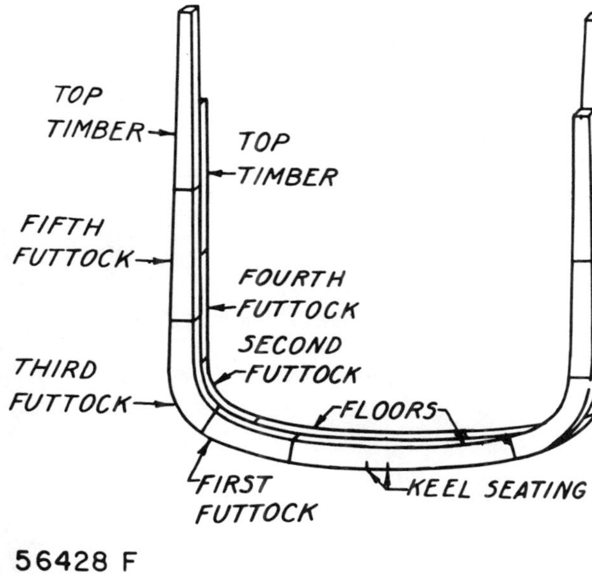

Figure 3-15.—Sawn frame for round-bottom ship, showing component wood parts.

<u>3.5.1.2.2 Sawn Frames for V-Bottom Vessels</u>.—Sawn frames for V-bottom boats are made of lumber usually not more than 2 inches thick by 6 or 8 inches wide. A typical V-bottom frame is shown in figure 3-16. Such frames are completely assembled before installation in the vessel. Principal parts of these frames are the bottom or floor members, the topside members, and wood or plywood gussets that strengthen joints at keel and chine.

Boards of suitable size are molded in the usual manner with plywood templates. They are then ripped out with a bandsaw or jigsaw, and the ends are cut with a circular saw to the angle necessary to produce the desired joint. Beveling is accomplished either as with heavy sawed futtocks or on a shaper by means of wood jigs. By the latter method, a jig is made up for each frame member that is to be beveled, as no two beveled frames are likely to be exactly the same.

Members are assembled into the complete frame by means of a scrive board (figure 3-17). This is a horizontal platform on which the frame lines are raised; it is equipped with slotted guides used to position the members to the body plan of the mold loft. The members are laid on the scrive board to coincide with the proper outline, and adjoining edges of each member are recut as necessary so that the sections are in proper alinement. The members are then removed to an assembly table and joined together by means of gusset plates, or "biscuits," which are glued and screwed in pairs over adjoining members. The screws are not removed, hence must be of nonferrous metal or galvanized or coated steel. The glue, which should be resorcinol resin, is applied with a mohair paint roller or stiff-bristle brush. Temporary spalls are fastened to the topside members until the glue has set.

White oak, mahogany, Sitka spruce, and baldcypress are the principal woods used for sawed frames of V-bottom vessels. Mahogany is preferred because of its light weight, freedom from checking, decay resistance, ease of working, low shrinkage and ability to hold its shape and size, and availability in the desired sizes. White oak is second choice, primarily because of its strength and because of its ability to hold fastenings. Sitka spruce is used because of its high strength-weight ratio, usually as a substitute for mahogany. Baldcypress is used because the all-heart grade combines strength with high resistance to decay. In high-speed boats of this type, the floor members are often made of a strong wood such as oak, or may be laminated, especially along the planing surface of the bottom, where stresses are most severe at high speeds. Topside members may be made of a weaker wood, such as mahogany or Sitka spruce. Gussets are generally Douglas-fir plywood.

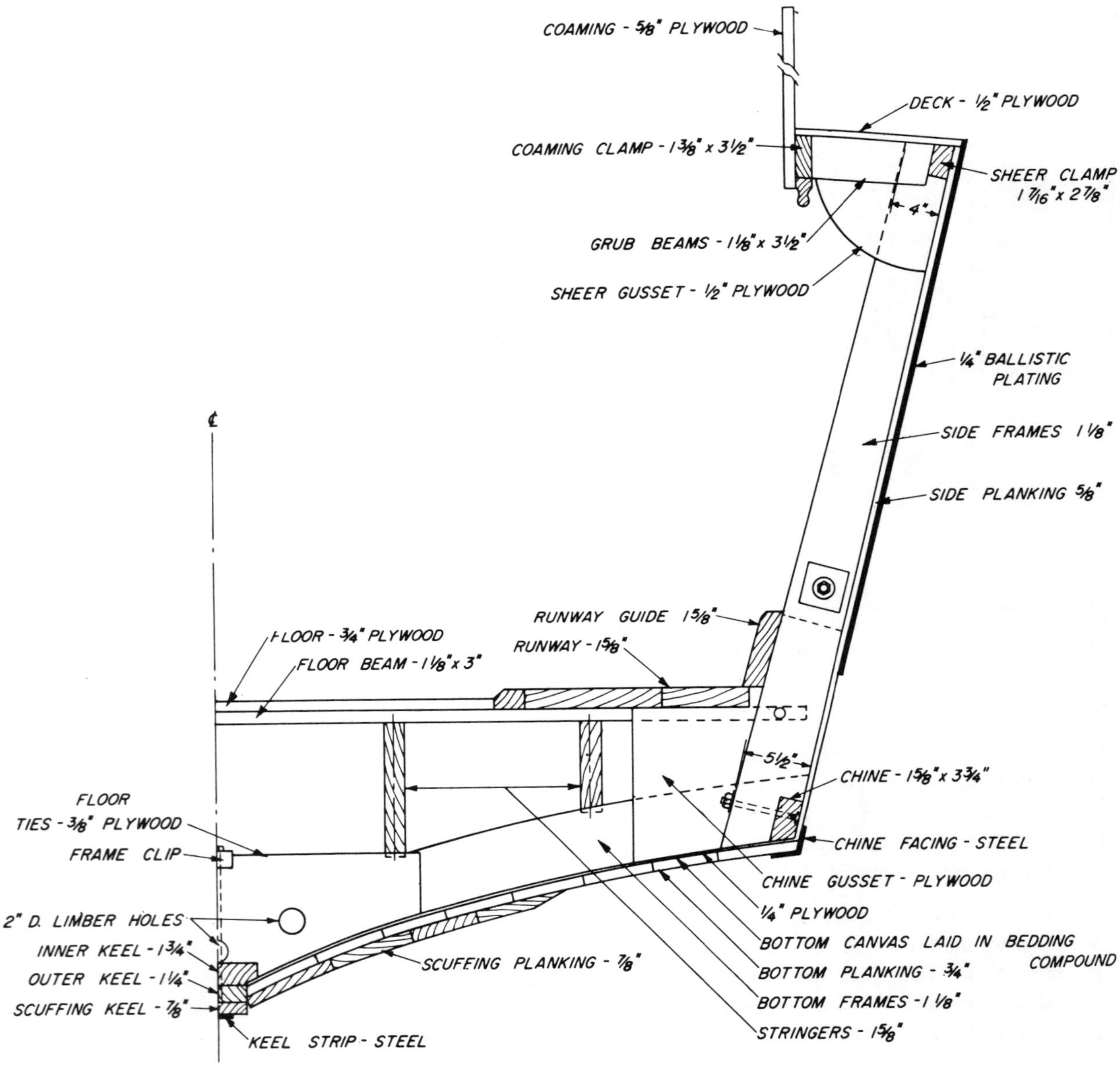

Figure 3-16.—Sawn-frame construction for V-bottom boat (LCVP).

Figure 3-17.—Scrive board used to assemble parts of a sawn V-bottom frame.

3.5.1.2.3 Bent Frames for Round-Bottom Vessels.—A bent frame consists of two sections butt-jointed together at the keel (figure 3-18) with the floor side-bolted to them. Bent frames are usually smaller in cross section than sawn frames. For certain types of boats, however, they have adequate strength and are therefore preferred because of the obvious economy of lumber possible through their use as well as their lighter weight. Good bending technique is essential to the production of bent frames. For such ships as minesweepers, subchasers, and some smaller boats, bent frames are frequently made up in two pieces, one nested in the other, where the radius of curvature is small. Where it is necessary to bend a frame to a sharp curvature and equipment to produce large end pressures is lacking, the frame can be slit from one end to a point beyond the curve; or a two-piece frame can be used.

White oak is used almost exclusively for bent frames because of its excellence for bending. Rock, winged, and cedar elm are good bending woods but are inferior to white oak in strength and decay resistance.

3.5.1.2.4 Bent Frames in V-Bottom Boats.—In some types of V-bottom patrol vessels, bent frames ranging from 1 to 1-1/2 inches in cross section are used in pairs, with filler blocks glued between them (figure 3-19). These are sometimes called web frames. Usually the bent members of such frames are steamed, then bent on a jig, and assembled into complete frames during the assembly of the boat. This is done because the longitudinals of the bottom and topsides pass between the bent members, filler blocks being fitted neatly in the openings between longitudinals. A variation of this type of bent frame uses a pair of bent members and filler blocks extending on each side only to a point just above the turn of the bilge, the topsides being one-piece members sawed to shape. In both, plywood gusset plates are used at the chine to strengthen the frame joint at that point.

Bent V-bottom frames are made of white oak because of its bending qualities. The filler blocks are of Sitka spruce, which is selected for its light weight. Other lightweight woods, such as Port-Orford-cedar and Alaska-cedar, could be used to advantage, as their heartwood has superior decay resistance.

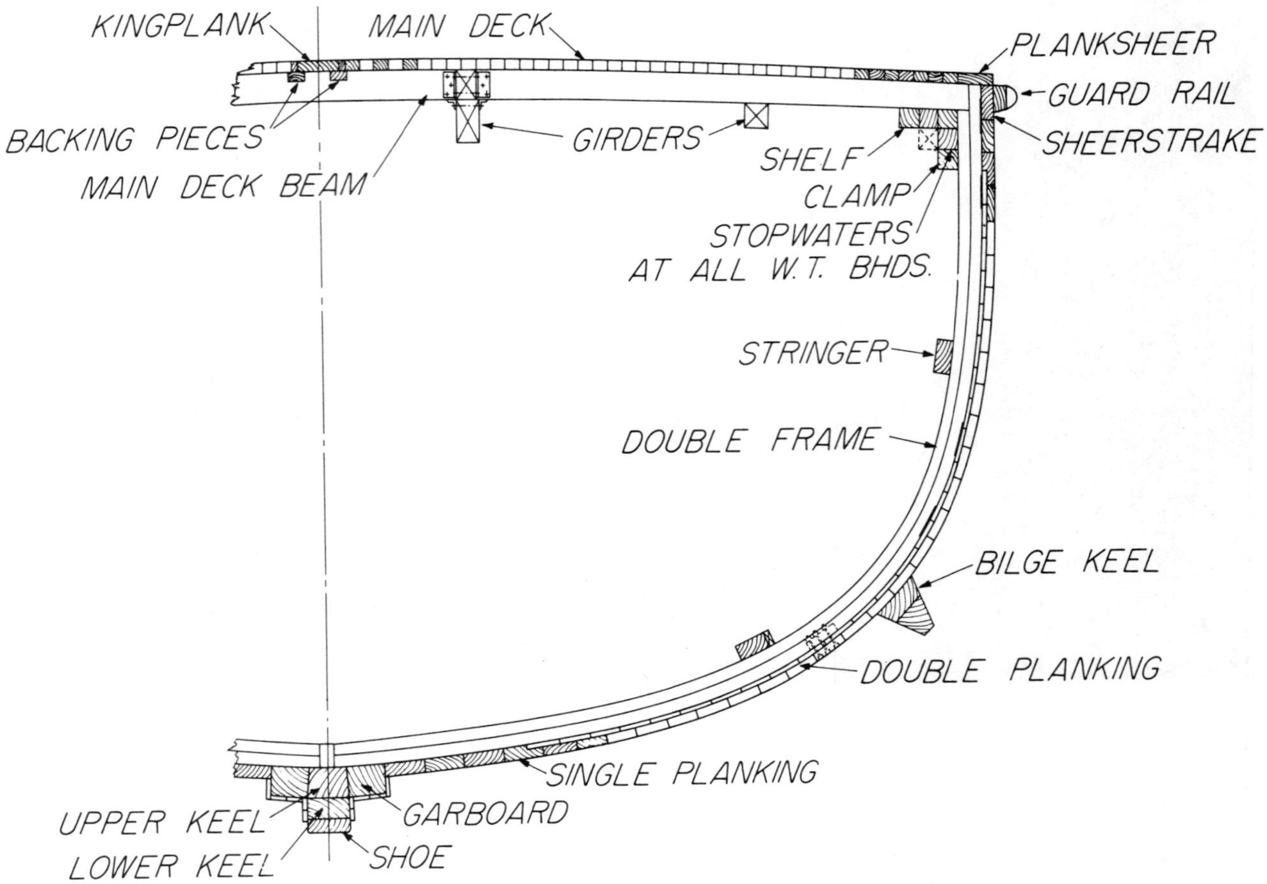

Figure 3-18.—Round-bottom bent frame, doubled for ease of bending.

3.5.1.2.5 Laminated Frames.—Procedures for making laminated frames are given in Section 2.2. Some shipyards have set up their own cutting, scarfing, gluing, forming, curing, and related equipment for laminating. This represents a considerable plant investment. Other shipyards purchase laminated parts from commercial laminators, so that only assembly is necessary at the shipyard. When low-cost facilities are available for shipment of laminated parts, this may prove more economical for the shipbuilder than to undertake lamination of the parts at his own yard. Circumstances, of course, vary at different yards, and each builder must hence decide upon the economics of the process as applied to his own business. Not to be overlooked, of course, are the possibilites of adapting laminating to private ship and boat construction.

Laminated frames are usually bandsawed to very nearly the correct bevel, so that little if any adzing is necessary for the attachment of planking, even for large ships.

White oak is the preferred species for laminated frames of large vessels. However, preservative-treated red oak is acceptable. Douglas-fir, southern yellow pine, and western larch are other suitable species, but western larch is rarely used. Frames for small vessels can be laminated of white oak, mahogany, Douglas-fir, southern yellow pine, and western larch, the heartwood of which has moderate decay resistance. Other species, such as red oak, can serve equally well if pressure treated with preservative.

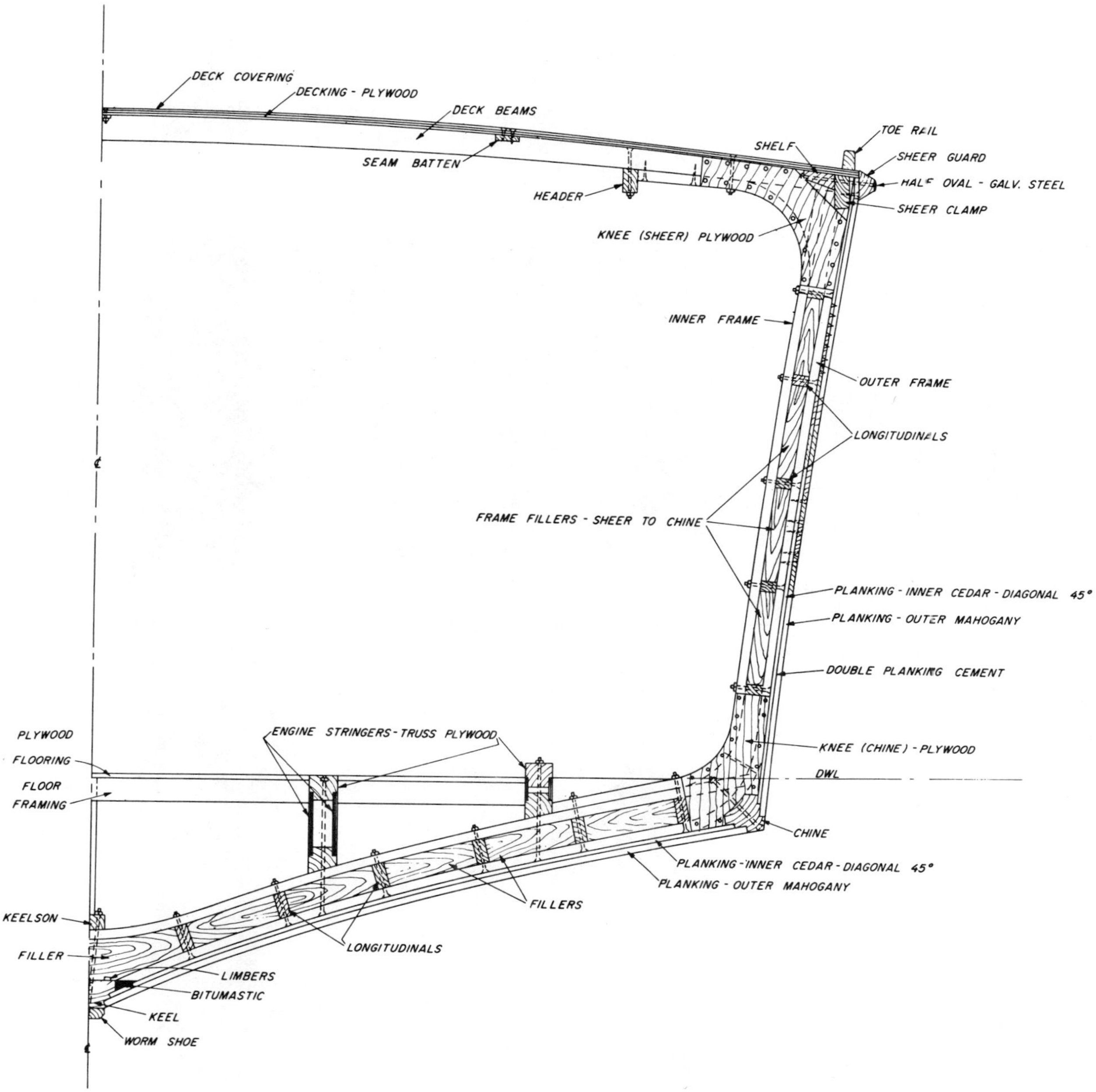

M 96096 F

Figure 3-19.—Web frame with bent inner, outer frames and filler blocks.

<u>3.5.1.3 Assembly of Frames in the Hull.</u>—Mounting of frames is one of the most important jobs involved in the assembly of the hull. Frames must be accurately positioned, centered, and plumbed; otherwise, serious difficulties will result and much work must be done over again. For large ships, cranes or other overhead handling devices are necessary for hoisting frames into position on the keel. For the smaller, lighter-weight hulls, positioning jigs are generally used (figures 3-6, 3-8), and the keel is, of course, laid when the frames are in position.

Some builders prefer to attach the first garboards to the keel before erecting the frames. Not only is this more convenient with respect to handling and fastening the garboards of such large vessels as the MSO minesweepers, which are laminated in one piece and, therefore, very long and difficult to handle because of their twist; but a wider base for the seating of the frames is thereby provided.

The usual practice in the erection of round-bottom frames is to start amidships and work fore and aft. However, this procedure can be varied to suit shipyard facilities. As a rule, the stem is left off until all frames except those that must be attached to it are in position; this facilitates hoisting and positioning of frames. The transom, on the other hand, is one of the first parts to be mounted (figure 3-20). Some builders use clamps to hold frames in place while supporting ribbands, spalls, and shoring are set up.

Figure 3-20.—Transom mounted on horn timber of MSO minesweeper. Note clamps.

Bow frames and cant frames are either butted against the keel-stem assembly or notched into it (figures 3-21, 3-22).

All frames, of course, should be checked for fairness when installed (figure 3-23).

Bent frames of round-bottom ships can be installed by either of two general methods. One method is to preform the frame on a jig or bending form before it is installed in the hull. The other is to bend the frame at the time it is installed by forcing it against ribbands erected for the purpose. Bending in place is successful only with frames of small cross section or where very moderate bends are required.

3.5.1.4 Stems, Sternposts, Transoms.—Key parts of any hull are the stem, sternpost, and transom. Together with the keel, they absorb the principal fore-and-aft stresses to which a hull is subjected in the water. Stem and transom are particularly vital in this respect in modern ship and boat designs, since the ends of most planking strakes, as well as the clamps, deck girders, chines, and similar longitudinals, are fastened to them directly or through an intermediate member, such as a breasthook. In many smaller vessels the sternpost is eliminated. The sternpost also is eliminated on some of the larger ships, such as the MSO minesweepers, especially when twin screws are employed. Where the sternpost is retained, however, it remains an integral part of the framework, supporting the stern, and is usually one of the largest members in the hull (figure 3-24).

Figure 3-21.—Cant frames notched to stem of laminated MSO minesweeper.

Essentially, stems and sternposts are simple to shape. The necessary operations of siding and beveling can be performed on a bandsaw. Conventional solid-wood stems consist generally of two parts, the stem proper, which may be sharply beveled for a cutwater, and the stem apron, or inner stem (figure 3-25). Such stems are connected to the keel by a more or less complicated assembly of fore-foot, forward deadwood, and knee. Although the stem proper is straight, the curved forefoot and sometimes the knee must be carefully made to avoid weakening cross grain. Cross grain should be held to a minimum when curved members are sawed out of a single piece of timber. This is usually done by laying out the template or pattern on the timber in such a way that the grain runs in the longest direction of the template (figure 3-26).

Laminated stems have replaced conventional solid-wood stems in many of the larger Navy ships. The advantages are readily apparent; a stronger, and often smaller, stem assembly results – a stem assembly that is free of weakening joints because it often includes apron, forefoot, deadwood, and knee of conventional stem (figure 3-27).

Figure 3-22.—Bow frames seated in precut notches in laminated stem of MSO minesweeper.

Figure 3-23.—Checking frames of a minesweeper for fairness.

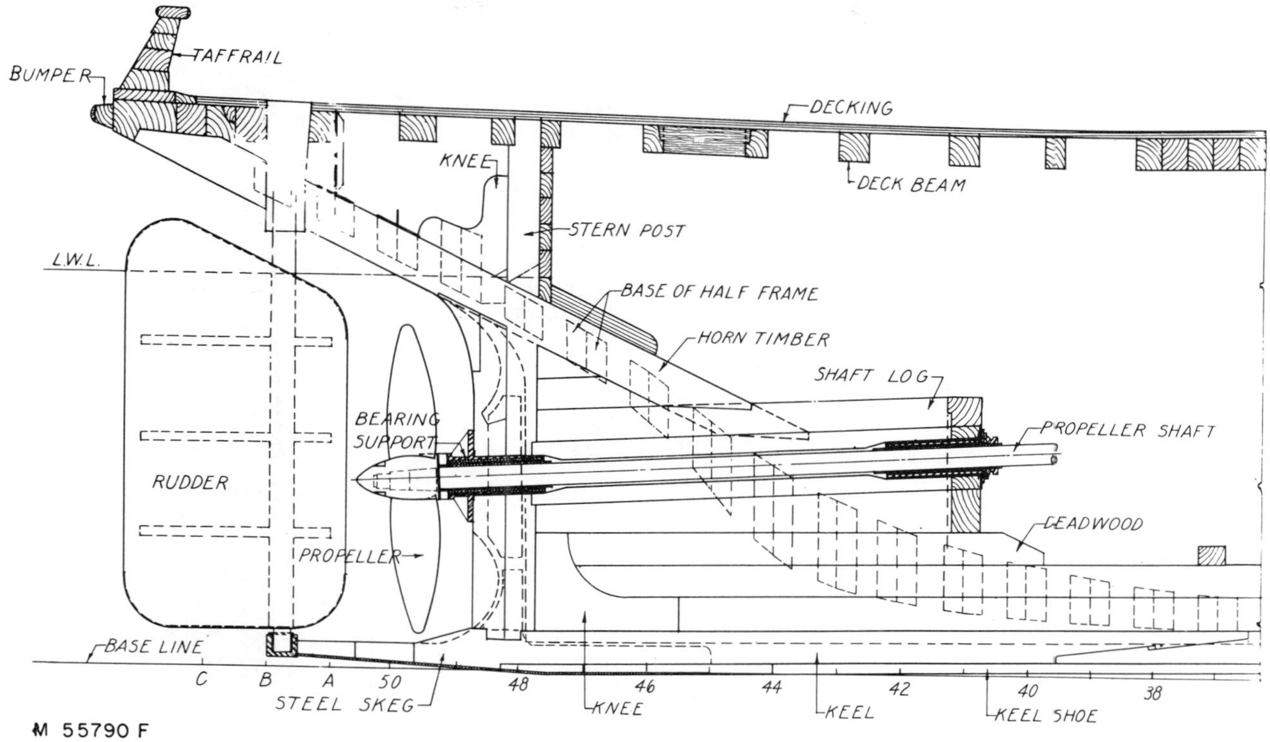

Figure 3-24.—Stern construction of a World War II YT harbor tug, showing sternpost and related framing.

In a stem keel assembly the rabbets are sometimes rough-cut on the keel before assembly. It is good practice, however, to assemble all parts of the stem and forefoot prior to the cutting of the rabbets. This work is done on small vessels with chisels; on larger vessels rabbets may be rough-cut with a portable circular saw and finished with adz or chisel.

The stem proper takes various curves. In almost all cases the stem is beveled so that its forward face is usually narrower than its thickness at the point where the rabbet is cut; this narrower width of face being determined by the width of the stem iron. In high-speed vessels, the beveling on the sides of the stem is determined by the angles described by the planking and the centerline of the vessel. High-speed vessels have very sharp cutwaters; that is, the angle of entrance of the stem to the water is very sharp. Beveling is accomplished with a bandsaw or sometimes a shaper; or it may be done entirely by hand. In some vessels the stem is also sided to variable thickness, thicker at top and bottom than at midlength.

In small craft, such as lifeboats, whaleboats, and motor launches, the stem may be beveled before it is assembled, thus providing a rabbet that can easily be finished with a chisel to the proper angles for planking after assembly to the keel. Faying surfaces of a two-part stem are always coated with bedding compound or white lead. Figure 3-28 shows two methods of fabricating a two-part curved stem from solid wood. Scarf joints should always be staggered as in A.

White oak heartwood is the preferred species for stems, both solid and laminated, with Douglas-fir, southern yellow pine, or western larch suitable alternates in large vessels. Mahogany is sometimes used in small craft.

Stopwaters always should be installed in all joints of the stem, keel, and sternpost assemblies. The inconspicuous and often overlooked stopwater is a vital part of a vessel's construction. If not properly placed, it can be the source of mysterious, hard-to-locate leaks.

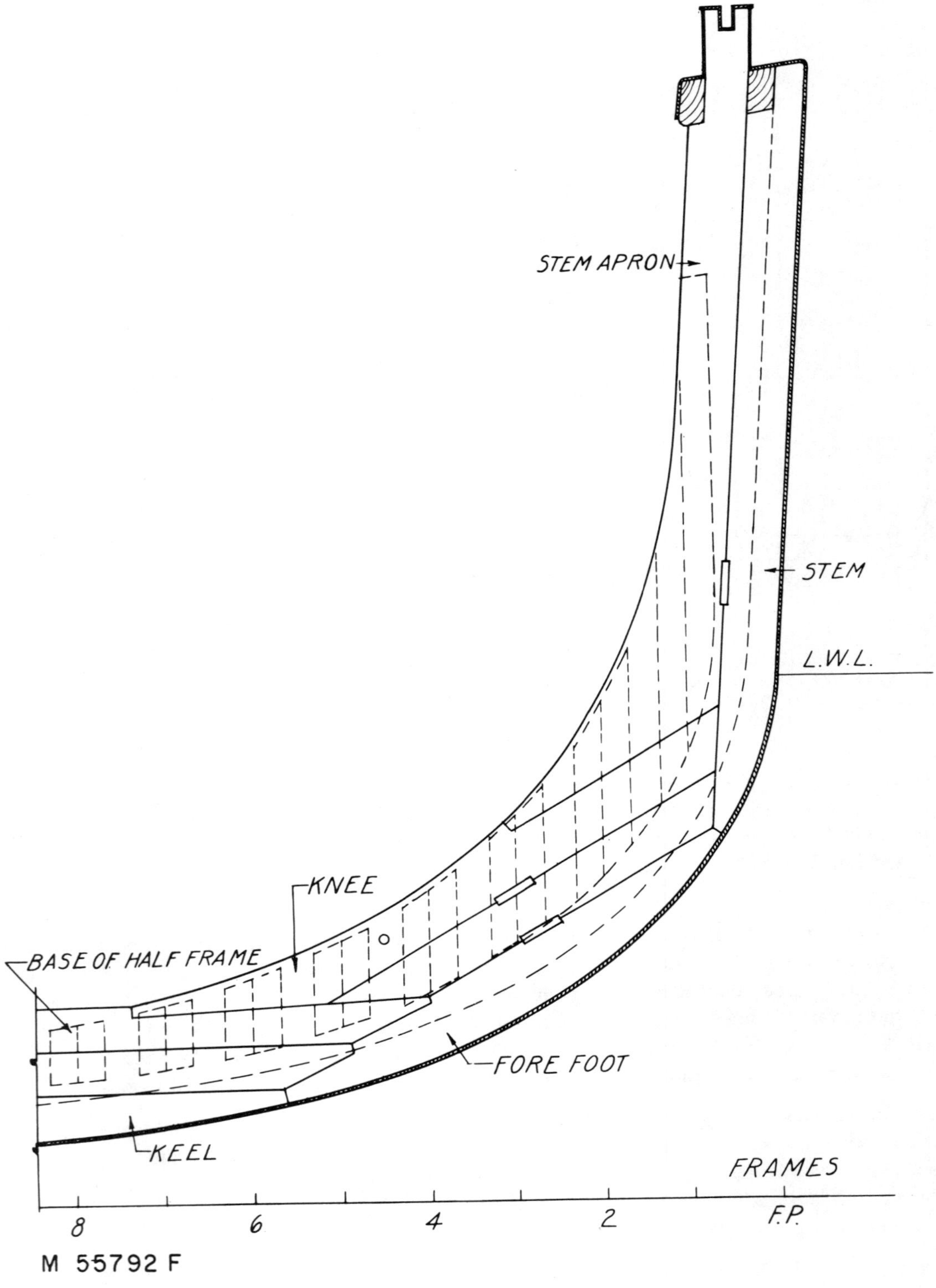

Figure 3-25.—Conventional solid wood stem construction for round-bottom ship, showing keel-connecting parts.

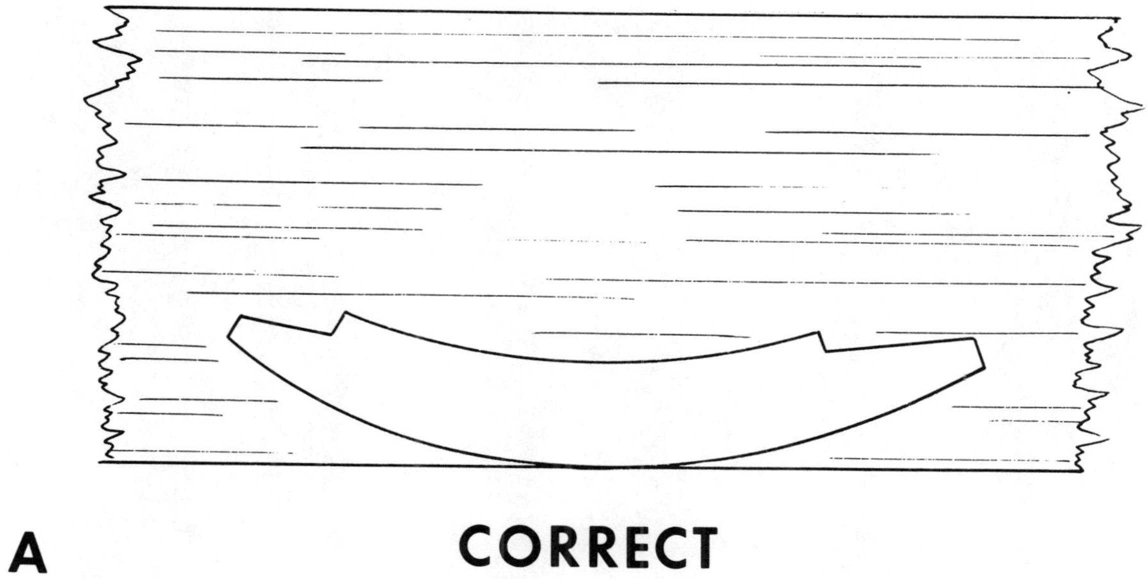

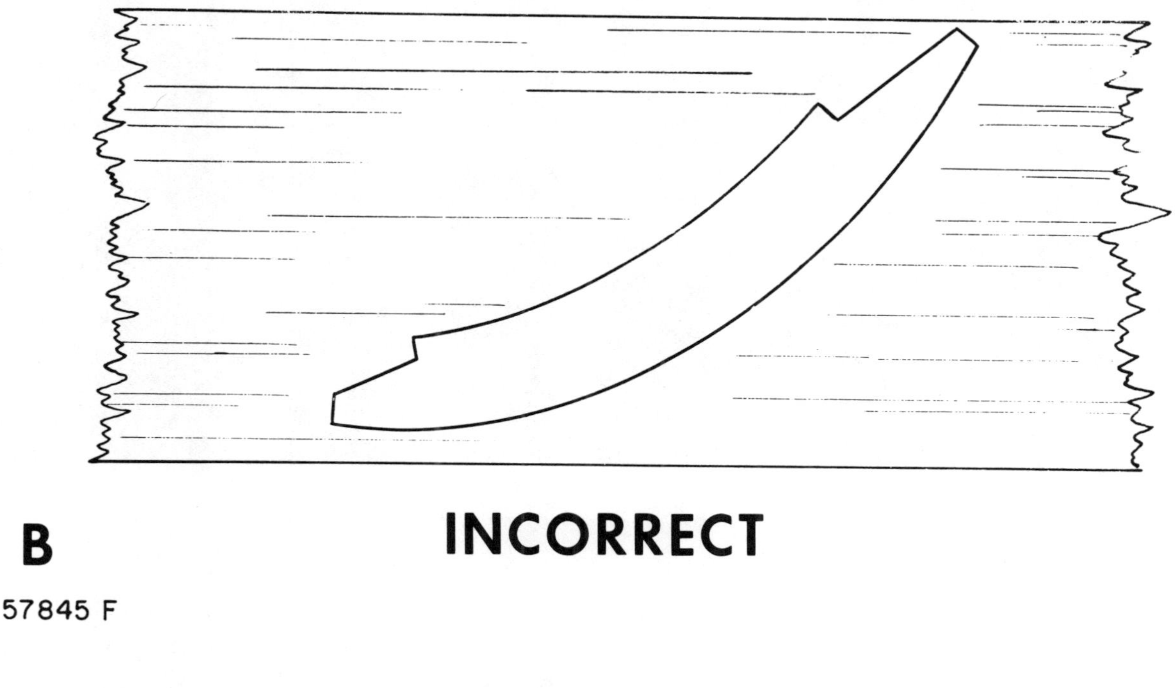

Figure 3-26.—Correct and incorrect methods of positioning template on straight-grained timber. A, Correct because grain of timber follows length of template. B, Incorrect because the piece will be unnecessarily weakened by cross grain.

Figure 3-27.—Stem-keel construction of an MSO minesweeper.

Generally speaking, stopwaters should be placed in every timber joint in the keel assembly where the joint passes from the outside of the hull to the inside at the rabbet. Stopwater should be placed at the exact point in the rabbet where the joint and the calking intersect (figure 3-29).

Holes for stopwaters should be bored clean and smooth. Stopwaters should fit holes snugly but not so tightly as to incur risk of splitting when they are driven. They should be made of a soft wood, such as white pine, and should not be painted.

If, after a vessel has been planked and launched, a stopwater leak should develop, the condition can be corrected by placing a stopwater in the offending joint a little outside the rabbet and calking the joint from the rabbet out through the emergency stopwater. This procedure, however, is not recommended as a regular practice in ship and boat building. Two stopwaters often are used, the joint between them being calked.

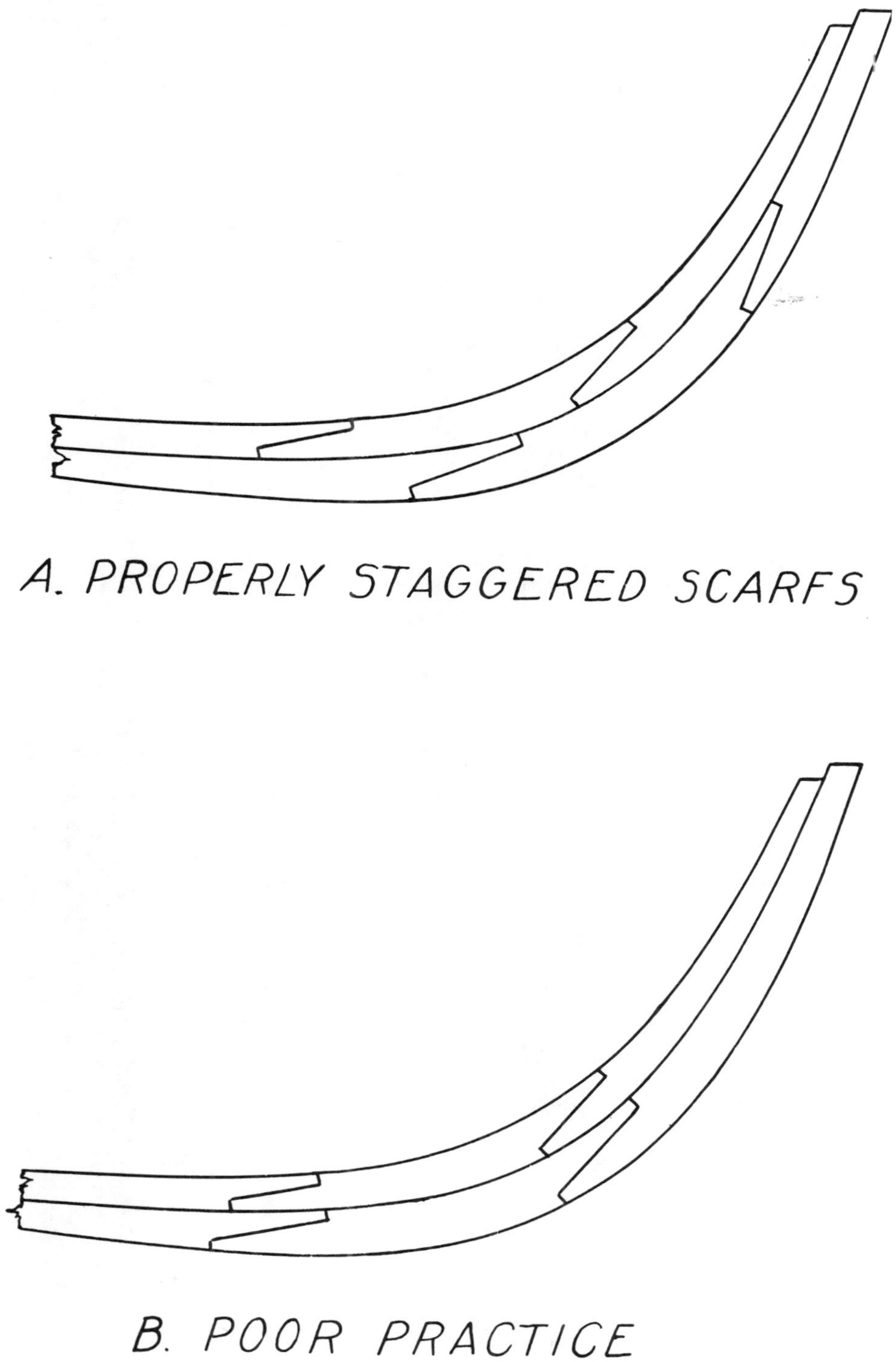

M 58165 F

Figure 3-28.—Two-part solid wood stem of a small boat, showing bolted scarf joints between keel, forefoot, and stem. A, Properly staggered scarfs; B, nested scarfs, which weaken the structure.

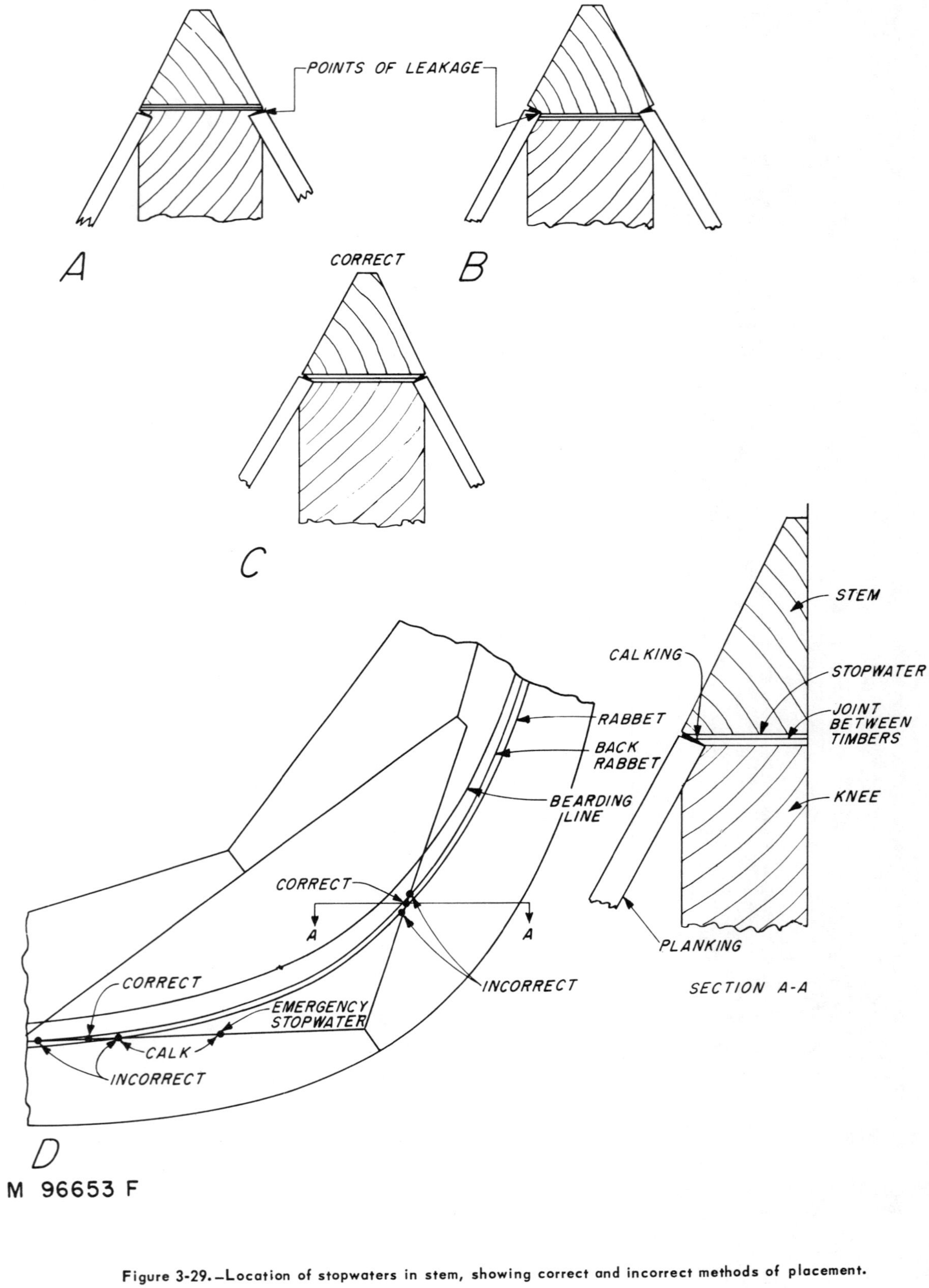

Figure 3-29.—Location of stopwaters in stem, showing correct and incorrect methods of placement.

Sternposts of heavy vessels may be the largest single piece of timber in the ship. It sometimes is necessary to make the sternpost of two or more timbers that are through-bolted together in order to provide a member big enough to accommodate the shaft and its fittings. Large sternposts are generally bored with a boring bar to provide for the installation of the stern bearing and sleeve of the propeller shaft. Sometimes sternposts are "swelled" at the shaft hole; that is, they are reduced in their thwartship dimension above and below the stern bearing. Such sternposts are employed in tugs, cargo and passenger vessels. Sternposts are shaped with bandsaw, adz, and plane. White oak heartwood is preferred for sternposts, for its combination of high holding power for fastenings and resistance to decay. Douglas-fir, southern yellow pine, and western larch are suitable alternates in large vessels. Mahogany sometimes is used in small boats.

Sternposts of small vessels, such as double-ender open boats, are sometimes laminated (figure 3-30). Preference is to laminate sternposts, knees, shaftlogs, and similar parts separately for ease of repair.

Many vessels are designed without sternposts, or with only rudimentary ones, especially vessels with more than one propeller. In fact, many small single-screw boats, such as personnel craft, do not require sternposts. The shaftlogs are placed entirely inside the vessels, the after bearing for the propeller shaft being a strut. The stern of such hulls is of transom construction.

The usual transom in Navy vessels consists of a heavy framework covered on its after side generally by two thicknesses of solid wood or plywood, between which a canvas or muslin fabric membrane is laid in bedding compound. The framework of the transom must be built to take great loads and stresses, especially on such craft as minesweepers (figure 3-31). Severe stresses are imposed when vessels are docking or in contact with wharves or other vessels. Wood for transoms must have high holding power for fastenings and good shock resistance.

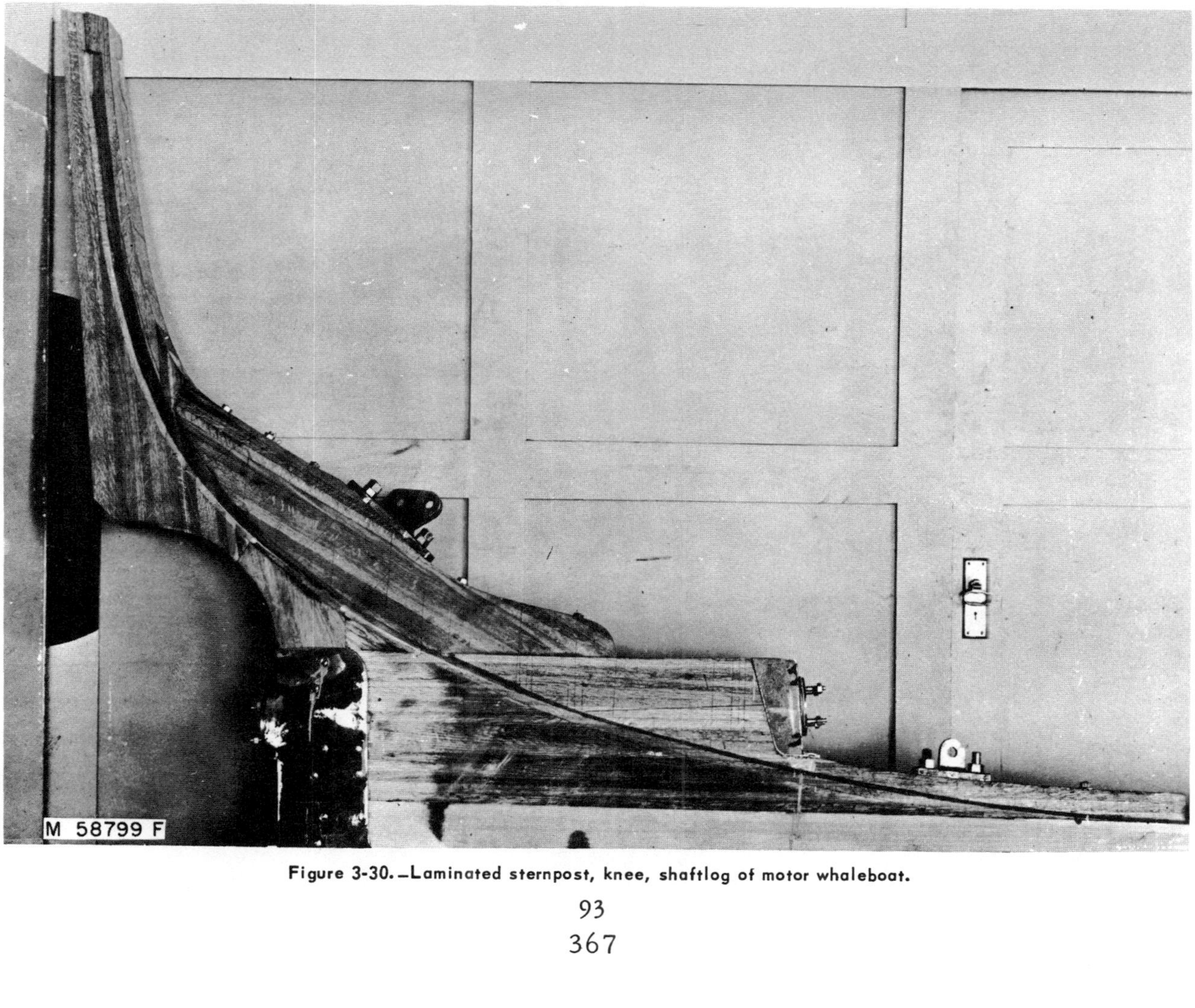

Figure 3-30.—Laminated sternpost, knee, shaftlog of motor whaleboat.

Figure 3-31.—Transom construction of MSO minesweeper.

The transom frame, or fashionpiece, may be either curved or rectangular in shape. Curved fashionpieces are made in the same way as other frames, and may consist of sawn futtocks, steam-bent solid wood, or laminated wood. Bent transom frames are formed over a jig and allowed to condition before they are installed. Complete framework of an MSO minesweeper transom is shown in figure 3-20. The framework of a rectangular transom is shown in figure 3-32.

Transoms must be securely fastened to keels. Fastening usually is done with a wood or metal knee. Figure 3-33 shows a metal knee to which the transom of the MSO minesweeper is bolted.

The species preferred for transom frame construction in large vessels is white oak. Southern yellow pine and Douglas-fir are alternate species sometimes used. In laminated transom frames, red oak pressure-treated with a good preservative is a suitable substitute for white oak heartwood. Mahogany is used for lightweight craft. Sitka spruce is sometimes used for the vertical stiffeners in transoms because it combines light weight with high strength.

The transom is commonly planked with the same wood as the rest of the hull, usually Douglas-fir, southern yellow pine, or white oak in larger ships. Transom planking of craft less than 100 feet long is commonly of mahogany, baldcypress, Port-Orford cedar, or Atlantic white-cedar. Douglas-fir boat-hull plywood is used on plywood-planked craft.

Figure 3-32.—Rectangular transom framing of Vosper PT boat.

3.5.1.5 Keelsons.–Keelsons are employed in round-bottom ships to add longitudinal or "backbone" strength to the structure to resist hogging and sagging stresses. The number of keelsons present depends upon the strength considered necessary by the designer for the particular type of vessel under construction. However, many are employed to make up the complete keelson assembly; the principal shaping factor involved is to finish all faying surfaces true to ensure tight joints. This finishing involves beveling to fit the inner curvature of frames fore and aft and is accomplished by accurate adzing or machine beveling by bandsaw or stationary or portable router.

The keelson assembly of a conventional sawn-frame ship consists of keelson, sister keelson, and rider keelsons. In ships, this makes up a massive bulk of wood resting on the floor timbers of the frames directly above the line of the keel and extending from stem to stern (figure 3-34). The keelson assembly is fastened to the keel and to the floor timbers and contributes largely to the longitudinal stiffness of the boat. Modified forms of keelsons, such as the arched and trussed, have been used.

Although the keelson is a prominent member in large ships, it usually is omitted in smaller craft. Longitudinal stiffness in smaller boats is obtained through the use of girders and minor longitudinal stringers distributed over the bottom of the boat.

Figure 3-33.–Metal knee connecting transom to horn timber of minesweeper.

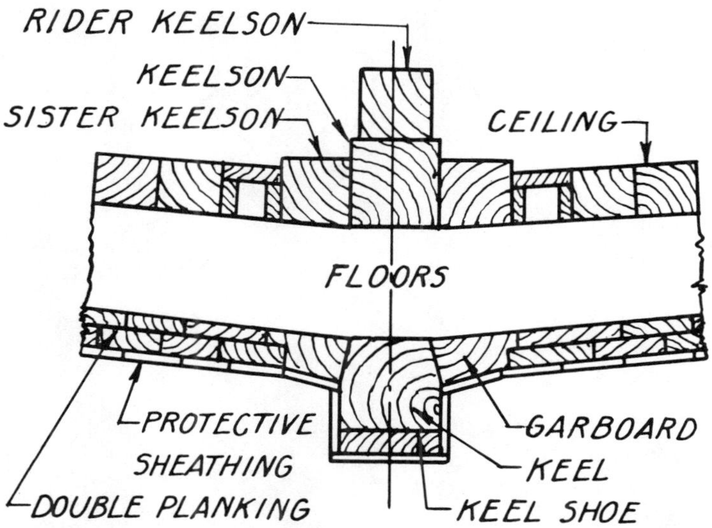

Figure 3-34.–Keelson assembly of ATR rescue tug.

The timbers used in solid wood keelsons should be as long as possible, to reduce the number of bolted scarf joints necessary to connect them. After the timbers are dressed to size, they are scarfed, laid in place, and bolted together. The scarf joints must be staggered if more than one keelson is used. In heavy timbers, the total length of scarf should be at least six times the depth of the piece. In light members, the length of the scarf may be as much as 12 times the depth of the piece.

Laminated keelsons are made full length, thus eliminating bolted scarf joints between timbers. Another advantage is that they can be made up to approximately the shape desired to conform to hull curvature and frame bevels, thus largely eliminating shaping operations.

Because of the length of keelsons in ships, it is customary to tighten bolts in staggered sequence; that is, every fourth bolt or so is turned down along the length of the keelson before intermediate bolts are tightened. Such tightening avoids the possible buckling and misalinement of keelson and frames that might result from tightening all bolts in sequence (figure 3-35). It also is considered good practice to begin tightening at midships and work fore and aft.

White oak, southern yellow pine, and Douglas-fir are the preferred species for keelsons because of their stiffness and ability to hold fastenings. Douglas-fir is substituted for white oak more frequently in keelsons than in keels, because it is more readily available in large sizes and long lengths. Western larch, although not used, has properties that indicate it would be a suitable wood for this purpose. White oak and Douglas-fir are preferred for laminated keelsons.

3.5.1.6 Clamps, Chines, Other Longitudinals.—Shelves, clamps and other longitudinals add stiffness to the hull and serve as support for deck beams and to connect frames and bulkheads. In ships they consist of heavy timbers made up of solid wood parts scarf-jointed together lengthwise, or of board-thickness pieces laminated together to the required cross-sectional size and length. Clamps usually are supported by one or more lodgers in heavy ships (figure 3-36), but in light, high-speed craft the clamp usually is a single member. The assembly of clamp, shelf, and lodgers supplants to some extent the hanging and lodger knees formerly used at the outboard ends of deck beams. In many high-speed boats, the sheer clamp provides a bearing for the deck beams. In large ships, shelf members bolted to the sheer clamp provide a wider base for deck beams.

Figure 3-35.—Laminated keelson of MSO minesweeper, showing through bolts for floor girders.

Shelf and clamp timbers are finished to dimension, scarfed (figure 3-37), and put in place on the boat. As in the keelson assembly, it is important to use the longest available lengths and to stagger the scarfs. Lamination, of course, eliminates much scarfing. The various timbers of the assembly are fastened together, and to the frames, by through bolts and drift bolts. After the deck beams and waterways have been put in place, fastenings are put through them into the clamp and shelf assembly.

The shelves, clamps, and lodgers near the bow of the ship are sometimes required by the lines of the design to take a considerable bend. At such places it often has been found necessary to use a series of bolt-laminated thinner pieces in lieu of one large solid piece. Steaming and prebending large solid members sometimes is necessary before installing them. At bow and stern, the clamp sometimes is beveled to fit the curvature of the frames. Sheer clamps and chines of boats often are steamed and bent over a form before installation, or they may be steamed or boiled and bent in place while still hot (figure 3-38). Prebending over a form is preferred with heavy members, because bending in place has a tendency to pull the hull out of shape. Such members as cover boards and waterways of smaller craft are generally bent or sawed to shape, but sometimes they are steamed and prebent.

The principal longitudinal members in addition to the keel, are the sheer clamps, shelves, engine girders, stringers, hogging clamps, and chines. Frequently, in vessels of the smaller, lighter types, including V-bottoms, longitudinal members are made up of two or more parts that are glued and screwed or bolted together (figure 3-39). This applies also to their scarfing. Resorcinol glue is best suited for use between the various parts of such multiple-piece longitudinals. Often the longitudinals are preformed, the glue setting while the pieces are held in jigs prior to their installation.

Figure 3-36.—Erecting clamp-lodger assembly of MSC 141-foot minesweeper.

Figure 3-37.—Mounting scarfed shelf member of MSC minesweeper.

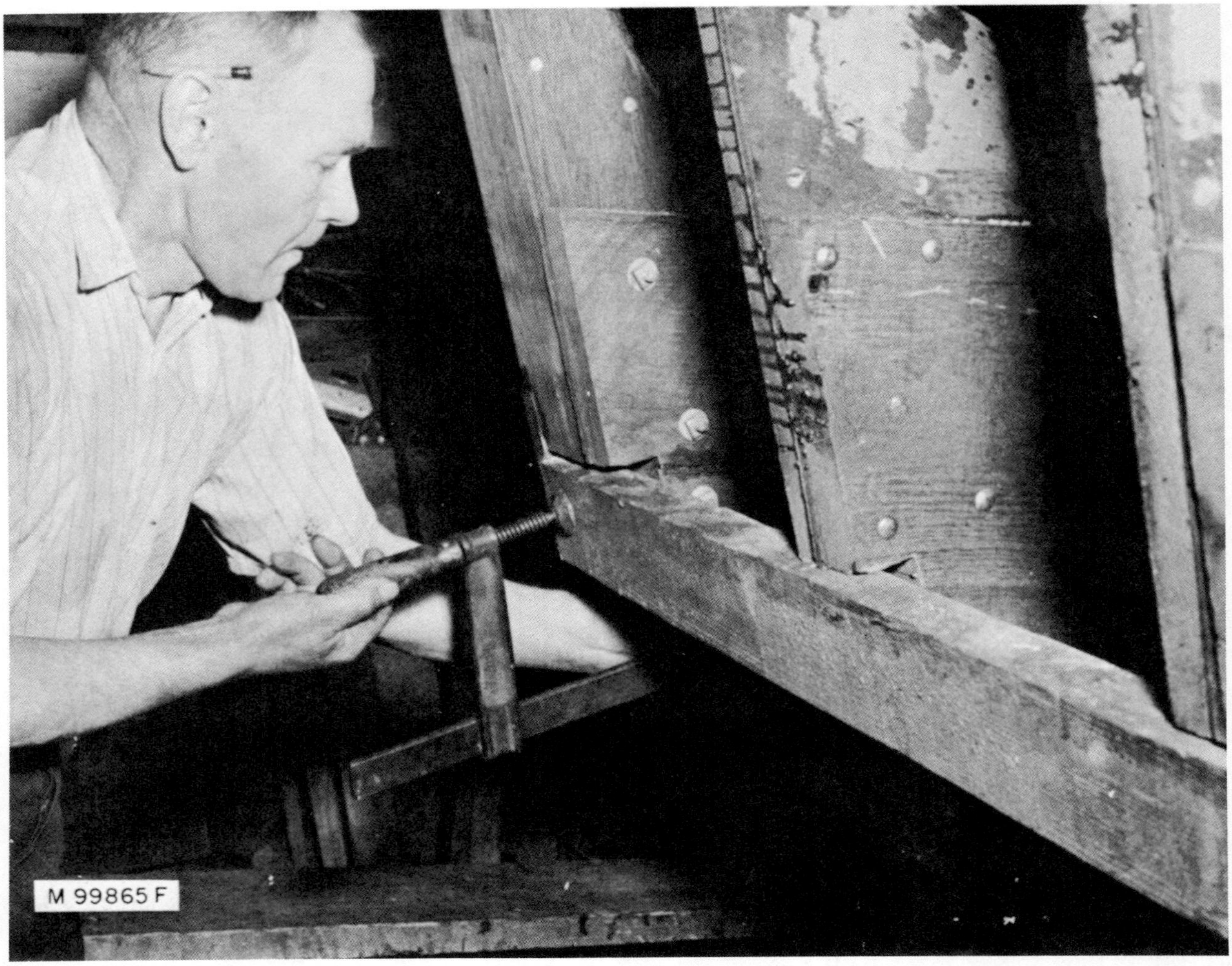

Figure 3-38.—Clamping chine of LCVP in place.

Longitudinals that are used to secure rigidity often pass through web frames (figure 3-19) and are secured to the latter by screws or bolts. This method sometimes is employed in light or high-speed craft of both the round-bottom and V-bottom types. In general practice with vessels not using web frames, longitudinals pass inside of the thwartship framing and are secured to such frames by bolting, with the exception of keels, sheer clamps, and chines.

Girders used in boats vary from simple solid-wood members tapered to fit the contour of the hull as the girder extends forward and aft, to complex I- and box-beam types and laminated members built to the shape of the required contours. Generally, girders are placed with the wide dimension of their cross section vertical. A box beam with solid wood flanges and plywood webs is illustrated by figure 3-40. Stiffeners of solid wood are fixed between the flanges. Such girders are glued, using a resorcinol glue, and screwed. Girders may be simply through-bolted to the frames, or – particularly when of considerable depth – they may be held in position by means of uprights which are bolted to the floors, frames, or both, and given added rigidity by means of transverse plywood stiffeners that may serve also as gusset plates for the floor (figure 3-40).

Longitudinal girders of solid or modified I-beam type sometimes are made in special jigs applicable to one particular operation. The glue is applied either by brush or by mechanical spreader, depending upon the size and shape of the member. Pressure usually is applied by C-clamps or other types of clamps. Resorcinol glue (Section 2.1) is best for this operation, both from the standpoint of durability and that of allowable assembly time. Where a large number of clamps or screws must be drawn up within a limited period, the factor of assembly time becomes important. Facilities for curing the glue in assemblies of this type, such as a heated chamber or kiln, can be employed to great advantage.

Figure 3-39.—Sheer clamp used on a V-bottom boat.

3.5.1.7 Deck Beams, Carlings.—The purpose of deck beams and carlings is to form a framework upon which decking is laid and secured. These members, together with bulkheads and decking, resist by far the greater part of the thwartship stresses to which a vessel is subjected. They must also be capable of sustaining the required vertical loads without objectionable deflection. Deck beams and carlings may be made of solid wood or of glued, laminated material. Deck beams (figures 3-41 and 3-42) may be sawed to shape of solid timbers or laminated.

The principal shaping operation involved in producing deck beams is cambering. The beams shown in figure 3-41 were sawed to the proper camber, as may be seen by their slight curvature. The designer of the vessel indicates the amount of camber required for the deck. The builder lays out the beam camber by means of a beam mold. A fair curve is imperative if the decking is to be properly installed in any vessel. Cambering is done by bandsawing, or with a jointer or shaper. Sometimes beams are built up of two or more pieces or laminated on a mold of proper camber.

Beams should be securely fastened to the clamps and shelves. Sometimes steel or wood knees are used to join deck beams and frames. Typical methods of fastening are shown in figure 3-43.

Often in small boats the deck beams are joined directly to the frames by means of a plywood gusset that is screwed or bolted and usually also glued to both members.

Glue-laminated deck beams used in certain light, high-speed boats are assembled and bent to the proper camber on jigs similar to those used to bend other curved, laminated members (Section 2.2).

Figure 3-40.—At center, a type of girder, of box-beam construction, used on a V-bottom boat.

Carlings are short beams forming a portion of the framing about deck openings; they also are called headers when they support the ends of interrupted deck beams. Carlings are molded and sided square with a surfacer or jointer, and the deck beams are secured by having their ends notched or dovetailed into headers.

Species used for deck beams and carlings include white oak, ash, Douglas-fir, southern yellow pine, western larch, and Sitka spruce. These woods are used because of their resistance to bending stresses and their decay resistance, with these exceptions: Sitka spruce and ash have low decay resistance, and spruce is in addition comparatively low in bending strength. Consequently, spruce and ash should be used only when protected from wetting and when well away from bilge water. In small boats, especially fast ones, Sitka spruce and mahogany are used for beams, primarily because of their high strength-weight ratio. Deck beams in both large and small boats can use laminated construction to advantage, because lamination makes possible a beam of practically straight grain and also affords a saving in material where the curvature of the beam is appreciable.

Figure 3-41.—Sawed-to-shape deck beams being installed in MSC minesweeper.

Figure 3-42.—Boring bolt hole in shelf of MSC minesweeper for attachment of deck beam.

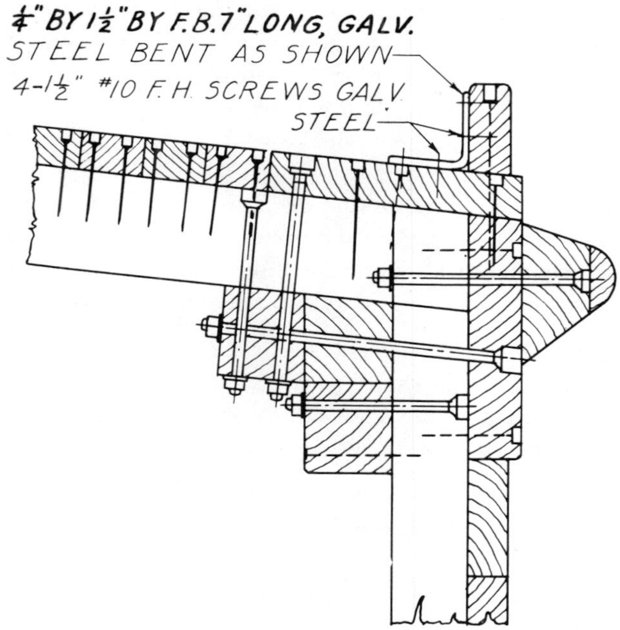

TYPICAL FRAME "A"

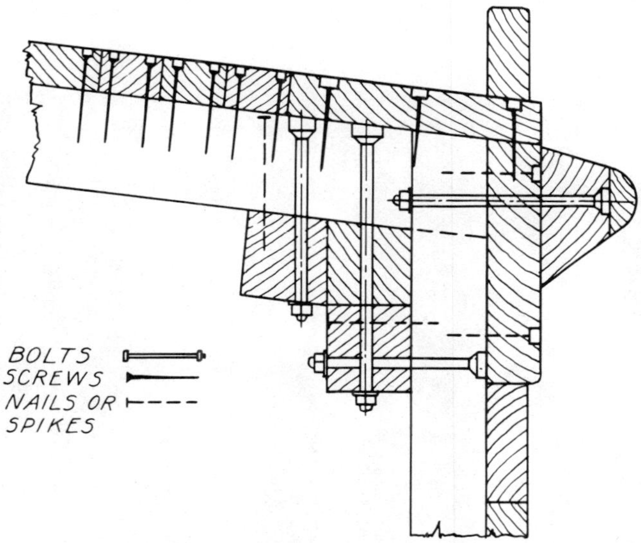

TYPICAL FRAME "B"

M 55796 F

Figure 3-43.—Typical sections of alternate frame and deck beam joints of a YMS minesweeper, showing arrangement of fastenings. Typical frames A and B are alternates.

3.5.1.8 Shaftlogs, Rudder Blocks.—A shaftlog is a structural member through which a propeller shaft passes. Its after end is generally secured to the sternpost. Its forward end should be secured to a header. Its purpose is generally to carry a metal sleeve which permits sea water to pass along the shaft as far inboard as the stuffing box, since nearly all modern vessels have water-lubricated stern bearings (figure 3-24). A shaftlog forms a portion of the after deadwood of the vessel. Solid wood shaftlogs of large vessels may be made of a solid timber bored with a boring bar, or may consist of two or four timbers bolted together. In the latter cases the shaft hole is cut out before the sections are bolted together, after which a boring bar may be used to finish the hole to the prescribed dimensions. All faying surfaces must, of course, be surfaced true so that they are in contact at all points, and they are painted with white lead when the parts are bolted together. Frequently wood splines are installed between the sections at stopwaters.

In most modern twin- or multiple-screw vessels, the shaftlogs are usually placed entirely within the skin of the ship and are often laminated. Such shaftlogs are sufficiently big to compensate for the loss of strength when the planking is pierced by the shaft hole. They are designed, when necessary, to receive and hold in place planking and frames that may be severed.

Even in such ships as the MSO minesweepers – which are designed to take advantage of modern fabricating methods that permit use of smaller members – massive wood parts are essential at points of great stress. Among these parts are rudder shaft blocks, engine foundations, and bases for special equipment needed to perform the functions for which the vessel is built (figure 3-44). Such blocking generally is laminated in the same manner as shaftlogs and carved to conform to the hull curvature.

Teak, Douglas-fir, and white oak are preferred for shaftlogs. Southern yellow pine and western larch are acceptable substitutes. For the large laminated shaftlogs, Douglas-fir is well suited because lower gluing temperatures are needed at the glue lines than for white oak (Section 2.2.8.2). These woods combine the necessary decay resistance, hardness, and ability to hold fastenings, can stand rough usage to a high degree, and are available in the large sizes required.

Figure 3-44.—Rudder shaft block mounted in MSO minesweeper.

3.5.1.9 Knees, Breasthooks, Riders, Pointers.—Knees, breasthooks, riders, and pointers are used to strengthen the vessel at certain joints and at other places of possible weakness. Knees are commonly used at such critical points as the juncture of sternpost or stem with the keel or deadwood, at the joints between deck beams and shelves, and occasionally to brace stanchions. They also are placed horizontally against deck beams in the way of hatches. Breasthooks are employed to attach the stem or apron to the shelves or clamps. Pointers are timbers used to brace the bow or stern, usually extending diagonally across the ceiling from the shelf forward to the stem or aft to the stern. Riders are bracing diagonals similar to pointers, but extend across the ceiling from the keelson to the lower deck beams.

Knees often form an approximate right angle, but may vary greatly depending upon the angle formed by the members joined (figure 3-45). Occasionally, knees are made of stock possessing grain curved to approximately the angle desired, which is usually obtained from stumps of trees. Such stock minimizes the inherent weakness of cross grain which results from sawing knees out of straight-grained wood. Material with natural crook is, however, increasingly difficult to procure, and most sawed knees are made from straight-grained stock. Lamination is usually employed for the larger knees (figure 3-45).

Knees are produced on a bandsaw; the cutting operation is, for the most part, a simple one. Heavy knees used on ships usually are dubbed to fit the angles occurring between deck beams and frames or ceiling at the bow or stern of the ship; such dubbing must be carefully done to ensure proper fit both fore and aft and vertical.

The breasthook as used on a heavily timbered vessel is ordinarily a flat knee fitted to the acute angle formed by shelves and stem apron and beveled to fit snugly against the shelf timbers or clamps. For smaller vessels, such as the PT, it is designed with a seat to fit around three sides of the stem, and the deck clamps are bolted to its arms. Figure 3-46 shows a typical glued breasthook used on a V-bottom boat. Such breasthooks should be sawed so that the grain in alternate pieces will run in about the same direction as that of either port or starboard longitudinal to avoid the weakening effects of cross grain.

These breasthooks are glued of wood in three or more layers; each layer is band sawed to shape in two pieces edge jointed together by glue during the gluing process. After the glue has cured, they are trimmed to final shape on a router or shaper and sanded smooth.

Riders and pointers must be carefully worked and bent to the inner curvature of the hull in order to lie snugly against the ceiling when bolted in position. They are usually dubbed to fit with an adz.

Knees with natural crook are preferred but are no longer available in quantities sufficient to meet the demand. The preferred species for natural knees is tamarack. White oak natural knees are sometimes available.

The preferred species for sawn knees are white oak, southern yellow pine, Douglas-fir, and Western larch. Where knees are sawed, they generally are made from the woods that are used in other parts of the ship. The laminated knee is superior to both the sawn and the natural knee because it has less cross grain than either of the other two and can be made in the sizes and dimensions required for any particular vessel. White oak or Douglas-fir is used.

3.5.1.10 Bulkheads.—Bulkhead is a term applied to any of the partition walls that subdivide the interior of a ship into compartments or rooms. The various types of bulkheads are distinguished by the addition of a word or words explaining the location, use, kind of material, or method of fabrication, such as transverse, longitudinal, and watertight. Bulkheads may serve several important purposes; they may support decks, resist stresses that tend to twist the hull, resist athwartship stresses, and – when they extend longitudinally – they aid in resisting hogging and sagging stresses. They usually serve to divide the hull into a number of watertight compartments for protection against water and fire.

In round-bottom vessels, bulkheads may be put into position early in the construction period or may be installed when hull construction is well along. Bulkheads sometimes are used to help form the shape of the hull by acting as supports for the longitudinals or for the ribbands. This method is commonly used in small boat construction and sometimes in large vessels. Bulkheads may be completely prefabricated be-

Figure 3-45.—Laminated knee as used in MSO minesweeper.

fore being put in place. For boats requiring ceiling, watertight bulkheads must be installed after the ceiling is in place; such bulkheads usually are built in place. Exterior-type Douglas-fir plywood is widely used for bulkhead faces, especially in small boats, being glued and screwed to a timber framework. In large vessels wood bulkheads are constructed by using a timber framework and a facing of two layers of tongued and grooved boards laid diagonally (figure 3-47), with fabric and bedding compound or white lead between the layers for watertightness. Calking compound is laid between the frame and bulkhead planking to keep water out of the bulkhead, where it might collect and set up a decay hazard (figure 3-48).

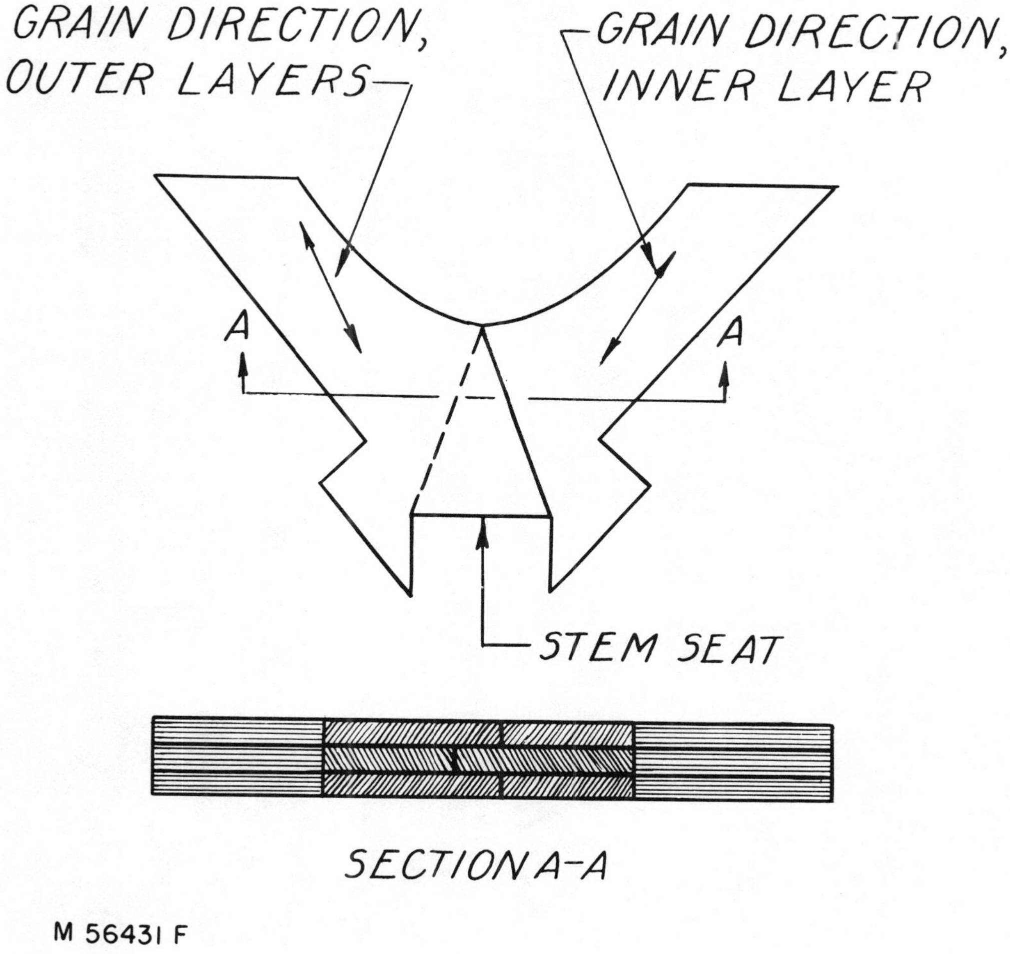

Figure 3-46.—Glued breasthook, each layer is sawed of two pieces of oak edge glued together.

In the lighter vessels, bulkheads are generally preassembled and installed complete. They usually consist of a light framework of solid wood and a double skin of plywood glued and screwed to this framework, sometimes with one or more bulkhead doorways. Plywood is laid in sections on the framework – to which glue has first been applied – and screwed down to exert gluing pressure on the joint between plywood and framework. The edges then are trimmed off with a portable router and the necessary openings for doors, longitudinals, girders, electrical conduits, and other equipment are cut out.

Figure 3-47.—Diagonal planking on MSO minesweeper bulkhead.

Considerable effort must be expended to make bulkheads watertight. The bulkhead itself is made entirely watertight by means of canvas set in white lead placed between reversed diagonal layers of wood. The watertightness of its margins is far more difficult to secure. A watertight division in the hull of a wood ship, especially when that ship is built with ceiling, becomes an installation involving a large amount of effort. Wood bulkheads of vessels with ceiling – such as the World War II ATR tugs – are, as a rule, secured in position by members known as chines, which are fayed and fastened to the ceiling and to the bulkhead proper. Chines generally are placed on both sides of the bulkhead. It is possible to obtain a high degree of watertightness if the following steps are carried out (figure 3-49), bearing in mind that no one step can be improperly or carelessly executed, and that such work must begin in the earliest stages of the construction of the vessel:

Figure 3-48.—Calking on bulkhead frame.

In vessels built with double-futtock frames, stopwaters must be installed between the frames and in the way of all butts in the futtocks. (The futtocks referred to are those in the immediate vicinity of the bulkhead. Designers almost always locate bulkheads directly in the way of a futtock frame.) Where the futtock frame lands on the keel, calking cotton laid in a thick sludge of bedding compound or white lead should be installed athwartship. The strands of calking cotton should be carried upward on the outer faces of the frames after dubbing as each succeeding plank goes on, whether the planking begins from the lower portion

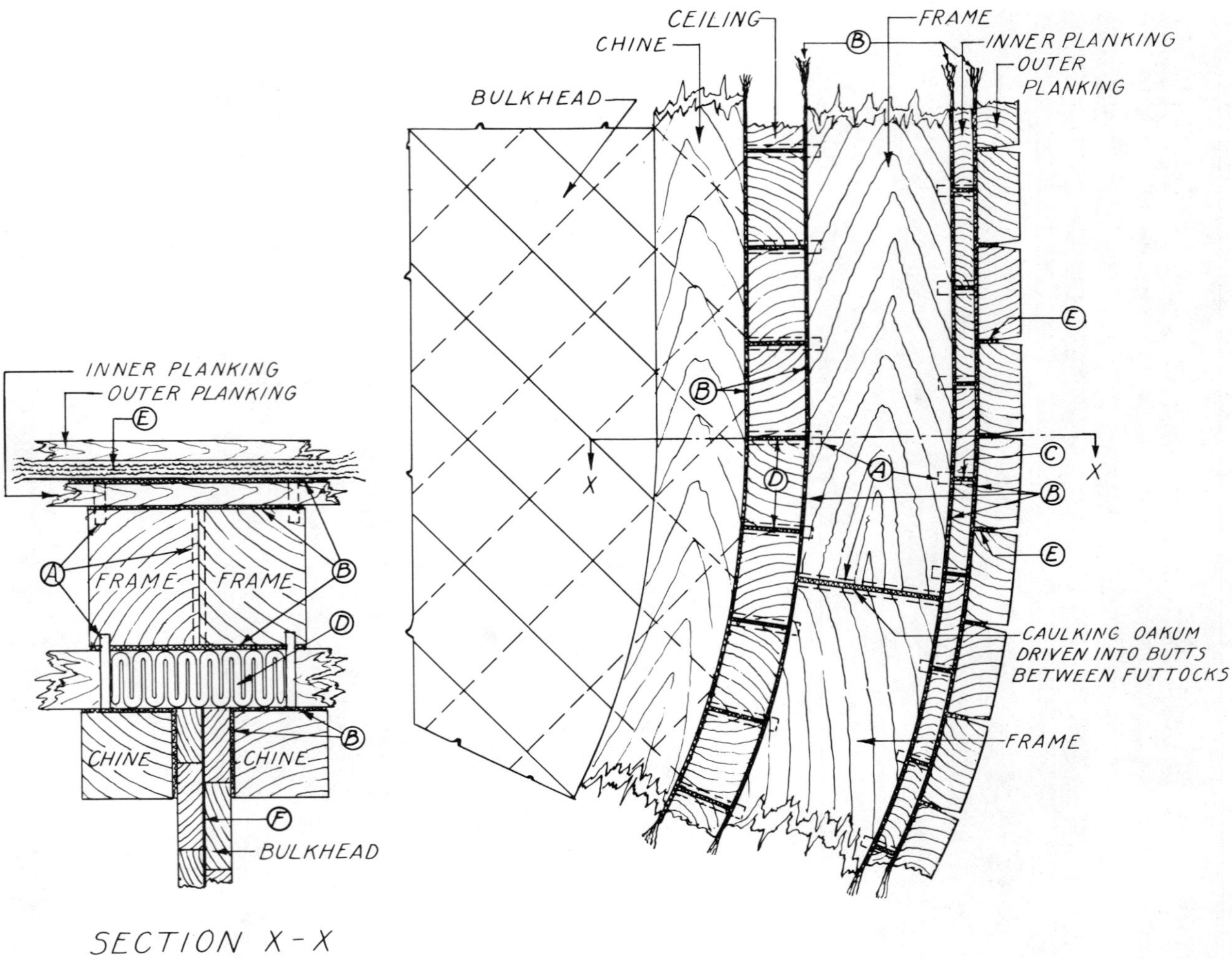

M 57846 F

Figure 3-49.—Bulkhead construction of sawed-futtock framed 165-foot ATR vessel, indicating method used to secure watertightness at intersections of bulkhead with skin of ship. A, Stopwater; B, calking cotton (not rolled) soaked in white lead, installed vertically between members; C, same as B, except that cotton is laid horizontally between the strakes of inner planking; D, same as B, except that cotton is laid zigzag horizontally between the strakes of ceiling; E, same as B, except that cotton is placed horizontally along seams between the outer planks in such a way as not to fill the outgaged parts of the seams. The outgaged parts are later calked with cotton and oakum. F, Canvas membrane laid in white lead.

of the vessel upward or from the deck edges downward (figure 3-50). As each plank is placed in position against the plank already fastened, a strand of calking cotton loaded with white lead should be placed between the inner face of the plank and the point where the out-gage of the plank ends. This method is used on vessels having either single or double planking.

Certain vessels, such as the World War II 110-foot subchasers, have steel watertight bulkheads that are secured to the faces of frames by means of lag screws and bolts. A canvas gasket soaked in white lead is placed around the margin of such bulkheads, between them and the frames. Often the boundaries of such bulkheads are formed by angle bars curved and fitted to suit the shape of the vessel and forming collars over the fore and aft members – such as clamps and stringers – which pass through the bulkheads. Such angle bars are in almost all cases welded to the steel bulkhead.

Figure 3-50.—Strands of cotton (see arrow) on bulkhead frame set in white lead over canvas membrane fcr watertightness.

Bulkhead frames for large boats are commonly made of the following species: white oak, southern yellow pine, Douglas-fir, and western larch. These woods are used because of their strength, ability to hold fastenings, and decay resistance in all-heart grades. The requirements for bulkhead frames in small boats are not so exacting, because they are seldom subjected to heavy loads. A variety of species is therefore used; Sitka spruce and mahogany because of their light weight and low shrinkage; Douglas-fir, southern yellow pine, western larch, and white oak because of their ability to hold fastenings.

Bulkheads frequently are planked with Douglas-fir, southern yellow pine, western larch, or baldcypress. The principal reason for using these species for planks is their availability and decay resistance. Some bulkheads are planked with Douglas-fir plywood.

3.5.1.11 Hull Planking.—Hull planking forms the shell of a wood ship. The purpose of planking is to produce strength and rigidity and to keep water out of the ship. It generally is the main strength member of the vessel, especially in such craft as cargo and passenger vessels, tugs, barges, minesweepers, and patrol craft. The stresses to which planking is subjected are similar to those to which the outer surfaces of a box girder are subjected. When a vessel undergoes hogging and sagging stresses, the planking receives and resists most of these bending stresses with their attendant horizontal shear.

The character of the planking varies with the type of vessel. Planking may be thick or thin, single or double, calked or uncalked, and laid fore and aft or at an angle. The thickness may range from a single shell of three-eighths-inch plywood to a total plank thickness of 6 inches or more (figure 3-51). Heavy-duty vessels have thick planking, either single or double, with the seams of the outer layer always calked.

Figure 3-51.—Double fore-and-aft planking of MSC minesweeper.

With all vessels, butts in the various strakes of the planks must be well staggered. Where double planking is employed, the seams between the various strakes of the inner planking must never be placed in line with the seams of the outer planking. Outer planking of large vessels is almost always laid in the fore and aft direction, but inner planking may be diagonal, as in the MSO minesweepers.

Types and combinations of planking used are:

Single planking.—a, Fore and aft strakes of solid wood; b, diagonally laid solid wood planks; c, plywood sheets.

Multiple planking.—a, Inner and outer shells of fore and aft strakes of solid wood; b, one or more inner shells of diagonally laid solid planks and outer shell of fore and aft solid planks; c, both inner and outer shells of diagonally laid solid planks; d, inner shell of plywood and outer shell of fore and aft solid planks; e, inner shell of plywood and outer shell of diagonal laid solid planks; f, inner and outer shells of plywood.

The choice of the type of planking is made by the designer. Many factors control the selection of planking; for instance, it is not feasible to use plywood shells on a craft with a pronounced double-curvature bottom except on small boats where veneer can be applied in thin strips and bag molding can be employed. Double-shell planking formed of solid planks tends to create greater resistance to horizontal shear, especially if diagonally laid (figure 3-52). Likewise, the placement of either shell in a diagonal position with respect to the keel tends to give fore and aft rigidity.

Where two shells are used on light, small craft, they are separated by a layer of fabric treated with white lead or bedding compound. The shells are fastened together in the bays between the frames by rivets, screws, or clinch nails, all of which are called "quilt fastenings." Double-shell planking sometimes is

Figure 3-52.—Laying diagonal planking on a PT boat.

calked only at the keel, stem, transom, and chine. Where calking is not employed, the joints are coated with seam compound and watertightness is obtained by firm fastening and good fitting.

The garboard is the plank which is placed against the keel (figure 3-53). Especially on the heavier types of craft, it is thicker than any other plank on the vessel. Next above it may be what is known as the "broad," sometimes also called garboards. Sometimes two broads are used. These are often not as thick as the first garboard but nevertheless thicker than the rest of the planking. Garboards and broads are face fastened to the frames and often edge fastened to the keel.

The watertightness and strength of the whole vessel depend to a great extent upon the proper installation of the planking. Sometimes the garboards and sheer strakes are laid first. Sometimes the garboards are temporarily fastened so that they may be removed after their adjoining planks have been fitted and fastened to help keep the vessel clear of debris during the succeeding building operations. Garboards that are edge fastened cannot, of course, be removed for such purposes.

The sheer strake usually is thicker than the balance of the planking – except the garboards and broads – to obtain greater strength to resist hogging and sagging stresses at a point where they may be especially severe. That portion of the planking which lies between the sheer strake and the broads almost always is of uniform thickness. It is customary to face fasten this planking.

Figure 3-53.—Garboard of MSO minesweeper, laminated to proper twist and bolted to keel and frames.

The more seams in the shell of a vessel, the less swelling and shrinking occurs on any one seam; hence, narrow planking is to be preferred. It is obvious, however, that planking that is too narrow does not give sufficient surface to permit installation of the required number of fastenings. Almost always it is necessary to vary the width of the planking at every frame throughout the length of the vessel unless she has a parallel midbody. Certain vessels – especially those with considerably more draught aft than forward – require the installation of planks known as "stealers," which are not continuous strakes throughout the vessel's length. They begin generally at the rabbet on the sternpost and are nibbed into other strakes at a point a few frames forward of the sternpost.

Planks are cut to shape and finished as far as possible before they are installed on the vessel. They are shaped according to a predetermined series of dimensions which ensures their proper taper throughout their length, the proper outgaging of the calking edges, and the hollowing of the inboard face if the particular strake happens to come at the turn of the bilge. The hood ends are also outgaged to receive calking. Butt joints are finished on the vessel. Where futtock frames are used, plank butts are located on them. Where steam-bent or laminated frames are used, butt joints in the planks fall in the frame bays and are reinforced and backed by butt blocks which are bolted, riveted, or screwed to the two planks so connected (figure 3-54). On large vessels, several planking gangs work simultaneously, but care is always taken that the work will proceed at the same rate on both sides. The last planks to go on the vessel are called "shutters." Large vessels sometimes have more than one shutter strake on a side.

Most planks must be bent into position to follow the curved surface of the hull. This often calls for preliminary steaming or boiling of the plank. After steaming, the plank is hurried to the vessel and put in place as quickly as possible while still hot. It is clamped to the frames to which it is to be fastened and progressively shoved into position against the planking already installed by means of wedges, clamps, or screws (figure 3-55). It then is fastened permanently in place. The garboards usually have to undergo a severe twist at the stem and sometimes at the stern. For this reason garboards are often made of oak, although the rest of the planking may be of some softwood species, because oak becomes more plastic when hot and wet than do softwoods and therefore will take a greater twist. Heavy garboards are often laminated to proper shape.

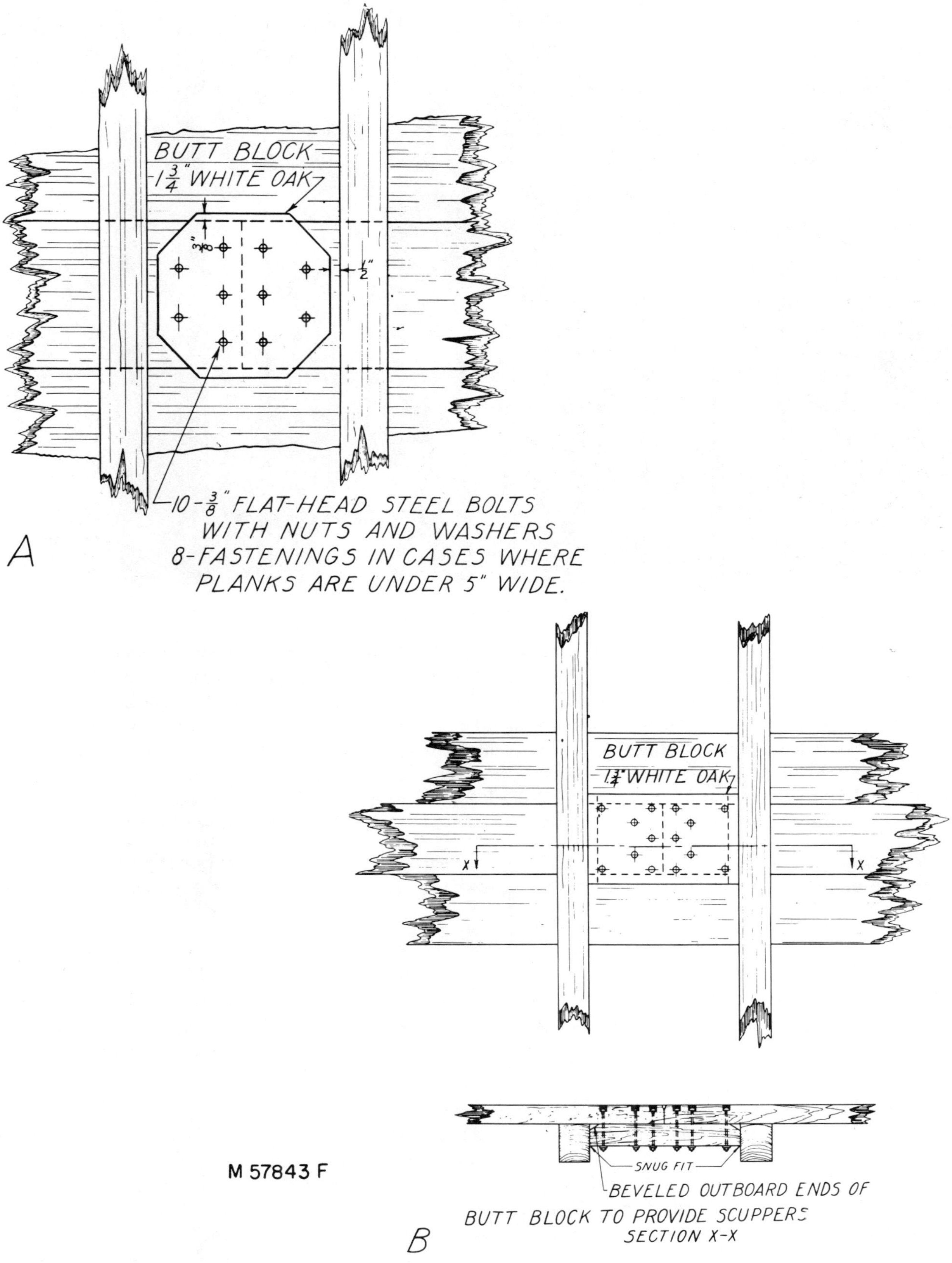

Figure 3-54.—Typical frame-bay butt blocks on YT harbor tug. A, Octagonal wood block overlapping planks above and below butt joint; B, wood block fitted snugly between frames.

Figure 3-55.—Forcing planking into place with C-clamp and wedges.

Planks are fastened to each frame with ship spikes, nails, drift bolts, through bolts, or a combination of these (figure 3-56). Special attention must be paid to the fastenings at the hood and butt ends of the planks. Where the thickness of the planking permits, the heads of the fastenings are driven into a counterbored hole deep enough to receive a wooden bung, which should be of the same wood as the plank in which it is set and so placed that its grain parallels that of the plank in order to ensure its swelling and shrinking in the same manner as the plank. The bung is dipped in white lead before being driven into the counterbored hole (figure 3-57). After installation, that portion of the bung that extends beyond the face of the plank is trimmed off. In thin planking, the head of the fastening is flushed over with white lead putty or bedding compound.

In double-planked hulls, the inner shell is sometimes sanded, especially in the way of bulkhead frames, to remove minor roughness before the canvas that goes under the outer planking at these frames is installed (figure 3-58).

Diagonal planking on V-bottom hulls requires spiling than does fore and aft planking. Frequently, the inner shell is laid diagonally and the outer shell fore and aft. Where the number of skilled workers is limited in a plant – especially when a large number of boats of identical design are to be built – diagonally laid inner and outer shells as a rule facilitate construction. Care must be exercised, when laying diagonal planking, to shape the planks so that their slope when laid on the hull will not become too nearly horizontal or vertical.

Plywood as a boat-planking material has several obvious advantages. It greatly reduces the number of seams necessary to calk or otherwise seal against water. It speeds fabrication considerably because the large sheets in which it is manufactured are more quickly fastened in place than are a greater number of solid wood planks. Plywood is more uniform in strength along and across the grain and is much less subject to splitting than is solid wood of equal thickness, and it distributes concentrated loads better to its supporting members because it is in large sheets. Plywood is more resistant to puncture by sharp objects than is solid wood of an equal thickness. These advantages make flat boat-hull plywood suitable for planking of hulls that do not require an appreciable amount of double curvature. Plywood has certain disadvan-

Figure 3-56.—Lag screws used for fastening planking of MSO minesweeper.

tages, however, and these should be taken into account when considering it for a particular type of boat. For example, while it shrinks and swells more uniformly than does solid wood, plywood shrinks or swells somewhat more than does solid wood in the longitudinal direction, as well as in its thickness. Swelling and shrinking, consequently, have caused difficulty in some types of vessels in which plywood has been tried, especially those that are in the water for prolonged periods during which the moisture content of the plywood increases markedly. The resultant swelling may cause buckling between frames and at stern joints. For such boats as landing craft (LCVP) or deckboats, which are out of the water a large part of the time, serious increase in moisture content and consequent swelling are not so likely to occur. Flat plywood cannot be bent to appreciable double curvature, a further limitation to its use as hull planking.

Plywood, when used for boat planking, is cut to shape and fastened to the frames, stem, transom, and keel, generally with screws (figure 3-59). Where edges of plywood sheets land on framework such as chines, clamps, and fashion pieces, they are laid in bedding compound. If the hull is built with two shells of plywood placed under solid planking, it is customary to fasten the inner shell in place with nails, either galvanized or copper. The outer shell, whether it is of solid wood or plywood, is generally fastened to the frames with screws or copper rivets (figure 3-60). Butt joints in plywood are backed by blocks, also made of plywood, which generally are fastened to the plywood planking by means of rivets; sometimes screws are employed. Bedding compound is spread between the butt blocks and the plywood shell.

Mahogany and Douglas-fir plywood are commonly used for planking. Whatever species is used, it should conform to the requirements and meet the tests for boat-hull plywood as given in Volume III, Section 4.5.

Figure 3-57.—Installing wood plugs in counterbored holes smeared with white lead.

Molded plywood hulls have been used to a limited extent for a few of the smaller types of boats, such as deckboats and shallow-draft assault boats (section 1.3). The entire shell of such boats is a unit built of several layers of veneer strips glued together with a waterproof glue. The hull is assembled over a mold built to produce the proper contours, and the assembly is then placed in a bag and treated in a retort with heat and pressure. The product is light, strong for its weight, and, when phenol-formaldehyde glues are used, the glue bonds are practically unaffected by exposure to water. Surfaces possessing severe and compound curvature can be readily produced, but for sections of single curvature the method is uneconomical. Since the whole shell is a single piece, there are no joints or seams to be made watertight.

Douglas-fir, southern yellow pine, Alaska yellow cedar, and, in some instances, white oak, which combine high strength with ability to hold fastenings and resistance to decay in all-heart grades, are the species commonly employed for planking in large boats and ships. Douglas-fir, southern yellow pine, and Alaska yellow cedar are used more extensively than white oak because they are more readily available and are easier to obtain in long lengths of good quality. They are lighter in weight and do not swell and shrink so much as white oak. White oak, however, excels in resistance to impact, in bending strength, in the ease in which it can be bent, and in its ability to hold fastenings. Where a vessel is planked entirely with white oak, it has been found advantageous to keep it in the water – except for overhauling and painting – because it tends to shrink more than one planked with Douglas-fir, southern yellow pine, or Alaska yellow cedar. Alaska yellow cedar, although not so strong in most respects as Douglas-fir or southern pine, is much more resistant to decay. In small boats, the species commonly used for planking are mahogany, Alaska yellow cedar, Port-Orford-cedar, Atlantic white-cedar, western redcedar, baldcypress, and Douglas-fir and mahogany plywood. These wood are preferred because of their light weight, stability, and heartwood decay resistance. The cedars absorb less water, thus are more suitable for light and high-speed boat construction. On the other hand, baldcypress is used principally for its decay resistance, but it absorbs exceptional amounts of water.

Figure 3-58.–Sanding inner shell of double-planked vessel with a portable disk sander

3.5.1.12 Ceiling.–Ceiling is fore-and-aft planking attached to the inner faces of the frames (figure 3-61), generally from the keelson to the deck clamps. Its main purpose is to strengthen the hull against both fore-and-aft and athwartship stresses, particularly hogging and sagging. Ceiling greatly interferes with ventilation of the frame bays, however, and for this reason is dispensed with in such ships as minesweepers, as well as in smaller hulls using laminated or bent frames. In hulls with such frames, which are free of strength-reducing joints found in sawed-futtock frames, ceiling is not so important from a strength standpoint. Where ceiling is used, as in the World War II sawed-futtock framed YM and YT tugs, some ventilation is provided by means of holes bored through the ceiling between frames up near the clamps. In general, however, such provision is inadequate.

Ceiling is generally calked only in the way of the watertight bulkheads where it also is stopwatered, hence it requires outgaging only at those points. It is face fastened to the frames with drift bolts, and sometimes with through-bolts that pass through the frames and, in some cases, through the planking as well. In certain vessels of the heavier types, the ceiling strakes located at and near the turn of the bilge are thicker than the rest. Scarf joints are sometimes employed (figure 3-62) to join ends of ceiling planks; they strengthen the hull against stresses such as those induced by hogging and sagging. To ensure access to limber holes and, in some cases, to the strainers in the bilge drainage system, the first strake of ceiling at either side of the keelson – termed the limber strake – is removable.

Figure 3-59.—Clamping plywood planking in place on LCVP.

Figure 3-60.—Clamping LCVP bottom planking tight edgewise over canvas membrane.

Figure 3-61.—Bow view of ceiling being installed in World War II YT tug.

Figure 3-62.—Drilling holes for edge bolting of ceiling at a scarf joint.

Figure 3-63.—Preservative-treated worm sheathing being installed.

Ceiling planks are shaped beforehand to a planking layout and are generally installed in the vessel prior to the outer planks, thereby providing a strong, rigid structure over which the planking proper is later placed. The installation of ceiling must progress at the same rate on both sides in order to prevent any possible distortion of the vessel. Frequently ceiling installed near the bow or the stern requires steaming or boiling before it can be bent into position.

Where the shell of the vessel is pierced by sea chests or any type of sea connection, it is necessary to provide a proper filler between the ceiling and the planking. Frequently this filler is let into the adjacent frames, thereby providing a rigid, continuous structure through which the flanged sleeves used in the sea connections may be installed.

Ceiling requires the use of woods that combine high strength – especially impact strength – and ability to hold fastenings with decay resistance and stability. Softwoods combining these properties to the best degree are Douglas-fir, southern yellow pine, and western larch; the first two are used almost exclusively for the ceiling of larger vessels. Oak is often used in combination with either southern yellow pine or Douglas-fir, especially where bends are sharp and difficult to make or where high resistance to impact is desired. Douglas-fir and southern yellow pine are used because of their relative lightness and their stability. Western larch is seldom used, but its properties are similar to those of Douglas-fir and southern yellow pine. Occasionally, woods that are low in decay resistance, such as Sitka spruce and white pine, are used for ceiling. These woods, because of their comparatively low strength, ability to hold fastenings, and decay resistance, are not the equal of any of the species previously discussed.

3.5.1.13 Sheathing.–Sheathing is a protective coating or shell of wood, metal, or both, that is placed over the underbody of a vessel. Its purpose is to provide protection against marine organisms (figure 3-63), ice, or damage on landing craft due to grounding. Sheathing also is used for the protection of vessels engaged in the handling of heavy metal equipment, such as submarine net buoys, motor boat boom fittings, and ground tackle.

Wood sheathing, to give adequate protection to the hull against ice and marine organism attack, should begin at a point above the load water line when the vessel is at her deepest draft and completely enclose the hull, including the keel and all deadwood, the sternpost, the wood skeg if any, and all other exposed parts of wood. The thickness of this sheathing varies from approximately 1/2 inch to 2 inches, according to the size of the vessel. The keel shoe should be sufficiently wider than the keel to cover the bottom edges of the sheathing fastened to the keel. If it is not, the shoe itself also should be protected by sheathing. Care must be taken to use moderately seasoned sheathing so that swelling of the wood after launching will not cause the sheathing fastenings to loosen. Swelling becomes severe where very dry lumber is used; hence lumber in the range of 15 to 20 percent moisture content is most suitable because it will neither shrink unduly while the ship is on the ways nor swell severely after launching.

The principal requirement for wood sheathing used as protection against marine organisms is ability to take preservative treatment. The sapwood of southern yellow pine is preferred for this use because it can be easily and thoroughly treated. Sapwood of other species, such as Douglas-fir, could be used but is not readily obtainable. If a species of low compressive strength is used for protective sheathing it will be necessary to use a species with high compressive strength across the grain for the keel shoe. This is essential because all vessels must be borne upon keel blocks while dry docked.

When sheathing is used for protection against ice alone, it should extend high enough up the sides of hull to form a belt of sufficient width to afford adequate protection against floating ice at any draft. This may require spiling of the sheathing strakes. Sheathing for this purpose is secured to the shell of the vessel generally by means of screws, the heads being slightly below the surface of the sheathing. A metal cutwater should either be placed over the stem iron or butted and welded to the aft face of the stem iron. This cutwater should extend vertically across the width of the sheathing belt and sufficiently far aft to provide the proper cutting action when ice fields are entered, thereby protecting the forward edges of the wood sheathing. The portion of wood sheathing that is placed under this metal cutwater should be tapered.

Ice sheathing requires wood capable of resisting severe abrasive action and of holding fastenings well. The species preferred for this purpose is Australian ironwood. White oak has given satisfactory service. Other species having properties that indicate they could be used satisfactorily are red oak, hickory, pecan, and rock elm. Black locust and live oak also have the required properties but are not readily available. The properties of the foregoing woods responsible for their selection are high toughness and hardness. White oak is used almost exclusively for keel shoes for the same reason that it is used for protection against ice.

Assault craft designed to run at full speed onto beaches often have a protective sheathing of hardwood on the fore part of the hull (figure 3-16) and a hardwood shoe on the fore part of the keel. This sheathing is attached by screws, and the forward ends are beveled.

Assault landing craft require sheathing that is capable of resisting severe abrasive action. Woods to meet this requirement must have a combination of high hardness and toughness; consequently, the wood preferred for this use is white oak. Other woods having the desired combination of hardness and toughness are rock elm, hickory, pecan, and red oak. These woods are low in decay resistance, but decay resistance is of secondary importance in boats of this type. Preservative-treated red oak combines decay resistance with hardness and toughness. Douglas-fir and southern yellow pine are not the equal of the hardwoods mentioned in toughness and hardness. They, however, are the toughest and hardest of the softwoods that are readily available and can be used as substitutes for the hardwoods mentioned.

The Navy has a large number of small and moderate-sized craft made of wood upon the sides of which it is necessary to place heavy metal sheathing extending from a point well below the turn of the bilge to the gunwale. This sheathing is secured to the hull by means of bolts that pass through both the planking and the frame. The use of metal sheathing creates a decay hazard that must be taken into consideration when selecting wood for planking. Where wood is sealed between two comparatively moisture-proof barriers, only highly decay-resistant woods should be used and sapwood should be rigidly excluded. Pressure-treated wood should be used in preference to a naturally decay-resistant wood. When pressure-treated wood is used, sapwood usually is superior to heartwood because of its ease of treatment. Nails, bolts, or other fastenings used to attach metal sheathing to wood hulls should be as nearly as possible of material similar to that of the metal sheathing in order to decrease the bimetallic potential. Even the use of galvanized nails in zinc-annealed sheets is considered unsatisfactory.

3.5.1.14 Guards or Fenders.—A guard or fender is a fore-and-aft timber running along the length of a vessel generally at deck level. Where a vessel has considerable freeboard, a second or even a third guard may be fitted below the one at the main deck edge. Sometimes additional guards are placed below the main deck at the stern or bow. The purpose of guards or fenders is to absorb blows, rubbing, or chafing when a boat comes in contact with a dock or with another vessel. The outer faces of most guards or fenders carry metal bars, which may be flat or half oval in cross section. A guard should have a cross-sectional dimension on its inboard face larger than that of the metal-covered outboard face to ensure a good attachment and to permit a generous bevel – especially on the under side – to protect it from being ripped off when the vessel is rolling against obstructions. Sometimes it is necessary to steam and prebend guards of large cross section. They are usually shaped on the inboard face to fit the hull contour at bow and stern.

Guards generally are built of one piece though on very large vessels they may be built up of several pieces. The sections of guards are fastened together by scarf joints rather than by plain butts. Scarf joints of guards should be so designed that the forward section of guard should overlap the after section. Guards frequently are tapered as they approach the bow or stern of the vessel, unless the vessel has a counter stern, such as that of a tug. Before installation, the inboard faces of guards are well coated with white lead. Guards are secured to the hull wherever possible by through bolting, and should be reinforced by parallel longitudinals within the hull itself if a clamp is not located at that point.

V-bottom craft frequently have guards located at the chines so that the chine knuckles may be properly protected. On high-speed vessels, chine guards have another purpose, that of keeping down spray from the bow wave. Small boats frequently do not have metal-shod guards.

Woods preferred for guards are white oak, southern yellow pine, and Douglas-fir because of their high resistance to impact, compressive strength, decay resistance, and ability to hold fastenings.

3.5.1.15 Decks and Decking.—Decks are those portions of a ship's structure which lie in a more or less horizontal position and correspond to the floors or flat roofs of a building. The purpose of decks is to cover the hull and create platforms on which other structures may be placed and on which personnel may walk. They are important structural members because they resist athwartship stresses, add greatly to the fore and aft stiffness of a vessel, and, together with the planking of the sides of the hull, resist hogging and sagging tendencies. Those decks which are exposed to the action of the elements, called weather decks, necessarily must be made and kept watertight to preserve their own structure and protect the parts of the vessel that are below them.

Figure 3-64.—Deck of MSO minesweeper. Note scarfed margin pieces and the adjacent square sectioned decking.

In general, decking is secured to deck beams. Decking of large craft usually consists of comparatively narrow planks that either run parallel to the centerline (figure 3-64) or are curved to conform to the curvature of the vessel's sides. Curved planks along the outer rim of the deck are called margin pieces (figure 3-64). It is desirable for the cross section of any one plank to be square or nearly so, because with wider planking the effects of swelling and shrinking are concentrated on fewer seams between planks, thereby exerting great stresses on the calking and fastenings and causing more leaks. Also, square cross sections

Figure 3-65.—Laminated king plank of MSO minesweeper nibbed for ends of deck margin pieces at bow.

permit turning the planks to achieve the desired vertical grain on the upper face. Excessive swelling due to the use of planking that is too wide has been known to cause an entire deck to buckle and rip loose from its beams. Edge-grained material is preferred for decking because wood shrinks and swells less across edge grain than across flat grain with changes in moisture content. Wood with low shrinkage properties is preferred.

The ends of margin pieces, unless curved to fit the ship's sides, are nibbed into a cover board or lock strake and are supported on their undersides by small intercostals placed between the deck beams. In some ships (figure 3-65), a king plank is used at the bow to take the ends of margin pieces.

Figure 3-66.—Laying metal-sheathed plywood decking.

Decks are fastened to their supporting beams by means of spikes, screws, or bolts. The planks are outgaged for calking. The joints between the various elements of a strake of decking are generally square butts that land upon supporting deck beams. It is customary to cover the heads of all fastenings with wood bungs set in bedding compound or white lead. Bungs should be made of the same wood as the deck planking; their grain should be parallel to the grain of the plank in which they are placed.

Plywood, often preservative-treated, is used on the decks of certain craft, either in single or double sheets, with fabric loaded with a bedding compound or white lead between them. Sometimes plywood is placed under fore-and-aft deck planking of solid wood. Double sheets of plywood are riveted together. Plywood weather decks of some ships are faced with metal (figure 3-66). Deck beams should be carefully surfaced for plywood decking for uniform bearing (figure 3-67).

Figure 3-67.—Sanding deck beam for plywood decking.

Plywood can be easily bent to the camber of deck beams – and on most vessels of light types to the slight sheer curvature as well – despite the fact that some double curvature is thereby required. Butts between adjoining sheets of plywood decking are frequently secured by means of a butt block.

Plywood usually is screwed to the deck beams (figure 3-68). Screws also are used when solid wood planking is installed over a plywood underlayer of deck. Because it is not feasible in such cases to use square decking, deck planks laid over plywood should be kept as narrow as possible, and one or more continuous shallow grooves should be cut on the underside of each plank to prevent cupping. Screws used for fastenings should not pass through such grooves.

The margins of plywood decks sometimes land in a rabbeted edge of the cover boards. These edges are set in white lead or bedding compound and screwed fast to the cover board. In other cases, the marginal edges of the plywood butt against the cover boards and are supported by intercostals placed between the deck beams. In still other cases, plywood portions of the deck are covered with canvas set in white lead or marine glue, the canvas extending over the marginal joint and being locked in place by a square spline that is forced into a groove in the cover board. Such a lock is generally employed when a vessel has a bulwark or toe rail that is not readily removable. In some vessels, the canvas is carried completely across the decks, lapped over the sides, and secured by means of copper tacks until the guards can be installed over its edges.

Figure 3-68.—Screwing plywood to deck beams of MSO minesweeper.

During the past few years, considerable interest has developed in substituting glass cloth for canvas. Glass cloth is a square weave cloth that weighs about 10 ounces per square yard. Polyester resin is generally used as a bonding agent. Limited experience with the glass cloth indicates a fair degree of success in correcting leaky conditions with plywood decks, provided proper precautions are taken to assure that the wood is dry and free from decay and careful workmanship is employed. The use of glass cloth is not advocated on planked decks.

Because of its multiple functions, decking material must have many properties, including resistance to abrasive action, compressive strength, resistance to impact, toughness, low shrinkage and swelling, resistance to decay, uniform texture, ability to hold fastenings, and ability to take and hold paint, especially on walkways and in the way of boat-handling gear and gun mounts. Plywood decking often is pressure treated with water-borne preservatives.

For plank decking of heavy-duty ships of all kinds, Douglas-fir and southern yellow pine are the most commonly used species. On weather decks, southern yellow pine may exude resin when exposed to hot sunlight – a factor to be considered when selecting a species for decking for a particular type of ship. To reduce swelling to a minimum, plank decking should be edge-grained. A surface more resistant to abrasion and slivering also is provided by edge-grained material. Narrow planks have less effect on tightness of seams than do wide ones because the total shrinking and swelling is divided among more seams.

Edge-grained Douglas-fir is used almost exclusively for decking in large ships. Douglas-fir is used because it is available in the sizes and quality needed and is readily available in edge-grain form. Mahogany is preferred for small boats because of its stability, its availability as clear stock in large sizes, and, to a large extent, its appearance. Teak has long been the preferred wood for ship decking but is scarce and costly.

Leaky decks on wood ships are of frequent occurrence. The leaks are most commonly found at junctions of the deck with coamings or deck houses. Leaks also are found in the seams, but not so frequently as at the junctions of deck and superstructure. In some ships – especially those built on unprotected ways – more trouble is experienced with leaks than in ships built under cover. Use of unseasoned lumber in decks also is conducive to leaks because of the openings between boards that form when the lumber shrinks. A remedy that has some merit when a ship is operating in a hot, dry climate is to wet down the decks to keep them swelled and tight.

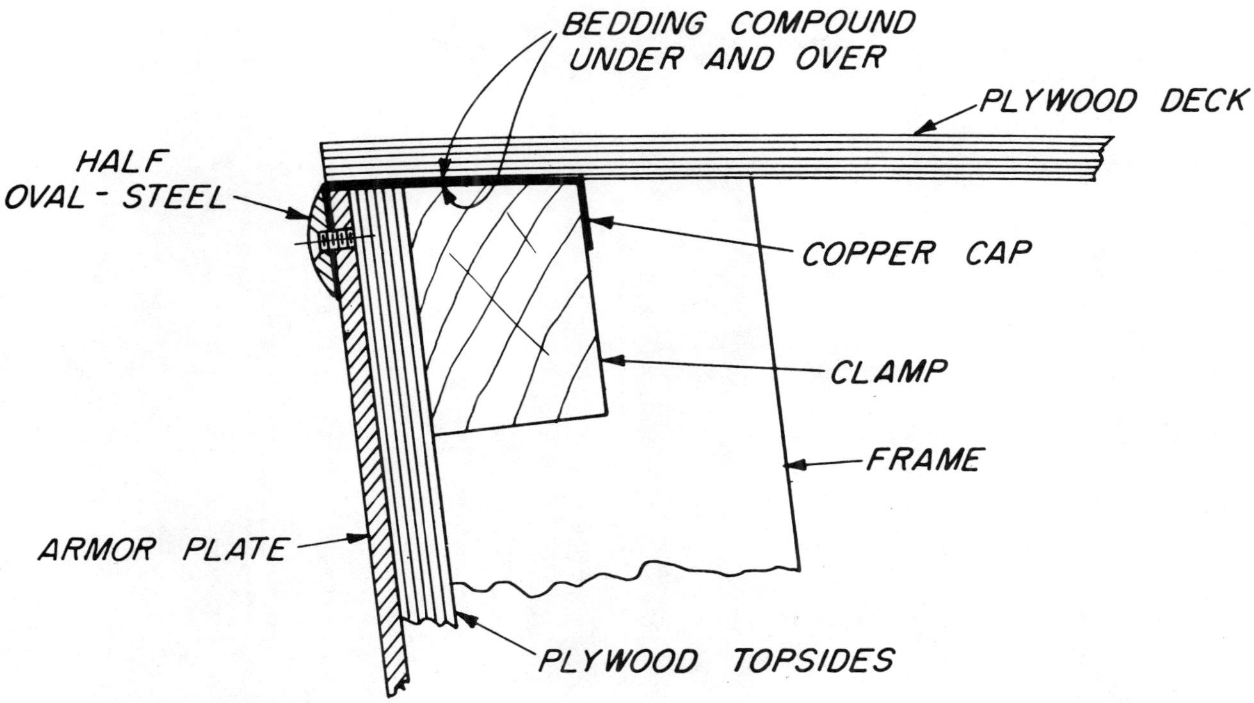

M 96452 F

Figure 3-69.—Copper flashing used to protect sheer clamp assembly from leakage through deck edge.

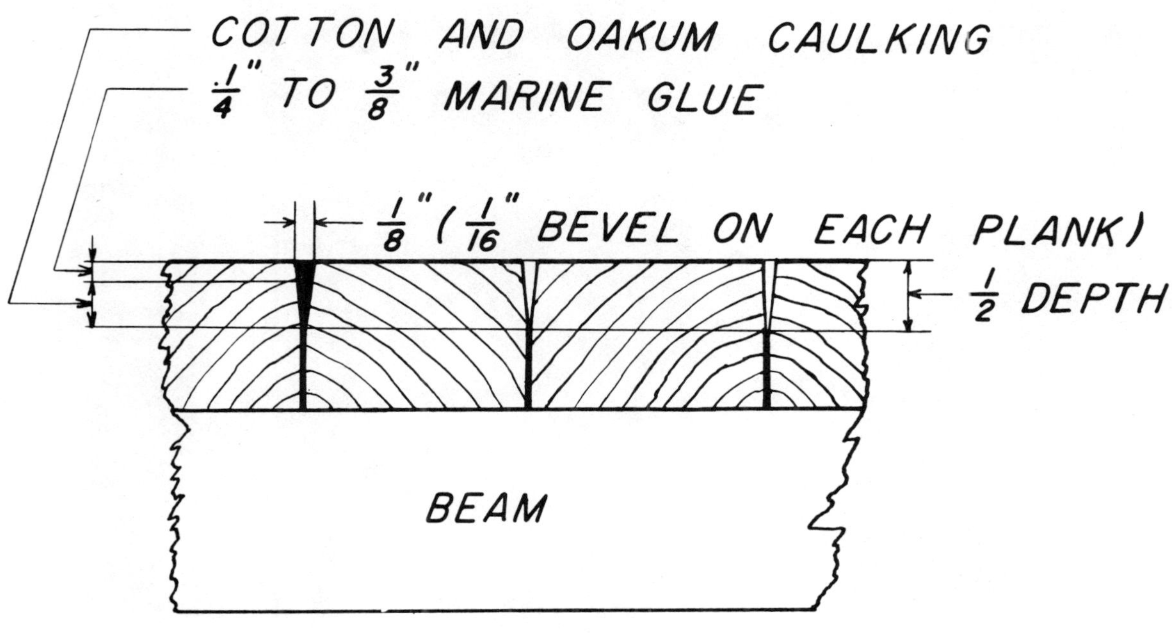

M 55795 F

Figure 3-70.—Typical outgaging required for a heavy deck.

Figure 3-71.—Driving the first thread of cotton into a seam.

Plywood has greater resistance to splitting under concentrated or impact loads than has solid wood of equal thickness; hence, it provides a more serviceable deck on light craft. When well-nailed or glued to each deck beam, plywood contributes much greater rigidity to the structure than does solid wood. Plywood affords speed of construction, does not require highly skilled craftsmen for calking, and, therefore, lends itself to quantity production. Douglas-fir plywood is used extensively, particularly in light craft, because of the ease with which it can be laid in place and because of its ready availability. Mahogany plywood is used where appearance is important.

While plywood decking is relatively free of seams that can spring leaks, a few critical points have developed on some plywood-decked craft, notably the sheer. Figure 3-69 shows a suggested use of metal flashing to avoid the moisture "pockets" that sometimes develop at this point. Canvas set in bedding compound is an alternate to metal flashing.

3.5.1.16 Calking.—Calking is a process of forcing cotton or cotton and oakum into specially beveled, or outgaged, seams between planks or other wood surfaces by means of a calking iron driven with a mallet. The purpose of this operation is to make the seams watertight.

Calking is an operation that depends first of all on properly outgaged planking (figure 3-70). In the event that seams have been improperly outgaged, it may be necessary to use an opening iron; such practice, however, should be avoided because it may result in torn and mashed wood fibers along the seams. The actual calking operation begins with the driving of one or two threads of calking cotton tightly into the bottom of the calking seam (figure 3-71). After this cotton has been well set in, one or more threads of oakum are driven into the seam. The calking is then set firmly home by means of a heavy "horsing" iron driven into the seam with a heavy mallet, or "beetle" (figure 3-72). Where heavy deck planking is employed, several successive horsing operations may be necessary as calking is fed into the seams in order to set it firmly in place and drive it deeply enough to allow sufficient seam space for seam compound, marine glue, or pitch. A power-driven horsing iron has been used to some extent to calk decks of large vessels (figure 3-73).

After the calking operation is finished, the seam should be cleared of all foreign material; this often is done with an air jet. The space between the calking and the planks then is filled with pitch, marine glue, or seam compound – as soon as practicable after the calking has been driven in – to protect the oakum and cotton from decay and moisture (Section 3.7.6).

Decay fungi frequently attack the calking cotton in old seams because of loss of seam compound above the cotton and exposure to rain or snow water. Cotton is almost pure cellulose and is easily rotted unless a preservative is used. Copper naphthenate-asphaltum naphtha solution imparts good fungus-resistant properties to the cotton. A less satisfactory application from an absorption standpoint is obtained by brushing or spraying the preservative into the seams after calking but before filling with marine glue. However, the calking operation using the preservative-treated cotton is more difficult due to a slight stickiness of the cotton.

Light vessels with thin hull planking or thin decking often are calked with a calking wheel rather than calking irons, because the wheel forces the calking in place by uniform pressure rather than by impact blows. Such a wheel can be successfully used only when seams are accurately beveled and with parallel edges. The amount of calking inserted should be carefully controlled, especially in lightly constructed vessels, because it is quite possible to "calk off" a plank from its frames if too much is driven in and forced beyond the outgage bevel.

3.5.1.17 Ratproofing.—With the exception of small, open craft, provision is made for ratproofing in nearly every Naval vessel. Care must be taken to install all barriers that are intended to prevent the entrance or nesting of rats in such a way that ventilation is impaired as little as possible. Wire screening with maximum mesh of 3/8 inch generally is employed to block intrusion of rats as it permits air movement for ventilation purposes. Solid barriers, such as solid intercostals and solid panels in lieu of grillwork or screening, should be avoided. The installation of any ratproofing barrier which prevents a free circulation of air tends to create conditions within the enclosure that are conducive to the development of wood-destroying fungi.

Figure 3-72.—Horsing oakum into a seam.

3.5.1.18 Rudders —Wood rudders – which are nearly always fitted with metal stocks – are used on the larger wood vessels as well as on many smaller tugs and commercial vessels.

Except in small craft, such as whaleboats, gigs, and power dories, rudders are made of more than one piece, generally by means of a series of vertical pieces joined by through bolting or drifting. Frequently the metal stock extends through the length of the rudder and is designed to permit – by means of a flange – the removal of the rudder without removing the upper portion of the rudder stock from the vessel, thus allowing the removal of propellers from single-screw craft. The rudder must be sufficiently sturdy to permit the vessel to turn whether it is going ahead or astern, and with the rudder placed in its most extreme position to port or starboard. Naval vessels, which must be much more maneuverable than commercial or pleasure craft, are sometimes run astern for considerable intervals. When a ship is under full power astern and the rudder is hard over in either direction, the stresses exerted against the rudder are much more severe than when the ship is going ahead.

When a vessel is operating in ice fields, the rudder is subjected to additional hazards due to the impact of ice thrown against it by the action of the screws. Such stresses are especially severe when the vessel is going astern.

Figure 3-73.—Power-driven horsing iron.

Primary requirements for rudders are ability to hold fastenings and strength. Good decay resistance also is requisite, because part of the rudder is usually above the water line. Because of the action of ice on a rudder, wood must be able to withstand rough usage, the best criteria of which are its hardness and toughness. White oak is generally preferred, especially for the larger vessels, because it combines the required properties in the highest degree. Dense southern yellow pine, however, is used as a substitute for white oak in large vessels. Sapwood of southern yellow pine can be used if properly treated with preservatives. Mahogany is used in smaller vessels and boats because of its light weight and decay resistance.

3.5.1.19 Masts and Spars.—Masts are vertical members, usually pole-shaped, which are used on powered vessels to support wireless and radar equipment, to act as derricks, and to support certain running and signal lights and crow's-nests. On sailing vessels, the masts carry the sails, together with gaffs, booms, spreaders, yards, and other equipment. Spars include booms, gaffs, signal yards, and the like.

Masts and spars are of two general classifications; (1) the solid type and (2) the hollow type. Few modern vessels have the solid type.

The hollow spar is usually made up of four or more sections to permit staggering of the scarf joints along the length of the spar and to obtain balanced construction. Each section is first assembled to the full length of the spar by scarf gluing. The pieces are next glued together until two half spars are formed. The halves are then hollowed, if necessary, and joined, and the whole is then rounded to the desired cylindrical shape.

The steps in the gluing of a hollow spar of four pieces are illustrated in figure 3-74. Gluing pressure is applied with retaining clamps. It has been reported that hollow spars glued with casein glue, when well protected with paint or varnish, have performed satisfactorily through years of service. Cold-setting, urea-resin glue has been used, but a more water-resistant glue – such as low-temperature phenol resorcinol or equivalent – is preferred.

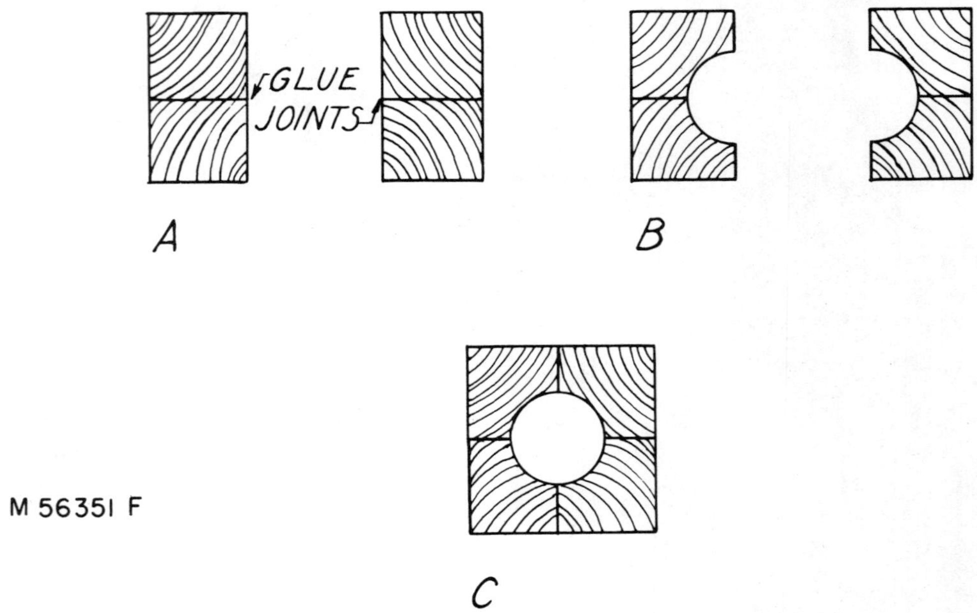

Figure 3-74.–Steps in gluing hollow spars. A, Two sections of spar are glued together as shown; B, sections are hollowed out; C, sections are glued together ready for turning.

On smaller craft, such as patrol boats and rescue craft, the masts are made either in the form of a hollow rectangle or a T-section. Where reduction in weight is required, masts are made of two or more pieces and hollowed out before assembly. Laminated wood spars bonded with phenolic-resin glues have been installed on certain vessels (figure 3-75).

Masts and spars are required to resist high bending and compressive stresses. In addition, the weight of the spar is important, especially for sail and small boats, and, consequently, woods with a high strength-weight ratio are preferred. Where the masts and spars are used for lifting heavy loads, the preference is for southern yellow pine and Douglas-fir because these woods combine high stiffness, compression, and bending strength. For masts in smaller boats, especially in sail boats, the preference is for spruce, principally Sitka spruce because of its comparatively light weight and high strength.

3.5.1.20 Superstructure.–Superstructure is that portion of the ship which is above the uppermost complete deck. It includes pilot houses, galleys, bridges, trunk cabins, and the like. Its purpose is to provide shelter for equipment and personnel. The effect of the elements upon the superstructure is often as severe as it is upon almost any other part of the ship because the superstructure is subjected to impact from boarding seas, the effects of heavy rolling, wetting and drying, and the action of ice. Although the superstructure does not add to the strength of the vessel, it is imperative that all structures exposed to the weather and the action of the sea be strong and as fully watertight as possible.

Wood superstructures are of two types – those built with a framework covered with solid wood in the form of staving and panels, and those covered with plywood or fiberboard. In small vessels the tops of superstructures generally are covered with tightly stretched and painted canvas placed over plywood or deck planking; in the larger vessels, decking is laid with calked seams similar to the other weather decks of the vessel. Superstructure should be placed on sills carrying back rabbets which project above the level of the deck. A series of long hold-down bolts, generally extending from the roof of the superstructure alongside of or through the studs and through the decks and deck beams below, is installed to secure the house to the deck.

Figure 3-75.—Laminated spar glued with a low-temperature-setting phenol type resin glue.

In large vessels, port lights and watertight metal doors are often used in deck houses to obtain complete watertightness and to provide sufficient strength to resist the action of the sea. Smaller vessels generally use wood doors and sash so built as to secure as much strength and watertightness as is possible.

The wood requirements for superstructures are similar in many respects to those for houses. However, decay resistance and strength – especially impact and bending strength – are much more important. For framing, therefore, the heavier and denser softwoods, such as Douglas-fir, southern yellow pine, and western larch, are preferred. These woods combine decay resistance with high stiffness, strength, and nail- and screw-holding power, and meet the requirements better than any other woods. The requirement for coverage likewise is somewhat more exacting than that for housing. The preference, therefore, is for the stronger and more highly durable woods, such as cypress and redwood. Western redcedar, one of the principal woods used for coverage in housing, has the necessary decay resistance but is low in strength.

Douglas-fir plywood is extensively used for superstructure coverage. It can be bent to moderate curvature in place or, where sharper bends are necessary, as for bridge cowling on MSO minesweepers, it can be bent on a mandrel before installation.

Certain fiberboards with high water resistance also are sometimes used, with or without overlays of resin-treated paper.

It is good practice to complete the superstructure in some sheltered area and then place and secure it in position on the vessel itself. Checking and other effects of shrinking and swelling are minimized and the structure is kept dry until protective paint, canvas, and other materials can be applied.

3.5.1.21 Joinery.—Portable furniture seldom is used on Navy vessels. Berths, desks, wardrobes, closets, racks, bins, chairs, and benches, while not ordinarily subjected to the action of seawater, should be well built and secured by mechanical means or by waterproof gluing.

The selection of woods for built-in furniture of ships and boats is, as with other furniture, largely a matter of appearance. The woods that are commonly used for furniture are suitable for boat furniture. The woods must be easy to work and must take paint or natural finishes well. Natural color and grain also are largely desired. Mahogany, walnut, and elm are selected for their appearance and ability to take natural finishes and for their low shrinkage and swelling. Where paint finish is to be used, the preferred wood is yellow-poplar. However, yellow birch is sometimes used and it is an excellent wood as a base for enamels.

Plywood, principally Douglas-fir and mahogany, is extensively used because of its availability in large flat sheets suitable for covering various types of surfaces and because the plywood construction swells and shrinks under varying atmospheric conditions much less than does solid wood.

3.5.1.22 Plywood Parts.—Fabrication of plywood parts of vessels ranges from simple routing out of gusset plates or biscuits to the shaping and joining of several panels into large sections for planking and decking and the formation of bent and molded parts.

The technique of bending and molding plywood parts is discussed in Section 1.3. Bent plywood is used for such parts as ventilator cowlings, deck light hoods, curved sections of deck cabin roofs, and fuel tank beds. Figure 1-8 shows a flat plywood panel being bent over a wood mandrel after preliminary soaking in hot water.

For ordinary work where strength is a principal requirement and appearance is relatively unimportant, the general practice is to use Douglas-fir or other softwood plywood. Where grain figure and appearance under natural finishes are important, hardwood plywood is generally selected. This may be yellow birch, mahogany, black walnut, or other furniture woods. Occasionally, hardwood plywoods are used because of special toughness and hardness requirements, as where severe weathering and other wear can be expected.

All plywood used in boatbuilding should be bonded with waterproof glues and otherwise meet requirements for marine use (Volume III, Section 4.5.1). Phenolic-resin glues, cured with heat, best meet the requirements.

3.6 PRESERVATIVE TREATMENT DURING ASSEMBLY.

3.6.1 General.—Certain parts of a ship inevitably are left exposed to attack by decay while in service unless given preservative treatment during the assembly operation. Even the use of pressure-treated lumber throughout the vessel, unless it is completely penetrated by the preservative, is no guarantee against the occurrence of such exposed points at joints and fittings; necessary cutting and shaping operations may result in the trimming off of the treated outer layers of wood, exposing the untreated interior of the timber. To provide some measure of protection at these vital points, preservatives may be applied by brush, spray, oil can or, if possible, by dipping the part in a solution of chemical.

If the parts are to be subsequently painted, following dipping or cold soaking, precautions should be taken (Volume II, Section 2.112) regarding bleeding through of copper naphthenate.

3.6.2 Critical Points for Decay in Ships.—Experience has shown that the points at which decay is more likely to occur in ships are the frames, especially half frames and cant frames butted against keel and deadwood; ill-ventilated parts between the stem and forepeak bulkhead, and between the stern and after-peak bulkhead, including such members as aprons, breasthooks, deadwood, knightheads, stems and stern-posts, pointers, hawse timbers, and transoms; the tops and ends of deck beams, the under side of decking in contact with beams, the contact sides of waterways and shelves, and contact points between all timbers as at scarf joints, butt joints, and other faying surfaces. At all such points of contact water is likely to collect due to leakage, faulty deck drainage, or some other cause. In double planked ships decay often is found at the interface between the outer and inner planking.

Large sawn frames are especially vulnerable to decay because, being large, they may contain sapwood edges and numerous butt joints at which – particularly below waterline – moisture may collect, due to leakage, in sufficient quantity to promote decay. Even when treated before installation, frames are exposed to decay by dubbing, resawing of the butt joints for better fit, and the drilling of numerous holes for fastenings.

3.6.3 Preservative Treatment.—The relative efficiency of the various methods of treating wood with preservatives is discussed in Volume II, Section 2.13. The builder should guide himself accordingly to get the best results possible, using pressure-treated material where possible and applying the methods most effective for other parts as they are feasible.

Naval contracts for many types of wood ships and boats require that all exposed parts be given at least a brush or spray treatment with an oil-borne toxic chemical (Volume II, Section 2.11). The effectiveness of this treatment in preventing the growth of decay fungi in ships is not known, particularly when applied to such naturally decay-resistant woods as white oak and mahogany. It is known that superficial applications of preservative generally offer only limited protection and the value of any treatment rests in the amount of preservative applied and the penetration secured, as well as in the character of the preservative. Hasty brushing and sparse application have dubious worth. The solution should be flowed on freely, not spread like paint. When brush treatment is used, the principles given in Volume II, Section 2.133, should be adhered to.

On members to be bent, the wood preservative should be applied after steaming. Requirements for preservative treatment of laminated oak members are given in Volume II, Section 2.14.

With chlorinated phenol preservatives, a hazard exists from skin irritation and certain protective garments should be worn by the user, such as neoprene gloves.

Butt ends of timbers – such as those of half frames at the keel – should be given especially liberal applications so that the deepest possible absorption of the chemical into the end grain will be obtained. This also is true at end-grain joints of deadwood, breasthooks, shelving, and other parts. Liberal application should also be given to dubbed surfaces and parts season-checked, chipped, or otherwise damaged so that moisture-collecting pockets are formed. Preservative should be squirted in quantity into fastening holes with a pressure lubricating can.

Faying surfaces of scarf joints usually are coated before being bolted. Since such joints expose end grain, they should be well coated with preservative before the bedding compound is applied. The toxic chemical should be allowed to dry thoroughly before being covered. Similar care should be observed in applying preservative to other joint surfaces that are painted, for, while the paint serves as a partial moisture barrier, it does not exclude moisture permanently, merely slowing its entry, and in time sufficient moisture will pass through the paint film to permit growth of decay unless a toxic preservative has first been applied to the wood.

With some forethought, small parts of even large vessels, such as butt blocks, knees, small deadwood, and plugs, can be given a more effective treatment than brushing. In most yards dipping probably is most convenient. Dipping in oil-borne preservatives does not cause swelling difficulties. Many more parts of the smaller vessels, such as floors, bent-frame filler blocks, breasthooks, knees, and gussets, can be effectively treated by the dipping process when oil-borne chemicals are used. Such treatment is especially desirable for such nondurable woods as spruce when used, for example, as filler blocks in double bent frames of V-bottom craft.

In some yards, copper naphthenate is liberally applied with a spray gun to various parts of a vessel during construction to reduce checking while the vessel is still on the ways. In such cases, spray application is reported to be more effective than brush application in treating the under and vertical sides of beams and similar parts and in forcing the preservative into checks, knot and bolt holes, and similar recesses in the wood.

3.7 PAINTING AND FINISHING.

3.7.1 General.—Painting and finishing include the application of paints, varnishes, sealers, fillers, and other materials applied to wood. Paints and other finishes are applied to protect wood from weathering, wood boring and fouling organisms, and fire, to improve appearance, for camouflage purposes; for sanitation and cleanliness, to reduce wear on decks, passageways, and other flooring; and to obtain nonskid walking surfaces. The term "finishing" is commonly used in the paint industry to describe the painting or varnish system employed, including fillers, sealers, and any other type of coating. In paint terminology, painting consists of the application of an opaque film, and varnishing the application of a transparent film. A sealer is generally a clear varnish product that is thinned, containing a large amount of transparent pigment.

3.7.2 Materials for Finishing Wood Boats and Ships.—The materials commonly used for wood craft are broadly classified as lacquers, oil paints, and varnishes. All the foregoing, when properly made for the purpose, are suitable for application directly on wood surfaces. In general, lacquers should not be combined with nonlacquers in a finishing system.

3.7.2.1 Lacquers.—Lacquers are commonly made with a cellulose ester of a low-viscosity type, together with a plasticizer and resin in suitable organic solvents and volatile thinners. Pigments are incorporated in the clear lacquers to make lacquer enamels. Lacquers dry chiefly by evaporation of the solvents and, therefore, may be made to dry very rapidly. As a rule, lacquers are applied in somewhat thinner coatings than are corresponding varnishes, so that two or more coats may be required where one coat of varnish or enamel would suffice. It is generally conceded that lacquer finishes on woods are not so durable as varnishes.

3.7.2.2 Oil Paints.—The oil paints for general marine purposes are commonly composed of drying oils, driers, volatile thinners, and pigments. Ordinarily paints contain opaque pigments such as white lead, zinc oxide, titanium oxide, alone or in combination, and transparent pigments (commonly called extenders) such as magnesium silicate or silica. The drying oil used is invariably a raw linseed oil.

3.7.2.3 Varnishes (Oleoresinous Products).—Varnish consists of resin blended with drying oil by heat according to a suitable cooking schedule; when the product has partly cooled, enough volatile thinner is added to produce the correct viscosity for application. Much less volatile thinner is required in varnishes than for lacquers. An enamel is made by incorporating a suitable quantity of hiding pigment in a varnish. Most finishes for ship and boat use are made with alkyd resins, but many are also made with phenolic resins. Still others are made for specific purposes with phenolic-modified alkyd resins.

For the drying oil, tung oil is usually preferred because it makes fast-drying varnishes, particularly in conjunction with the phenolic resins, but various satisfactory alternates for tung oil now are available and widely used. The properties of a varnish depend also on the proportions of resin and drying oil. Varnishes are called "long in oil" or "short in oil" according as they are made with much or little oil for a given amount of resin. A varnish is of "10-gallon length in oil" if it is made with 10 gallons of drying oil per 100 pounds of resin, "20-gallon length" if there are 20 gallons of oil per 100 pounds of resin, and so on. In general, a varnish dries faster and has better moisture-excluding effectiveness the shorter it is in oil; but, when too short, it lacks plasticity and durability. The length in oil, therefore, is commonly adjusted to give the fastest drying consistent with the required degree of durability for the use to which it is to be put. If unduly fast drying is demanded, it can be attained only by sacrificing durability.

According to their function in finishing systems, some special-purpose paints used in wood boats and ships contain fire-retardant or antifouling ingredients.

3.7.3 Antifouling Paints.—Certain specially made paints are designed to resist growth of barnacles on ship bottoms and to protect the bottoms of wood boats and vessels from attacks by marine borers (Volume II, Section 2.2). The attachment of marine growths, such as barnacles and other organisms of the sea, seriously affects the speed and fuel consumption of a ship. In tropical waters, marine borers present an even more serious problem in preserving wood craft.

Antifouling compositions must have a toxic effect and should be sufficiently durable to avoid the need of too frequent repainting. To produce the required toxic effect, the toxic elements must leach slowly from the paint film. A paint film that is hard and impervious to water may render the toxicants ineffective. It is the practice to make the period between launching and application of the last coat of antifouling paint as short as possible. This is done to prevent formation of a hard surface film and also because cuprous oxide, when exposed to air, turns to cupric oxide, a less soluble and consequently less toxic substance.

The compounds that have been found most effective in antifouling properties are mercuric and cuprous oxides. Mercuric oxide effectively prevents the growth of certain fouling organisms, while cuprous oxide is more effective with others. A combination of both compounds is often used in bottom paints for wood boats and ships.

None of the antifouling paints in present use provide permanent protection against barnacles or marine borers. Inasmuch as their effectiveness is dependent upon a slow leaching or dissolving of the toxic ingredients, frequent repaintings – usually at 6-month or shorter intervals – are required. Any mechanical damage to the paint film – such as that caused by grounding of the vessel – should be repaired as quickly as possible, this is especially true of keel surfaces.

3.7.4 Fire-Retardant Paints.—Paints of good fire-retarding effectiveness (Volume II, Section 2.5) can be made with ammonium phosphate, borax, or sodium silicate. A major weakness of coatings containing these compounds, however, is their inability to retain their effectiveness after exposure to water. They are consequently unsuited for many ship uses.

3.7.5 Preparation of Wood Surfaces for Painting.—The preparation required of wood surfaces prior to finishing depends on the species of wood to be finished and on the appearance of the desired finish. If a finish of mirrorlike smoothness is required, it is necessary to apply a coating of substantial thickness that can be sanded soon after it is dry to provide a perfectly level foundation for enamel or lacquer enamel. Sanding is accomplished through the application of a sanding surfacer. Ease of sanding is achieved through the incorporation of a high proportion of pigments so that the coating becomes porous as soon as the volatile thinner evaporates. The pigments used, such as magnesium silicate and china clay, are largely of low opacity and of relatively large particle size to provide easy and quick sanding. Enough opaque pigment, such as titanium dioxide, is commonly incorporated to give the coating good opacity.

Hulls and framing of wood Naval craft can be made acceptably smooth by planing or sanding, and personnel quarters and working spaces should be sanded down to a smooth, even texture prior to application of finishes. A glazing compound is sometimes applied to hull planking before it is painted, to cover the seams.

3.7.6 Seam Compounds.—Seam compounds are special types of putties designed to fill and level off countersunk screw holes, bolt holes, and seams to reduce skin friction. In addition, they protect the calking (Section 3.5.1.16). The types employed above the water line often differ from those used below the water line. Some of the latter contain ingredients which make them resistant to marine borers. Those employed above the waterline are usually high-grade putties, consisting mainly of whiting or whiting and white lead or other pigments mixed to a paste consistency with semidrying oils. A priming coat of paint should be used prior to application of seam compounds.

Properties of seam compounds that should be considered in their selection are the ease of application, adherence of the compound to the wood, ability to remain plastic, and resistance to weathering.

A class of seam compounds now widely used differs markedly from older conventional seam compounds in that it is a liquid when applied, This material has a polysulfide base. It does not expand and flow out of seams when exposed to the hot sun nor does it become brittle in very cold weather. It is used chiefly on decks. (See BUSHIPS Instruction 9140.9.)

Materials used for filling deck seams, often referred to as "marine glues," are not glues but fillers; they consist usually of petroleum asphalt.

3.7.7 Wood Fillers.—Wood filler is essentially a highly pigmented sealer designed to fill and level off the large pores in hardwoods having pores as large or larger than those in birch. These include ash, chestnut, elm, khaya, mahogany, oak, and walnut. The pigments commonly are silica and magnesium silicate or similar pigments of low opacity. Filler may be furnished in the consistency of paste to be thinned before application, or it may be furnished ready for application. Filler may be applied by brushing, spraying, or mopping; but, in any case, after much of the volatile has evaporated and before the rest has hardened. It is wiped with a rag, burlap, or moss across the grain of the wood to pack the large pores and remove any excess. A good filler functions also as a sealer and may be applied directly on bare wood.

3.7.8 Methods of Applying Paints.—Paint is applied at practically all stages of the construction of the boat. Under conditions where painting is done at various stages of construction, the paint usually is applied by brush. Where a special floor area or shop is provided for painting, or where boats and ships are built in the open, the spraying method of application often is used.

3.7.8.1 Spraying.—Spraying is a rapid method of application that is especially effective when applying fast-drying products on large surfaces. Spraying must be done only where there are facilities to exhaust the mist and fumes which may be exploded or ignited by fire or sparks. Boats, in most cases, are complicated structures with numerous spaces inaccessible for spraying equipment and sometimes spaces should not be sprayed because of their proximity to machinery or other equipment and because such space cannot be properly ventilated during spraying sufficiently to protect the workmen. Furthermore, different types of construction work are usually in progress simultaneously with the painting, and the fumes and mist connected with spraying would be objectionable. In most states, detailed laws govern equipment and safety precautions for industrial spray finishing. Technical details about the equipment needed for specific operations are obtainable from manufacturers of spray equipment.

3.7.8.2 Brushing.—Brushing is the most common means of applying finishes to wood vessels. Many finishers believe that certain coatings can be applied on wood more effectively by brushing than by spraying; this view, however, is not accepted generally.

3.7.9 General Directions for Applying Finishes.—The surface should be dry and entirely free from smudge, dust, or grease before paint is applied. Certain paints are designed for application on wet surfaces, but it is essential even with these that the surface be clean. Seam compounds and putties should be applied after the priming coat. Ordinary window putty is not suitable for marine use; the variety generally used is white lead putty. Plastic cements, also are used for special purposes. Painting should not be done in cold or freezing weather. At temperatures in the range of 35° to 50° F., painting may be facilitated by putting less oil and more turpentine in the paint than would be used in warm weather. No coat of paint should be applied until the previous coat is thoroughly dry.